Series: Environmental and Agricultural Nanotechnology

Series Editor: Kamel A. Abd-Elsalam

Volume

Soil Nanotechnology

Books in the Series

- Biopolymer Nanoparticles: A Novel Approach for Eco-conscious Environmental Applications
  Ayat F. Hashim, Rabab M.A. El-Sherif, Farah K. Ahmed and Sabu Thomas (eds.), 2026
- Green Nanotechnology for Stress-Resilient Crops and Sustainable Agroecosystems
  Jen-Tsung Chen (ed.), 2026
- Soil Nanotechnology
  Akhtar Hameed and Muhammad Zahid (eds.), 2026
- Smart Nanomaterials for Environmental Remediation
  Megha Barot and Azamal Husen (eds.), 2026

Series: Environmental and Agricultural Nanotechnology

Series Editor: Kamel A. Abd-Elsalam

Volume

Soil Nanotechnology

Editors

Akhtar Hameed

Institute of Plant Protection
MNS University of Agriculture
Multan, Pakistan

Muhammad Zahid

Department of Chemistry
University of Agriculture Faisalabad
Faisalabad, Pakistan

CRC Press
Taylor & Francis Group
Boca Raton London New York

CRC Press is an imprint of the
Taylor & Francis Group, an **informa** business
A SCIENCE PUBLISHERS BOOK

First edition published 2026
by CRC Press
2385 NW Executive Center Drive, Suite 320, Boca Raton FL 33431

and by CRC Press
4 Park Square, Milton Park, Abingdon, Oxon, OX14 4RN

CRC Press is an imprint of Taylor & Francis Group, LLC

Library of Congress Cataloging-in-Publication Data (applied for)

ISBN: 978-1-032-87829-4 (hbk)
ISBN: 978-1-032-90502-0 (pbk)
ISBN: 978-1-003-55835-4 (ebk)

DOI: 10.1201/9781003558354

Typeset in Times New Roman
by Prime Publishing Services

Preface to the Series

Welcome to the "Environmental and Agricultural Nanotechnology" book series, a specialty forum revealing the breakthrough promise of nanotechnology in two of the most vital areas of human endeavor: environment and agriculture. As Kamel A. Abd-Elsalam, Series Editor from the Plant Pathology Research Institute, Agricultural Research Center (ARC), I'm pleased to present this series of publications by leading figures in the field.

Nanotechnology is a promise technology, with creative solutions to the majority of the globe's most contentious problems. It presents innovative alternatives in environmental applications to prevent pollution, protect resources, and effectively address environmental effects. Consider new nanomaterials with previously unavailable pollutant absorption properties from air, water, and soil, or new sensors with specific environmental detection and cleansing. These are no longer concepts on paper anymore; they are becoming things and hard facts in a rush to redefine our way of environmental stewardship.

Parallel to the concerns, nanotechnology is also going to revolutionize agriculture, ushering in new avenues of food production and sustainable land management. This includes higher yields from crops through improved delivery of nutrients, optimization of nutrient use efficiency, enhanced fertility of the soil, and development of climate change-resilient agro-systems. The combination of the nanotechnological advances is the door to a future where farming is not only productive but considerably more sustainable.

This book series is to be a comprehensive treatise, taking the reader through the fundamentals of and concepts in nanoscience, nanomaterials, and their unique properties in relation to environmental and agronomic applications. We cover the specifics of the application of nanotechnology in controlling environmental pollution, e.g., its worth in contaminant decontamination and environmental monitoring. Moreover, we detail how nanotechnology can be applied in sustainable agricultural use and resource management, such as the application of targeted agrochemical delivery as well as crop adaptation to climate change.

Through these volumes, we hope to impart a deep appreciation of recent advances and envision future breakthroughs. This volume is a must-read for researchers, academics, policymakers, and entrepreneurs interested in leveraging nanotechnology for a sustainable and green future. In our opinion, by bridging the gap between frontier nanoscience and practical use in the environment and agriculture, we can contribute importantly to global efforts for sustainable development and higher living standards.

Preface

Soil health and sustainability have always been at the core of agricultural productivity and environmental conservation. With rapid advancements in technology, soil science is now embracing nanotechnology as a revolutionary tool to address critical challenges related to soil degradation, contamination, and nutrient management. "Soil Nanotechnology" brings together cutting-edge research and applications that highlight the transformative role of nanomaterials in soil systems. This book provides an in-depth exploration of how nanotechnology is reshaping soil science, from enhancing fertility to mitigating environmental pollutants. The chapters are designed to cover fundamental principles, emerging trends, and potential challenges associated with the application of nanomaterials in soil ecosystems. Beginning with an introduction to the principles and applications of soil nanotechnology, the book delves into specific topics such as the impact of nanomaterials on soil health and productivity, their role in remediation and pollution control, and their use in soil fertility enhancement. A crucial aspect of soil nanotechnology is its ability to improve soil monitoring and assessment through nanosensors. These sensors offer real-time data on soil conditions, enabling precise nutrient management and sustainable agricultural practices. The book also discusses how nanoparticles influence soil biogeochemistry, plant-soil interactions, and microbial communities, thereby affecting overall soil health and ecosystem stability. Additionally, the incorporation of Geographic Information System (GIS) and remote sensing technologies with nanotechnology paves the way for site-specific and efficient nutrient management strategies. The environmental and ecological implications of engineered nanomaterials are also a key focus. Topics such as soil nanoparticle interactions, their fate and transportation, and their potential effects on soil microbes are critically examined. Furthermore, the role of nanotechnology in soil erosion control, carbon sequestration, and climate change mitigation highlights its potential in promoting long-term sustainability. Given the increasing interest in nano-biochar and nanomaterial-based soil amendments, this book provides insights into their applications for improving crop production and land rehabilitation. The regulatory and environmental considerations of soil nanotechnology are also discussed, ensuring a balanced understanding of both the benefits and challenges associated with its widespread adoption. As soil science moves toward a more technology-driven future, this book serves as a comprehensive guide for researchers, academicians, policymakers, and industry professionals interested in the convergence of nanotechnology and soil management. By addressing both the scientific advancements and ethical considerations, "Soil Nanotechnology" aims to pave the way for sustainable agricultural and environmental solutions. We hope this book will inspire further research and innovation in this dynamic field, leading to enhanced soil health and productivity for future generations.

Contents

CHAPTER-1

Introduction to Soil Nanotechnology

Principles and Applications

Akhtar Hameed,[1,*] *Rana Binyamin,*[1] *Abdul Rehman,*[2] *Hira Akhtar,*[1] *Hafiz Muhammad Usman Aslam*[3] and *Faizan Ali*[4]

1. Introduction to Soil Nanotechnology

1.1 History of Nanotechnology

The concept of nanotechnology initially came into being in the medieval times when numerous colors were imparted in glass using gold nanoparticles. Richard Adolf Zsigmondy made the first observation and the size measurement of nanoparticles in 1914, used the term "nanometer" to define a particle with a size of 1/1,000,000th of a millimeter and established the first system to classify the particles based on their size in nanometer range. However, Richard Feynman, a Nobel Laureate, established the foundation of nano-science by introducing the idea of nanotechnology in his famous talk titled "There's plenty of room at the bottom". Even though he did not use the term "nanotechnology", he gave the idea of manipulation of matter at atomic level to develop "nanoscale" machines. Although studies on nanoscience have been going on for many years, the development of nanoscale material was not possible due to limited technological advancement. The term "nanotechnology" was first used by a researcher, Norio Taniguchi from the University of Tokyo, who explored it using a "top down" approach, while this term was introduced to the world by Eric Drexler in 1986 in his famous book "Engines of Creation" (Tarafdar and Adhikari 2015).

1.2 Definition and Concepts

"Nanotechnology" is the invention of Richard Feynman, who received the Nobel Prize in Physics in 1965. In 1959, during an American Physical Society meeting at Caltech, he gave the idea of the manipulation of matter at atomic level. It is the edge of material sciences that comprises the substances having unique properties in comparison to their larger macroscopic counterparts

[1] Institute of Plant Protection, MNS University of Agriculture, Multan 66000, Pakistan.
[2] Department of Plant Pathology, University of Agriculture Faisalabad 38000, Pakistan
[3] Department of Plant Pathology, San Luis Valley Research Center, Colorado State University, USA.
[4] Department of Agricultural Science & Engineering, College of Agriculture, Tennessee State University, Nashville, TN 37209, USA.
* Corresponding author: akhtar.hameed@mnsuam.edu.pk

DOI: 10.1201/9781003558354-1

(Hulla et al. 2015). It is the use of tiny particles ranging from 1 to 100 nanometers in size, known as nanoparticles, with a great potential to be used in medical field, agriculture and other industries due to their extremely small size and large surface area (Ghidan et al. 2017). For the identification of plant pathogens, it is introduced as fourth resourceful tool for cellular and molecular biology (Sharon et al. 2010).

2. Types of Nanomaterials

Nanomaterials are categorized into five major types (mentioned in Figure 1) depending upon their unique properties, composition, structure and functionality (Saleh 2020).

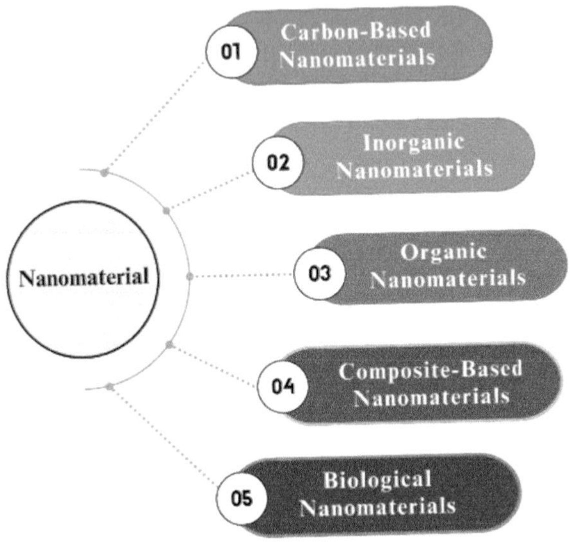

Figure 1. Classification of nanomaterials.

2.1 Carbon-based Nanomaterials

Carbon-based nanomaterials are composed entirely of carbon and exist in various forms and structures. These include Carbon Nanotubes (CNTs), which are wire-like hollow cylindrical tubes; Graphene (Gr), made up of layers of exfoliated graphite; Fullerenes, which have a specific number of carbon atoms in their structure (e.g., C60, which are ellipsoid or spherical); carbon nanofibers, which are long and thin nano-fibers; and carbon black, which exists as particulate matter (Sannino 2021).

2.2 Inorganic Nanomaterials

Particles made up of elements other than carbon are classified as inorganic nanomaterials (Ealia and Saravanakumar 2017). These materials are primarily composed of metallic oxides as well as pure and precious metals. Additionally, silicon and ceramics like semiconductors are also included in this category (Sannino 2021). The most commonly used metals for the synthesis of nanomaterials are zinc (Zn), aluminum (Al), copper (Cu), iron (Fe), cobalt (Co), and cadmium (Cd). Among the most frequently produced nanomaterials are iron oxide (Fe_2O_3), zinc oxide (ZnO), aluminum oxide (Al_2O_3), cerium oxide (CeO_2), silicon dioxide (SiO_2), and titanium oxide (TiO_2) (Ealia and Saravanakumar 2017).

2.3 Organic Nanomaterials

Apart from carbon-based or inorganic nanomaterials, there are also nanomaterials and nanostructures made from organic materials. These include dendrimers, liposomes, micelles, and polymers, which are typically formed through weak interactions and self-assembly techniques to form complex structures (Sannino 2021).

2.4 Composite-based Nanomaterials

These are complex nanostructures that exhibit multiphase characteristics with at least one component having nanoscale dimensions. These structures can integrate nanoparticles of different shapes and compositions or incorporate nanoparticles within bulk materials such as hybrid nanofibers or more complex frameworks like metal-organic frameworks. Nanocomposites can encompass any combination of metal-based, carbon-based, or organic-based nanomaterials with bulk materials of different types, including ceramics, metals, or polymers (Sannino 2021).

2.5 Bio Nanomaterial

Among biological nanomaterials, DNA and viruses are commonly encountered. For example, DNA is an organic molecular nanowire made of repeating molecular units. Similarly, viruses also exhibit similar structures (Sannino 2021).

3. Synthesis Methods

Numerous methods have been employed to synthesize nanoparticles, which can be categorized as chemical, biological, or physical techniques. These methods are further divided into top-down and bottom-up approaches (Usman et al. 2020).

3.1 Top Down Approach

This approach initiates with a bulk material which is gradually broken down, bit by bit, to produce fine nanoparticles. Common techniques used in this process are laser ablation, arc discharge, ball milling, micromachining, wire explosion, inert gas condensation, ion-sputtering, and lithography (scanning probe, nanosphere, electron beam, photo, soft, and block copolymer) (Usman et al. 2020).

3.2 Bottom Up Method

The bottom-up approach involves assembling or merging atoms and molecules to make a variety of nanoparticles (Dhand et al., 2015). Common techniques used in this process are electrochemical oxidation or reduction, sonochemical synthesis, photochemical synthesis, thermolysis, chemical reduction, solvothermal or hydrothermal synthesis, microemulsion, interfacial methods, microwave-assisted synthesis, atomic layer deposition, vapor phase chemical decomposition condensation, arrested precipitation, solvated metal atom dispersion, sol-gel fabrication, and biological synthesis methods (such as using plant extracts, yeast, fungi, bacteria) (Usman et al. 2020).

4. Characterization of Nanoparticles

A thorough physicochemical characterization of nanoparticles is crucial due to their applications in diverse fields. It helps to analyze their attributes such as shape, size, structure, morphology, surface area, composition, intensity, and stability. Moreover, purity, thermal, and electrical conductivity of nanoparticles are also evaluated through characterization. For the study of the size and shape of nanoparticles, the advanced analytical techniques are Transmission Electron Microscopy (TEM), High-Resolution Transmission Electron Microscopy (HRTEM), X-Ray Diffraction (XRD), Scanning Electron Microscopy (SEM), Environmental Scanning Electron Microscopy (ESEM), Atomic Force Microscopy (AFM), Dynamic Light Scattering (DLS), Photon Correlation Spectroscopy (PCS), Condensation Particle Counter (CPC), and Fourier-Transform Infrared (FTIR) Spectroscopy. Among all of them, XRD, SEM, and TEM are the most extensively used techniques to analyze surface morphology. For a detailed study of magnetic properties, Superconducting Quantum Interference Device (SQUID), Electron Paramagnetic Resonance (EPR), and Vibrating Sample Magnetometer (VSM) techniques are employed (Dikshit et al. 2021).

5. Analytical Techniques in Soil Nanotechnology

Numerous analytical techniques have been used to analyze the behavior and properties of nanoparticles. Microscopy, spectroscopy, and other advanced techniques are included among them. Each of them is concerned with specific properties of nanoparticles, as mentioned in Table 1.

5.1 Microscopy

Optical microscopy is limited to observing particles at the micron level because of its low resolving power. Therefore, to achieve nanoscale resolution for the observation of particles, different approaches must be employed. In this context, the currently employed techniques are scanning electron microscopy (SEM), atomic force microscopy (AFM), scanning tunneling microscopy (STM), and transmission electron microscopy (TEM). These enable the visualization of nanoparticles and allow for highly magnified images (Anees et al. 2024). The selected methods for the identification of shape, size, morphology, and distribution are SEM and FESEM, which are commonly used and highly efficient (Azizi et al. 2014, Habeeb Rahuman et al. 2022). In SEM, there are thermionic emission guns or field emission guns, which facilitate the production of high-resolution images of samples with features within a range of approximately 1 to 5 nm by scanning the surface of the sample using a high-energy beam of electrons. During such scanning, X-rays are also emitted and used in energy-dispersive X-ray (EDX) spectroscopy for nanoparticle characterization (Mohan and Renjanadevi 2016). Therefore, surface morphology and size distribution can be visualized by TEM (Thokchom et al. 2023), whereas HREM assists in determining the structure and morphology of nanoparticles (Sharma et al. 2013). In TEM, the electron beam illuminates the sample, leading to the formation of transmitted electrons. Subsequently, these electrons are magnified and transmitted through an objective lens onto a photographic plate (Anees et al. 2024). On the other hand, AFM has the ability to define both quantitative and qualitative information such as morphology, surface roughness, and particle size of nanoparticles (Baer et al. 2010).

5.2 Spectroscopy

Spectroscopic techniques reveal aspects of the nanoparticles and how they interact with electromagnetic radiation in order to determine size, shape, and concentration (Stampoulis et al. 2009). FTIR, Raman spectroscopy, and UV-visible spectroscopy are some of the most common spectral techniques employed in the analysis of nanoparticles (Anees et al. 2024). Raman spectroscopy is a technique that involves non-destructive inelastic scattering of monochromatic light waves. The technique offers a perspective on molecular vibrations and structural information of nanoparticles (Slavov et al. 2024). Likewise, FTIR confirms the functional groups on the surface of the nanoparticles (González-González et al. 2022) and their composition (Anees et al. 2024), while the synthesis of nanoparticles is confirmed through UV-visible spectroscopy (Mozdbar et al. 2023). In this technique, a UV-visible spectrophotometer tool directs light of a particular wavelength in the UV-visible range to the sample. Such light is absorbed by the nanoparticles. This absorption is dependent on the amount of sample in solution—the higher the concentration, the higher the extent of absorption. Concavity and other attributes show that high absorption is a sign of high sample concentration (Anees et al. 2024).

5.3 Other Advanced Techniques

Advanced analytical techniques are used to characterize nanomaterials in the soil system and to understand their interaction with soil. In this context, DLS, PCS, XRD, CPC, and zeta potential are some frequently used methods. X-ray diffraction (XRD) is used to determine the size and structure of crystalline nanomaterials as well as to analyze their atomic structure and physical properties (Habeeb Rahuman et al. 2022). Meanwhile, DLS analysis determines hydrodynamic particle size based on light scattering (Ali et al. 2020). PCS is also a type of DLS that assesses fluctuations in scattered light intensity to provide details on dimensions and size distributions of nanomaterials (Gun'ko et al. 2003). Similarly, CPC condenses nanoparticles using light condensation techniques and then subsequently counts and detects them (Rezaei and Johnson 2021), while zeta potential analyzes both surface charge and particle size, providing insight into the stability of nanoparticles (Yahaya Pudza et al. 2020).

Table 1. A brief summary of analytical techniques, along with their functions.

Techniques	**Functions**
Microscopy	
AFM	Provides 3D image and detailed surface analysis with atomic scale resolution
SEM and FESEM	Provides high resolution images of shape, morphology, size and surface morphology
TEM	Provides detailed image of average size and surface morphology
HRTEM	Determines nanoparticles' structure and detailed surface morphology
Spectroscopy	
Raman spectroscopy	Analyzes molecular vibrations and provides structural information of nanoparticles
FTIR	Determines composition and presence of functional groups on nanoparticles' surface
UV-Visible spectroscopy	Confirms nanoparticles synthesis and determines concentration based on light absorption
Other Advanced Techniques	
XRD	Analyzes atomic structure and determines size and structure of nanoparticles
DLS	Measures hydrodynamic particle size by analyzing light scattering fluctuations
PCS	Analyzes size distribution of small particles in suspension and focuses on nanoparticles' dimensions
CPC	Detects and counts airborne nanoparticles by enlarging them for light scattering measurements
Zeta potential	Analyzes surface charge, particle size and stability

6. Scope and Importance

Since the beginning of time, agriculture has been the foundation of the global economy, especially in developing countries where 60% of the population is dependent on it for their survival (Sharma et al. 2022). But currently, agriculture is facing multiple issues including a continuously changing climate, urbanization, reduction of arable lands, low production, and rising input costs, which contribute to malnutrition and poverty. The primary causes of farmers' economic losses are the increasing cost of inputs and declining crop yield. These issues are mainly attributed to imbalanced nutrient levels and reduced soil fertility (Rajput et al. 2022).

In light of these rising concerns, significant changes are needed in the global agricultural production system (Usman et al. 2020). In response to this, nanotechnology is emerging as the sixth revolutionary technology of the modern era (Elizabath et al. 2019). It is an interdisciplinary science already used in various fields. However, it has recently begun making its way into the agriculture sector, bringing new ideas and concepts to increase productivity and reduce poverty. Its promising potential to advance the agriculture sector by improving input efficiency and addressing agricultural issues has also been highlighted in previous research (Usman et al. 2020). Despite being in its early stages, this innovative technology holds the potential to reshape agriculture in the coming years by using a variety of techniques including genetically engineered plants, controlled and smart delivery systems of agrochemicals, nano-array-based gene technologies for gene expression in plants, early disease detection, and soil protection (Usman et al. 2020).

7. Applications of Nanotechnology in Soil Science

7.1 Soil Remediation

The use of nanotechnology-based *in situ* techniques has significantly garnered attention in soil remediation. The methods that remediate soil contaminants are reduction reaction, immobilization, and adsorption (Cecchin et al. 2017), illustrated in Figure 2. Inorganic soil contaminants are removed by the adsorption method, while organic contaminants are removed using reduction reaction and degradation methods (Fajardo et al. 2012).

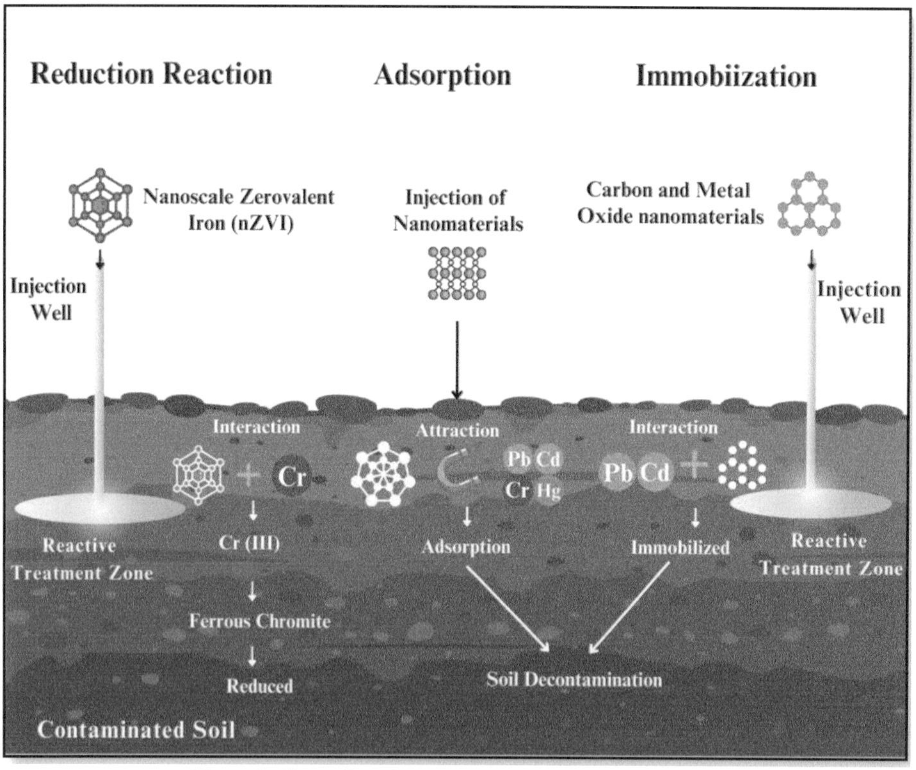

Figure 2. Illustration of nanoparticles remediating the soil.

7.1.1 Reduction Reaction

In the reduction reaction, nanoscale zero-valent iron (nZVI) is more prominently used to decontaminate organic compounds and heavy metals from soil. It also holds potential to decontaminate freshwater and groundwater as well (Qian et al. 2020). As nZVI is nano-sized with a large surface area, it has high remediation efficiency and can make direct contact with the contaminants. It is introduced into contaminated soil through injection wells. Its adsorption and reduction potential convert harmful pollutants such as chromium (Cr) into less harmful forms. In a previous study, it was demonstrated that the introduction of nZVI through injection wells in soil converted Cr (VI) to Cr (III). Then Cr (III) led to the formation of ferrous chromite (Rabbani et al. 2016). Similarly, it was reported in another study that the remediation capacity of the combined application of nZVI and biochar was higher than the individual application of nZVI (Qian et al. 2020).

7.1.2 Immobilization

Immobilization is relatively cheaper to implement when dealing with decontamination of the soil and is considered as environmentally friendly (Usman et al. 2020). One of the key stages of immobilization is the selection of nanomaterials, which is made based on the type of pollutant and soil conditions. For these reasons, metal oxide and carbon-based nanomaterials have been selected for the adsorption of organic and inorganic pollutants existing in the soil (Qian et al. 2020). Among the selected nanomaterials, carbon based materials are known to exist in different forms such as graphene, fullerenes, carbon nanotubes which are also used as an adsorbent to retain the soil contaminants. They collect the contaminated soil depending on the hydrophobicity of the pollutants and their capacity to adsorb the pollutants (Guerra et al. 2018). Nevertheless, metal oxide nanoparticles have been applied in the removal of soil through the surface complexation technique. The surface complexation techniques often employ metal oxides like iron (III) oxide and titanium dioxide (TiO_2). Among these metal oxides, iron (III) oxide possesses high degree of adsorption and hence can be quite efficient in the removal of metals such as arsenic and cadmium (Qian et al. 2020).

7.1.3 Adsorption

Adsorption for the removal of heavy metal ions such as Hg, Cd, Cr, and Pb has attracted a lot of attention due to its simplicity as well as being relatively cheap to implement. Wei et al. (2013) reported that adsorptive materials remove contaminants by attracting them toward their internal structure and surface. The effectiveness of an adsorption material depends upon its ability to attract contaminants. Therefore, nanomaterials, with their extremely small size and large surface area, hold high adsorption capacity, which enables them to remediate the soil (Shan et al. 2009).

7.2 Fertilizer Delivery System

Nanofertilizers are primarily nutrient carriers with a size range of 30 to 40 nm, equivalent to one-billionth of a meter. Due to their large surface area, they have the potential to hold nutrient ions for a long time and release them gradually as required by crops (Subramanian et al. 2015). Currently, three main approaches of nanotechnology are used for crop protection and fertilization. The first approach is the reformulation of fertilizers and supplements to the nanoscale, which makes them highly efficient. The second is the integration of nano-additives with bulk materials to improve their performance and functionality, while the third approach is the coating of fertilizers with nanomaterials, in which fertilizers are encapsulated with nanoporous substances and nanosized thin films that facilitate the smart and controlled release of fertilizers into the soil (Mastronardi et al. 2015).

In addition to these advancements, it is also reported by Subramanian et al. (2005) that nanocomposites and nanofertilizers are highly efficient for the controlled release of nutrients from fertilizer granules. The controlled release of nanofertilizers not only enhances nutrient use efficiency but also reduces nutrient losses in soil and their accumulation in the environment. Moreover, it is possible to deliver these nanofertilizers to root zones of plants more effectively and in a timely manner with high precision. It is observed that nanotechnology-based nitrogenous and phosphatic fertilizers are highly soluble in soil, reduce losses, and increase nutrient use efficiency (Subramanian et al. 2015).

7.3 Soil Moisture Retention

In rainfed areas, agriculture is completely dependent on traditional approaches to conserve water and soil, which include check dams, construction of reservoirs, and steep terracing. Coarse-structured soils often have a weak structure and coarse texture, resulting in low soil water retention capacity. Therefore, to enhance water holding capacity and soil structure—particularly in soils with coarse structure—it is imperative to explore effective solutions. In this context, nanocarbon presents a promising solution. It helps to reduce water loss through percolation and retain more water for extended periods in the plant root zones. Additionally, nanocarbon, derived from coconut shells and containing naturally occurring carbon, contributes to controlling carbon emissions in the environment. Consequently, the use of nanocarbon in agriculture is gaining traction and support. Incorporation of nanocarbon into sandy soils has shown significant improvements in water holding capacity and infiltration rates. Analysis of soil water characteristics revealed that nanocarbon not only enhanced water retention but also increased the percentage of fine pores in the soil. In order to test four different levels of nanocarbon content across five different thicknesses within the soil profile, an infiltration experiment was conducted. The results demonstrated that cumulative infiltration was positively influenced by both nanocarbon content and the thickness of the nanocarbon mixture layer. The combination of more than 0.5% content and a mixture layer thickness within the range of 3 to 5 cm led to a highly positive effect on water infiltration and soil performance. Moreover, the water retention in soil also increased, indicating a reduction of water losses through percolation in deeper soil layers and improved plant growth (Zhou and Chen 2017). Similarly, in arid and desert environment, the use of superabsorbent polymers is gaining traction to manage water losses as they have shown promising outcomes in improving plant growth, fertilizer retention, minimizing plant mortality rate and the use of irrigation water (Nge et al. 2004). Similarly, the soil treated with double-coated slow-release fertilizers and superabsorbents demonstrated reduced soil water transpiration as compared to untreated soil (Xu et al. 2024).

7.4 Pest and Disease Control

A broad range of attackers, including insects, pests, pathogens, and weeds, impact crops. These attackers can be categorized into various types, such as leaf senescence accelerators, stand reducers, light stealers, tissue consumers, assimilate sappers, and photosynthetic rate reducers (Savary and Willocquet 2020). Conventional pesticides are used to protect crops from these threats. However, a significant drawback is that most pesticides are lost in the environment through spray drift or runoff, leading to off-target accumulation. Estimates suggest that only 0.1% of applied pesticides effectively reach the targeted pest, while the majority either photodegrade, accumulate, or pollute the environment. In this context, the use of nanomaterials presents an efficient and innovative solution to address these issues. Nanomaterials offer encapsulation of pesticides to prevent their degradation, accumulation, and wastage. The encapsulation of nanomaterials on active ingredients improves their physicochemical properties and enables their smart and slow release into the environment. To achieve these coatings, the active ingredients are trapped or loaded within nanosized polymeric structures to form nanopesticides. This nanotechnology-based pesticide delivery system is a promising strategy for pest and disease control, as nanoformulations have the potential to convert insoluble chemicals into soluble forms and allow for controlled and targeted delivery of active ingredients. In this way, it minimizes the required amount of chemicals for disease control and reduces their bioaccumulation in the environment. In addition, these nanopesticides are highly stable and resistant to leaching, washing, evaporation, photodegradation, and microbial degradation (Chowdappa and Gowda 2013).

8. Nanotechnology for Sustainable Agriculture

8.1 Enhancing Crop Productivity

Nanotechnology plays a key role in improving various aspects of agriculture associated with crop production, such as photosynthesis, seed germination, disease control, crop yield, and inducing disease resistance in plants (Shang et al. 2019). Recent reports demonstrate that when commercial fertilizers are applied as foliar sprays in combination with metal nanoparticles such as Zn, Fe, Cu, nano-N, nano-K, and nano-P, plant growth, economic yield, and nutrient use efficiency were significantly higher compared to other treatments (Rahman et al. 2021). Similarly, it is reported by field trials that the nutrient use efficiency of nanofertilizers is notably higher in comparison with conventional fertilizers. For example, when nitrogenous nanofertilizers were applied at half the amount of conventional nitrogenous fertilizer, the crop yield was 7.9 tons/ha compared to 7.3 tons/ha with conventional fertilizer (Carmona et al. 2022). Additionally, the application of nanotechnology-based fertilizers such as nitrogen, phosphorus, and potassium resulted in higher starch content, harvest index, economic yield, and nutrient use efficiency than the control treatment (Abd El-Azeim et al. 2020). The use of nanoparticles in tomato plants has led to an improved cost-benefit ratio compared to traditional fertilizers (Rahman et al. 2021).

8.2 Precision Farming

Precision farming is a modern approach to farm management. It utilizes advanced tools to improve agricultural practices. In precision farming, farmers use nanosensors that enable them to apply precise amounts of agrochemicals at targeted points. In this way, the inputs are efficiently applied, resulting in economic benefits. Farmers can remotely detect the requirements of crops and manage the farm accurately. Moreover, nanosensors, in combination with nanotechnology-based smart delivery systems, empower precision agriculture. It enhances the efficiency of agricultural resources such as agrochemicals, nutrients, and water. Similarly, more advanced precision agriculture is achieved by integrating these nanomaterials with the Global Positioning System (GPS) and satellite imaging. This technology enables the remote detection of crop pests and stress conditions at earlier stages, allowing farm managers to automatically adjust the use of resources, pesticide applications, and irrigation levels while minimizing wastage. Moreover, the use of fertilizers encapsulated with nanomaterials is increasing due to their ability to be used in lesser amounts with minimal environmental impacts. Similarly, in dry and semi-arid lands, the storage and protection of water is another of its achievements. Further benefits of nanotechnology extend to the production of genetically modified resistant crops and alternative fuel production methods (Mastronardi et al. 2015). An integrated use of all these technologies in agriculture holds potential to push agriculture toward sustainability.

8.3 Sustainable Resource Management

The use of nanotechnology offers innovative advancements in agriculture for sustainable resource management. Precision agriculture, which uses advanced technologies to improve crop yield, is significantly influenced by the potential of nanotechnology. Nanotechnology has introduced manipulation of materials and structures at the nanoscale as a new dimension in precision farming. Nanosensors for real-time monitoring of soil and plant health are another key factor. They provide information on environmental and plant status, which is useful for effective and accurate crop management. In addition, it provides social advantages by reducing the consumption of agrochemicals and their wastage and ensures smart delivery. It creates a relationship between nanotechnology and precision agriculture that enables the measurement of field variability, decreases input costs, and helps in generating maps for variable rate application in terms of precipitation. We can also conclude that this technology not only increases crop yield but also encourages the adoption of agricultural practices that ensure sustainability (Yadav et al. 2023).

9. Environmental Implications of Soil Nanotechnology

9.1 Nano-ecotoxicology and Environmental Impacts

Nanotechnology is an advanced technology that can be applied to agriculture as well as other fields. Still, there are some other aspects of this technological advancement that have negative implications that must be discussed (Yadav et al. 2023). One of the main concern is that they are very small in size, hence they could be delivered either through water or air, which leads to environmental contamination. When nanoparticles are in the environment, they are a threat to ecology. For example, they can disrupt the balance of micronutrients and microorganism essential for soil fertility (Tang et al. 2023). Although nanoparticles enable the smart release of pesticides in the environment, their dispersal in the environment can contribute to the loss of biodiversity, soil and water pollution, and resistance in pests and pathogens. Previous studies demonstrated that nanopesticides are harmful to bees which are important plant pollinators (Haris et al. 2023). Additionally, their accumulation in plants and animals adversely impacts human health, as these particles are consumed through their feed (Tiede et al. 2009). Exposure to nanoparticles has been linked to respiratory diseases, cardiovascular issues, and neurological damage (Pietroiusti et al. 2018). Direct skin exposure to nanoparticles can lead to systemic toxicity, as these particles enter the bloodstream (Larese et al. 2009) and produce reactive oxygen species (ROS) that damage cellular membranes (Fu et al. 2014). Moreover, substantial investment is required for research and development in nanotechnology, making it expensive. It is unaffordable for small-scale growers, resulting in an unequal distribution of the benefits of nanotechnology in society (Yadav et al. 2023). Last but not least is the knowledge gap regarding the environmental and ecosystem impacts of nanotechnology, which justifies significant investment to mitigate potential hazards of these tiny particles to humans and wildlife (Jafari et al. 2017).

9.2 Soil Health and Microbial Diversity

Nanoparticles are also found naturally in the soil matrix either in the form of primary nanoparticles or aggregates/agglomerates. However, their artificial introduction into the soil raises significant concerns, as they do not degrade easily and accumulate in soil over the long term. This accumulation poses a serious threat to the conservation of the microbial community in soil, which is crucial to maintaining soil fertility. The effect of nanoparticles on the soil ecosystem is influenced by several factors, including their concentration, soil type, and enzymatic activity. It has been demonstrated that a higher concentration of nanoparticles negatively affects the activity of dehydrogenase enzyme, which is essential for plant health. Moreover, soil fertility and plant nutrition are also affected through imbalances of essential nutrients and disruption of the soil's self-cleaning ability. Accumulated nanoparticles can also inhibit the activity of beneficial microbes in soil. Soil factors such as pH, texture, structure, and organic matter content are important in influencing the microbial community and the impact of pollutants' toxicity on microorganisms. The mobility of soil contaminants can be altered by these interactions. Additionally, the soil's ability to sustain microbial communities has also declined. Therefore, the testing of nanoparticles across different types of soil is necessary to understand their environmental impacts (Javed et al. 2019).

10. Future Prospects of Soil Nanotechnology

10.1 Emerging Trends

Nanotechnology holds potential to ensure precision in agriculture through its innovative solutions. It offers monitoring of soil conditions, insect infestations, reduced environmental impacts, and optimized use of resources (Yadav et al. 2023). Additionally, nanotechnology can be used to mitigate challenges associated with climate change and food security by developing new crop varieties that are resistant to environmental stress (Haris et al. 2023). Precision agriculture integrated with nanotechnology has led to a technological revolution in agriculture. It is promoting sustainable agriculture and reshaping traditional farming practices. Now, the direct delivery of chemicals to plants is possible through the use of nanosensors, nanochips, and nanomaterials. The pesticides, herbicides, and fertilizers encapsulated with nanomaterials ensure that agrochemicals are delivered to plants at the right time and place, minimizing wastage and environmental impacts. Moreover, two commercially applied nanofertilizers are immune system enhancers and plant growth regulators (Zaim et al. 2023). These products are capable of offering all essential nutrients to plants, including zinc, iron, and silica, by stimulating resistance in plants as well as nutrient absorption, leading to enhanced growth and yield (Tyagi et al. 2023).

10.2 Policy and Regulatory Considerations

The existing knowledge gaps on the effects of nanotechnology on ecosystems clearly point to the need for significant investment to reduce risks to human beings and wildlife. For these reasons, detailed studies and trials must be conducted to assess their effects on the environment and other facets. Unfortunately, the current data available on the risks associated with nanotechnology-based products is very limited, which is why more research should be conducted to establish the potential dangers posed by nanotechnology. Further, a regulatory framework must be developed across organizations and ministries for food nanotechnology (Jafari et al. 2017). Consequently, before implementing regulatory guidelines for their use in agriculture, comprehensive assessments and risk management protocols must be established (Amenta et al. 2015).

11. Conclusion

Nanotechnology is a revolutionary science with the ability to transform agriculture. It provides solutions that may enable the enhancement of soil fertility, optimal utilization of resources, and increased yields. The combination of nanomaterials and nanotechnology tools has been found to boost soil fertility by restoring it; this can address problems of nutrient deficiency through the application of nanofertilizers. However, the development of appropriate regulatory frameworks and further research is needed for such implementation to be safe and fruitful. With advancements in this field, additional investigations into soil nanotechnology will be necessary for sustainable agriculture aimed at addressing food security concerns.

12. Acknowledgement

The authors gratefully acknowledge the support of the Institute of Plant Protection, Muhammad Nawaz Sharif University of Agriculture, Multan for the support in this publication.

References

Abd El-Azeim, M., Sherif, M., Hussien, M., I. Tantawy and S. Bashandy. 2020. Impacts of nano- and non-nanofertilizers on potato quality and productivity. Acta Ecol. Sin. 40: 388–397.

Ali, S.K., H. Wani, C. Upadhyay, K.S. Madhur, I. Khan, S. Gul et al. 2020. Synthesis of CdS nanoparticles by chemical co-precipitation method and its comparative analysis of particle size via structural and optical characterization. Indones. Phys. Rev. 3: 100–110.

Amenta, V., K. Aschberger, M. Arena, H. Bouwmeester, F.B. Moniz, P. Brandhoff et al. 2015. Regulatory aspects of nanotechnology in the agri/feed/food sector in EU and non-EU countries. Regul. Toxicol. Pharmacol. 73: 463–476.

Anees, A., R. Gupta, P. Phanindra, O.A. Fabiyi, U.K. Thera, T.T. Bello et al. 2024. Green nanoparticles synthesis and applications. pp. 89. In: Advanced Nanotechnology in Plants: Methods and Applications.

Azizi, S., M.B. Ahmad, F. Namvar and R. Mohamad. 2014. Green biosynthesis and characterization of zinc oxide nanoparticles using brown marine macroalga Sargassum muticum aqueous extract. Mater. Lett. 116: 275–277.

Baer, D.R., D.J. Gaspar, P. Nachimuthu, S.D. Techane and D.G. Castner. 2010. Application of surface chemical analysis tools for characterization of nanoparticles. Anal. Bioanal. Chem. 396: 983–1002.

Carmona, F.J., A. Guagliardi and N. Masciocchi. 2022. Nanosized calcium phosphates as novel macronutrient nano-fertilizers. Nanomaterials. 12: 2709.
Cecchin, I., K.R. Reddy, A. Thomé, E.F. Tessaro and F. Schnaid. 2017. Nanobioremediation: Integration of nanoparticles and bioremediation for sustainable remediation of chlorinated organic contaminants in soils. Int. Biodeterior. Biodegrad. 119: 419–428.
Chowdappa, P. and S. Gowda. 2013. Nanotechnology in crop protection: status and scope. Pest Manag. Hortic. Ecosyst. 19: 131–151.
Dhand, C., N. Dwivedi, X.J. Loh, A.N.J. Ying, N.K. Verma, R.W. Beuerman et al. 2015. Methods and strategies for the synthesis of diverse nanoparticles and their applications: a comprehensive overview. RSC Adv. 5: 105003–105037.
Dikshit, P.K., J. Kumar, A.K. Das, S. Sadhu, S. Sharma, S. Singh, P.K. Gupta et al. 2021. Green synthesis of metallic nanoparticles: applications and limitations. Catalysts. 11: 902.
Ealia, S.A.M. and M.P. Saravanakumar. 2017. A review on the classification, characterisation, synthesis of nanoparticles and their application. IOP Conf. Ser. Mater. Sci. Eng.
Elizabath, A., M. Babychan, A.M. Mathew and G.M. Syriac. 2019. Application of nanotechnology in agriculture. Int. J. Pure Appl. Biosci. 7: 131–139.
Fajardo, C., L. Ortíz, M. Rodríguez-Membibre, M. Nande, M. Lobo and M. Martin. 2012. Assessing the impact of zero-valent iron (ZVI) nanotechnology on soil microbial structure and functionality: a molecular approach. Chemosphere. 86: 802–808.
Fu, P.P., Q. Xia, H.-M. Hwang, P.C. Ray and H. Yu. 2014. Mechanisms of nanotoxicity: generation of reactive oxygen species. J. Food Drug Anal. 22: 64–75.
Ghidan, A.Y., T.M. Al-Antary, N.M. Salem and A.M. Awwad. 2017. Facile green synthetic route to the zinc oxide (ZnO NPs) nanoparticles: effect on green peach aphid and antibacterial activity. J. Agric. Sci. 9: 131–138.
González-González, R.B., L.T. González, M. Madou, C. Leyva-Porras, S.O. Martinez-Chapa and A. Mendoza. 2022. Synthesis, purification, and characterization of carbon dots from non-activated and activated pyrolytic carbon black. Nanomaterials. 12: 298.
Guerra, F.D., M.F. Attia, D.C. Whitehead and F. Alexis. 2018. Nanotechnology for environmental remediation: materials and applications. Molecules. 23: 1760.
Gun'ko, V.M., A.V. Klyueva, Y.N. Levchuk and R. Leboda. 2003. Photon correlation spectroscopy investigations of proteins. Adv. Colloid Interface Sci. 105: 201–328.
Habeeb Rahuman, H.B., R. Dhandapani, S. Narayanan, V. Palanivel, R. Paramasivam, R. Subbarayalu et al. 2022. Medicinal plants mediated the green synthesis of silver nanoparticles and their biomedical applications. IET Nanobiotechnol. 16: 115–144.
Haris, M., T. Hussain, H.I. Mohamed, A. Khan, M.S. Ansari, A. Tauseef et al. 2023. Nanotechnology – A new frontier of nano-farming in agricultural and food production and its development. Sci. Total Environ. 857: 159639.
Hulla, J., S.C. Sahu and A.W. Hayes. 2015. Nanotechnology: history and future. Hum. Exp. Toxicol. 34: 1318–1321.
Jafari, S.M., I. Katouzian and S. Akhavan. 2017. Safety and regulatory issues of nanocapsules. pp. 545–590. *In*: Yildirim, A.S.K. [ed.]. Nanoencapsulation Technologies for the Food and Nutraceutical Industries. Elsevier.
Javed, Z., K. Dashora, M. Mishra, V.D. Fasake and A. Srivastva. 2019. Effect of accumulation of nanoparticles in soil health – a concern on future. Front. Nanosci. Nanotechnol. 5: 1.
Khan, S.T. 2020. Interaction of engineered nanomaterials with soil microbiome and plants: their impact on plant and soil health. Sustain. Agric. Rev. 41: Nanotechnology for Plant Growth and Development. 181–199.
Larese, F.F., F. D'Agostin, M. Crosera, G. Adami, N. Renzi, M. Bovenzi and G. Maina. 2009. Human skin penetration of silver nanoparticles through intact and damaged skin. Toxicol. 255: 33–37.
Mastronardi, E., P. Tsae, X. Zhang, C. Monreal and M.C. DeRosa. 2015. Strategic role of nanotechnology in fertilizers: potential and limitations. pp. 25–67. *In*: Johnson, S. [ed.]. Nanotechnologies in Food and Agriculture.
Mohan, A.C. and B. Renjanadevi. 2016. Preparation of zinc oxide nanoparticles and its characterization using scanning electron microscopy (SEM) and X-ray diffraction (XRD). Procedia Technol. 24: 761–766.
Mozdbar, A., A. Nouralishahi, S. Fatemi and F.S. Talatori. 2023. The impact of Carbon Quantum Dots (CQDs) on the photocatalytic activity of TiO2 under UV and visible light. J. Water Process Eng. 51: 103465.
Nge, T.T., N. Hori, A. Takemura and H. Ono. 2004. Swelling behavior of chitosan/poly (acrylic acid) complex. J. Appl. Polym. Sci. 92: 2930–2940.
Pietroiusti, A., H. Stockmann-Juvala, F. Lucaroni and K. Savolainen. 2018. Nanomaterial exposure, toxicity, and impact on human health. Wiley Interdiscip. Rev. Nanomed. Nanobiotechnol. 10: e1513.
Qian, Y., C. Qin, M. Chen and S. Lin. 2020. Nanotechnology in soil remediation– applications vs. implications. Ecotoxicol. Environ. Saf. 201: 110815.
Rabbani, M.M., I. Ahmed and S.J. Park. 2016. Application of nanotechnology to remediate contaminated soils. pp. 219–229. *In*: Singh, M. [ed.]. Environmental Remediation Technologies for Metal-Contaminated Soils.

Rahman, M.H., M.N. Hasan, S. Nigar, F. Ma, M.S.A. Aly and M.Z.H. Khan. 2021. Synthesis and characterization of a mixed nanofertilizer influencing the nutrient use efficiency, productivity, and nutritive value of tomato fruits. ACS Omega. 6(41): 27112–27120.

Rajput, V.D., T. Minkina, S.K. Upadhyay, A. Kumari, A. Ranjan, S. Mandzhieva et al. 2022. Nanotechnology in the restoration of polluted soil. Nanomaterials. 12(5): 769.

Rezaei, M. and M.S. Johnson. 2021. Airborne nanoparticles: Control and detection. pp. 85–133. *In*: Wood, T.S. [ed.]. Air Pollution Sources, Statistics and Health Effects.

Saleh, T.A. 2020. Nanomaterials: Classification, properties, and environmental toxicities. Environ. Technol. Innov. 20: 101067.

Sannino, D. 2021. Types and classification of nanomaterials. pp. 15–38. *In*: Longoria, P. [ed.]. Nanotechnology: Trends and Future Applications.

Savary, S. and L. Willocquet. 2020. Modeling the impact of crop diseases on global food security. Annu. Rev. Phytopathol. 58(1): 313–341.

Shan, G., S. Yan, R. Tyagi, R.Y. Surampalli and T.C. Zhang. 2009. Applications of nanomaterials in environmental science and engineering. Pract. Period. Hazard. Toxic Radioact. Waste Manag. 13(2): 110–119.

Shang, Y., M.K. Hasan, G.J. Ahammed, M. Li, H. Yin and J. Zhou. 2019. Applications of nanotechnology in plant growth and crop protection: A review. Molecules. 24(14): 2558.

Sharma, B., M. Jasper, C. Lemckert and M. Muminovic. 2022. Patterns of rural community livelihood and managing natural resources: A systematic literature review. Int. J. Sustain. Econ. Soc. Cult. Context. 19(1): 35.

Sharma, P., G. Darabdhara, T.M. Reddy, A. Borah, P. Bezboruah, P. Gogoi et al. 2013. Synthesis, characterization and catalytic application of Au NPs-reduced graphene oxide composites material: An eco-friendly approach. Catal. Commun. 40: 139–144.

Sharon, M., A.K. Choudhary and R. Kumar. 2010. Nanotechnology in agricultural diseases and food safety. J. Phytol. 2(4).

Slavov, D., E. Tomaszewska, J. Grobelny, N. Drenchev, D. Karashanova, Z. Peshev et al. 2024. FTIR spectroscopy revealed nonplanar conformers, chain order, and packaging density in diOctadecylamine- and Octadecylamine-passivated gold nanoparticles. J. Mol. Struct. 138827.

Stampoulis, D., S.K. Sinha and J.C. White. 2009. Assay-dependent phytotoxicity of nanoparticles to plants. Environ. Sci. Technol. 43(24): 9473–9479.

Subramanian, K.S., A. Manikandan, M. Thirunavukkarasu and C.S. Rahale. 2015. Nano-fertilizers for balanced crop nutrition. Nanotechnol. Food Agric. 69–80.

Tang, R., D. Zhu, Y. Luo, D. He, H. Zhang, A. El-Naggar et al. 2023. Nanoplastics induce molecular toxicity in earthworm: Integrated multi-omics, morphological, and intestinal microorganism analyses. J. Hazard. Mater. 442: 130034.

Tarafdar, J. and T. Adhikari. 2015. Nanotechnology in soil science. Soil Sci.: An Introduction. Chapter: Nanotechnology in Soil Science. 775–807.

Thokchom, B., S.M. Bhavi, M.B. Abbigeri, A.K. Shettar and R.B. Yarajarla. 2023. Green synthesis, characterization and biomedical applications of Centella asiatica-derived carbon dots. Carbon Lett. 33(4): 1057–1071.

Tiede, K., M. Hassellov, E. Beeitbarth, Q. Chaudhry and A.B. Boxall. 2009. Considerations for environmental fate and ecotoxicity testing to support environmental risk assessments for engineered nanoparticles. J. Chromatogr. A. 1216(3): 503–509.

Tyagi, P.K., A. Arya, S. Ramniwas and S. Tyagi. 2023. Recent trends in nanotechnology in precision and sustainable agriculture. pp. 1256319. *In*: Front. Media SA, eds. Vol. 14.

Usman, M., M. Farooq, A. Wakeel, A. Nawaz, S.A. Cheema, H. ur Rehman et al. 2020. Nanotechnology in agriculture: Current status, challenges and future opportunities. Sci. Total Environ. 721: 137778.

Xu, F., S. Zhang, Q. Wu, C. An, X. Li, X. Chen and Z. Dong. 2024. A biodegradable multifunctional pectin–montmorillonite fertilizer coating: Controlled-release, water-retention and soil-cementation. Int. J. Biol. Macromol. 134118.

Yadav, A., K. Yadav, R. Ahmad and K.A. Abd-Elsalam. 2023. Emerging frontiers in nanotechnology for precision agriculture: advancements, hurdles and prospects. Agrochemicals. 2(2): 220–256.

Yahaya Pudza, M., Z. Zainal Abidin, S. Abdul Rashid, F. Md Yasin, A. Noor and M.A. Issa. 2020. Eco-friendly sustainable fluorescent carbon dots for the adsorption of heavy metal ions in aqueous environment. Nanomaterials. 10(2): 315.

Zaim, N.S.H.B.H., H.L. Tan, S.M.A. Rahman, N.F. Abu Bakar, M.S. Osman, V.K. Thakur et al. 2023. Recent advances in seed coating treatment using nanoparticles and nanofibers for enhanced seed germination and protection. J. Plant Growth Regul. 42(12): 7374–7402.

Zhou, B. and X. Chen. 2017. Effect of Nano-carbon on water holding capacity in a Sandy soil of the loess plateau. Earth Sci. Res. J. 21(4): 189–195.

CHAPTER-2

Impacts of Nanomaterials on Soil Health and Productivity

Najma Majeed,[1] *Ahmad Mahmood,*[2] *Shakeel Ahmad,*[1] *Muhammad Imran,*[1,*] *Muhammad Zeeshan Gulzar*[3] and *Akhtar Hameed*[4]

1. Introduction

Nanoparticles have become more commonly used in industrial applications, as well as consumer and medical items, over the previous decade, and this trend is expected to continue. Nanoparticles can be introduced into soils directly through fertilizers or plant protection chemicals, or indirectly by application to land or wastewater treatment products such as sludges or biosolids. Nanoparticles can infiltrate aquatic systems directly by industrial discharges, wastewater treatment effluent dumping, or indirectly via soil surface runoff. Researchers have utilized laboratory experiments to begin to understand the effects of nanoparticles on waterways and soils, and this account covers that research and the translation of those findings to environmental environments (Batley et al. 2013). Soil, which is a natural resource on the earth's surface, is very significant in agricultural production as well as the maintenance of the environment. It provides plants with nutrition, water and soil, which helps them establish their root system. This chapter highlights the correlation between the fertility and quality of soil and crop productivity, which is in turn defines food access and price (Doran and Zeiss 2000). Soil health is the functional capacity of soils as living ecosystems at any given time. Its conservation is vital to the continued productivity and sustainability of agriculture in the future (Bünemann et al. 2006). Soil is an important resource for food security; it supplies nutrient requirements for crops. Changes such as erosion and salinization are known to decrease agricultural production, thus affecting yields and food security (Bationo et al. 2007). Organic matter and trace minerals in soils improve the nutrient status, while soils with low physical and chemical quality result in low nutrient density. Climate change, variability, and extreme weather conditions can accelerate the rates of soil degeneration which ultimately accelerate the food insecurity. Soil health requires sustainability when it comes to the use of soil in order to support agriculture and other farming practices (Lal 2016). The global food requirement is now rising due to a growing population, which is anticipated to reach 9 billion in the year 2050. As such, the world faces a big challenge in terms of food production (Duro et al. 2020). Factors that can influence food supply include energy consumption, farming, and inadequate distribution, mostly to the developing regions.

[1] Department of Soil and Environmental Sciences, Muhammad Nawaz Shareef University of Agriculture, Multan.
[2] Department of Climate Change, Muhammad Nawaz Shareef University of Agriculture, Multan.
[3] CAS-Key Laboratory of Crust-Mantle Materials and the Environments, School of Earth and Space Sciences, University of Science and Technology of China, Hefei 230026, PR China.
[4] Institute of Plant Protection, MNS University of Agriculture, Multan 66000, Pakistan.
* Corresponding author: m.imran@mnsuam.edu.pk

DOI: 10.1201/9781003558354-2

In order to solve these issues, farmers are now using new technologies and practices such as nano modified products and precision farming to increase production, quality of crop, soil health, water efficiency and water quality, to feed the ever-increasing population, thus enhancing food security.

1.1 Nanomaterials and their Unique Properties

Nanomaterials (NMs) are defined as materials that contain structural features with one or more dimensions being less than one micrometer. A true nanomaterial is defined by structures at the atomic or molecular scale (around 0.2 nm). While some bulk crystals have atomic planes spaced at the nanometer level, their overall macroscopic size means they are generally not considered nanomaterials. The applications of nanotechnology are useful to address environmental concerns and discourse since nanomaterial has an ability to capture greenhouse gases and degrade toxic substances. At this scale, materials possess unique characteristics not present at the macroscopic scale. These properties result from the high surface area-to-volume ratio and quantum confinement effects, which dominate at the nanoscale (Baig et al. 2021). Characterized by at least one dimension less than 100 nanometers, NMs possess optical, magnetic, or electrical characteristics that influence fields such as electronics, medicine, and other industries through their use in nanotechnology. Nanomaterials may be natural or synthesized, and they exhibit unique physical and chemical properties. Some common applications include healthcare products, cosmetics, ecological remediation, and air treatment. In the health sector, nanoparticles are used for drug delivery, while carbon nanotubes are developed for bacterial sensors. These materials are also used in aerospace—such as carbon nanotubes for flexible aircraft wings—and in nanowires like zinc oxide nanowires for flexible solar cells and water treatment. The nanomaterials also assist in preservation processes within the environment including air purification, and ozone layer preservation (Mekuye and Abera 2023). Nanomaterials are materials with at least one dimension in the size range of 1–100 nm, and their properties differ dramatically from those of bulk materials. These features are mostly due to the enormous surface-to-volume ratio and quantum confinement phenomena that are unique to NMs. NMs, for example, have size-related properties, such as quantum confinement, which causes matter, such as quantum dots, to emit light differently depending on its size. Furthermore, nanoparticles exhibit improved mechanical characteristics, elasticity, and heat conductivity, as seen in graphene and carbon nanotubes. Their increased surface reactivity also makes them excellent catalysts for a variety of chemical processes. Because nanostructures have different physical and chemical properties than microstructures and macroscale materials, they have a wide range of applications in medicine, agriculture and food industries, biotechnology (Arcos 2023), environmental pollution detection and control, water treatment, and electronics. These features are inextricably linked to the increased atom fraction and surface energy of nanoscale materials compared to bulk substances (El-Kady et al. 2023).

1.2 Applications in Agriculture and Environmental Sciences

NMs are increasingly being used in agriculture and environmental sciences because they offer potential solutions to the gaps in productivity, sustainability, and ecological health. In agriculture, nanoparticles are used to increase crop development, pest control, and delivery via nano-fertilizers and nano pesticides, minimizing the use of chemical fertilizers and pesticides, therefore lowering pollution. Furthermore, smart sensors evaluate the health of soils and crops in order to improve farming techniques (Neme et al. 2021). Nanoparticles such as titanium dioxide and graphene oxide are used in environmental research to remove pollutants from water and air. They are also used in advanced filters to eliminate undesirable pollutants, ensuring ample clean water supplies. Furthermore, nanoparticles are used to trap greenhouse gases and degrade toxic compounds, illustrating the importance of nanotechnology in resolving environmental concerns and promoting sustainability (Gogos et al. 2012).

1.3 Increasing Use of Nanomaterials in Various Applications

NM applications are quickly developing across all domains of technology because of their huge surface area, quantum size effects, and better mechanical, optical, and chemical capabilities. Nanomaterials such as graphene and carbon nanotubes (CNTs) have emerged as key materials for designing and manufacturing high-speed, miniaturized, and highly efficient electronic devices such as transistors, sensors, and energy storage devices. Nanomaterials are used in the medical industry for drug delivery, which directs the drug to specific targets in the body, limiting the drug's impact on healthy tissues. The cosmetics sector is also leveraging the benefits of NMs, particularly their enhanced efficacy and stability in sunscreens and anti-aging creams. Furthermore, NMs are increasingly being applied to environmental uses, in treatment of water where the NMs remove various pollutants, and also in cleaning of air where the nanomaterials degrade pollutants. These applications illustrate how nanomaterials are likely to revolutionize many industries over the years to come (Batley et al. 2013).

1.4 Interaction between Nanomaterials and Soil Ecosystem

The behavior of nanomaterials in much of the world, and specifically in soil ecosystems, is still under extensive research. This is because, when used, nanomaterials largely interfere with the physicochemical and biological processes in soil. They can alter physical characteristics of soils, including structure, nutrient cycling, microbial distribution, and plant growth. However, the type and intensity of these interactions are a function of one or many factors, such as the properties of nanomaterials, the nature of the soil, and environmental conditions. Knowledge of these interactions is important to evaluate the potential dangers and opportunities of nanomaterials in agricultural and environmental fields (Rajput et al. 2023). The absorption of nanomaterials into the terrestrial biosphere, particularly in soils, is a complex and critical topic due to their ability to significantly alter soil system functionality. Nanomaterials introduced into soil systems can affect soil minerals, organic matter, and microbes in a variety of ways, potentially influencing soil structure, fertility, and microbial activity. Certain nanomaterials, such as silver nanoparticles, have antimicrobial characteristics that may influence soil microbial populations and their role in nutrient cycling and organic matter decomposition. Conversely, some nanomaterials may improve soil fertility by enhancing nutrient availability to plants, as seen in nanoscale fertilizers that promote plant growth. However, doubts remain about the long-term impacts of nanomaterials on soil biota and the possibility of bioaccumulation, which might lead to toxicity in soil-dwelling organisms and, as a result, affect organisms further up the food chain. As a result, while the use of nanoparticles in agriculture has potential benefits, further research into their effects on soil ecosystems is required to counteract any negative effects of nanotechnology (Tourinho et al. 2012).

1.5 Purpose of Writing this Book Chapter: Strength and Powerful Points

The goal of this book chapter is to provide a thorough assessment of current knowledge about nanomaterials' applications in agriculture and environmental research, with a focus on their behavior in soil systems. It provides an overview of the existing literature on nanomaterials to better understand the potential benefits and drawbacks of using them to improve agricultural productivity and environmental sustainability. Furthermore, it aims to fill knowledge gaps and analyze potential dangers, thereby offering specific directions for future research and policy formulation. The purpose of this chapter is to explain how these two components interact and how nanomaterials can alter the physical and chemical properties of soil, as well as the microbial populations that live in it. As a result, recognizing the chapter's strengths and key points will help readers understand the importance of this research area and its contribution to environmental advancement (Ragab et al. 2022).

2. Nanomaterials and their Types

Nanomaterials are of great importance today due to the tremendous innovations arising from their unique physical, chemical, and biological properties, which make them outperform conventional materials in terms of their bulk characteristics. Given that there are millions of nanomaterials used across industry and reported in scientific literature, it is logical to classify them according to several key parameters. In recent years, nanomaterials have garnered significant attention from researchers due to their size and enhanced physical, biological, and chemical characteristics compared to bulk materials. These NMs are grouped according to their size, chemical constitution, morphology and origin. Some of the manufactured NMs are being synthesized from various sources and categorized according to their types. Many NMs have been mass-produced to meet the needs of a wide range of industries. NMs are formed from two main sources: synthetic sources and naturally occurring nanoparticles. This chapter discusses the types and classifications of NMs, as well as the various types of nanomaterials isolated from natural and manmade sources.

2.1 Classification of Nanomaterials with Respect to Different Parameters

Nanomaterials are materials that exhibit nanoscale dimensions and are utilized in a diverse range of applications, including drug delivery and imaging. These materials can be classified into four primary categories: zero-dimensional (0D), one-dimensional (1D), two-dimensional (2D), and three-dimensional (3D). Carbon-based nanoparticles are renowned for their durability, electrical conductivity, and adaptability, whereas metal-based nanomaterials are used in catalysis, electronics, and medicine. Ceramic-based nanomaterials—composed of oxides, carbides, and nitrides—are employed in photocatalysis, coatings, and drug delivery due to their chemical stability and hardness. Polymeric nanostructures, made from organic polymers, can be designed to transport drugs, proteins, or DNA to specific cells or tissues (Buzea et al. 2007). The aerospace, automotive, and biotechnology industries use composite nanomaterials, which are made up of several nanoparticles.

2.2 Physicochemical Properties that Influence the Interaction of Soil and Nanomaterials

Chemical and physical properties such as size, shape, surface area, surface charge, chemical nature, and many other factors are influential parameters that determine the toxicity of nanomaterials and influence their interaction with soil (Ali et al. 2021), as summarized in Figure 1.

The physical and mechanical properties of the soil, such as structural organization, bulk density, porosity, permeability, texture, temperature, moisture content and others, are of interest because they are correlated to productiveness of specific plant species and root development (Bayat et al. 2019). Soil and nanomaterial interaction is a dynamic process involving soil physical and chemical characteristics, which influence the behavior of nanomaterials in the environment in terms of fate, transport, and bioavailability. These properties can affect the adsorption, agglomeration, dissolution, and transformation of NMs in soil (Suazo-Hernández et al. 2023). For instance, NM mobility in the soil can be influenced by soil properties such as high organic matter content and high CEC, which may enhance adsorption of nanomaterials, and low soil pH, which enhances the dissolution of particular nanomaterials. Nanotechnology is among the emerging novel research areas of the present time. The contribution of nano related applications to various research areas in soil science cannot be overemphasized, as the subject has been explored in various fields (Ghormade et al. 2011). NMs are implemented in the management of farm practices for the following reasons: their small size, high reactivity, and large specific surface area. In addition, the presence of NMs also influences properties of the soil structure. Furthermore, it is suggested that the concentration and size of ENPs may exert influence on the physical characteristics of soil (Komendová et al. 2019) many studies have reported its adverse effect on biota, but the potential influence on the properties and functions of environmental compartments has largely been ignored. In this work, the effect

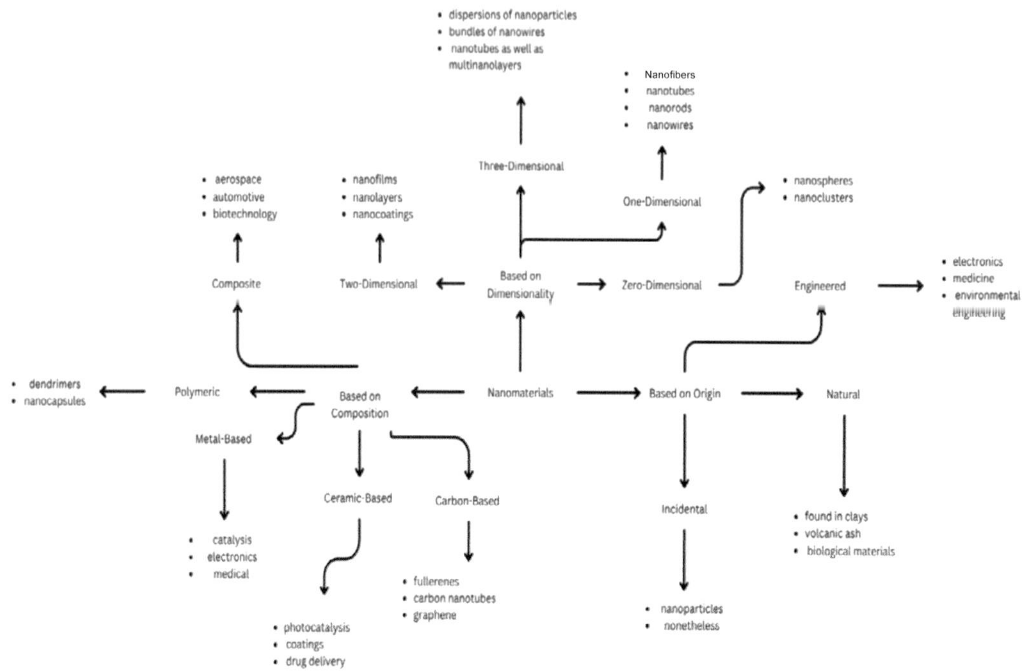

Figure 1. Physicochemical Properties that Influence the Interaction of Soil and Nanomaterials.

of Pt nanoparticles on the functions and properties of model soil organic matter has been studied. Using differential scanning calorimetry and molecular modeling, the effect of a wide range of 3 nm Pt nanoparticles concentrations on water holding capacity, the strength of water binding, the stability of water molecule bridges and the content of aliphatic crystallites was studied. It was found that strong hydration of the nanoparticles influences the 3D water structural network and acts as kosmotropic agents (structure-forming. The cation exchange capacity (CEC) of the soil depends on surface charge as well as surface area. In agricultural soils, CEC is an important parameter since it characterizes the ability of the soil to adsorb nutrients, i.e., to be fertilized (Das et al. 2018). A few studies have reported findings on CEC in the presence of ENPs. Soil pH is a key component in determining soil fertility and overall health (Fajardo et al. 2015). The pH values obtained in soils with ENPs vary. In certain investigations, changes in soil pH were caused by the indirect impact of ENPs (Ben-Moshe et al. 2013). The presence of ENPs causes fluctuations in soil pH, which are determined by matrix parameters and ENP type. The soil Eh indicates oxidation-reduction reactions and is affected by oxygen (O_2) concentration, precipitation, temperature, and organic matter content (Zhang et al. 2020). Eh is an essential parameter in agronomy because it regulates the functioning of the soil-plant-microorganism system as well as the solubility of nutrients and pollutants. Over time, ENP deposition in soils increases, improving particle rigidity and causing changes in pH, EC, Eh, and SOM. ENPs can chemically interact with nutrient particles to form various complexes or precipitates. There is limited published data on the specific effects and influence of ENPs on soil characteristics; therefore, there are potential advantages along with issues and disadvantages of ENPs that need to be investigated further.

2.3 Discussion of Challenges and Consideration for Nanoparticle Synthesis

The most challenging aspect is gaining adequate control over particle size, shape, and surface characteristics, which often influence their performance and applications. Also common among synthesis methods is the use of hazardous chemicals that could pollute the environment and cause adverse health effects. Another factor is achieving repeatability and feasibility in nanoparticle formation processes, which is important when considering their utility for large-scale production. Additionally, understanding the atomic-level processes involved in nanoparticle formation and growth is critical for controlling synthesis parameters and generating novel materials. These are challenges that require ongoing efforts to harness nanotechnology for overcoming the above-mentioned limitations and achieving optimal utilization of nanoparticles.

Regularity of size and shape is especially important because these factors demonstrably affect the properties and potential uses of nanoparticles. Other factors that should be monitored include temperature, pH, and reaction time, as these influence synthesis outcomes However, the direct toxicity and disposal of toxic chemicals during synthesis also form a major concern. Another challenge is achieving economic scalability while maintaining quality and reducing costs for large-scale applications (Zhang et al. 2014).

NMs can be regarded as particles of the century and have attracted significant attention. These nanomaterials have demonstrated progress across a number of sectors due to their size-dependent physicochemical properties and have proven to be some of the most inspiring materials in the technical world. Nanomaterials with minuscule size bear high reactivity owing to the large surface plasmon resonance, which significantly enhances their analytical capabilities (Baker et al. 2017).

3. Entry Routes of Nanomaterials into Soil

Nanotechnology is a rapidly growing field, and engineered nanoparticles (ENPs) are being used in a variety of industries, including cosmetics, bioremediation, and water purification. As a result, ENPs are likely to be released into the environment during their manufacture and use, potentially contaminating food and water sources. Nanoparticles may also occur naturally in the environment, originate from water bodies, or be released into the environment as waste. Consequently, concerns have been raised about the possible effects of indirect human exposure to ENPs. To identify the factors and processes that are likely to influence the behavior and impacts of nanoparticles in the environment, it is essential to thoroughly explain the probable pathways of exposure to ENPs in environmental media. Global trends in NM production and application indicate that 49% occurs in the United States, 30% in the European Union, and the remaining in other regions. From the European Union, the UK forms the largest market contributing to nearly one-third of the amount (Aitken et al. 2006). Some of the companies reported that they were preparing or incorporating NMs in their products. Among them, however, only some of them (such as manufacturers of paints and varnishes, cosmetics, catalysts, polymer compositions, etc.) were using NMs to any noticeable extent. However, with more new applications and markets that are yet to be developed for NM-based products, this scenario is expected to change in the future.

The primary current applications of NMs in the UK included catalysts, lubricants and fuel additives, paints, pigments and coatings, conductive inks and printing materials, cosmetics and personal care products (e.g., sunscreens), drug delivery/bio-nanotechnology, functional coatings (e.g., on glass, textiles), hydrogen storage and fuel cell applications, nanoelectronics and sensor devices, optics and optic devices, security and authentication applications, and structural material. An additional study found that NMs have potential applications in construction materials (including coatings inside drinking-water pipes), detergents, food processing, paper manufacturing, agrochemicals and plant protection products, plastics (including food packaging), and weapons and explosives.

3.1 Agricultural Inputs (Use of Nanomaterials as Nano-fertilizers and Nano-pesticides)

Nanotechnology is viewed as a unique method for addressing a variety of issues in agriculture and related industries. The agricultural system is challenged with difficulties such as the development of microbial resistance owing to excessive pesticide usage and the deterioration of soil health due to excessive fertilizer application, both of which have negative environmental consequences. A considerable amount of applied fertilizers and insecticides is lost to the environment and not utilized by plants, reducing the effectiveness of traditional approaches. Nanosensors provide early-stage disease identification, allowing for rapid management actions and increased output. Various disease control products, such as nano-pesticides, nano-fungicides, and nano-bactericides, are currently in use to protect crops against various biological stresses induced by microorganisms (Rani et al. 2020).

3.2 Anthropogenic Nanomaterials

NMs have been created by humans for millennia, as byproducts of simple combustion (with sizes as small as several nm) and food cooking, and more recently, chemical manufacturing, welding, ore refining and smelting, combustion in vehicle and airplane engines (Rogers et al. 2005), combustion of treated pulverized sewage sludge, and combustion of coal and fuel oil for power generation (Linak et al. 2000). While engineered nanoparticles have been on the market for some time and are commonly used in cosmetics, sporting goods, tires, stain-resistant clothing, sunscreens, toothpaste, food additives, and so on, these nanomaterials, along with newer, more deliberately fabricated nanoparticles such as carbon nanotubes, make up a small minority of environmental nanomaterials. The manufacture of man-made nanoparticles ranges from the well-established multi-ton-per-year production of carbon black (for vehicle tires) to the microgram quantities of luminous quantum dots (for biological imaging). Diesel and automotive exhaust are the principal sources of air nanoparticles and microparticles in metropolitan areas. Prior to the publication of these results (Evelyn et al. 2003), they were assumed not to exist in the environment, and their presence was attributed solely to engineering by material scientists. Nanoparticles account for 20% of particle mass but more than 90% of the number of diesel-generated particles. The application of nanoparticles in cosmetics is not new. Black soot and mineral powders have been utilized as cosmetics for thousands of years in ancient Egypt, and some of them are still used today. Because of recent advances in nanotechnology, engineered NMs have gained popularity in the cosmetics sector for a variety of reasons. Many cosmetic and personal care products contain nanomaterials (Serpone et al. 2001). Personal care items include deodorants, soap, toothpaste, shampoo, and hair conditioner, as well as sunscreen and cosmetics.

NMs can be synthesized by many methods, including: gas-phase processes (flame pyrolysis, high-temperature evaporation, and plasma synthesis); vapor deposition synthesis (electron, thermal, and laser beam evaporation); colloidal or liquid-phase methods, in which chemical reactions in solvents lead to the formation of colloids; and mechanical processes such as grinding, milling, and alloying (Seaton et al. 2009).

3.3 Industrial Sources

Industrial sources of nanomaterials pose significant challenges in terms of disposal and potential accidents due to their unique properties and potential environmental impacts. Nanomaterials from industrial activities can enter waste streams through various pathways. For example, nanomaterials used in coatings, composites, plastics, light bulbs, and glass and ceramics are often released to incinerators and landfills. The disposal of nanomaterials is often not regulated stringently. In the U.S., manufacturers producing less than 100 kg of nanomaterials annually are exempt from notifying the EPA and establishing contingency plans, making it difficult to track and evaluate the risks associated with nanowaste (Holder et al. 2013). Accidental exposure to nanomaterials during industrial processes can occur through inhalation, dermal contact, or other routes. Activities like machining, sanding, drilling, or cleaning systems containing nanomaterials can aerosolize these particles, leading to potential health risks for workers (Judith and Vasudevan 2022).

3.4 Natural Sources

Natural NMs are formed in the soil, water, atmosphere, and space by physical, chemical, and biological processes. In general, natural nanomaterials are discovered in the atmosphere (including the entire troposphere; nanoparticles are suspended or moved by air even at high levels), hydrosphere, lithosphere, and biosphere (micro- and macro-organisms up to humans) (Tan et al. 2018). NMs arise naturally through physicochemical processes. Natural sources of NMs include forest fire products, volcanic ash, ocean spray, radioactive decay, and the weathering of metal- or anion-containing rocks (Barhoum et al. 2022). Nanoparticles and nanostructured materials are assumed to have formed from meteorites created during the Big Bang event, which resulted in the development of the cosmos and Earth. Nanodiamonds collected from meteorites are thought to be the most abundant type of pre-solar grain (Dai et al. 2002). Nanomaterials found naturally on Earth have long been linked to the emergence of life, from the first cells to humans. Nanomaterials and their derivatives have been created and used by humankind since prehistoric times. Advances in nanomaterial characterization techniques (Barhoum and García-Betancourt 2018) allow for better assessment of all types of nanomaterials and their origins in Earth systems. Long-term effects on the Earth's environment and human health can be estimated.

3.5 Transport Mechanisms

Agricultural soils are one of the locations where engineered nanomaterials (ENMs) accumulate. Soil exposure to ENMs occurs through the intentional use of nano-agrochemicals, as well as incidental contamination from industrial waste release, irrigation with wastewater or gray water, amendment with ENM-loaded sludge (soil conditioning to stimulate plant growth), or atmospheric fallout (Medina-Velo et al. 2017). Nanomaterials, due to their unique properties, exhibit distinct transport mechanisms compared to larger particles. These mechanisms are crucial in understanding their behavior in various environments, including biological systems and engineered materials. Nanomaterials can be transported through various mechanisms, significantly impacting their environmental distribution, bioavailability, and potential effects on human health and ecosystems. The primary transport mechanisms of nanomaterials include diffusion, advection, sedimentation, and aggregation (Nie et al. 2022).

The stability of ENMs in aquatic environments is an important element in determining their transit and ultimate fate. Large ENM aggregates will soon precipitate out, limiting transport and bioavailability (Lin et al. 2010). Soil serves as an essential sink for ENMs after they are released into the environment, as well as a potential source of ENMs in groundwater. Understanding ENM transport behavior in natural soil systems is critical for determining their possible influence on the food chain and groundwater (He et al. 2007). In most cases, columns packed with porous media (e.g., glass beads, silica sand, model soils) were utilized to investigate the filtration of various ENMs supported or not by dispersants. Particle transport in porous media is largely determined by the pace at which stationary grain surfaces catch or filter them. Particle filtration theory can be utilized to better explain the movement of ENMs in soil (Schrick et al. 2004).

Dissolution may be a significant fate for nanoparticles, particularly metal-based nanomaterials in the environment and inside organisms. A few studies have investigated the dissolution of metal-based nanoparticles in culture media and their contribution to nanotoxicity (Wang et al. 2009). Although the exact contribution of nanoparticle dissolution to nanotoxicity varies between studies, it is widely agreed that dissolution can play an important role in determining the toxicity of various nanomaterials to organisms (Auffan et al. 2009). As a result, understanding the dissolution of ENMs is critical for determining their fate in the environment and the underlying mechanism of nanotoxicity.

Recent research has investigated the toxicity of ENMs to aquatic organisms, with the bulk of findings indicating that the organisms ingested ENMs. Although terrestrial creatures may primarily absorb nanoparticles by inhalation or ingestion, aquatic organisms mostly absorb ENMs through regular feeding behavior and/or water filtration through the gills. Engineered nanomaterials consumed with food may pass through the digestive tract before being expelled or disseminated throughout the body (Bystrzejewska-Piotrowska et al. 2009). In biological systems, a wide range of proteins, enzymes, polysaccharides, and other compounds have been shown to rapidly adsorb to ENM surfaces and hence have a major impact on the toxicity and fate of ENMs in the organism's body, warranting more investigation.

4. Techniques for Studying Nanomaterials in Soil

Studying nanomaterials in soil involves various techniques to understand their presence, distribution, and effects on the soil environment. It also analyzes chemical composition, optical qualities, mass, electrical properties, and other physical or chemical characteristics. There are numerous nanometrological techniques used to gather morphological data on nanoparticles, such as SEM, TEM, AFM, XRD, and EDX, as well as various other techniques to study the electrical behavior of nanoparticles. However, studying nanomaterials using these techniques presents its own set of challenges, such as inherent properties of nanomaterials that may impede accurate study of their features, insufficient production at the laboratory scale, and the multidisciplinary nature of the field, all of which affect access to characterization tools required for specific applications. It is also vital to investigate the usability of data obtained from characterization tools, the limitations of these techniques, and the tool's capabilities while keeping the application in mind (Bano and Munawar 2023).

Various characterization approaches can be classified based on their application. For example, microscopy-based methods for surface, morphology, and size data (TEM, FESEM, AFM, etc.); magnetic techniques for studying specific types of groups (SQUID, VSM, etc.); and structural or elemental properties (X-ray spectroscopy, XRD, UV-Vis spectroscopic techniques) are summarized in Table 1.

4.1 Emerging Techniques for Studying Nanoparticle-Soil-Microbe Interactions

The influence of ENMs in soil is highly dependent on soil type and ENM characteristics (Gottschalk and Nowack 2011). Depending on their bioavailability, ENMs may interact strongly with charged particles in the soil, be soluble in soil water, or be absorbed by diverse living organisms in the soil (Abbas et al. 2020). Aside from the soil, ENMs can enter other ecosystems such as sediments

Table 1. Some key parameters and the instruments used to measure them.

Analytical Techniques		
Techniques	**Uses**	**References**
Transmission Electron Microscopy	It provides high-resolution images of nanomaterials. Allows direct observation of their morphology, and internal structure	(Jayawardena et al. 2021)
Atomic Force Microscopy	Provides detailed surface topography	
Scanning Tunneling Microscopy	To examine the surface morphology and topography of nanomaterials	(Zhou et al. 2018)
X-ray Diffraction Spectroscopy	To determine the crystalline structure and phase composition of nanomaterials. It provides information on the crystallite size, lattice parameters, and the degree of crystallinity.	(Yan et al. 2018)
Raman Spectroscopy	It is highly sensitive to changes in molecular structure and is used to study the chemical composition, phase transitions, and stress or strain within nanomaterials.	(Farber et al. 2019)
FTIR	Identifies the functional groups and chemical bonds To understand surface modifications, and chemical interactions	(Bechtel et al. 2020)
X-ray Photoelectron Spectroscopy	Crucial for elemental analysis and surface chemistry studies	(Stevie and Donley 2020)

and aquatic bodies. ENMs' physicochemical properties, such as size, morphology, chemical characteristics, and other surface properties, play an important role in determining their fate in soil because they significantly alter the particles' electrical, optical, and catalytic behavior.

Another study found that adding TiO_2 and polystyrene ENMs to the rhizosphere of lettuce seedlings reduced the population of rhizospheric bacteria, which inhibited shoot and root growth (Kibbey and Strevett 2019). When soils were irrigated with ENM-containing wastewater, the population of cyanobacteria and a variety of unknown Archaea increased, whereas A. thaliana's life cycle was significantly reduced (Khodakarami et al. 2022). Carbon nanoparticles also altered the microbial ecology of the rice rhizosphere and caused environmental toxicity (Hao et al. 2018).

ENMs progressively accumulate in the soil via many pathways, and many of them have microbicidal properties, at least at higher levels. As a result, these ENMs may hinder important microbial biogeochemical processes. These processes include ammonification, denitrification, nitrogen fixation, phosphate solubilization, and plant growth-promoting (PGP) activities, all of which are essential for soil fertility and ecosystem health. Several studies have been published on the effect of ENMs on microbe-mediated cycling of a range of critical nutrients, including carbon and nitrogen (Klaine et al. 2008). ENMs have been shown to influence a variety of different processes and PGP activities, including nitrification, denitrification, potassium and phosphate solubilization, and microbial plant disease defense. The soil contains a variety of natural nanomaterials, including clay, organic compounds, and numerous metal and metal oxide nanoparticles. However, numerous ENMs, as described earlier, are also eliminated.

However, as previously noted, some ENMs are disposed into soil and water as a result of the widespread use of nanomaterial-based products (Tratnyek and Johnson 2006). Approximately 60–90% of ENMs are disposed of in landfills, with soil accounting for the remaining 10–25%. This indicates that tons of these ENMs reach the soil each year. SiO_2, titania (TiO_2), zinc oxides (ZnO), iron, and alumina (Al_2O_3) are the most common ENMs generated in varied situations. The dynamic multimodal fate and transport model (nanoFate) predicted the time-dependent accumulation of ENMs in various settings at high concentrations.

Unfortunately, the concentration of several highly manufactured ENMs, such as TiO_2 and ZnO, frequently surpasses the lowest hazardous threshold. Notably, nanomaterial interactions with the soil ecosystem cannot be generalized since soil, soil microbiome, and plant microbiome differ from one another. Several studies have looked into the effects of ENMs on soil-dwelling organisms such as bacteria and plants. Microorganisms play essential roles in the biogeochemical cycling of nutrients and help in the mobilization and fixation of nutrients for plants, thereby improving total soil fertility and plant health (Khan et al. 2021).

4.2 Impacts of Nanomaterials on Soil Physical, Chemical and Biological Properties

NMs can also reduce the overall mass or volume provided by the soil in two ways: either through direct filling of the interparticle space in the soil or indirectly through enhancement of the biomass or soil humus content through the force with which the particle bonds (Sharma and Kumar 2024). For instance, they can enhance the water retention capacity of a particular soil by either providing a larger surface area and increased hydrophilicity to the soil particles or by developing water swelling hydrogels. This can assist plants in avoiding water loss and handling drought stress conditions. Furthermore, there is a substantial reduction in soil liquefaction along with a decrease in soil settlement, as evidenced by (Arora et al. 2022).

NMs have the ability to improve the quality of the soil as well as crop yield through modulation of other soil processes. Nanomaterials in the soil can affect the pH, CEC, and nutrient/contaminant content in the soil. Soil pH tests and electrochemical methods are employed to determine variation in soil pH or pH level resulting from association with nanomaterials. NMs have the potential to enhance the efficiency of nutrients such as fertilizers, pesticides, and other inputs through their controlled release and by minimizing their pollution effects (Raliya et al. 2016). NMs can also impact the redox potential and chemical reactions within soil in areas of nutrient cycling and contaminant biodegradation.

These alterations in biological properties of the nanomaterials enable interaction with microbial activity that subsequently affects physical and chemical characteristics of the soil such as enzyme activity and soil respiration. Nanotechnology and NMs have the potential to be used in agriculture for improving the properties of soil and crop production by managing bacterial activity and nutrient transformations. Some impacts of NMs can also be observed on the activity and multitude of soil microorganisms, which in turn influences the decomposition of organic matters of the soil and the availability of nutrients. Enzyme assays are used to measure the changes on the relevant enzymes of the soil, in connection with the nanomaterials. Soil CO_2 efflux is determined using techniques known as respirometry wherein rates of respiration are obtained to assess microbial nutrient transformations and soil quality (Liu et al. 2021).

4.3 Impacts of Nanomaterials on Soil Health and Productivity

NM can have an impact on soil health and productivity, altering plant development and crop yields. Among the many inorganic pollutants that may wind up in soil, NMs are one of the most commonly used (Gottschalk and Nowack 2011). NMs may change the microbial ecology of the soil or plant, which may have a direct or indirect impact on plant growth. The soil conditions under which crops are cultivated are often regarded as the most significant factor for the plant. Thus, soil stresses such as salt, drought, compaction, acidity, suboptimal root zone temperature, nutrient availability, soil types, and soil biota functionality can all have a negative impact on plant performance in the future. These stresses (biotic-abiotic) have a direct impact on agriculture; nevertheless, they can be managed in a sustainable fashion by applying modern advances, particularly nanotechnological techniques and nano-enabled products (Rajput et al. 2021).

A recent study found that applying carbon-based NMs increased *Z. mays* growth by enhancing nutrient uptake while also boosting soil fertility by stimulating soil enzymes. Metal-based NMs such as Fe, Cu, Co, and ZnO promoted growth in *Glycine max* under drought stress conditions. Se-based NMs alleviate high-temperature stress in *Sorghum bicolor* (Djanaguiraman et al. 2018). A wide range of microbes, including *Brevibacterium frigoritolerans*, *Bacillus thuringiensis*, and *Bacillus velezensis*, have been shown to reduce NaCl stress through root secretion. The combined application of Si-Zn NMs and plant growth-promoting microorganisms minimizes the salt's impact on plant growth (Osman et al. 2021). The use of ZnO-based NMs and biofertilizers together has been shown to protect *Carthamus tinctorius* from salt stress by enhancing antioxidant enzyme activity and lowering malondialdehyde and proline levels (Yasmin et al. 2021). These are a few studies summarized in this review that demonstrate how nanotechnology may manage several types of soil stress that directly or indirectly affect plant development and yield quality. Nanofertilizers have a huge surface area and particles that are smaller than the pores in a plant's root system and leaves, increasing their penetration into the plant and nutrient utilization efficiency (Hong et al. 2021). The usage of NM-encapsulated fertilizers has been found to boost nutrient bioavailability and absorption in crop plants (Khan and Rizvi 2017). The capacity of zeolite-based nanofertilizers to gradually provide nutrients to agricultural plants increases crop nutrient supply during the growing season while decreasing nutrient loss due to volatilization, leaching, denitrification, and soil fixation (Soltys et al. 2020). Nanotechnology enables the achievement of sustainable development goals by addressing major concerns such as malnutrition, poverty, and food security through improved soil health and sustainable crop production. The potential of nanotechnology propels a new green revolution by reducing the risks involved with farming. Despite its numerous applications, nanotechnology has not yet been completely employed, notably in agriculture and associated sectors. To better understand the interaction between plants and microorganisms and to reduce the pressures that the soil imposes on plants, NMs can be applied to agriculture (Rajput et al. 2023).

5. Future Prospects and Research Directions

Soil nanotechnology research seeks to improve soil fertility, health, and agricultural productivity by controlling nanoparticles. It aims to solve challenges related to environmental sustainability, nitrogen management, and soil quality. Significant progress has been made in developing NM fate and uptake models over the last decade; however, as previously discussed, more work is needed to produce models suitable for NM risk assessment as well as to advance our fundamental understanding of, and ability to predict, NM fate and uptake. Interdisciplinary collaboration among soil scientists, agronomists, chemists, and environmental scientists is critical for future studies. The study of nanomaterials in soil ecosystems is still in its early stages, with major knowledge gaps. Understanding long-term fate, interactions, bioaccumulation potential, and standardized detection and quantification methodologies are among the top research goals.

5.1 Development of Advanced Models for Predicting Nanoparticle Behavior in Soil

This study reviews current models for predicting the environmental fate of NMs and their uptake by cells and organisms. It underlines the need to understand nano-specific processes such as aggregation, sedimentation, and dissolution, as well as the dynamic nature of organismal NM uptake. The uptake of NMs by organisms is influenced by the physicochemical properties of the NMs themselves, as well as the properties of the environment and the organisms involved. Numerous studies have documented the uptake of various NMs, with evidence shown by BioConcentration Factors (BCFs) greater than one. Future efforts should focus on parameterization, calibration, and validation using known data rather than expanding to more sophisticated models. Collaboration between experimentalists and modelers is recommended for rapid progress. Additionally, several articles review the current state-of-the-art regarding NM uptake and accumulation in organisms, outlining the routes of uptake and discussing the specific factors that govern this process (Darlington et al. 2009). However, there is still limited information available on modeling the uptake and accumulation of NMs in organisms.

The bulk of NMs used in goods today are essentially nanoscale versions of particles that have been used for decades, such as silica, titania, and ceria, or new nanoscale materials, such as quantum dots and carbon nanotubes, developed to have specific qualities. However, they are relatively simple materials when compared to many of the sophisticated materials that will be available in the future, which will be multi-component (for example, core-shell, segmented, layered materials with complex geometries and the possibility for evolution or self-healing). These new NMs are being created to reach unique features and functions, which may have an impact on NM transformation processes and rates when compared to existing NMs (Baalousha et al. 2016). The models in the manuscript are based on spherical particles and may not be applicable to non-spherical particles. Sharp edges or points can promote dissolution, aggregation, or uptake, resulting in crystal structural flaws or surface spaces. To avoid potentially inappropriate data, careful consideration of NM geometry and constituent compositions is required.

5.2 Exploration of Nanomaterial Applications for Soil Remediation and Improvement

NMs could be effectively used for the purpose of reclamation and modification of soils, thereby providing unique solutions to some of the greatest ecological problems. Because of their physicochemical properties including surface area, reactivity and possibility of surface functionalization, they are very effective in the adsorption of contaminants in soil. For instance, some nanomaterials have the ability to adsorb heavy metals, reduce organic pollutants and ameliorate detrimental chemicals, hence rejuvenating the fertility of the soil. Furthermore, nanomaterials demonstrate significant potential in enhancing critical soil quality parameters, including water retention capacity, nutrient availability, and structural stability, which are fundamental to agricultural productivity (Attia and

Elsheery 2020). For instance, graphene oxide-based carbon nanomaterials enhance the ability of soil to aerate as well as enhance water retention, thus enhancing root development and crop yields. Nevertheless, as with other nanotechnology applications in environmental science, the use of nanomaterials for improving the contaminated soil properties and remediation is still in its infancy. More research needs to be conducted in order to determine the long term effects of such nanomaterials on the soil and its microbial communities (Pan and Xing 2012). Effective NM based technologies that are safe to apply on soils could completely transform this sector and improve soil management and environmental conservation for agricultural purposes.

5.3 Assessment of Long-term Environmental and Human Health Risks

Evaluating the survival implications of nanomaterials, as well as their impact on the environment and human health, is critical given the growing use of nanotechnology in industries such as agriculture, medicine, and consumer products. Because of their small size, nanomaterials have the ability to pass through membranes in living organisms, which can pose a risk (Hristozov and Malsch 2009). Long-term exposure to nanomaterials may cause bioaccumulation, concentration, and chronic toxicity in both biota and humans. Previous research suggests that nanomaterials can accumulate in the environment, influencing soil, water, and air matrices in ways that are not well understood. Furthermore, research is needed into the pathways through which nanomaterials may pose health hazards to people, such as potential carcinogenic, neurotoxic, and endocrine-disrupting effects. As a result, it is critical to build comprehensive risk assessment models that take into account the advantages and potential drawbacks of nanomaterial use in both short- and long-term scenarios to avoid negative environmental and human health effects (Jamuna and Ravishankar 2014).

5.4 Policy and Regulatory Frameworks for Nanomaterial Use in Agriculture

Nanotechnology offers new strategies for developing quality agricultural products. However, there are some potential ill effects that need to be mitigated. Researchers should focus on green synthesis to make it more reliable and eco-friendly. Nano-priming of seeds can help maintain sustainable agriculture by creating nano-primed plants that change at the molecular level without harming the environment or plants. Primary screening and monitoring of nanoparticle delivery are crucial. Regulatory systems for handling nanotechnologies should be updated to ensure safety measures and regulations are effective (Amenta et al. 2015). Developing strong policy and regulatory frameworks for the use of nanomaterials in agriculture is critical to ensuring their safe and sustainable use. As nanotechnology advances, its application in agricultural practices—such as increasing soil fertility, improving crop protection, and remediating contaminated soils—has shown great promise. However, the unique features of nanoparticles present regulatory issues, as conventional safety assessments may fail to capture their environmental and health implications. Effective regulations should cover the whole lifecycle of NMs, from manufacture and application to disposal, while promoting transparency and public participation in decision-making. Regulations should be based on thorough scientific research, with standardized testing techniques and risk assessment models tailored to the unique properties of NMs (Kumari et al. 2023).

6. Conclusion

NMs are increasingly being integrated into agricultural systems, offering potential benefits such as soil fertility improvement, crop protection, and soil treatment. However, their interaction with the soil ecosystem is crucial for responsible innovation. Advanced analytical techniques are essential for understanding how nanomaterials affect soil properties and productivity. Challenges include interaction with soil components, mobility, bioavailability, and persistence. To ensure safe use, knowledge gaps must be identified, research priorities set, models developed, and long-term environmental and human health risks assessed. Establishing robust policy and regulatory frameworks is crucial for managing the risks and benefits of nanomaterials in agriculture. The current evaluation looks at how to deal with various soil stressors and grow crops in a sustainable manner. The studies provided here demonstrate the potential of NMs in soils to improve microorganisms or agriculturally significant microbes and boost their function, hence improving pollutant biodegradation or reducing soil pressure. Global soil health degradation is becoming a major issue in supporting human needs, notably food security. The inclusion of NMs into biological processes aids in the removal of polluted soil. The incorporation of nanotechnology into several approaches has created new opportunities for improving soil health and agricultural productivity.

7. Acknowledgement

We (Authors) are highly thankful to the University of Agriculture Faisalabad and Yangzhou University for their support in this publication. Authors also want to thank Professor Dr Abdul Wakeel from Soil Fertility and Plant Nutrition Laboratory, University of Agriculture Faisalabad, Pakistan, and Dr. Akhtar Hameed, Assistant Professor (Plant Pathology), Institute of Plant Protection, MNSUAM, for providing us this opportunity. We are highly thankful to our families who sacrificed their good times for the accomplishment of this milestone.

References

Abbas, Q., B. Yousaf, M.U. Amina, M.A.A. Ali, M.A.M. Munir, A. El-Naggar, J. Rinklebe et al. 2020. Transformation pathways and fate of engineered nanoparticles (ENPs) in distinct interactive environmental compartments: A review. Environ. Int. 138: 105646.

Aitken, R.J., M.Q. Chaudhry, A.B.A. Boxall and M. Hull. 2006. Manufacture and use of nanomaterials: current status in the UK and global trends. Occup. Med. 56: 300–306.

Ali, S.S., R. Al-Tohamy, E. Koutra, M.S. Moawad, M. Kornaros, A.M. Mustafa et al. 2021. Nanobiotechnological advancements in agriculture and food industry: Applications, nanotoxicity, and future perspectives. Sci. Total Environ. 792: 148359.

Amenta, V., K. Aschberger, M. Arena, H. Bouwmeester, F. Botelho Moniz, P. Brandhoff et al. 2015. Regulatory aspects of nanotechnology in the agri/feed/food sector in EU and non-EU countries. Regul. Toxicol. Pharmacol. 73: 463-476.

Arcos, D. 2023. Nanomaterials in Biomedicine 2022. Int. J. Mol. Sci. 24: 9026.

Arora, S., A. Kumar, S. Ogita and Y.-Y. Yau. 2022. Biotechnological Innovations for Environmental Bioremediation. Springer Nature Singapore, Singapore.

Attia, T.M.S. and N.I. Elsheery. 2020. Nanomaterials: Scope, applications, and challenges in agriculture and soil reclamation. pp. 1–39. *In*: Hayat, S., J. Pichtel, M. Faizan and Q. Fariduddin [eds.]. Sustainable Agriculture Reviews 41: Nanotechnology for Plant Growth and Development. Springer Int. Publ., Cham.

Auffan, M., J. Rose, J.Y. Bottero, G.V. Lowry, J.P. Jolivet and M.R. Wiesner. 2009. Towards a definition of inorganic nanoparticles from an environmental, health and safety perspective. Nat. Nanotech. 4: 634–641.

Baalousha, M., G. Cornelis, T.A.J. Kuhlbusch, I. Lynch, C. Nickel, W. Peijnenburg and N.W. van den Brink. 2016. Modeling nanomaterial fate and uptake in the environment: current knowledge and future trends. Environ. Sci. Nano 3: 323–345.

Baig, N., I. Kammakakam and W. Falath. 2021. Nanomaterials: a review of synthesis methods, properties, recent progress, and challenges. Mater. Adv. 2: 1821–1871.

Baker, S., T. Volova, S.V. Prudnikova, S. Satish and N. Prasad M.N. 2017. Nanoagroparticles emerging trends and future prospect in modern agriculture system. Environ. Toxicol. Pharmacol. 53: 10–17.

Bano, K. and A. Munawar. 2023. Analytical techniques for characterization of nanomaterials. pp. 28–51. *In*: Munawar, A. [ed.]. Modeling and Simulation of Functional Nanomaterials for Forensic Investigation. IGI Global.

Barhoum, A., M.L. García-Betancourt, J. Jeevanandam, E.A. Hussien, S.A. Mekkawy, M. Mostafa et al. 2022. Review on natural, incidental, bioinspired, and engineered nanomaterials: history, definitions, classifications, synthesis, properties, market, toxicities, risks, and regulations. Nanomaterials. 12: 177.

Barhoum, A. and M.L. García-Betancourt. 2018. Physicochemical characterization of nanomaterials: size, morphology, optical, magnetic, and electrical properties. pp. 279–304. *In*: Barhoum, A. and A.S.H. Makhlouf [eds.]. Emerging Applications of Nanoparticles and Architecture Nanostructures. Elsevier.

Bationo, A., B. Waswa, J. Kihara and J. Kimetu. 2007. Advances in integrated soil fertility management in sub-Saharan Africa: challenges and opportunities. Nutr. Cycl. Agroecosyst. doi: 10.1007/s10705-007-9096-4.

Batley, G.E., J.K. Kirby and M.J. McLaughlin. 2013. Fate and risks of nanomaterials in aquatic and terrestrial environments. Acc. Chem. Res. 46: 854–862.

Bayat, H., Z. Kolahchi, S. Valaey, M. Rastgou and S. Mahdavi. 2019. Iron and magnesium nano-oxide effects on some physical and mechanical properties of a loamy Hypocalcic Cambisol. Geoderma. 335: 57–68.

Bechtel, H.A., S.C. Johnson, O. Khatib, E.A. Muller and M.B. Raschke. 2020. Synchrotron infrared nano-spectroscopy and -imaging. Surf. Sci. Rep. 75: 100493.

Ben-Moshe, T., S. Frenk, I. Dror, D. Minz and B. Berkowitz. 2013. Effects of metal oxide nanoparticles on soil properties. Chemosphere. 90: 640-646.

Bünemann, E.K., G.D. Schwenke and L.V. Zwieten. 2006. Impact of agricultural inputs on soil organisms—a review. Soil Res. 44: 379–406.

Buzea, C., I.I. Pacheco and K. Robbie. 2007. Nanomaterials and nanoparticles: sources and toxicity. Biointerphases 2: MR17-71.

Bystrzejewska-Piotrowska, G., J. Golimowski and P.L. Urban. 2009. Nanoparticles: their potential toxicity, waste and environmental management. Waste Manag. 29: 2587–2595.

Dai, Z.R., J.P. Bradley, D.J. Joswiak, D.E. Brownlee, H.G.M. Hill and M.J. Genge. 2002. Possible in situ formation of meteoritic nanodiamonds in the early Solar System. Nature. 418: 157–159.

Darlington, T.K., A.M. Neigh, M.T. Spencer, O.T.N. Guyen and S.J. Oldenburg. 2009. Nanoparticle characteristics affecting environmental fate and transport through soil. Environ. Toxicol. Chem. 28: 1191–1199.

Das, P., S. Barua, S. Sarkar, N. Karak, P. Bhattacharyya, N. Raza, K.-H. Kim and S.S. Bhattacharya. 2018. Plant extract-mediated green silver nanoparticles: efficacy as soil conditioner and plant growth promoter. J. Hazard. Mater. 346: 62–72.

Djanaguiraman, M., N. Belliraj, S.H. Bossmann and P.V.V. Prasad. 2018. High-temperature stress alleviation by selenium nanoparticle treatment in grain sorghum. ACS Omega. 3:2479–2491.

Doran, J.W. and M.R. Zeiss. 2000. Soil health and sustainability: managing the biotic component of soil quality. Appl. Soil Ecol. 15: 3–11.

Duro, J.A., C. Lauk, T. Kastner, K.-H. Erb and H. Haberl. 2020. Global inequalities in food consumption, cropland demand and land-use efficiency: a decomposition analysis. Glob. Environ. Change. 64: 102124.

El-Kady, M.M., I. Ansari, C. Arora, N. Rai, S. Soni, D.K. Verma et al. 2023. Nanomaterials: A comprehensive review of applications, toxicity, impact, and fate to environment. J. Mol. Liq. 370: 121046.

Evelyn, A., S. Mannick and P.A. Sermon. 2003. Unusual carbon-based nanofibers and chains among diesel-emitted particles. Nano Lett. 3: 63–64.

Fajardo, C., M. Gil-Díaz, G. Costa, J. Alonso, A.M. Guerrero, M. Nande et al. 2015. Residual impact of aged nZVI on heavy metal-polluted soils. Sci. Total Environ. 535: 79–84.

Farber, C., J. Li, E. Hager, R. Chemelewski, J. Mullet, A.Yu. Rogachev et al. 2019. Complementarity of Raman and infrared spectroscopy for structural characterization of plant epicuticular waxes. ACS Omega. 4: 3700–3707.

Ghormade, V., M.V. Deshpande and K.M. Paknikar. 2011. Perspectives for nano-biotechnology enabled protection and nutrition of plants. Biotechnol. Adv. 29: 792–803.

Gogos, A., K. Knauer and T.D. Bucheli. 2012. Nanomaterials in plant protection and fertilization: current state, foreseen applications, and research priorities. J. Agric. Food Chem. 60: 9781–9792.

Gottschalk, F. and B. Nowack. 2011. The release of engineered nanomaterials to the environment. J. Environ. Monit. 13: 1145–1155.

Hao, Y., C. Ma, Z. Zhang, Y. Song, W. Cao, J. Guo et al. 2018. Carbon nanomaterials alter plant physiology and soil bacterial community composition in a rice-soil-bacterial ecosystem. Environ. Pollut. 232: 123–136.

He, F., D. Zhao, J. Liu and C.B. Roberts. 2007. Stabilization of Fe–Pd nanoparticles with sodium carboxymethyl cellulose for enhanced transport and dechlorination of trichloroethylene in soil and groundwater. Ind. Eng. Chem. Res. 46: 29–34.

Hong, J., C. Wang, D.C. Wagner, J.L. Gardea-Torresdey, F. He and C.M. Rico. 2021. Foliar application of nanoparticles: mechanisms of absorption, transfer, and multiple impacts. Environ. Sci.: Nano 8: 1196–1210.

Hristozov, D. and I. Malsch. 2009. Hazards and risks of engineered nanoparticles for the environment and human health. Sustainability 1: 1161–1194.

Jamuna, B.A. and R.V. Ravishankar. 2014. Environmental risk, human health, and toxic effects of nanoparticles. In: Nanomaterials for Environmental Protection. John Wiley & Sons, Ltd. pp. 523–535.

Jayawardena, H.S.N., S.H. Liyanage, K. Rathnayake, U. Patel and M. Yan. 2021. Analytical methods for characterization of nanomaterial surfaces. Anal. Chem. 93: 1889–1911.

Judith, J.V. and N. Vasudevan. 2022. Synthesis of nanomaterial from industrial waste and its application in environmental pollutant remediation. Environ. Eng. Res. 27.

Khan, M.R. and T.F. Rizvi. 2017. Application of nanofertilizer and nanopesticides for improvements in crop production and protection. pp. 405–427. *In*: Ghorbanpour, M., K. Manika, and A. Varma [eds.]. Nanoscience and Plant-Soil Systems. Springer International Publishing, Cham.

Khan, S.T., S.F. Adil, M.R. Shaik, H.Z. Alkhathlan, M. Khan and M. Khan. 2021. Engineered nanomaterials in soil: Their impact on soil microbiome and plant health. Plants (Basel). 11: 109.

Khodakarami, L., S. Pourmanafi, A.R. Soffianian and A. Lotfi. 2022. Modeling spatial distribution of carbon sequestration, CO_2 absorption, and O2 production in an urban area: Integrating ground-based data, remote sensing technique, and GWR model. Earth and Space Sci. 9: e2022EA002261.

Kibbey, T.C.G. and K.A. Strevett. 2019. The effect of nanoparticles on soil and rhizosphere bacteria and plant growth in lettuce seedlings. Chemosphere. 221: 703–707.

Klaine, S.J., P.J.J. Alvarez, G.E. Batley, T.F. Fernandes, R.D. Handy, D.Y. Lyon et al. 2008. Nanomaterials in the environment: Behavior, fate, bioavailability, and effects. Environ. Toxicol. Chem. 27: 1825–1851.

Komendová, R., J. Žídek, M. Berka, M. Jemelková, V. Řezáčová, P. Conte and J. Kučerík. 2019. Small-sized platinum nanoparticles in soil organic matter: Influence on water holding capacity, evaporation and structural rigidity. Sci. Total Environ. 694: 133822.

Kumari, R., K. Suman, S. Karmakar, V. Mishra, S.G. Lakra, G.K. Saurav et al. 2023. Regulation and safety measures for nanotechnology-based agri-products. Front. Genome. Ed. 5.

Lal, R. 2016. Soil health and carbon management. Food and Energy Security. 5: 212–222.

Holder, A.L., E.P. Vejerano, X. Zhou and L.C. Marr. 2013. Nanomaterial disposal by incineration. Environ. Sci.: Processes Impacts. 15: 1652–1664.

Lin, D., X. Tian, F. Wu and B. Xing. 2010. Fate and transport of engineered nanomaterials in the environment. J. Environ. Qual. 39: 1896–1908.

Linak, W.P., C.A. Miller and J.O.L. Wendt. 2000. Comparison of particle size distributions and elemental partitioning from the combustion of pulverized coal and residual fuel oil. J. Air Waste Manag. Assoc. 50: 1532–1544.

Liu, Y., Y. Li, B. Pan, X. Zhang and H. Zhang. 2021. Application of low dosage of copper oxide and zinc oxide nanoparticles boosts bacterial and fungal communities in soil. Sci. Total Environ. 757: 143807.

Medina-Velo, I.A., J.R. Peralta-Videa and J.L. Gardea-Torresdey. 2017. Assessing plant uptake and transport mechanisms of engineered nanomaterials from soil. MRS Bulletin. 42: 379–384.

Mekuye, B. and B. Abera. 2023. Nanomaterials: An overview of synthesis, classification, characterization, and applications. Nano Select. 4: 486–501.

Neme, K., A. Nafady, S. Uddin and Y.B. Tola. 2021. Application of nanotechnology in agriculture, postharvest loss reduction and food processing: food security implication and challenges. Heliyon. 7: e08539.

Nie, D., C. Liu, M. Yu, X. Jiang, N. Wang and Y. Gan. 2022. Elasticity regulates nanomaterial transport as delivery vehicles: Design, characterization, mechanisms and state of the art. Biomaterials. 291: 121879.

Osman, H.S., S.M. Gowayed, M. Elbagory, A.E.-D. Omara, A.M.A. El-Monem, U.A. Abd El-Razek et al. 2021. Interactive impacts of beneficial microbes and Si-Zn nanocomposite on growth and productivity of soybean subjected to water deficit under salt-affected soil conditions. Plants. 10: 1396.

Pan, B. and B. Xing. 2012. Applications and implications of manufactured nanoparticles in soils:

Ragab, S.M., L. Turoop, S. Runo and S. Nyanjom. 2022. The effect of foliar application of zinc oxide nanoparticles and Moringa oleifera leaf extract on growth, biochemical parameters and in promoting salt stress tolerance in faba bean. AJB. 21: 252–266.

Rajput, V.D., A. Kumari, S.K. Upadhyay, T. Minkina, S. Mandzhieva, A. Ranjan et al. 2023. Can nanomaterials improve the soil microbiome and crop productivity? Agriculture. 13: 231.

Rajput, V.D., A. Singh, T. Minkina, S. Rawat, S. Mandzhieva, S. Sushkova et al. 2021. Nano-enabled products: Challenges and opportunities for sustainable agriculture. Plants. 10: 2727.

Raliya, R., J.C. Tarafdar and P. Biswas. 2016. Enhancing the mobilization of native phosphorus in the mung bean rhizosphere using ZnO nanoparticles synthesized by soil fungi. J. Agric. Food Chem. 64: 3111–3118.

Rani, A., K. Rani, J. Tokas, Anamika, A. Singh, R. Kumar, H. Punia et al. 2020. Nanomaterials for agriculture input use efficiency. pp. 137–175. *In*: Kumar, S., R.S. Meena and M.K. Jhariya [eds.]. Resources Use Efficiency in Agriculture. Springer, Singapore.

Rogers, F., P. Arnott, B. Zielinska, J. Sagebiel, K.E. Kelly, D. Wagner et al. 2005. Real-time measurements of jet aircraft engine exhaust. J. Air Waste Manag. Assoc. 55: 583–593.

Schrick, B., B.W. Hydutsky, J.L. Blough and T.E. Mallouk. 2004. Delivery vehicles for zerovalent metal nanoparticles in soil and groundwater. Chem. Mater. 16: 2187–2193.

Seaton, A., L. Tran, R. Aitken and K. Donaldson. 2009. Nanoparticles, human health hazard and regulation. J. Royal Soc. Interface. 7: S119–S129.

Serpone, N., A. Salinaro and A. Emeline. 2001. Deleterious effects of sunscreen titanium dioxide nanoparticles on DNA: Efforts to limit DNA damage by particle surface modification. Nanoparticles and Nanostructured Surfaces: Novel Reporters with Biological Applications. SPIE. pp. 86–98.

Sharma, R. and V. Kumar. 2024. Nano-enabled agriculture for sustainable soil. Waste Manag. Bull. 2: 152–161.

Soltys, L., I. Myronyuk, T. Tatarchuk and V. Tsinurchyn. 2020. Zeolite-based composites as slow-release fertilizers (Review). Phys. Chem. Solid State. 21: 89–104.

Stevie, F. A. and C. L. Donley. (2020). Introduction to x-ray photoelectron spectroscopy. Journal of Vacuum Science & Technology A, 38(6), 063204. https://doi.org/10.1116/6.0000412.

Suazo-Hernández, J., N. Arancibia-Miranda, R. Mlih, L. Cáceres Jensen, N. Bolan and M.L. Mora. 2023. Impact on some soil physical and chemical properties caused by metal and metallic oxide engineered nanoparticles: A review. Nanomaterials. 13: 572.

Tan, K.X., A. Barhoum, S. Pan and M.K. Danquah. 2018. Chapter 5 - Risks and toxicity of nanoparticles and nanostructured materials. pp. 121–139. *In*: Barhoum, A. and A.S.H. Makhlouf [eds.]. Emerging Applications of Nanoparticles and Architectured Nanostructures. Elsevier.

Tourinho, P.S., C.A.M. van Gestel, S. Lofts, C. Svendsen, A.M.V.M. Soares and S. Loureiro. 2012. Metal-based nanoparticles in soil: Fate, behavior, and effects on soil invertebrates. Environ. Toxicol. Chem. 31: 1679–1692.

Tratnyek, P.G. and R.L. Johnson. 2006. Nanotechnologies for environmental cleanup. Nano Today. 1: 44–48.

Wang, H., R.L. Wick and B. Xing. 2009. Toxicity of nanoparticulate and bulk ZnO, Al2O3 and TiO2 to the nematode *Caenorhabditis elegans*. Environ. Pollut. 157: 1171–1177.

Yan, N., N. Xia, L. Liao, M. Zhu, F. Jin, R. Jin et al. 2018. Unraveling the long-pursued Au144 structure by X-ray crystallography. Sci. Adv. 4: 7259.

Yasmin, H., J. Mazher, A. Azmat, A. Nosheen, R. Naz, M.N. Hassan et al. 2021. Combined application of zinc oxide nanoparticles and biofertilizer to induce salt resistance in safflower by regulating ion homeostasis and antioxidant defense responses. Ecotoxicol. Environ. Saf. 218: 112262.

Zhang, W., J. Long, J. Li, M. Zhang, X. Ye, W. Chang et al. 2020. Effect of metal oxide nanoparticles on the chemical speciation of heavy metals and micronutrient bioavailability in paddy soil. Int. J. Environ. Res. Public Health. 17: 2482.

Zhang, W., H. Xiao and K.L. Parkin. 2014. Apoptosis in MCF-7 breast cancer cells induced by S-alkenylmercaptocysteine (CySSR) species derived from Allium tissues in combination with sodium selenite. Food and Chemical Toxicology 68: 1–10.

Zhou, Q., S. Kaappa, S. Malola, H. Lu, D. Guan, Y. Li et al. 2018. Realspace imaging with pattern recognition of a ligandprotected Ag374 nanocluster at submolecular resolution. Nat. Commun. 9: 2948.

CHAPTER-3

Nanotechnology for Soil Remediation and Pollution Control

Hamna Bashir,[1] *Naila Ghafoor,*[2] *Muhammad Ashir Hameed,*[1] *Saad bin Zafar,*[1] *Muhammad Mahroz Hussain*[3,4,*] and *Shengsen Wang*[4,*]

1. Introduction

Water, air and terrestrial pollution pose a significant risk to both humans and the environment, soon after the industrial revolution. Human activities, whether in the air, water, or soil, are causing potential disruptions to every aspect of the ecosystem, leading to substantial difficulties. Industrial, residential, and agricultural processes, which are essential for human survival, substantially contributed to the entry of pollutants into the environment without any check and balance on daily basis throughout the globe (Kumar and Bharadvaja 2019). Mekonnen and Hoekstra (2016) revealed that around 800 million individuals from developing countries lack access to clean drinking water. The proliferation of water-intensive sectors, such as pharmaceuticals, food production, agriculture, textiles, tanning, fertilizer manufacturing, mining, and steel production, has led to a significant increase in water pollution. This, in turn, has become a key contributing factor to the spread of waterborne infections (Huerta-Fontela et al. 2010, Hussain et al. 2020). Surface water degradation is caused not only by industries but also by untreated wastewater from agricultural and urban areas. This wastewater contains toxic pesticides, detergents, and traces of organic chemicals (Schwarzenbach et al. 2010). The existence of enduring organic pollutants, metal(loids), dioxins, dyes, and plastics can cause detrimental effects on marine ecosystems and the organisms that rely on them (Luo et al. 2014). A survey conducted revealed that the use of improved modes of disinfecting the households, industries, swimming pools, etc., has released suspended dangerous chemicals and acidic effluents into water, affecting approximately 33% of the globe. These processes have also led to the accumulation of toxic compounds and metal ions in plants, animals, fish, crabs, lobsters, squids, and oysters (Taylor et al. 2017).

Furthermore, human activities are also known to influence the depletion of the ozone layer by emitting large quantities of various greenhouse gases such as carbon dioxide, hydrogen sulfide, methane, nitrous oxide, ozone, along with chlorofluorocarbons (Anderson et al. 2016). The process of industrialization has also led to the increased use of automobiles, excessive reliance on coal and

[1] Institute of Soil and Environmental Sciences, University of Agriculture, Faisalabad-38040, Pakistan.
[2] Department of Zoology, Wildlife and Fisheries. University of Agriculture, Faisalabad 38040, Pakistan.
[3] HAM Organics (PVT) Limited, Nankana Sahib, Punjab, Pakistan.
[4] College of Environmental Science and Engineering, Yangzhou University, Jiangsu China.
* Corresponding authors: hmahroz@gmail.com; wangss@yzu.edu.cn

DOI: 10.1201/9781003558354-3

gasoline, which in turn has raised the levels of aerial ozone, lead, PM2.5 and PM10 particulate matter, CO, SO_2, and NO_2. These pollutants have impacted various compartments of the Earth, including land, water, and air. The toxic components engulfing them also have negative impacts on their fundamental structures and balances (Kampa and Castanas 2008). Gaseous substances such as water vapor, humidity, chlorofluorocarbons, and CO_2 have a significant detrimental impact on human health. Consequently, there has been a notable increase in the number of individuals suffering from asthma, cardiovascular ailments, and respiratory disorders, including cancer. Furthermore, both short- and long-term exposure results in a rise in premature fatalities and a reduced lifespan, particularly among the youth and the elderly, rendering them the most susceptible demographics. It exerts significant effects on humans, agricultural practices, soil, and groundwater.

Chemicals, oil refineries, leather, textiles, steel, automobile, household, biomedical, mining, and electronics industries are significant contributors to alterations in soil quality and composition (Steffan et al. 2018). The textile industry annually produces a significant amount of solid waste, consisting of yarns, textiles, and chemicals. Furthermore, the presence of solid sludge, which includes both organic and inorganic matter, is a significant challenge in managing environmental pollution. This sludge includes heavy metals such as lead, lethal biochemical dyes, and micronutrients in the form of phosphate and sulfate, as reported by Bhatia and Devraj (2017). Agriculture, too, poses threats to humans and other living organisms, particularly through the use of fertilizers and pesticides and their seepage into water tables. A study conducted by Aktar et al. (2009) explored that over one million individuals are afflicted with chronic illnesses as a direct result of exposure to pesticides and fertilizers. Fertilizers containing nutritive elements cause the proliferation of microalgae in still water bodies upon the arrival of these chemicals. This occurrence, known as eutrophication, poses a danger to the survival of aquatic life forms since oxygen and sunlight from the surface cannot penetrate deeper into the water bodies (Glibert 2017). Pesticide contamination can lead to a range of adverse effects in humans, including nausea, dizziness, allergies, headaches, tremors, and unconsciousness (Xu et al. 2019). Regular exposure to chemical fertilizers and pesticides can hinder the process of carbon mineralization and the symbiotic fixation of nitrogen. It modifies the communication system between rhizobia and legumes by blocking the activity of nitrogenase and impacting the processes of ammonification, assimilation, nitrification, and denitrification. The populations of non-target macrofauna, including wasps, earthworms, insects, nematodes, and flies, might experience significant decreases due to the impact of fertilizers and other contaminants (Naveed et al. 2014). Lead, mercury, arsenic, cadmium, and chromium, which are classified as heavy metals, are extensively utilized in the electronic equipment and electroplating sectors (Thundiyil et al. 2008).

Exposure to high levels of heavy metals, particularly arsenic, can cause both immediate and long-term health problems, including damage to the central nervous system, kidneys, blood cells, high blood pressure, cancer, lung illness, internal organ damage, seizures, and loss of consciousness (Hussain et al. 2021b).

Several procedures have been investigated and implemented for cure and retreat, but they have yielded inadequate or unsatisfactory outcomes. Numerous current cleanup techniques possess significant disadvantages: they are costly and time-consuming, require extensive space, and produce additional contaminants (Baruah and Dutta 2009). The repeated ineffectiveness of traditional treatment approaches and the exploration of alternative technologies have resulted in the utilization of nanotechnology and engineering concepts to supplant current technologies (Madima et al. 2020). Nanomaterials counterpoint these problems and exhibit definite activities targeted toward specific zones. Mineral and magnetic nanoparticles provide a simple process, coupled with the low cost of pollution removal in the ecosystem (Panda et al. 2021). When they are used to selectively adsorb specific pollutants from water, magnetic adsorbent nanoparticles can then be easily recovered from wastewater (Mashkoor et al. 2020). Due to their desirable characteristics—including relatively simple and inexpensive synthesis, eco-friendliness, reusability, recyclability, and availability of necessary raw materials, energy, and water—nanomaterials are widely applied in numerous environmental applications. Such features are crucial for environmental remediation, including the purification or removal of pollutants such as wastewater, pesticides, fertilizers, dyes, and heavy metals from polluted areas (Singh and Mehata 2019).

2. Environmental Remediation Technologies

The choice of contaminant treatment technologies depends on the type and concentration of pollutants, the degree and extent of contamination, the geographical distribution of the contaminants, the cost implications of the treatment processes, the intended goal of the treatment technology, as well as the prevailing policies. Initial assessment includes the location of the land, geology, and the impacts of the land on human health, as reported by Kumar and Bharadvaja (2019). Conventional pollution treatment primarily employs three methods: chemical, biological, and physical (Guo et al. 2021). Most treatment procedures transform one type of pollutant into another. Membranes are employed in wastewater treatment to effectively separate salts and other large particles. These separated substances are subsequently transformed into salts and solids, which may be disposed of without the need for further treatment. Another instance involves the utilization of activated carbon for the process of adsorption. Once the adsorption process is complete, the solid adsorbent becomes ineffective and, if not properly disposed of, can lead to soil contamination Chemical approaches employ noxious substances as coagulants, flocculants, peroxides, Fenton's reagent, etc., which are also irreversible (Peng and Guo 2020). Biological treatments utilizing microorganisms such as bacteria, fungi, and plants are the only methods capable of ecologically acceptable degradation and desalination of complex contaminants into carbon dioxide and water.

3. Remediation of Air, Water and Soil Contaminants using Physical Methods

Various types of scrubbers are employed based on the specific pollutant being targeted, including Venturi scrubbers, wet scrubbers, dry scrubbers, and adsorbers (Endres et al. 2018). Prior to implementing additional air pollution control measures, it is recommended to employ bag filters for the filtration of larger particles (PM > 2.5 μm), as suggested by Liu et al. (2019). Sediment chambers, scrubbers, filters, cyclones, and electrostatic precipitators are examples of physical approaches used to reduce air pollution. Scrubbers play an important role in the separation of numerous pollutants produced during combustion, including particulate matter, polycyclic aromatic hydrocarbons, heavy metals, carbon dioxide, SOx, and NOx. Sodium hydroxide is used in scrubber cleanup as a neutralizing agent for SOx, hydrogen sulfide, and sulfuric acid. Naphtha, as per Üresin et al. (2015), emits sulfates, bisulfates, and carbonates, which are non-toxic. Multicyclone systems, which often consist of 10 to 15 individual units operating in parallel, are highly effective at removing large particles (e.g., dust, pollen, and allergens) larger than 10 micrometers (μm). Their operation relies on centrifugal force, not centripetal force. The process begins when high-speed, particle-laden gas is forced tangentially into the cyclones. This creates a powerful vortex. The resulting centrifugal force flings denser particles outward toward the cyclone walls. The particles slide down the walls and are collected in a hopper at the bottom, often forming a sludge or dry waste, while the cleaned gas exits through the top. It is important to note that while efficient for large particles, cyclones are less effective at capturing fine particles and aerosols smaller than 2.5 μm (PM2.5).

The resolution of water pollution issues has consistently posed difficulties due to the variability of pollutant concentrations, the presence of non-point source pollutants, the creation of secondary pollutants, the absence of cost-efficiency, and the inadequate management of residual waste. Traditional treatment options for contaminated water include physical, biological, and chemical methods. Physical treatment functions as a barrier that effectively segregates contaminants from wastewater. A brief summary of physical methods used to remediate the air, water, and soil is given in Table 1.

Table 1. Heavy metal removal approaches: Sustainable environmentally friendly techniques with disadvantages. Reproduced and modified from (Kaur and Roy 2021).

Approach	Advantage	Disadvantage	Mechanism	Reference
Conventional methods	Good controllability and resilience to high amount of heavy metals present	Low metal removal efficiencies, high operational costs, production of waste by products that are difficult to manage, high energy requirements lead to secondary pollution, reducing soil fertility	Chemical precipitation, ion exchange, chemical oxidation, reduction, reverse osmosis, ultrafiltration	(Kaur and Roy 2021)
Application of nanotechnology in the removal of toxic anions	Extraordinary adsorption capacity and reactivity due to surface effect, small size effect, quantum effect, and macro quantum tunnel effect	Agglomeration, few real life applications, use of toxic chemicals, lack of extensive study, experimentation and documentation	Single stage-treatment method, chelation, metal adsorption	
Phytoremediation based removal of heavy metals	Cost-effective, efficient, easy maintenance, and eco-friendly	Leaching of metals into groundwater, besides persistence of soil amendments, produces waste, requires wide area, proper set-up and location, lengthy process	Phytodegradation, phytoaccumulation, phytoextraction, phytovolatilization, phytostabilization	
Microbes assisted remediation of heavy metals	Reduces energy consumption and cost, inhibits pathogens, eliminates odor and improves air quality	Low in process, limited due to presence of non-biodegradable contaminants and may also cause toxicity of microbes	Biosorption and bioaccumulation, passive and active mechanisms	

This process is dependent on the size of the particles that exist in the wastewater stream. The filters are primarily categorized as fine and coarse, with pore diameters ranging from 10 microns to 10 inches. Grids are frequently employed as obstacles to effectively eliminate large-sized particles and trash (> 1 mm) from incoming wastewater streams (Ali et al. 2019a). Methods such as sand tanks, sedimentation tanks, and primary clarifiers are employed to eliminate fine suspended particles, floating sludge, and sludge. Activated carbon adsorption is a widely used physical treatment technique for eliminating color and various impurities from wastewater (Sandhyarani 2019). The process of filtration, utilizing both sand and activated carbon filters, actively contributes to the separation of solid particles from wastewater. The method of separation relies on two primary filter categorizations: surface filters and depth filters, which are determined by the distribution of pore sizes throughout the medium and surface. The limitations of physical processing include substantial capital investment, increased spatial requirements, elevated expenses for routine maintenance, and a significant need for energy.

4. Contaminants Removal using Chemical Approaches

Coagulation, sedimentation, flocculation, chemical leaching, oxidation, fixing, purification, and advanced oxidation are used in chemical methods. Advanced oxidation methods include the Fenton process, photolysis, ozonation, UV/peroxides, metal oxide nanocatalyst photocatalysis, wet chemical oxidation, and sonochemistry. Superior results have been achieved with these approaches over standard chemical procedures (Čehovin et al. 2017). These treatment systems are efficient and have minimal pollutant oxidation energy. Frequently, a mixture of these strategies is employed to attain optimal outcomes. Advanced oxidation methods aim to decrease the number of microorganisms and the concentration of chemicals in wastewater. The photolytic destruction of peroxide generates hydroxyl and ionic free radicals, which effectively eliminate leftover chemicals and microbes in water by breaking them down. According to Karci et al. (2013), this method has demonstrated a 78–95% efficiency in reducing the total organic carbon burden over a 90-minute timeframe. This method is ideal for removing phenol, bacteria, viruses, fungi, chemicals, dyes, endocrine-disrupting substances, and heavy metals from sewage. Semiconductor nanocatalysts including ZnSe, Cu-NiO, ZnO, TiO_2, GaAs, GaN, InP, ZnS, CdS, and CdSe increase wastewater oxidation (Jadhav and Biswas 2018). External energy sources such as visible or ultraviolet light stimulate ions from the valence band to the conduction band. As a result, hydroxyl radicals are released, which then target and break down contaminants. While many chemical approaches are user-friendly, they are not devoid of limitations. The primary drawbacks of this approach include the utilization of ecologically harmful substances, challenges in scaling up applications, lack of cost-effectiveness, and the requirement for additional treatment using nanoporous membranes or biological substances.

4.1 Biological Treatment Methods

Biological treatment is an ecologically sound procedure that assists in the decomposition and demineralization of contaminants into water and carbon dioxide. The process employs microbes and plants to eliminate or break down contaminants. There is no alternative method that can guarantee the thorough breakdown and removal of contaminants. It efficiently eliminates essential nutrients including phosphate, nitrates, nitrites, and ammonia. Biological filtration techniques are crucial in decreasing biological oxygen demand, suspended particles, and dissolved pollutants, as well as eliminating micronutrients from wastewater. This biochemical reaction is completed through two inherent respiratory mechanisms, specifically aerobic digestion and anaerobic digestion (Kumar and Bharadvaja 2019). In the aerobic process of treatment known as activated sludge, oxygen serves as the electron acceptor, while in anaerobic digestion, oxygen is not present. In order to perform both operations, it is necessary to optimize the reactor configuration and abiotic factors such as flow rate, pH, temperature, aeration, and feed water concentration. The biochemical treatment process employs suspended and attached growth reactors, operating in both batch and continuous modes. *Zooglea, Arcobacter, Acinetobacter, Trichosporon*, and *Fusarium* are often encountered microorganisms in aerobic treatment procedures. They are also known as "flocculating bacteria" (Romauld et al. 2019, Hussain et al. 2021a, Hussain and Farooqi 2021). Aerobic treatment is a process that degrades organic materials within a very short time, usually 10–20 days. On the other hand, anaerobic treatment processes are relatively slow, performing their tasks with longer solids retention time (SRT) and hydraulic retention time (HRT), and can reach up to 30 days (Alepu et al. 2016). *Methanobacterales* are species of mesothermic bacteria which constitute a separate type of microorganisms called the Archaea. They are believed to be involved in the degradation of pollutants, and in absorbing and producing useful energy, referred to as "biogas" from wastes. Anaerobic processes exhibit greater susceptibility to abiotic stimuli, such as variations in pH, temperature, and nutrient concentrations in the influent waste. This characteristic is widely regarded as a significant drawback of the process. Tropical countries are generally more conducive to biological activities compared to temperate countries (Zang et al. 2015). Nevertheless, this drawback can be mitigated by augmenting the solids and hydraulic retention durations, as well as linking several reactors in a sequential manner. The processed wastewater is subsequently directed to a secondary clarifier to eliminate suspended particles and activated sludge.

5. Application of Nanotechnology-Based Methods for the Remediation of the Environment

Nanotechnology has progressively infiltrated common applications in the past few decades. Nanotechnology offers a potential remedy for soil, water, and air pollution, according to Guerra et al. (2017). Nanoparticles possess a diverse array of characteristics, including high porosity, high reactivity, good agglomeration, large surface area, and improved physical, mechanical, optical, and

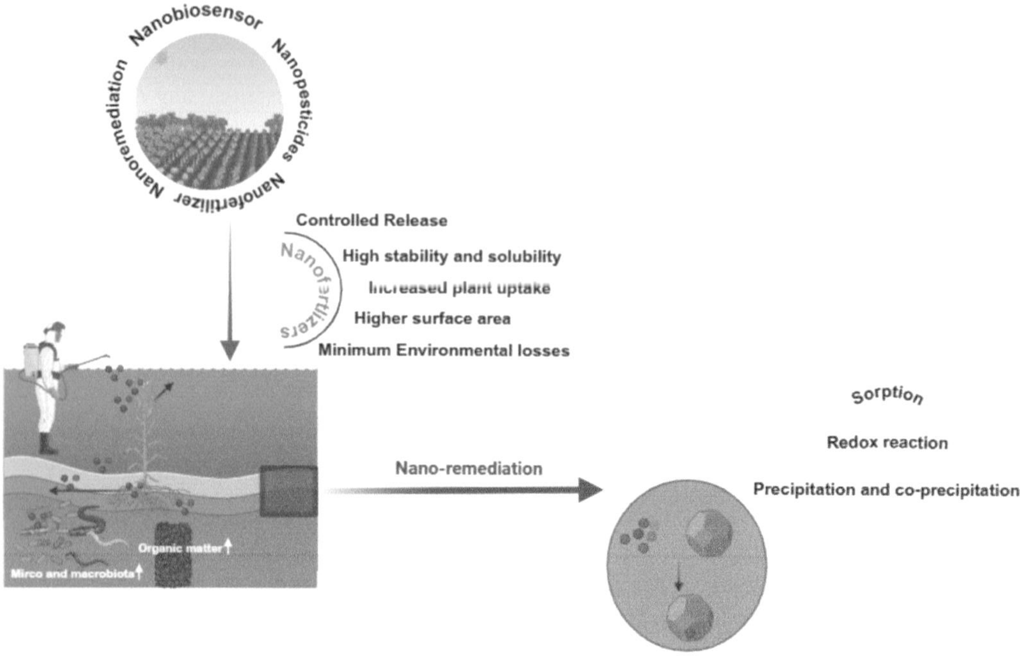

Figure 1. Possible mechanism of implication of nanoparticles in soil remediation and fertilizations.

electrical capabilities in comparison to bulk materials (Borji et al. 2020). Nanoparticles possess distinctive surface characteristics, such as chemically functionalized surfaces, that enable them to selectively target particular pollutants. Its mechanism has been illustrated in Figure 1.

Furthermore, modifying the physical and surface characteristics enhances their qualities in comparison to conventional treatment methods, consequently augmenting therapeutic effectiveness (Guerra et al. 2017). It is crucial to acknowledge that in traditional treatments, the materials utilized for restoration are, in fact, sometimes the source of contamination. Bioremediation is a technique that utilizes living organisms to eliminate pollutants from the environment. It achieves this by minimizing or completely removing the need for additional waste processing, resulting in zero or near-zero post-processing waste. More recently, researchers have used variants of the concepts availing nanotechnology to design and produce nanocomposites with unique properties such as selectivity, specificity, biodegradability, and reusability. These are particularly valuable when used to direct synthetic materialization toward specific environmental goals. Advances in bio-nanotechnology now enable the development of nanocomposites with mechanisms of zero toxicity and enhanced environmental compatibility. Unlike other synthesis techniques that use chemical reduction and capping agents, this process leaves no chemical residues (Rasli et al. 2020). A wide range of nanomaterials—including metals, metal oxides, non-metals, semiconductors, polymers, and ceramics—are effective for environmental remediation. Their utility stems from their strong adsorbing and absorbing characteristics, making them highly suitable for applications in membrane filtration and catalysis to remove pollutants. These nanostructures include tubular, allotropic, rod-like, quantum dot, wire, and spherical nanoparticles. Consequently, nanomaterials are very effective in addressing some of the most significant environmental issues due to such features. Optimization and formulation play a crucial role in achieving significant scalability and commercialization while minimizing byproduct emissions. Table 2 provides a summary of nanoparticles along with their remediation potential.

Table 2. Nanoparticles' potential to remediate environment and mechanisms.

Remediation potential	Nanoparticles' types	References
Remediation of chlorinated organic solvents, dyes, pesticides, heavy metals, and antibiotics contaminated water bodies.	Iron (Fe)	(Bashir et al. 2022, Crane et al. 2015, Klimkova et al. 2011)
Increase the rate and efficiency of remediation		
Efficient removal of heavy metals and polycyclic aromatic hydrocarbons (PAHs) from the environment.	Silica (Si)	(Ali et al. 2019b, Nazaralian et al. 2017, Petrunin et al. 2015)
Adsorption of heavy metals and organic pollutants and the elimination of harmful bacteria from contaminated water bodies. Photocatalyst degradation of pollutants	Carbon Nanotubes (CNTs)	(Gangupomu et al. 2014, Madima et al. 2020, Mashkoor et al. 2020, Shashanka et al. 2022, Yan et al. 2015)
Antibacterial, antifungal, and antiviral property. Disinfection of contaminated water.	Silver (Ag)	(Jadhav and Biswas 2018)

5.1 Nanomaterials and their Role in Environmental Remediation

The progress made in nanotechnology, materials science, catalysis science, and biological sciences, as well as the intersections between these disciplines, provide economically efficient, socially acceptable, and eco-friendly answers to numerous environmental challenges. Nanomaterials are defined by the presence of at least one external dimension in the 1–100 nm range. This nanoscale confinement can lead to unique electronic properties. For instance, in one-dimensional nanomaterials, electron transport is constrained along a single axis. Furthermore, the ratio of surface area to volume experiences exponential growth when compared to bulk materials. As a result, the physical and chemical characteristics of these materials experience significant alterations in comparison to larger quantities. Nanomaterials with adjustable surface properties, strong structural integrity, high ability to capture pollutants, improved ability to selectively target specific pollutants, a greater capacity to break down naturally, and increased ability to be reused have been developed in order to decrease pollution in the environment.

5.2 Synthesis and Characterization of Nanomaterials

There exist three distinct methodologies for the production of nanomaterials: biological, chemical, and physical. The eco-friendly qualities of biosynthetic approaches make them preferable over thermodynamic methods for the production of nanomaterials. Nanomaterials are categorized into distinct classes according to criteria such as dimensionality, shape, magnetism, etc. (Kolahalam et al. 2019). Following the synthesis process, analyses are conducted to examine various material qualities such as morphology (shape and size), structure (amorphous or crystalline), bonding properties, and electromagnetic characteristics. This is done so that a confirmation could be made if the nanomaterial of the desired properties has been synthesized. For the purpose of identifying the crystallinity of the synthesized elements, techniques such as such as X-ray diffraction (XRD) are employed, while thermogravimetric analysis is used for phase transition. The morphology and physical characteristics of nanomaterials are evaluated using atomic force microscopy (AFM) and transmission electron microscopy (TEM). SEM (scanning electron microscopy) is used to determine surface characteristics, morphology, and dimensions. Magnetic properties are identified with the help of Electron Paramagnetic Resonance (EPR) spectroscopy, Vibrating Sample Magnetometer (VSM), and Superconducting Quantum Interference Device (SQUID). Energy-dispersive X-ray spectroscopy (EDS) is a microanalytical technique used alongside SEM and TEM to detect elemental composition. Further characterization tools include FTIR, XPS, Raman, and XAS, which are used to evaluate chemical bonding and connection patterns in the sample, as highlighted by Reddy et al. (2012) and Bashir et al. (2022).

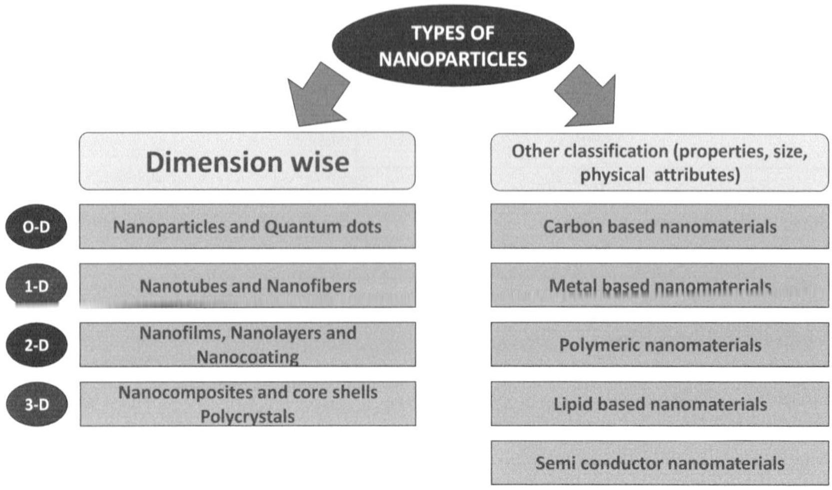

Figure 2. Classification of nanoparticles according to size, dimensions and physical attributes.

5.3 Categories of Nanomaterials and their Remediation Efficacy

In recent times, there has been extensive exploration of several categories of nanomaterials, including both synthetic and naturally occurring ones, with the anticipation of further additions to this roster in the coming years (Figure 2). Hence, it is imperative to categorize them suitably. In 1995, Gleiter initially categorized nanomaterials according to their chemical characteristics, specifically nano polymers and oxides of metal. The diameters of the nanoparticles were considered, a factor that Gleiter overlooked. Nanomaterials are categorized according to their geometric properties, such as zero-dimensional (0D), one-dimensional (1D), two-dimensional (2D), and three-dimensional (3D) structures (Baruah and Dutta 2009, Bashir et al. 2022, Panda et al. 2021, Rasli et al. 2020, Sandhyarani 2019, Xu et al. 2019).

6. Other Nanomaterials

6.1 Nanosorbents

Nanosorbents refer to nanoscale organic or inorganic molecules that possess a high surface area and are capable of absorbing contaminants, making them highly useful in the remediation of contaminated locations. Zeolite is a highly prevalent nanomaterial. Zeolites are aluminosilicate minerals that contain water molecules in their structure. They are fundamentally crystals. Their surface is charged through electrostatic pressure with cationic elements like silver and zinc on their surface (Čehovin et al. 2017, Jadhav and Biswas 2018, Luo et al. 2014, Reddy et al. 2012, Schwarzenbach et al. 2010). One explanation for the great adsorption capacity of nanoscale zeolites is their high surface area to volume ratio. They are quite efficient in terms of cation exchange and are used as catalysts. El-Sayed (2020) further elucidated that zeolites provide notable chemical and mechanical stability. Pollutants are removed via the nanocomposite because the zeolite nanopores contain infused silver ions, which possess antibacterial, antiviral, and other beneficial properties. They are also utilized in anaerobic breakdown, aerobic water purification, denitrification, composting, and biological water treatment (Behravesh et al. 2020).

Additional nanosorbents include polymeric ones. These composites have more surface area than traditional adsorbents (Zare et al. 2018). There are several types of nanoadsorbents depending on their materials, including graphene and metal oxides. These polymer nanocomposites are ecologically sustainable and can undergo rapid degradation. Nanocatalysts and redox-active nanoparticles are nanomaterials that possess catalytic characteristics. Due to their high surface-to-volume ratio, they exhibit a high level of reactivity (Zare et al. 2018). They are employed for the purpose of eliminating radioactive materials and heavy metals. These substances are classified as non-metals, including zero-valent nanoparticles like nanosilver, as well as nanometal equivalents including titanium oxide.

Certain oxides exhibit superparamagnetic properties, enabling the effortless elimination of impurities under weak magnetic fields, such as nanomagnetite (El-Sayed 2020). Nanoparticles composed of metal atoms in their zero-valent state, such as silver nanoparticles (AgNPs), possess antibacterial properties, making them suitable for use as disinfectants. They adhere to the cellular surface of bacteria and infiltrate cells by altering the structure of the membranes. This results in the generation of detrimental free radicals that can damage cells. Silver undergoes a chemical reaction with the thiol groups present in enzymes, leading to their destruction (Lu et al. 2016). Iron nanoparticles exhibit exceptional adsorption capabilities. They show robust oxidation and precipitating characteristics, though their reducing capabilities are limited. In the absence of oxygen, water or protons cause the oxidation of zero-valent iron, resulting in the release of ferrous iron and hydrogen. This is followed by a redox reaction between the pollutants and nanoscale zero-valent iron, resulting in the formation of iron (III) and iron hydroxide. Iron hydroxide facilitates the removal of impurities. Zero-valent iron nanoparticles also remove organic molecules by oxidizing various organic contaminants via the discharge of hydroxyl radicals (Bashir et al. 2022). Zero-valent iron nanoparticles can remove dyes, heavy metals, radioactive elements, and both organic and inorganic substances. Zinc removes impurities faster than iron due to its higher standard reduction potential (Fu et al. 2014).

The efficacy of zero-valent zinc nanoparticles in removing substances is restricted to halogenated organic chemicals, specifically carbon tetrachloride. Lu et al. (2016) found that carbon tetrachloride degrades at a higher rate with zero-valent zinc than with zero-valent iron nanomaterials. Metal oxide nanoparticles exhibit high instability during photocatalytic reactions. Titanium oxide possesses a significantly wide band gap; hence, the activation of titanium oxide nanoparticles necessitates exposure to UV irradiation, resulting in the liberation of hydroxyl radicals. These radicals are detrimental to cellular health and impair bacterial, fungal, and algal cell structure and activity (Foster et al. 2011). UV light creates reactive oxygen species that interact with pollutants and thoroughly degrade polycyclic aromatic hydrocarbons, organic substances like chlorine, chemicals, dyes, and heavy metals. NZVI reacts with various redox-sensitive pollutants and other electron-accepting substances, releasing H^+ ions into the pore water. This acidifies the sediment, subsequently mobilizing heavy metals (HMs). They are photocatalytic, photostable, biochemically stable, and accessible. Iron oxide nanoadsorbents include Fe_3O_4 (magnetic magnetite), Fe_2O_3 (magnetic hematite), and Fe_2O_3 (non-magnetic). Due to their compact size and large surface area, polluted sites can remove numerous heavy metals (Singh et al. 2010).

7. Potential of Nanoparticles in Remediation of Heavy Metal(liods)

7.1 Nano Zero-valent Iron (NZVI)

Non-simatic and simatic NZVI are porous materials and have particle sizes ranging from 10 – 100 nm with core-shell structures. The core material generally comprises of zero valent iron or metallic iron and its main active material includes Rare-Earth-Permanent Magnets (REPMs). Iron interacts with the surface by oxidation and the subsequently generated layer comprises Fe oxides (Crane et al. 2015). The octahedral inner shell is considered as electron donor that contributes to reduction of given molecule, while the outer shell of this structure is considered electron acceptor that promotes adsorption and surface complexation reactions. Examining prior findings, it has been pointed out that due to the peculiar architecture of NZVI, it is exceptionally feasible for the reduction of heavy metals (Crane and Scott 2012). As reported by Li et al. (2017), NZVI is emerging as a potential purification agent, as it has been found effective in the purification of arsenic, copper, and other toxic essential metallic elements. Two main methods are widely used to synthesize NZVI: the physical method and the chemical method. Ball milling is one of the physical treatment techniques, along with gas condensing and ultrasonic-assisted procedures. On the other hand, controlled chemical co-precipitation, pulsed electrodeposition, liquid flame spraying, gas-phase reduction, and liquid-phase reduction are examples of chemical treatment techniques. Another preparation technique of NZVI is high-temperature catalytic reduction on goethite or hematite using hydrogen gas at a temperature of 200–600°C, as explained by Nurmi et al. (2005). Another method is by deconstruction of pentacarbonyl iron, as reported by Li et al. (2008). Among the described methods, such as hydrogenation, hydrocracking, pyrolysis, and catalytic cracking, the most widely used and severe method is reduction with sodium.

However, several factors can constrain and lower the effectiveness of NZVI, including surface alterations and compounding with other chemicals. To overcome these problems, various types of coating materials with excellent properties for steric stabilization have been explored (Hotze et al. 2010). These coatings are primarily made of anionic or nonionic surfactants and include polyelectrolytes intruding into biopolymers. Enhancers increase the stability and mobility of NZVI in such a manner that results in more efficient degradation of pollutants. In addition, stabilizers contain functional groups such as –OH and –COOH, which act as chelating groups for heavy metal ions and aid in their adsorption (Zou et al. 2016). Among the NZVI materials used in recent years, those modified with surface stabilizers have been the most popular for handling heavy metals as a form of environmental pollution.

Characterization investigations such as the SEM, FTIT and XRD probes have shown that the addition of SA was helpful in inhibiting the rate of aggregation and oxidative degradation of the NZVI (Bashir et al. 2022). Researchers also noted that the application of SA-NZVI was most efficient in the process of remediation of sediment contaminated with cadmium. This indicates that NZVI can react with various redox-sensitive pollutants and other electron-accepting substances. These reactions release H^+ ions into the pore water, lowering the sediment pH and subsequently mobilizing heavy metals (HMs). In another study, Geng et al. (2009) synthesized chitosan-stabilized Fe0 NPs to apply them for the removal of Cr(VI) by reduction of Cr(VI) present in the water samples. Thus, the researchers proved that FeO-chitosan nanoparticles work effectively in the reduction of Cr(VI). In the subsequent study, it has also been discovered that chitosan chelate with Fe(III) at higher degree and prevents cross linking of Fe(III)-Cr(III) laterite, thus increasing the rate of Cr(VI) reduction. In addition, initial chromium is in the oxidation state of the hexavalent Cr(VI) and the process of reduction to the trivalent Cr(III) can qualify for a pseudo first-order reaction. Nonetheless, there being several surface modification technologies, the following are some merits associated with this area of research. Just like any other field of research, surface modification also has some demerits in relation to the alteration of NZVI. For instance, in coating development, placing a layer over the NZVI that shields it seriously inhibits the rate at which the NZVI undergoes oxidation and, moreover, creates a zone free of any contact between the NZVI and the targeted pollutants. This also poses a hindrance to the diffusion of pollutants in the solution to the surface of the NZVI, perhaps removing the ability of the NZVI to neutralize ions in solution. It becomes challenging to compare its efficiency between surface-modified and unmodified NZVI, as its efficiency highly depends on the coat layer of NZVI, the conditions of modification, and the removal of impurities (Klimkova et al. 2011).

7.2 Supported NZVI

Ameliorative surface modifications were used to retard the formation of passivation layers and agglomerations by the scientists. In addition, they also prepared "carriers" which can serve to improve the transport and dispersion of NZVI. NZVI, when incorporated in a carrier, moves and stabilizes throughout the field of application or, in the case of field repairs, is preferred. The delivery vehicles are usually formed from materials with a sizeable surface area, high portability, and distinguished features. Graphene oxide reinforced with NZVI shows excellent pollutant adsorption properties, and there was a relatively high level of interaction between the loaded NZVI and the target pollutants (Sun et al. 2014). Materials used to well NZVI include chitosan, porous carbon layers, carbon, carboxymethyl cellulose, graphene oxide, bentonite, and biochar. According to literature findings, highly concentrated NZVI has demonstrated significant efficacy in the removal of heavy metals. Shang et al. (2017) effectively created NZVI on grass residual biochar to eliminate Cr(VI) from water. They demonstrated that the highest adsorption effectiveness (98.71%) was achieved at a pH of 2. Furthermore, the presence of humic acid and sulfate expedited the elimination of Cr(VI), but the presence of bicarbonate hindered the reaction. The act of loading materials onto the NZVI surface serves to avoid the aggregation of nanoparticles (NPs), resulting in an increase in surface area and the stabilization of sites for the removal of external contaminants.

8. Environmental Factors Affecting the Reactivity of Fe-Based NPS with HMS

Contemporary research concerning the reactivity of NZVI with HMs proves that this process is influenced by the characteristics of the solution. Thus, understanding NP–HM interactions and NP's relevant use for the achievement of the remediation objectives, it should be noted that both NP properties and potential factors affecting the behavior of the NPs in the polluted environment need to be considered. This section will highlight several parameters of the environment that influence the topic as follows: Some of these factors include oxidation/reduction potential (ORP) (Tandon and Singh 2016), co-existing cation/anion (Lv et al. 2011), pH (Vítková et al. 2017), and soil organic matter (Zhang et al. 2017). The power density of this cispectral manifold is four in the set pH scale 1. This means that the ability of the concentration of HMs and nature of the solution to affect the reduction of heavy metal content in the solution with NZVI is greatly determined by the solution's pH level. For instance, effects like a change in pH may impact the corrosion of NZVI and, in the process, impact the reaction rates as well as the engulfment capacities of the NZVI to the pollutant. In addition, if the quantity of solution pH is low, it indicates that a high quantity of hydroxide ions (OH^-) shall have been used, which in turn shall facilitate the enhancement of the reduction process of the NZVI. However, the aspect of pH plays an important role as it relates to the change in the surface charge of NZVI, and this in turn relates to the degree of protonation and hence strongly influences the sorptive characteristics of the NZVI for HMs.

8.1 Organic Matter (OM)

This study found that OM stands for organic matter, and this matter exists both in water and soil and affects the reactivity of NZVI (Nanoscale Zero Valent Iron). Multi-component interactions have been suggested by Wang et al. (2019). For example, the authors argued that due to the presence of OM, it becomes more effective at dissolving and forming complexes with HMs and other organic species. This leads to the process of passivating the Zero Valent Iron or, in some cases, the incorporation of more Zero Valent Iron (ZVI). If analyzed proportionally within the context of organic matter, the key constituents are humic and fulvic acids. They sorb NZVI particles preferentially, according to Tratnyek et al. (2001), and they can attenuate the reaction sites with HMs, thereby lowering the reactivity of NZVI.

Similarly, OM and stabilizers affect the surface of NZVI; because of this, the size as well as the stability of NZVI is disturbed (Zhao et al. 2016). Dissolved organic matter (DOM) and inorganic matter (HMs) can interact with metals to form coordination compounds that may alter some of the basic properties and, hence, have a different capacity to be adsorbed or immobilized (Peng et al. 2017). Furthermore, the decomposition of sediment OM can reduce the sediment pH, thereby affecting the reactivity of the metal ions of sample OM and OMNPs. However, different studies have indicated that there is a harmony between OM and HM immobilization that leads to a decrease in the activity and toxicity of such elements. Based on the information discussed, it was concluded that OM has a primary influence on the behavior of different nanomaterials as well as HM ions. It should therefore be taken into account while using nanomaterials for the remediation of HM-contaminated environments. Oxidation-reduction potential (ORP), or redox potential, as the name suggests, is the potential a system has to bring about oxidation/reduction reactions.

The use of ORP is very critical when describing the mobility, availability, redox activity, and toxicity of some HMs. Hence, the ORP of sediment moderates the behavior of numerous high-value elements such as chromium, selenium, cobalt, lead, arsenic, nickel, and copper. These elements impact the mobility and toxicity of pollutants independently (Tandon and Singh 2016). Taking a cue from what has been pointed out regarding OPR, NZVI equally works nobly in regard to the immobilization of pollutants based on the research. Equations (16) and (17) from the regression analysis confirm that NZVI is organized as a bi-electron donor system involving Fe0 and Fe2 + counterparts. This implies that NZVI can react with a number of redox-sensitive pollutants and other electron-accepting substances in the environment. The incorporation of H^+ ions into the pore water lowers the pH in the sediment, which in turn releases the heavy metals (HMs).

9. Possibilities for Commercial Applications

Due to their exceptional surface features, newly developed nanomaterials exhibit high efficiency in removing heavy metal ions. Nevertheless, clear disparities exist between industrial production and laboratory-scale settings. The commercial use of novel nanomaterials will be influenced by factors such as cost, toxicity, and feasibility for large-scale implementation. Currently, MOF materials are synthesized using costly metal precursors and organic ligands, which elevates costs for wastewater treatment applications. Thus, using cost-effective organic solvents as precursors is advisable. nZVI is primarily produced from inexpensive iron salts; its synthesis is simple, and the adsorbent is cost-effective. The synthesis conditions required for MXenes are severe, resulting in elevated production expenses. Hence, substituting etching with ball milling can result in cost reduction, necessitating further investigation into the manufacturing process of MXenes. g-C_3N_4 is an affordable adsorbent due to its inexpensive raw ingredients and straightforward manufacturing procedure.

As part of pollution remediation, it is unavoidable that new nanomaterials will be introduced into the environment. Their diminutive size, excellent mobility, elevated reactivity, and formidable ability to permeate may pose potential hazards to the surroundings. MOF materials consist of metal precursors that are coordinated with organic ligands. Research has demonstrated that the release of metal ions has a direct influence on the harmfulness of MOFs. Hence, it is advisable to choose non-toxic metal salts, such as iron salts, as metal precursors in order to prevent any potential toxicity. nZVI is mostly produced through the synthesis of iron salts, which are typically non-toxic. At the nanoscale, alterations in the characteristics of iron lead to toxicity in animals. Li et al. demonstrated that the presence of nZVI can disrupt the function of antioxidant enzymes and cause lipid peroxidation. The aforementioned research primarily concentrates on the toxicity of exposed nZVI. However, these conventional iron nanoparticles are prone to agglomeration and oxidation, resulting in inadequate dispersion stability. Consequently, their potential applicability is restricted. Indeed, modified nanoscale zero-valent iron (nZVI) is frequently employed in the field of environmental remediation. In their study, Phenrat et al. (2008) examined the neurotoxic properties of several forms of nZVI and determined that freshly prepared nZVI exhibited the highest level of toxicity, whereas surface-modified nZVI had the lowest level of toxicity MXenes are typically synthesized through the targeted removal of the A-element layer from the MAX phase using an appropriate etchant at ambient temperature. Hence, the establishment of a fluorine-free technique for synthesizing MXene, with the aim of its widespread use in commercial settings, holds significant significance. g-C_3N_4 is mostly derived from non-toxic urea and is inherently non-toxic, making it a commonly employed environmentally friendly substance.

Nanomaterials are extensively employed in tower, pilot, and industrial production for the purpose of pollutant removal, owing to their exceptional capabilities. These materials exhibit excellent adsorption performance and demonstrate consistent repeatability. Furthermore, the significant adsorption capability underscores their potential use in the treatment of wastewater. NZVI, which stands for nano zero-valent iron, has been utilized in practical scenarios as a very efficient adsorbent for the purpose of eliminating metal ions. Chlorinated hydrocarbon has been eliminated throughout Europe with nZVI. The pilot-scale experiments demonstrated a reduction in the quantity of chlorinated hydrocarbons by 40–80%. Similarly, the full-scale remediation work showed a decrease in the concentration of chlorinated hydrocarbon to approximately 20% of the initial value (Lefevre et al. 2016). News sources indicate that a booster manufacturer in my nation uses nZVI (nanoscale zero-valent iron) for the purpose of treating electroplating effluent.

Essentially, by optimizing the experimental settings, new nanomaterials can effectively fulfill the criteria of being inexpensive and non-toxic. The widespread implementation of nanomaterials for the removal of heavy metal ions in wastewater, including in pilot and industrial settings, has been successfully achieved. In the not-too-distant future, the synthesis of nanomaterials will be simplified and made more affordable due to technological advancements. If comprehensive study is conducted, the commercial utilization of novel nanomaterials may prove feasible, as indicated by these findings.

10. Environmental Fate of Nanomaterials

The environmental fate of nanoparticles remains incompletely investigated and comprehended. Multiple instances of confirmed toxicological impacts of nanoparticles on human beings have been reported, which can be directly attributed to the production and utilization of nanomaterials. Presently, there is a lack of comprehensive data regarding the bioavailability, transformation, and absorption of nanomaterials by both people and other organisms. Once they dock at the cell surface, they are capable of penetrating the cell through endocytosis and affecting cellular metabolism. The trophic level that is lower indicates an even higher predisposition to accumulate or magnify nanoparticles, which can potentially enter the human body through the food chain (Lowry et al. 2012). However, while there have been improvements in the evaluation of the risks of nanomaterials, several restrictions remain for future studies. As a result, anthropological studies could benefit from addressing potential uncertainties relating to the risks involved in the use of nanomaterials and nanodevices. Future emphasis should be placed on environmental impacts and safety issues. There are several factors operating in the environment that may lead to the degradation of nanomaterials and reduced competencies. Therefore, the best strategy to consider is to focus on the methods by which nanomaterials contribute to environmental degradation in some ways. This relates to the loss of protective shells, mechanical stress, chemical reactions involving oxidation and reduction processes, and the buildup and dissolution of the nanomaterial (Kahlon et al. 2018). Among the substrates of biological agents, some include biotransformation processes for the transformation of nanomaterials. Furthermore, consideration of the performance of electrochemical methods, the working environment with its potential pH and temperature, and the kind and concentration of the target pollutants must be studied and set to the right level. However, the density of nanoparticles should also be properly selected to obtain the highest impact on the concentration of the two components in the mixture. Therefore, in an endeavor to achieve better outcomes, all of the aforementioned constituents must be taken into consideration before employing nanoparticles to realize the intended goal of selective pollutant degradation.

11. Conclusions

Human activities are the main cause of environmental pollution and degradation. Conventional pollution control technologies require more energy and material consumption and may result in secondary pollution. Nanotechnology has emerged as a potential solution that can address several environmental problems. The properties of base materials change dramatically at the nanoscale, providing ideal activity against target contaminants. Attempts have been made to reduce environmental pollution by using nanomaterials with tunable surface functionality, structural stability, high adsorption capacity, enhanced selectivity and specificity, improved biodegradability, and reusability of synthetic materials. Nanoengineered materials have been used as catalysts, adsorbents, and additives. The fate of nanomaterials in the environment and the potential health effects on all organisms and the environment are not well understood so far and further research is required in this area to avoid potential risks in using these materials at the nanoscale. A better understanding of the operating conditions, such as pH and temperature of the environment, dosage, type and initial concentration of contaminants, can lead to good results. The synthesis of nanomaterials by physicochemical methods is expensive and environmentally challenging. Biosynthesis of nanomaterials can provide economical and environmentally friendly solutions for the production of nanomaterials, thus making their application in environmental remediation a sustainable approach.

12. Acknowledgement

Authors want to thank the University of Agriculture Faisalabad and Yangzhou University for the support in this publication. Authors also want to thank Professor Dr Abdul Wakeel from Soil Fertility and Plant Nutrition Laboratory, University of Agriculture Faisalabad, Pakistan.

References

Aktar, W., D. Sengupta and A. Chowdhury. 2009. Impact of pesticides use in agriculture: their benefits and hazards. Interdiscip. Toxicol. 2: 1–12.

Alepu, O.E., Z. Li, H. Ikhumhen, L. Kalakodio, K. Wang and G. Segun. 2016. Effect of hydraulic retention time on anaerobic digestion of Xiao Jiahe municipal sludge. Int. J. Waste Resour. 6: 1–5.

Ali, H., K.W. Kim, S.G. Bang, H.B. Chae, S.W. Shin and C.W. Park. 2019a. Numerical modeling of fluid structure interaction between sewage water flow and bar screen to improve the screening process. Water Environ. J. 33: 560–573.

Ali, S., M. Rizwan, A. Hussain, M.Z. ur Rehman, B. Ali, B. Yousaf et al. 2019b. Silicon nanoparticles enhanced the growth and reduced the cadmium accumulation in grains of wheat (*Triticum aestivum* L.). Plant Physiol. Biochem. 140: 1–8.

Anderson, T.R., E. Hawkins and P.D. Jones. 2016. CO_2, the greenhouse effect and global warming: from the pioneering work of Arrhenius and Callendar to today's Earth System Models. Endeavour. 40: 178–187.

Baruah, S. and J. Dutta. 2009. Nanotechnology applications in pollution sensing and degradation in agriculture: a review. Environ. Chem. Lett. 7: 191–204.

Bashir, H., I. Bibi, M.M. Hussain, N.K. Niazi and J. Iqbal. 2022. Nanoparticulate iron oxide minerals for arsenic removal from contaminated water. pp. 459–479. *In*: Niazi, N.K., M.M. Hussain and I. Bibi [eds.]. Global Arsenic Hazard: Ecotoxicology and Remediation. Springer.

Behravesh, S., N. Mirghaffari, A.A. Alemrajabi, F. Davar and M. Soleimani. 2020. Photocatalytic degradation of acetaminophen and codeine medicines using a novel zeolite-supported TiO2 and ZnO under UV and sunlight irradiation. Environ. Sci. Pollut. Res. 27: 26929–26942.

Bhatia, S. and S. Devraj. 2017. Pollution Control in Textile Industry. WPI Publishing.

Borji, H., G.M. Ayoub, M. Al-Hindi, L. Malaeb and H.Z. Hamdan. 2020. Nanotechnology to remove polychlorinated biphenyls and polycyclic aromatic hydrocarbons from water: a review. Environ. Chem. Lett. 18: 729–746.

Čehovin, M., A. Medic, J. Scheideler, J. Mielcke, A. Ried, B. Kompare et al. 2017. Hydrodynamic cavitation in combination with the ozone, hydrogen peroxide and the UV-based advanced oxidation processes for the removal of natural organic matter from drinking water. Ultrason. Sonochem. 37: 394–404.

Crane, R.A., M. Dickinson and T.B. Scott. 2015. Nanoscale zero-valent iron particles for the remediation of plutonium and uranium contaminated solutions. Chem. Eng. J. 262: 319–325.

Crane, R.A. and T.B. Scott. 2012. Nanoscale zero-valent iron: future prospects for an emerging water treatment technology. J. Hazard. Mater. 211: 112–125.

El-Sayed, M.E. 2020. Nanoadsorbents for water and wastewater remediation. Sci. Total Environ. 739: 139903.

Endres, S., F. Maes, F. Hopkins, K. Houghton, E.M. Mårtensson, J. Oeffner et al. 2018. A new perspective at the ship-air-sea-interface: The environmental impacts of exhaust gas scrubber discharge. Front. Mar. Sci. 5: 361303.

Foster, H.A., I.B. Ditta, S. Varghese and A. Steele. 2011. Photocatalytic disinfection using titanium dioxide: spectrum and mechanism of antimicrobial activity. Appl. Microbiol. Biotechnol. 90: 1847–1868.

Fu, F., D.D. Dionysiou and H. Liu. 2014. The use of zero-valent iron for groundwater remediation and wastewater treatment: a review. J. Hazard. Mater. 267: 194–205.

Gangupomu, R.H., M.L. Sattler and D. Ramirez. 2014. Carbon nanotubes for air pollutant control via adsorption: A review. Rev. Nanoscience Nanotechnol. 3: 149–160.

Glibert, P.M. 2017. Eutrophication, harmful algae and biodiversity—Challenging paradigms in a world of complex nutrient changes. Mar. Pollut. Bull. 124: 591–606.

Guerra, F.D., M.L. Campbell, D.C. Whitehead and F. Alexis. 2017. Tunable properties of functional nanoparticles for efficient capture of VOCs. ChemistrySelect 2: 9889–9894.

Guo, S., C. Xiao, N. Zhou and R. Chi. 2021. Speciation, toxicity, microbial remediation and phytoremediation of soil chromium contamination. Environ. Chem. Lett. 19: 1413–1431. doi:10.1007/s10311-020-01114-6

Hotze, E.M., T. Phenrat and G.V. Lowry. 2010. Nanoparticle aggregation: challenges to understanding transport and reactivity in the environment. J. Environ. Qual. 39: 1909–1924.

Huerta-Fontela, M., M.T. Galceran and F. Ventura. 2010. Fast liquid chromatography–quadrupole-linear ion trap mass spectrometry for the analysis of pharmaceuticals and hormones in water resources. J. Chromatogr. A 1217: 4212–4222.

Hussain, M., Z. Farooqi, W. Mohy-Ud-Din, F. Younas, M. Shahzad, M. Ghani et al. 2021a. Application of bioremediation as a sustainable approach to remediate heavy metal and pesticide polluted environments. Plant Environ. 1: 62–92.

Hussain, M.M., I. Bibi, N.K. Niazi, M. Shahid, J. Iqbal, M.B. Shakoor et al. 2021b. Arsenic biogeochemical cycling in paddy soil-rice system: Interaction with various factors, amendments and mineral nutrients. Sci. Total Environ. 145040. doi: https://doi.org/10.1016/j.scitotenv.2021.145040.

Hussain, M.M. and Z.U.R. Farooqi. 2021. Allelopathic bacteria as an alternate weedicide: Progress and future standpoints. pp. 211–230. *In*: Microbiota and Biofertilizers. Springer.

Hussain, M.M., J. Wang, I. Bibi, M. Shahid, N.K. Niazi, J. Iqbal et al. 2020. Arsenic speciation and biotransformation pathways in the aquatic ecosystem: The significance of algae. J. Hazard. Mater. 403: 124027. doi: https://doi.org/10.1016/j.jhazmat.2020.124027.

Jadhav, J. and S. Biswas. 2018. Hybrid ZnO: Ag core-shell nanoparticles for wastewater treatment: Growth mechanism and plasmonically enhanced photocatalytic activity. Appl. Surf. Sci. 456: 49–58.

Kahlon, S.K., G. Sharma, J. Julka, A. Kumar, S. Sharma and F.J. Stadler. 2018. Impact of heavy metals and nanoparticles on aquatic biota. Environ. Chem. Lett. 16: 919–946.

Kampa, M. and E. Castanas. 2008. Human health effects of air pollution. Environ. Pollut. 151: 362–367.

Karci, A., I. Arslan-Alaton, T. Olmez-Hanci and M. Bekbolet. 2013. Degradation and detoxification of industrially important phenol derivatives in water by direct UV-C photolysis and H2O2/UV-C process: A comparative study. Chem. Eng. J. 224: 4–9.

Kaur, S. and A. Roy. 2021. Bioremediation of heavy metals from wastewater using nanomaterials. Environ. Dev. Sustain. 23: 9617–9640.

Klimkova, S., M. Cernik, L. Lacinova, J. Filip, D. Jancik and R. Zboril. 2011. Zero-valent iron nanoparticles in treatment of acid mine water from in situ uranium leaching. Chemosphere 82: 1178–1184.

Kolahalam, L.A., I.K. Viswanath, B.S. Diwakar, B. Govindh, V. Reddy and Y. Murthy. 2019. Review on nanomaterials: Synthesis and applications. Mater. Today Proc. 18: 2182–2190.

Kumar, L. and N. Bharadvaja. 2019. Enzymatic bioremediation: a smart tool to fight environmental pollutants. pp. 99–118. *In*: Mishra, P. [ed.]. Smart Bioremediation Technologies. Elsevier, Amsterdam.

Lefevre, E., N. Bossa, M.R. Wiesner and C.K. Gunsch. 2016. A review of the environmental implications of in situ remediation by nanoscale zero valent iron (nZVI). behavior, tranoport and impacts on microbial communities. Sci. Total Environ. 565: 889–901.

Li, S., W. Wang, F. Liang and W.X. Zhang. 2017. Heavy metal removal using nanoscale zero-valent iron (nZVI): Theory and application. J. Hazard. Mater. 322: 163–171.

Li, X.Q., D.W. Elliott and W.X. Zhang. 2008. Zero-valent iron nanoparticles for abatement of environmental pollutants: materials and engineering aspects. pp. 309–330. *In*: Gupta, M.K. and A.J. Jadhav [eds.]. Particulate systems in nano-and biotechnologies. CRC Press, Boca Raton, FL.

Liu, Y., F. Liu, Z. Qi, C. Shen, F. Li, C. Ma et al. 2019. Simultaneous oxidation and sorption of highly toxic Sb (III) using a dual-functional electroactive filter. Environ Pollut. 251: 72–80.

Lowry, G.V., K.B. Gregory, S.C. Apte and J.R. Lead. 2012. Transformations of nanomaterials in the environment. ACS Publications.

Lu, H., J. Wang, M. Stoller, T. Wang, Y. Bao and H. Hao. 2016. An overview of nanomaterials for water and wastewater treatment. Adv. Mater. Sci. Eng. 2016(1): 4.

Luo, Y., W. Guo, H.H. Ngo, L.D. Nghiem, F.I. Hai, J. Zhang et al. 2014. A review on the occurrence of micropollutants in the aquatic environment and their fate and removal during wastewater treatment. Sci. Total Environ. 473: 619–641.

Lv, X., J. Xu, G. Jiang and X. Xu. 2011. Removal of chromium (VI) from wastewater by nanoscale zero-valent iron particles supported on multiwalled carbon nanotubes. Chemosphere. 85(7): 1204–1209.

Madima, N., S. Mishra, I. Inamuddin and A. Mishra. 2020. Carbon-based nanomaterials for remediation of organic and inorganic pollutants from wastewater. A review. Environ. Chem. Lett. 18: 1169–1191.

Mashkoor, F., A. Nasar and I. Inamuddin. 2020. Carbon nanotube-based adsorbents for the removal of dyes from waters: a review. Environ. Chem. Lett. 18(3): 605–629.

Mekonnen, M.M. and A.Y. Hoekstra. 2016. Four billion people facing severe water scarcity. Sci. Adv. 2(2): e1500323.

Naveed, M., P. Moldrup, E. Arthur, M. Holmstrup, M. Nicolaisen, M. Tuller et al. 2014. Simultaneous loss of soil biodiversity and functions along a copper contamination gradient: when soil goes to sleep. Soil Sci. Soc. Am. J. 78(4): 1239–1250.

Nazaralian, S., A. Majd, S. Irian, F. Najafi, F. Ghahremaninejad, T. Landberg et al. 2017. Comparison of silicon nanoparticles and silicate treatments in fenugreek. Plant Physiol. Biochem. 115: 25–33.

Noh, S.Y., J.E. Heo, S.H. Woo, S.J. Kim, M.H. Ock, Y.J. Kim et al. 2018. Performance improvement of a cyclone separator using multiple subsidiary cyclones. Powder Technol. 338: 145–152.

Nurmi, J.T., P.G. Tratnyek, V. Sarathy, D.R. Baer, J.E. Amonette, K. Pecher et al. 2005. Characterization and properties of metallic iron nanoparticles: spectroscopy, electrochemistry, and kinetics. Environ. Sci. Technol. 39(5): 1221–1230.

Panda, S.K., I. Aggarwal, H. Kumar, L. Prasad, A. Kumar, A. Sharma et al. 2021. Magnetite nanoparticles as sorbents for dye removal: a review. Environ. Chem. Lett. 19: 2487–2525.

Peng, H. and J. Guo. 2020. Removal of chromium from wastewater by membrane filtration, chemical precipitation, ion exchange, adsorption electrocoagulation, electrochemical reduction, electrodialysis, electrodeionization, photocatalysis and nanotechnology: a review. Environ. Chem. Lett. 18(6): 2055–2068.

Peng, Z., C. Xiong, W. Wang, F. Tan, Y. Xu, X. Wang et al. 2017. Facile modification of nanoscale zero-valent iron with high stability for Cr (VI) remediation. Sci. Total Environ. 596: 266–273.

Petrunin, M., L. Maksaeva, T. Yurasova, E. Terekhova, M. Maleeva, V. Kotenev et al. 2015. Formation of organosilicon self-organizing nanolayers on an iron surface from vapor phase and their effect on corrosion behavior of metal. Prot. Met. Phys. Chem. Surf. 51: 1010–1017.

Phenrat, T., N. Saleh, K. Sirk, H.J. Kim, R.D. Tilton and G.V. Lowry. 2008. Stabilization of aqueous nanoscale zerovalent iron dispersions by anionic polyelectrolytes: adsorbed anionic polyelectrolyte layer properties and their effect on aggregation and sedimentation. J. Nanopart. Res. 10: 795–814.

Rasli, N.I., H. Basri and Z. Harun. 2020. Zinc oxide from aloe vera extract: two-level factorial screening of biosynthesis parameters. Heliyon. 6(1).

Reddy, L.H., J.L. Arias, J. Nicolas and P. Couvreur. 2012. Magnetic nanoparticles: design and characterization, toxicity and biocompatibility, pharmaceutical and biomedical applications. Chem. Rev. 112(11): 5818–5878.

Romauld, S.I., R. Venkataraghavan, D. Yuvaraj, V.I. Devi and S. Hashika. 2019. Mycoremediation of hydrocarbon and its products using *Fusarium oxysporum*. Res. J. Pharm. Technol. 12(9): 4216–4224.

Sandhyarani, N. 2019. Surface modification methods for electrochemical biosensors. pp. 45–75. *In*: Electrochemical biosensors. Elsevier.

Schwarzenbach, R.P., T. Egli, T.B. Hofstetter, U. Von Gunten and B. Wehrli. 2010. Global water pollution and human health. Annu. Rev. Environ. Resour. 35: 109–136.

Shang, J., M. Zong, Y. Yu, X. Kong, Q. Du and Q. Liao. 2017. Removal of chromium (VI) from water using nanoscale zerovalent iron particles supported on herb-residue biochar. J. Environ. Manage. 197: 331–337.

Shashanka, R., G.K. Jayaprakash, M. Kumar and B.E. Kumara Swamy. 2022. Electrocatalytic determination of ascorbic acid using a green synthesised magnetite nano-flake modified carbon paste electrode by cyclic voltammetric method. Mater. Res. Innov. 26(4): 229–239.

Singh, B., M. Gräfe, N. Kaur and A. Liese. 2010. Applications of synchrotron-based X-ray diffraction and X-ray absorption spectroscopy to the understanding of poorly crystalline and metal-substituted iron oxides. pp. 199–254. *In*: Developments in Soil Science, vol 34. Elsevier.

Singh, M. K. and M. S. Mehata, 2019. Phase-dependent optical and photocatalytic performance of synthesized titanium dioxide (TiO2) nanoparticles. Optik 193: 163011.

Steffan, J., E. Brevik, L. Burgess and A. Cerdà. 2018. The effect of soil on human health: an overview. Eur. J. Soil Sci. 69(1): 159–171.

Sun, C., C. Ding, W. Cheng and X. Wang. 2014. Simultaneous adsorption and reduction of U (VI) on reduced graphene oxide-supported nanoscale zerovalent iron. J. Hazard. Mater. 280: 399–408.

Tandon, P.K. and S.B. Singh. 2016. Redox processes in water remediation. Environ. Chem. Lett. 14: 15–25.

Taylor, V., B. Goodale, A. Raab, T. Schwerdtle, K. Reimer, S. Conklin et al. 2017. Human exposure to organic arsenic species from seafood. Sci. Total Environ. 580: 266–282.

Thundiyil, J.G., J. Stober, N. Besbelli and J. Pronczuk. 2008. Acute pesticide poisoning: a proposed classification tool. Bull. World Health Organ. 86: 205–209.

Tratnyek, P.G., M.M. Scherer, B. Deng and S. Hu. 2001. Effects of natural organic matter, anthropogenic surfactants, and model quinones on the reduction of contaminants by zero-valent iron. Water Res. 35(18): 4435–4443.

Üresin, E., H.İ. Saraç, A. Sarıoğlan, Ş. Ay and F. Akgün. 2015. An experimental study for H2S and CO_2 removal via caustic scrubbing system. Process Saf. Environ. Prot. 94: 196–202.

Vítková, M., S. Rákosová, Z. Michálková and M. Komárek. 2017. Metal (loid) s behaviour in soils amended with nano zero-valent iron as a function of pH and time. J. Environ. Manag. 186: 268–276.

Wang, T., Y. Liu, J. Wang, X. Wang, B. Liu and Y. Wang. 2019. In-situ remediation of hexavalent chromium contaminated groundwater and saturated soil using stabilized iron sulfide nanoparticles. J. Environ. Manag. 231: 679–686.

Xu, J., C. Liu, P.-C. Hsu, J. Zhao, T. Wu, J. Tang et al. 2019. Remediation of heavy metal contaminated soil by asymmetrical alternating current electrochemistry. Nat. Commun. 10(1): 2440.

Yan, Q., J. Street and F. Yu. 2015. Synthesis of carbon-encapsulated iron nanoparticles from wood derived sugars by hydrothermal carbonization (HTC) and their application to convert bio-syngas into liquid hydrocarbons. Biomass Bioenergy 83: 85–95.

Zang, Y., Y. Li, C. Wang, W. Zhang and W. Xiong. 2015. Towards more accurate life cycle assessment of biological wastewater treatment plants: a review. J. Clean. Prod. 107: 676–692.

Zare, E.N., A. Motahari and M. Sillanpää. 2018. Nanoadsorbents based on conducting polymer nanocomposites with main focus on polyaniline and its derivatives for removal of heavy metal ions/dyes: a review. Environ. Res. 162: 173–195.

Zhang, N., Z. Fang and R. Zhang. 2017. Comparison of several amendments for in-site remediating chromium-contaminated farmland soil. Water Air Soil Pollut. 228: 1–10.

Zou, Y., X. Wang, A. Khan, P. Wang, Y. Liu, A. Alsaedi et al. 2016. Environmental remediation and application of nanoscale zero-valent iron and its composites for the removal of heavy metal ions: a review. Environ. Sci. Technol. 50(14): 7290–7304.

CHAPTER-4

Nanomaterials for Soil Fertility Enhancement

Akhtar Hameed,[1,*] *Rana Binyamin,*[1] *Subhan Ali,*[1] *Hira Akhtar,*[1]
Muhammad Waqar Alam[2] and *Nighat Raza*[3]

1. Introduction to Soil Fertility

1.1 Definition and Importance of Soil Fertility

The capability of soil to sustain agricultural plant growth by providing them required essential nutrients and suitable physical, biological and chemical conditions for plant growth is known as soil fertility (Ditta et al. 2020). Physical media include the size, shape, and texture of soil particles, while biological refers to soil microbiota, and chemical media refer to pH level, soil acidity, and chemical inputs to fulfill soil nutrition. All of these components make the soil healthy and productive when they are present in balanced quantities (Das et al. 2022). Plant growth, development, and production all depend upon the level of soil fertility. The more fertile the soil is, the healthier it will be, and the higher the production that will be achieved. Soil fertility can be improved through the addition of organic matter or inorganic fertilizers to soil. Organic matter may already be present in the soil in the form of fossils, which are decomposed by soil microbiota (Furey and Tilman 2021).

The effects of soil fertility are the most important in almost every aspect of life which is linked with the agriculture sector. The provision of enough food for the global population is the most prominent benefit which is achieved based on fertile soil, as high-yield production of food crops is attained in fertile soil (Di Falco and Zoupanidou 2017). Fertile soil can provide all the nutrients to plants for their growth and development. Cultivation of crops in fertile soil increases their production in terms of both quality and quantity, and farmers can gain more benefits. In the case of plant resistance, soil fertility is important for producing strong and healthy plants that have more stamina to resist pathogens (Wolf et al. 2023). Therefore, the use of pesticides is limited, which creates a positive impact on biodiversity. Soil fertility increases the water-holding capacity of soil, resulting in tolerance to drought conditions. Aeration in the soil is improved, enabling easy and balanced access of air and water to plants. Environmental sustainability, support for beneficial microbes, optimization of soil pH, and availability of soil organic matter are other important benefits attained from fertile soil.

[1] Institute of Plant Protection, MNS University of Agriculture, Multan 61000, Pakistan.
[2] Department of Botany, University of Okara, Pakistan
[3] Department of Food Science and Technology, MNS University of Agriculture Multan 61000, Pakistan
* Corresponding author: akhtarhameed@mnsuam.edu.pk

DOI: 10.1201/9781003558354-4

1.2 Factors Affecting Soil Fertility

As soil fertility is necessary for better growth and production of crops, some factors influence the capability of soil to provide optimum conditions to plants. These factors include soil pH, moisture content, bulk density, climate, soil biota, types of soil, nutrient retention capacity, cation exchange capacity, etc. These factors determine how much fertile the soil is (Neina 2019).

1.3 Current Challenges in Soil Fertility

Soil faces many challenges that result in the loss of fertility, ultimately affecting the cultivated crops in terms of quality and quantity. The loss of soil fertility has been a major challenge faced by agricultural land. The reasons behind this loss include water and air pollution, soil compaction, lack of tillage operations, extensive use of chemical pesticides and fertilizers, lack of proper irrigation, intensification of crop production, continuous cultivation of major crops, and depletion of soil microbiota, etc. (Shrivastava and Kumar 2015). The acidity of rain due to air pollution has increased, which destroys soil health by accelerating soil acidity. Water pollution concerns the addition of industrial waste into river water, which is often used to irrigate cultivated land. Heavy metal ions present in polluted water accumulate in the soil and disturb its quality. The reduced use of tillage operations converts the soil into hard pans, leading to compaction. The extensive use of pesticides and chemicals helps to solve the problem of crops temporarily but they have side effects on the soil and harm the soil microbiota. Due to these issues, the soil quality is degraded and then farmers use extensive chemical fertilizers to retain the soil fertility but they affect the environment and biodiversity. Human beings are found to be involved in the above major issues, and with the increase in population, these problems are increasing continuously. These alarming issues disturb not only soil life but also every life that exists on the earth, including human beings (Tudi et al. 2021).

2. Introduction to Nanomaterials

2.1 Definition and Types of Nanomaterials

With the advancement and modernization of the world, different advanced technologies have emerged to serve the globe, and nanotechnology is one of them. Nanotechnology has broad-spectrum applications in almost every sector such as biomedical, agriculture, industrial, and energy sectors. The nanoparticles (particles with a diameter ranging from 1–100 nm) used in nanotechnology are referred to as nanomaterials, which are generally defined as materials that have at least one external dimension with a size of 1–100 nm (Malik et al. 2023). Nanotechnology, by using different nanomaterials, has developed products with improved performance. These nanomaterials may be engineered, naturally present, or produced incidentally. These nanomaterials are synthesized using two approaches: the top-down approach and the bottom-up approach. They are classified based on the number of dimensions they possess or the material from which they are derived. The most common nanomaterials include nanorods, nanotubes, fullerenes, nanoribbons, nanofoam, etc. (Khan et al. 2019).

2.2 Potential Applications in Agriculture

Like their application in other fields, nanotechnology has made breakthroughs in agriculture with improved results in the production and protection of crops, summarized in table 1. The use of nanotechnology in the development of precision agriculture is termed nano-farming. Reduced and low-quality crop yields due to plant diseases are major challenges in agriculture, and a solution via nanotechnology is promising (Shang et al. 2019). Nanotechnology improves crop production, crop protection, soil enhancement, stress tolerance, and precision farming. Crop production is increased through the usage of nanofertilizers, while crop protection is improved through the development of

Table 1. Nanomaterials and their functions in agriculture.

Nanomaterials	Functions
Carbon nanotubes	Germination of seed
Nanonutrients	Plant/animal/human nutrition
Nanopesticides	Plant protection
Nanoscale carriers	Efficient delivery of fertilizer and pesticides
Nanosensor	Detection of plant diseases and nutrient deficiencies

nanopesticides and nanosensors, which help in the detection of plant diseases and their treatment. The major advantage of using these novel techniques is that they are site-specific. Nanomaterials are used to improve soil quality by converting degraded soil into healthy soil (Yadav et al., 2010). Stress tolerance is enhanced by using nanoparticles that aim to minimize abiotic stress like UV light, temperature, etc. These disciplines of agriculture are being focused on using nanotechnology on a somewhat larger scale. Precision farming is the basic need of recent times for the conservation of biodiversity by reducing the side effects of widely used chemicals on the ecosystem (Al-Khayri et al. 2023).

Several nanotechnology advancements have emerged in gene techniques for stress-induced gene expression in plants and animals. Future developments in nanomaterial applications depend heavily on the delivery of genes and medicinal compounds to particular cellular areas using microarrays (DNA chips or biochips), micro-electromechanical systems, microfluidics, and quality control. Prospects for agriculture (Day 2005), smart agrochemical delivery systems (e.g., pesticides and fertilizers), early disease and pollutant detection in agricultural commodities, and systems integration for food processing, packaging, monitoring, and natural resource management are all developing (Chau et al. 2007). Food and agricultural product production are major economic drivers, and they are influenced by each of these factors. Since nanotechnology is expected to benefit consumers, producers, farmers, the general public, and the ecosystem more broadly than other technologies, it is anticipated to be one of the major influences in the long run (Rashidi and Khosravi-Darani 2011, Raynolds 2004). Some of their uses in agriculture are illustrated in Figure 1.

Conversely, nanotoxicity poses serious risks to human health, agriculture, and the environment (Fischer and Chan 2007). Not all nanoparticles are harmful to human health since their toxicity depends on several factors, such as size, aggregation, composition, crystallinity, and surface functionalization. Due to their small size, ingested nanomaterials can penetrate biological tissues and organs, as well as the lymphatic and circulatory systems. Depending on their size and makeup, certain nanoparticles have the potential to cause irreversible harm to cells and organelles (Daima and Navya 2016). Their synthesis also influences the ecosystem due to the negative impact of different chemicals and metals used, while green synthesis imposes positive impacts. However, not much research has been done on the toxicological potential of nanoparticles in food, agriculture, and the environment (Nel et al. 2006).

3. Role of Nanomaterial in Plants

3.1 Nanoparticles: Synthesis and Characterization Techniques

Nanoparticles are defined as nano-objects with all external dimensions at the nanoscale, and the length of long and short axes is not significantly different, according to the International Organization for Standardization (ISO), a global federation of national standards organizations. Nanoparticles can be divided into three size ranges: larger than 500 nm, between 100 and 500 nm, and between 1 and 100 nm, even though the nanoscale is typically defined as 1 to 100 nm (Khan et al. 2019). Based on their size distribution, nanoparticles may have intense properties that are related

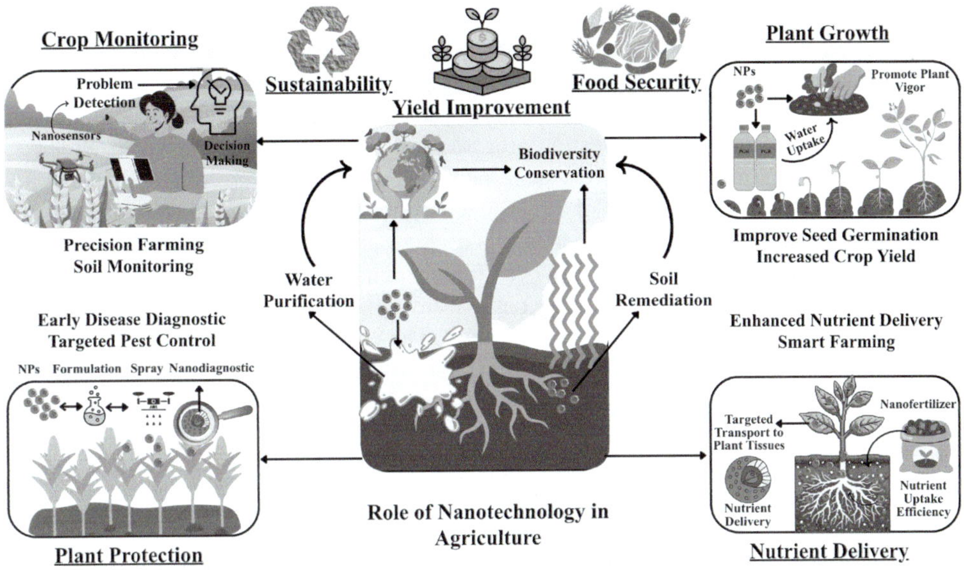

Figure 1. Applications of nanotechnology in agriculture.

to their size. They display surprising properties and produce quantum phenomena if they are small enough to confine their electrons. For instance, compared to those in slab form, gold nanoparticles in solution appear red and melt at much lower temperatures. According to Biliu and Coseri (2019), a high surface-area-to-volume ratio of nanoparticles enables significant changes in surface area and surface contact characteristics, such as catalytic surface-enhanced plasmon resonance. One of the following properties can be found in nanoparticles: metallic, dielectric, semiconductor, magnetic, or multifunctional—which combines multiple attributes. Applications or potential uses for them span numerous industries. Because nanoparticles are similar in size to viruses (10–100 nm) and cells (1–10 μm), they are useful in the life sciences and the environment (Buzea et al. 2007). Because of their high surface-area-to-volume ratio and strong interactions with surfactant molecules, nanoparticles can cling to biological entities without impairing their ability to function. Because of their small size and large surface area, nanoparticles have the potential to be a particularly sensitive tool for identifying a particular contaminant, which is a common source of contamination in environmental applications. There may be opportunities to alleviate environmental contamination through nanoparticle engineering (Khan et al. 2021). Following are the methods to synthesize nanomaterials.

3.1.1 Chemical Reduction

When producing zero-valent nanoparticles by chemically reducing aqueous salts of metals, like silver nitrate ($AgNO_3$) in the case of silver nanoparticle manufacturing, chemical reduction is an efficient wet chemical method (Rahme 2020). At least one reducing agent is used to provide metal ions with electrons and reduce them to zero-valent to minimize the precursor metal salt. Three often-used reductants are ascorbate, citrate, and borohydride. To maintain the stability of reduced nanoparticles, a stabilizing agent like cetyltrimethylammonium bromide is added. Reducing compounds, such as sodium citrate, can function as stabilizing agents in the synthesis of silver nanoparticles (Kakkar et al. 2012).

3.1.2 Co-precipitation

Under some circumstances, soluble chemicals are precipitated. Usually, nucleation occurs in solution suddenly when the component concentration hits supersaturation. Surface diffusion boosts nucleation, culminating in the formation of nanoparticles (Zhang et al. 2011). To achieve homogeneous nanoparticles, the nucleation process must be delayed during growth. Precipitation can be achieved using a variety of processes, including co-precipitation, micro-emulsion/inverse micro-emulsion, polyol, and others. For the production of Fe_3O_4 nanoparticles, co-precipitation is a simple method. Before NH_4OH was added, the mixture of two chloride salts, FeCl2 and FeCl3, with a 1:2 molar ratio of Fe^{2+} and Fe^{3+}, had been rapidly agitated and held at 70C resulting in black precipitate formation (Zhang et al. 2011). To remove any remaining impurities, Fe_3O_4 nanoparticles have been recovered after repeated purification using magnetic separation with ethanol and distilled water. By adjusting the pH and ion concentration, it is possible to change the size of nanoparticles in the solution.

3.1.3 Seeding

According to Frens (1972, 1973), the seeding procedure involves reducing the salt content to generate nanoparticles. To control the size and shape of the growing nanoparticles, stabilizers are employed in this procedure. Over time, this method was refined using different kinds of stabilizers, reductant agents, and seeding.

For instance, Perrault and Chan created a method in 2009 that lowers HAuCl4in an aqueous solution containing gold nanoparticles by using hydroquinone. The hydroquinone and gold nanoparticle seed can catalyze the reduction of gold ions into their surface. If the stabilizer is citrate, then the citrate method is usually employed to produce the seed nanoparticles. Nanoparticle, with a minimum size of 30–300 nm, can be created using this process (Bastús et al. 2014).

3.1.4 Microemulsion and Inverse Microemulsion

Microemulsions are isotropic and thermodynamically stable blends of "oil," water, and surfactant, or a mixture of the latter with a co-surfactant, making this an effective procedure for manufacturing nanomaterials. There are two main varieties of microemulsions: direct (oil dispersed in water) and reverse (water dispersed in oil). Micelles, which are tiny aqueous phase droplets, can also contain oil in the form of a combination of surfactants and may possess salts or other components.

A mechanism that yields nanomaterials can also occur when the micelles contact. According to Arab et al. (2018), the surfactants in "oil" regulate the growth of the nanomaterials. This method allows for the creation of Fe_3O_4 nanoparticles and their functionalization with a silica layer (Feltin and Pileni 1997). Fe_3O_4/Au, core-shell nanoparticles, which help to enhance biocompatibility and stop magnetic nanomaterials (MNMs) from oxidizing, can also be made using this process (Li et al. 2011). Additionally, the easiest method for creating multifunctional nanomaterials is the inverter microemulsion procedure, which creates the silica matrix in an aqueous phase.

3.1.5 Hydrothermal Method

According to Darr et al. (2017), the hydrothermal process forms nanoparticle crystals under high pressure and temperature by heterogeneous reactions involving insoluble chemicals at normal temperature and pressure. To develop crystals, an equipment called an autoclave—a steel pressure vessel filled with water and nutrients— is utilized. Hydrothermal synthesis is typically carried out at temperatures below 300°C. According to the nucleation theory, the critical state produces a favorable reaction field for the synthesis of nanoparticles by greatly supersaturating the medium and increasing the reaction rate. In supercritical water, this technique has been used to produce metal oxide and semiconductor nanoparticles (Holmes et al. 2003).

3.1.6 Sono-electrode Position

The useful method of sono-electrodeposition is used to create nanoparticles. It has been effectively used to produce FePt and CoPt metallic nanoparticles (Luong et al. 2011). Sono-electrodeposition is a technique that creates metallic nanoparticles by fusing the benefits of electrodeposition with mechanical vibrations from ultrasonic waves (Zhu et al. 2000). One of the primary drawbacks of traditional ways of producing silver nanoparticles, such as chemical reduction, is the unexpected presence of harmful ions in the final products.

The majority of the dangerous ions in the outcome, such as nitrate and thiosulfate, originate from the silver precursor. According to Irzh et al. (2007), silver acetate is a good silver precursor to use, but it's expensive and difficult to work with in everyday situations. A modified sono-electrodeposition method for producing silver nanoparticles in a safe solution was proposed by Tuan et al. (2011). The difference is that a silver plate was used as the cathode rather than silver salts, which prevented unexpected ions from the salt from occurring.

3.2 Types of Nanomaterials (NMs) Involved in Agriculture

3.2.1 Carbon-based NPs

The carbon nanotube (CNT) and fullerene subgroups of carbon-based NMs are two important subgroups. Fullerenes contain NMs with spherical hollow cages that resemble allotropic forms of carbon. Their electrical conductivity, high strength, structure, electron affinity, and flexibility have sparked a great deal of commercial interest. These materials consist of hexagonal and pentagonal carbon units that have undergone sp^2 hybridization. CNTs form tubular structures with a diameter of 1–2 nm despite their elongation. These have the appearance of graphite sheets stacked on top of one another. Thus, they are referred to as single-walled (SWNTs), double-walled (DWNTs), or multi-walled carbon nanotubes (MWNTs) depending on how many walls are present in the rolled sheets (Zhou et al. 2009). Carbon-based nanomaterials improve agricultural yield through the use of plant growth boosters and novel nanomaterial-based fertilizers. Plant protection and production can be controlled using biosensors and nanopesticides based on carbon nanomaterials. By using nanopesticides efficiently, soil fertility can be conserved.

3.2.2 Metal NMs

Metals make up the entirety of metal nanoparticles. Because of their well-known localized surface plasmon resonance (LSPR) characteristics, these NMs display a variety of electrical properties. A significant absorption band is displayed by Cu, Ag, and Au nanoparticles in the visible portion of the solar electromagnetic spectrum. Because metal nanoparticles have more features than other materials, such as size, shape, and facet control, they are used in many scientific fields (An and Somorjai 2012). Nano-products can provide new and better solutions for various agricultural problems such as detection and mitigation of infection long before symptoms are visually evident, nutrient deficiency or other health problems, delivery of nano-formulated pesticides, delivery of fertilizers to reduce nutrient loss into the soil by leaching, nano-herbicides for effective weed population control and post-implementation detoxification, and field sensing systems to monitor crop management (Das 2022).

3.2.3 Ceramics NMs

Ceramic nanoparticles are minuscule particles with special properties that are created by heat-treating and cooling inorganic, non-metallic materials. Amorphous, polycrystalline, dense, porous, and hollow are just a few of the forms they can take. They are renowned for their durability and resistance to heat. Coatings, batteries, and catalysts are just a few of the uses for ceramic nanoparticles (Lv and Mu 2014).

3.2.4 Polymeric NMs

Active compounds that are either surface-adsorbed onto the polymeric core or encapsulated inside the polymeric body can be found in polymeric nanomaterials (NMs) that range in size from 1 to 1,000 nm. These NMs, which are frequently organic in composition, are referred to in the literature as polymer nanomaterials (PNMs). Most of the time, they resemble nanospheres or nanocapsules (Khan et al. 2023). Polymeric nanoparticles applied to agricultural and food packaging may help to achieve the goal of increasing food security. Recent research has provided knowledge regarding the synthesis of novel polymeric nanoparticles that are affordable, biodegradable, safe, biocompatible and easily reproducible from natural sources (Khodakovskaya and Marmiroli 2023).

3.2.5 Nanoclays

Nanoclay is a nanomaterial made of layered mineral silicates with particles smaller than 100 nm in at least one dimension. Nanoclay, with its large surface area, high aspect ratio, capacity to increase mechanical and barrier properties and adjustable qualities, has become an appealing material for a range of fields, including polymer composites, nanoelectronics, biomedical engineering, wastewater treatment, and energy storage (Abulyazied and Ene 2021). One of the most significant advantages of nano clay is its capacity to improve the mechanical and physical properties of polymers, especially when used as a reinforcement agent. The incorporation of nanoclays into polymer matrices results in high-performance polymer nanocomposites with increased attributes, including stiffness, strength, toughness, and thermal stability. They have various types of nanoclays including montmorillonite and kaolinite. Montmorillonite is a nanoclay mineral from the smectite group with a layered structure. The chemical formula is $(Na,Ca)_{0.33}(Al, Mg)_2(Si_4O_{10})(OH)_2 \cdot nH_2O$, with n indicating the number of water molecules in the interlayer gap. The mineral has a high cation exchange capacity; therefore, it may exchange ions with the surrounding environment. Montmorillonite, with its exchangeable ions, can help buffer soil pH. It helps to keep the soil pH in a healthy range for plant growth, which is critical for nutrient availability and microbial activity (Uddin et al. 2024).

Kaolinite is a mineral that belongs to the kaolin group with the chemical formula $Al_2Si_2O_5(OH)_4$. The material has a layered arrangement that includes a tetrahedral layer and an octahedral sheet, with oxygen atoms shared across layers, eventually resulting in a tri-layer formation. The advantages include chemical inertness and affordable cost. Kaolin is used by farmers to protect agricultural products from insects, mites, fungi, and bacteria, as well as to prevent sunburn and heat stress. This natural clay has been utilized for many years in foods, cosmetics, and other products (Boraei et al. 2024).

3.3 Role of Nanomaterials in Soil Fertility Enhancement

3.3.1 Nanomaterial in Restoration of Soils

Soil is a naturally occurring substance that interacts on many scales to provide a wide range of ecosystem products and services. It is a liquid, solid, and gaseous phase domain mediated by organic carbon. It is known that soil organic carbon affects the functionality, quality, and health of the soil. For maintaining soil health, the nitrogen cycle, carbon transformations, and soil structure preservation are all essential (Ramesh et al. 2019). The primary factors influencing these traits are microbial activity and the metabolic process. If NM treatments can increase this activity, the soil's fertility and health may improve.

Since soil biodiversity is thought to be the main determinant of soil health, the interactions between soil animals and microbes determine soil fertility and productivity. Carbon nanotube and graphene oxide treatments of the soil reduced soil enzyme activity in the short term but had no discernible impact on microbial biomass (Ge et al. 2016). Reducing agricultural input wastes in various environmental contexts and improving management and conservation techniques are just a few of the ways that nanotechnology, one of the most significant inventions of the twenty-first century, can improve agricultural practices today and facilitate sustainable development.

When combined with different farming techniques and soil-healthy additions, nanotechnology can increase soil microbial and animal activity (Ramteke et al. 2023). An agricultural system that is more climate-resilient will benefit from restoring damaged soil and enhancing soil health. It will eventually help to achieve a steady and substantial revenue from agriculture using a sustainable method. The growing soil conditions of plants have a major effect on them. Due to their potential to impact plant performance, soil stress factors such as salt, dryness, acidity, suboptimal root zone temperature, nutrient availability, and appropriate soil biota functionality are crucial for agricultural productivity (Rajput et al. 2023). Nanotechnology-created variations in soil microbiota have the potential to improve agricultural productivity and soil health.

3.3.2 Enhancement of Soil Structure

The physical features of soil, including structure, texture, bulk density, and porosity, are influenced by nanomaterials in different ways. Water drainage and soil aeration can be improved by nanomaterials via the formation of nanoscale holes or channels between soil aggregates. Platinum nanomaterials (NMs) enhance the saturated hydraulic conductivity by tightening the soil structure and reducing the spacing between soil particles. Moreover, NMs may increase the amount of soil organic matter by fortifying the links between soil particles or by filling the spaces between soil particles to reduce the bulk density of the soil (Suazo-Hernández et al. 2023). Carbon nanotubes, with exceptional elastic qualities and high aspect ratios, may enhance the aggregation of sandy loam soil by 35%. The hydrological regimes of soil are enhanced by using NMs in different ways. For example, they can build water-absorbing hydrogels or improve the surface area and hydrophilicity of soil particles, which improves the soil's ability to store water (Dhanapal et al. 2024). Plants stressed by drought and water loss may benefit from this ability. Furthermore, Arora et al. (2019) observed a significant decrease in soil liquefaction along with a decrease in soil settlement. The use of chemical oxidation, sorption, or reduction mechanisms in soil to remediate polluted soils has drawn further attention to NMs. Nanomaterials are excellent remediation materials due to their increased chemical reactivity, large specific surface area, and selectivity (Arora et al. 2019).

3.3.3 Impact of Nanomaterials on Soil Microbial Communities

Compant et al. (2019) define the term "plant microbiome" as the collection of bacteria that reside on different plant components. The microorganisms that inhabit the leaves (phyllosphere), the rhizoplane (the outer layer of the roots), the rhizosphere (the small region of soil directly impacted by root secretions), and the endosphere (the interior of the plant system) are among them. Many factors, including the type of plant, its age and health, its secretions, the surrounding environment, physicochemical characteristics, and the amount of microorganisms in the soil, have an impact on the intricate relationship a plant has with its microbiota (Pang et al. 2021). Research has revealed that certain bacteria selectively colonize plant rhizospheres, which is an interesting finding. The microorganisms of a plant's rhizosphere, phyllosphere, and endosphere may be different even under identical weather conditions. By producing unique exudates, plants may also benefit from certain microbial populations for particular purposes (HuangXing-Feng et al. 2014).

Bacteria provide a host of well-known plant support functions, including tolerance and disease defense mechanisms, nutrient fixation, nutrient mobilization, micronutrient sequestration, and effector chemical production. In light of climate change, it is imperative to investigate the role and adaptability of the soil microbiome, which can be modified for improved soil and plant health. According to Dubey et al. (2019), soil microbiomes are becoming increasingly important for biogeochemical cycling. They contain the macro- and micronutrients necessary for plant and animal survival, as well as other essential elements. By enhancing the soil's natural ability to produce suppressiveness and resistance to disease and infestation, managing the soil microbiota can help avoid overusing pesticides. Two examples of microbes that help their hosts are mycorrhizal fungi and rhizobia, a diazotrophic bacterium that creates root nodules on legumes. Phosphate can be solubilized by resource-mobilizing bacteria like *Pantoea* spp. and *Pseudomonas* spp., which can then convert the phosphate into insoluble forms $Ca_3(PO_4)_2$.

Other compounds produced by various microorganisms, such as IAA, are known to promote plant growth. Additionally, siderophores or organic acids, which are micronutrients similar to zinc, can be delivered to plants by bacteria. NMs permit significant biological activity because of their unique surface features, which may be helpful for altering the physicochemical and biological aspects of the soil (Wang et al. 2013). Due to NMs' hydrophobicity and surface impacts on the biochemical and soil environments, a variety of interrelated mechanisms—including plant root exudates, soil organic matter, and rhizosphere bacteria—contribute to the mineralization and mobilization of nutrients. Cu-based NMs (10 mg/L) induced root exudation in hydroponic growing conditions. Plant roots are encouraged by TiO_2 and Fe_3O_4 (50–200 mg/kg) NMs to exude more in either alkaline or saline soil. CuO (500 mg/kg) NMs lower pH and mobilize nutrients to raise soil pH in acidic soil (Rajput et al. 2023b).

These investigations demonstrated the value of NMs in achieving soil pH neutrality, which improves soil quality and nutrient flow. NMs can improve phytostimulation by increasing the synthesis of carbohydrates, fatty acids, and amino acids, altering gene expression, regulating antioxidant activity, enhancing nutrient transport, and improving other activities. TiO_2-based NMs boost P and N content in *Oryza sativa*, whereas CuO- and ZnO-based NMs encourage *Glycyrrhiza glabra* phenols, anthocyanins, and other antioxidant compounds. Certain elements and mineral phases can be chelated and dissolved with the aid of siderophores. Increased microbial siderophore synthesis appears to expedite the breakdown of hematite (Fe_2O_3) NMs, with the plants subsequently absorbing the released Fe. Another study (Wang et al. 2020) found that using siderophore-mobilized Fe to mobilize Fe-doped, high aspect-ratio NMs greatly reduced toxicity. Copper, zinc, and manganese are just a few of the metals that siderophores can chelate. Siderophores have a strong propensity to chelate metals, which may facilitate NM degradation and encourage hetero-aggregation with soil minerals.

The link between the soil microbiome and plants is referred to as "bidirectional" since the bacteria feed on the carbon-rich materials that the plants produce while the microbes aid in the growth and development of the plants. Remarkably, research has shown that the microbiome of *Arabidopsis thaliana* affects both biomass and the blooming period (Krämer 2015). Another study (Rajput et al. 2023c) found that the rhizospheric microbial populations of *A. thaliana* are critical to the plant's defense against diseases. The bacterial community provides defense against wilt sickness. In-depth research on plant microbiomes may support sustainable agriculture by increasing crop yield and nutritional value while lowering the need for synthetic pesticides and fertilizers. Plants have a variety of microbiomes that can change in response to changing environmental conditions; therefore, it's critical to identify and describe the core microbiome that is unique to them and typically stable. These microbiomes can be altered to satisfy specific needs, such as increased growth, resistance to illness, and better agricultural quality (Kumawat et al. 2022). Such thorough research will contribute to agriculture becoming less dependent on agrochemicals and more sustainable.

Among the numerous inorganic contaminants that may find their way into soil, NMs are among the most widely utilized substances. NMs may modify the microbial ecology of soil or plants, which may have an impact on plant growth either directly or indirectly. For example, it has been observed that nanoscale ZnO and TiO_2 affect soil microbial populations. The bacteria that fix nitrogen and oxidize methane experienced a sharp reduction following NM treatment, whereas the bacteria that degrade refractory organic pollutants, such as those belonging to the *Sphingomonadaceae* family, saw a considerable increase in population. TiO_2, one of the common soil nanomaterials, did not affect arbuscular mycorrhizal root colonization in the wheat microbiome, despite variations in the populations of a particular group of microorganisms (Karimi and Fard 2017). Moreover, changes in the number of microorganisms may suggest that TiO_2 NMs have contaminated the soil. In another study that examined the effects of Ag NMs in more depth, it was discovered that these compounds significantly affected the microbial communities in the soil.

Ag NPs introduced at 0.01 mg kg^{-1} drastically reduced the populations of ammonia oxidizers and proteobacteria, although actinobacteria, acidobacteria, and bacteroidetes increased in density rapidly. According to Jiao et al. (2022), leucine was reduced by exposure to Ag NPs at 0.01–1.0 mg kg^{-1}. Nitrogen-fixer abundance, soil microbial biomass, and aminopeptidase activity were also affected. It was shown that the density of fast-growing bacteria dropped three to four times in soils treated with C60 fullerenes, which had an average diameter of 50 nm (Maurice 2023). TiO_2 injections and amine-modified polystyrene nanospheres were utilized to treat the rhizosphere of *Lactuca sativa* seedlings in a different study. This reduced the amount of rhizospheric bacteria present in the rhizosphere and impeded the growth of the plant (Rajput et al. 2023d). Wet soils containing wastewater significantly reduced the life cycle of *A. thaliana* nanoscale substances (Liu et al. 2018). Populations of cyanobacteria increased, and several newly unidentified archaea were discovered. In addition, carbon nanoparticles had negative effects on the ecosystem and changed the bacterial composition in the rhizosphere of *Oryza sativa*. In an investigation of the microbial community, the addition of CNTs to the soil did not affect *Solanum lycopersicum* plants (Patel et al. 2019).

Numerous research works have investigated the impact of nanomaterials (NMs) on microorganisms involved in the regulation of multiple vital elements, such as carbon and nitrogen. The enzymes acid phosphatase, glycosaminidase, β-glucosidase, and arylsulfatase are essential to the nitrogen cycle. Carbon in soil samples that have been exposed to Ag NMs shows lower activity. Denitrification is the microbiological process most sensitive to CuO nanoparticles, according to (Zhu et al. 2024). According to research by Klanjšček et al. (2017), high quantities of nano-CeO_2 hampered nitrogen fixation in leguminous *Glycine max* crops. Silver's well-known antibacterial properties have led to substantial research compared to other NMs. Ag-NMs, for instance, have been shown in a study to prevent the free-living nitrogen fixer *Azotobacter vinelandii* from multiplying. The ability of NMs to bind the bacterial exopolysaccharides and to resist steric repulsion stabilizes the NMs' suspension. It assists in converting inaccessible soil components into forms that are utilized by plants, hence reducing biotic and abiotic stress and pathogen activity and strengthening plant defenses. Similarly, mycorrhiza and rhizobia symbioses affect the microbial community and plant structure, the mineralization of organic materials, the cycling of soil nutrients, and the resilience and performance of ecosystems (Ambrosini et al. 2015).

3.3.4 Nanotechnology for Soil pH Regulation

Nanotechnology has offered a promising solution to problems related to pH regulation, soil bioremediation, and soil structure. With the development of nanosensors, the monitoring of soil pH has become easier with accurate and precise data in real time. These nanosensors are fixed in the soil to monitor and automatically regulate soil pH as per requirements. Some nanoparticle-based fertilizers, nanolime, and nanoclays present in soil may alter its pH level up to the desired range (Zain et al. 2023). These nanoparticle-based products maintain soil fertility and amendments. In addition to these activities, some nanoparticles influence microbial activity concerning soil pH. In other words, nanotechnology protects soil biodiversity by releasing pH-regulating agents in an organized way. Still, research on nanotechnology regarding soil pH regulation is being conducted worldwide to achieve the stability, fertility, and sustainability of soil (Sharma and Kumar 2024).

3.3.5 Nanomaterials for Nutrient Delivery and Uptake

Sluggish crop yields and precision farming require efficient ways to get pesticides into plant cells. Conventional methods, such as spraying, furrow irrigation, or sprinklers, are used to apply agrochemicals to crops. These methods can lead to over-application or nutrient loss, which may have detrimental long-term effects on soil quality (Sojka et al. 2007). Additionally, a very small amount—much less than the minimum concentration required—reaches the plants due to several processes, such as microbial degradation, photolysis, evaporation, drift, hydrolysis, and loss. Because of their large surface area and smaller particles relative to the pores of a plant's leaves and root system, nanofertilizers can maximize nutrient uptake in plants. It has been demonstrated that applying fertilizers enclosed in NMs increases the bioavailability and nutrient uptake of farming plants (Iqbal et al. 2019). By progressively supplying nutrients to agricultural plants, zeolite-based nanofertilizers increase crop nutrient availability during the growing season and reduce nitrogen loss through soil fixation, volatilization, leaching, and denitrification. In a study, the effects of TiO_2-based NMs at 500 and 750 mg/kg on *Oryza sativa* spp. and the nutrient availability in different soil textures, such as sandy, silty loam, and clay loam with silt, were examined. In comparison to other soil textures, plant chlorophyll levels, shoot and root length, and biomass increased when 500 mg/kg TiO_2 NMs were added to silty-clay loam soil (Rajput et al. 2023).

According to the study, the primary nutrients that lead to prolonged growth and higher biomass in plants exposed to TiO_2 NMs are Ca, Fe, and P, and TiO_2 NMs might improve the availability of these nutrients (Burke et al. 2015). Another study looked at the efficiency of hydroxyapatite nanomedicines (NMs) treated with urea for regulated nitrogen release. According to Ramírez-Rodríguez et al. (2020), the findings showed that after a brief initial release, nitrogen from nanofertilizers was released steadily over the next sixty days. Impacts on crop growth and soil physicochemical characteristics of both conventionally manufactured Ag NMs and made-in-the-USA Green Ag NMs (GS-NPs) were investigated using *Thuja occidentalis* plant leaf extract. Following the application of GS-NMs, the soil's pH started to become closer to neutral, and there was a significant increase in its ability to hold water, exchange cations, and supply N/P. The fact that soils treated with Green Ag-NMs significantly reduced nitrate offers more evidence of the mechanism behind improved N availability in laboratory conditions. Nitrogen availability was both leached and maintained in soil layers under the root zone (Farooq et al. 2022).

4. Nano-Fertilizers: Efficient Source of Nutrition for Plants

Agricultural inputs that use nanotechnology to increase nutrient delivery and enhance plant growth and productivity are referred to as nanofertilizers, nanoscale fertilizers, or simply nanofertilizers (Iqbal et al. 2019). Their benefits are illustrated in Figure 2. Compared to conventional fertilizers, these are made to provide nutrients to plants more effectively and specifically. Nanoparticles are the building blocks of nanofertilizers. Nanoparticles can be made from polymers, metals and metal oxides, and carbon-based compounds. The material utilized depends on the intended uses and characteristics of the nanofertilizers. Improving the availability, uptake, and utilization of nutrients by plants is the main objective of using nanofertilizers. Because of their small size, nanoparticles can enter plant tissues more quickly and increase the efficiency of nutrient absorption. Additionally, by preventing nutrient volatilization or leaching, nanofertilizers can improve nutrient utilization efficiency while lowering environmental losses (Kalia and Kaur 2018).

Moreover, plants can be guaranteed a steady supply of nutrients by using nanofertilizers that are designed to release nutrients gradually over time. Fertilizer application frequency is reduced, and nutrient waste is minimized thanks to this controlled-release mechanism. Furthermore, nutrients can be efficiently applied and targeted at particular plant tissues or organs with the help of nanofertilizers (Elemike et al. 2019).

4.1 Classification of Nanofertilizers Based on Requirement

Nanofertilizers deliver the needed nutrients in a regulated manner required for plant development and growth. Nanofertilizers are divided into two categories based on macronutrients and micronutrients.

4.1.1 Macronutrients

NFs are derived from large-scale, nanoscale macronutrients required by agricultural plants (Anjum et al. 2023). The macronutrients are N, P, K, Ca, Mg and S. Their requirements are related to the rise in the world's food consumption, which is a result of population growth. It is anticipated that by 2050,

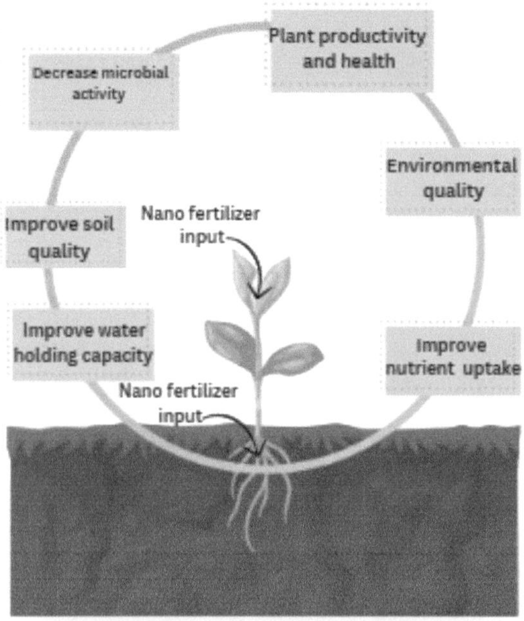

Figure 2. Benefits of nanomaterials in agriculture.

the demand for macronutrients will reach 263 Mt. The creation of nano-fertilizers for macronutrients may be crucial to achieving macronutrient requirements because of their higher use efficiency, which, in comparison to conventional fertilizers with their extremely low use efficiency of 20% or less, will lead to higher crop growth and output. Worldwide, the application of nanotechnology has enabled the production of macronutrient nano-fertilizers, which have been shown to significantly increase resource efficiency and improve crop growth and yield (Chatterjee and Ravishankar 2023). Nano-fertilizers may be a more cost-effective option than the current standard chemical fertilizers because their usage as a source of nutrients in agriculture also reduces cultivation costs.

4.1.1.1 Nitrogen (N)-NMs

For many plant functions in agriculture, nitrogen is the most important nutrient. Fertilizers high in nitrogen are more prone to contaminate and threaten groundwater. Nitrogen releases more slowly in its absorbed form than in its ionic form. The way slow-release nanofertilizers and sub-nanocomposites are coated and bound controls how much nutrient is released from the fertilizer capsule (Fen et al. 2022). By reducing nitrogen leaching loss, emissions losses, and continuing amalgamation by soil microbial communities, nanofertilizers offer a tremendous opportunity to benefit the economy, energy, and ecology. For instance, N was gradually released when urea was applied to zeolite chips. Similarly, N was released into the soil gradually and sustainably when urea-modified hydroxyapatite nanomedicines (NMs) were pressure-encapsulated into the softwood cavities of *Gliricidia sepium*. When compared to conventional nitrogen sources, this method's nitrogen supply was found to be ideal for up to 60 days, giving the plants a lot of N early in the crop's life and relatively little in the later stages (Meisinger et al. 2015).

4.1.1.2 Phosphorus (P)-NMs

Phosphorus is involved in many different metabolic processes and is a necessary component of several metabolites. Chemical fertilizers containing P have very low fertilizer use efficiency—only 20% of the fertilizer is available to crops, while the remaining fertilizer either runs off or is deposited in the soil, eutrophicating aquatic bodies (Karamesouti and Gasparatos 2017). To lower environmental risks and increase nutrient use efficiency (NUE), nanotechnology is essential. Wet chemical synthesis was used to create hydroxyl apatite [$Ca_5(PO_4)_3OH$] NMs, which were then compared to conventional phosphatic fertilizers and soybean crop performance in a greenhouse to increase P use efficiency. The growth rate (33%) and seed yield (20%) were much higher than with traditional phosphatic fertilizers (Amanullah et al. 2021). We used *Aspergillus* spp. to microbially synthesize phosphorus nanoparticles from tricalcium phosphate. The experiment, which was carried out in a dry area, showed that using nanomaterials greatly decreased the amount of fertilizer needed. When cluster beans and pearl millet were foliar-applied with 640 mg/hectare of nanophosphorus, the corresponding yield of 80 kg/ha phosphorus was obtained.

4.1.1.3 Potassium (K)-NMs

Utilizing carriers as the foundation material, potassium NMs have been created and assessed in controlled environments. The increased growth tendency of *Kalmi*, as well as its uptake and concentration of potassium (K) and phosphorus (P) in comparison to standard fertilizer treatments, indicate that nano-fertilizer has a bright future in agriculture.

4.1.1.4 Calcium (Ca)-NMs

Ca is essential for cell elongation and improves the structure of plant cell walls by producing calcium pectate. $CaCO_3$ NMs (20–80 nm) at 160 mg/L as Ca in Hoagland solution were assessed for 80 days as a source of Ca in peanuts grown in sand, and the results were compared with the control (no Ca) and a soluble source of Ca as $Ca(NO_3)_2$ (200 mg/L). When compared to the control and the soluble source of calcium [$Ca(NO_3)_2$], fresh biomass was significantly improved (Perveen et al. 2022). In addition, the roots and stems of the seedlings absorbed more calcium than the control group.

4.1.1.5 Magnesium (Mg)-NMs

Mg is a necessary part of chlorophyll, which helps with photosynthesis. It also aids in the synthesis of cell proteins and amino acids. Additionally, this facilitates the uptake and movement of phosphorus and increases plant tolerance to both biotic and abiotic stressors. The combined effects of foliar treatment with Mg-NMs and Fe-NMs (0.5 g/L) on the photosynthetic efficiency of the black-eyed pea (*Vigna unguiculata*) were examined by Delfani et al. (2014) (Singh et al. 2021). The application of Fe- and Mg-NMs together led to an 8% decrease in grain production despite increasing growth and yield, according to the results. In comparison to the control group, regular application of Mg-NMs increased Mg absorption in several plant tissues. It follows that Mg-NMs may improve Mg absorption.

4.1.2 Micronutrient Nanofertilizers

Due to soil overexploitation, widespread micronutrient shortages have become problematic. Even though plants only require very small amounts of these minerals for nourishment, their deficiency leads to hidden hunger (Dimkpa and Bindraban 2016). To overcome these deficiencies, the essential nutrients should be supplied via cutting-edge, practical methods like the application of nanoparticles, which have been demonstrated to enhance plant uptake patterns. A major component of plant physiology is micronutrients. Despite having a very low need (100 ppm), they are essential for some plant metabolic processes. Micronutrients include iron (Fe), boron (B), manganese (Mn), zinc (Zn), copper (Cu), molybdenum (Mo), chloride (Cl), and nickel (Ni), according to Hänsch and Mendel (2009). These are also used for applying various complex fertilizers with varying NPK concentrations to agricultural plants. Micronutrients are an excellent source of nutrients with minimal environmental impact. The availability of micronutrients to plants is significantly influenced by small changes in soil factors, such as pH, soil texture, and organic matter. This leads to extremely low fertilizer consumption efficiency of micronutrient-containing fertilizers. Micronutrient-containing nanofertilizers can be used to meet their needs. By altering the particle size, micronutrient fertilizers can be made more agronomically efficient. Both the specific surface area and the number of particles per unit weight of applied fertilizers are increased by decreasing the size of the particles. Zinc sulphate ($ZnSO_4$) (1.4 to 2 mm) was shown to be somewhat less effective than fine $ZnSO_4$ (0.8 to 1.2 mm), whereas granular zinc oxide (ZnO) (2.0 to 2.5 mm) was shown to be completely ineffectual (Mahdieh et al. 2018). Smaller granules increased Zn dispersion, while increasing surface contact area with Zn fertilizers improved Zn uptake. More investigation and analysis with a focus on nanomaterial particle size should be carried out to increase fertilizer production and efficacy.

4.2 Applications

Applying large amounts of urea, nitrate, ammonium salts, or phosphate compounds to plants is extremely risky and ineffective because the majority of these nutrients are lost as runoff and result in pollution (Mengel et al. 2001). Nanomaterials' clever system of distribution allows plants to absorb nutrients. These nano-based agriproducts like nanofertilizers and nanopesticides have their methods of application based on their releasing point to the target materials. The following methods describe their mode of application with the releasing methods.

4.2.1 Chemical

Fertilizers can be encapsulated in nanoparticles and released gradually, which could help meet crop and soil requirements. Because of its stability, the coating slows down the pace at which fertilizers dissolve and regulates its release over time. Potential uses for kaolin and polymeric biocompatible nanoparticles in agricultural nanotechnology. Corradini et al. (2010) investigated the gradual release of NPK fertilizers (urea, calcium phosphate, potassium chloride) using biodegradable polymeric chitosan NPs (~ 78 nm) (Srivastava et al. 2019).

4.2.2 Biofertilizers and Micronutrients

Biofertilizers' preparations are coated with polymeric NMs to create formulations that are resistant to desiccation. In horticultural crops, foliar micronutrient treatment increased the absorption of micronutrients. Peteu et al. (2010) discovered that plant development and soil health were enhanced by the use of nanomaterials to progressively release micronutrients in the soil (Chaudhary et al. 2023).

4.2.3 Seed Treatment

Enhancing the micronutrient delivery system can increase the efficiency of nutrient utilization. It has been determined that clay nanocomposite superabsorbent can be used as a coating or seed treatment before sowing. In the rain-fed ecosystem, nano clay composite can provide moisture to the crop's rhizosphere to help seedlings and seeds survive in water-stressed conditions, especially during the long dry spell that follows a rainstorm. Therefore, using nano clay composite superabsorbents may be a viable way to save semi-arid agriculture. Yadav et al. (2010) discovered that fragrant rice's grain output and quality were enhanced by Zn-enriched urea (Yadav et al. 2010). Zinc deficiencies in wheat, maize, sunflower, peanut, and soybean have been demonstrated to be remedied by seeds coated with leprosy-ZnP or teprosyn-Zn.

5. Safety and Environmental Considerations

5.1 Toxicity of Nanomaterials

Due to technological advancements, a substantial increase in the quality and yield of crops has been observed in the agriculture sector; however, serious concerns have also been raised owing to the toxicity of nanomaterials toward growth, root and shoot extension, seed germination, and biomass of plants. Once NPs are translocated inside plants, they can directly change the structure of cells, molecules, and membranes, and may also affect the protective system of plants. Furthermore, there may be unintended effects of NPs based on their properties, solubilization, and generation of reactive compounds. NPs stimulate the production of reactive oxygen species (ROS) that affect the growth and development of plants (Muzammil et al. 2023). Animals are also affected by the side effects imposed by the application of nanomaterials on soil organisms and other organisms that depend on the cultivation of that land. The heavy metal ions from metal oxide nanomaterials may accumulate in animal bodies, and metabolism is heavily affected, leading to serious problems in their health (Haghighizadeh et al. 2024).

5.2 Environmental Impacts

To enable the safe and effective use of nanotechnology in agriculture, major obstacles including the safety and environmental effects of nanofertilizers need to be overcome, because they provide plants with delayed and prolonged release of nutrients. Nanomaterials have the potential to increase crop yields and decrease environmental pollution (Fatima et al. 2020). Nanotechnology-based agricultural products are subject to regulation by several agencies, including the Food Standards Agency (FSA), the Department for Environment, Food and Rural Affairs (DEFRA), and the Health and Safety Executive (HSE) in the United Kingdom (Chavez-Hernandez et al. 2024). Reducing possible environmental and health risks in agriculture requires careful handling and application of nanomaterials. The FDA has recommended that the use of nanotechnology in food products be subject to risk assessment and evaluation.

In conclusion, to guarantee the safe and effective application of nanotechnology in agriculture, it is imperative to tackle the safety issues and environmental effects of nanofertilizers. To ensure the safe application of nanotechnology in agriculture, regulatory bodies from all over the world have created laws and regulations. The safe handling and use of nanomaterials in agriculture are critical to minimize possible dangers to the environment and human health. More research is needed to develop green, cost-effective, and safe nanomaterials for future applications, as well as to understand the toxicology and environmental implications of direct and indirect exposure to nanomaterials.

5.3 Regulatory and Policy Aspect

The use of nanotechnology in agricultural goods raises concerns about potential environmental and health risks. To address these concerns, regulatory authorities around the world have created rules and regulations to ensure the safe use of nanotechnology in agricultural goods. Here are some important regulations and guidelines that must be implemented by regulatory bodies for nanotechnology-based agricultural goods:

- Regulatory Oversight: Every country must possess its own structure to implement nanotechnology in agriculture.
- International Standards: To ensure the safety and quality of nanotechnology-based agricultural products, International Standards (ISO) must be established.
- Risk Evaluation: Before any nanotechnology-based agricultural product is approved and launched, regulatory organizations must demand a risk evaluation. It will assess the toxicity of utilized nanomaterials, the risk of environmental release, and the impact on human health.
- Research and Development: To ensure the safety of human health and the environment, regulatory bodies must promote the research and development of nanotechnology-based agricultural products.
- Labeling: Regulatory authorities must demand the labeling of nanomaterial-containing agricultural goods. It will allow consumers to make more educated judgments regarding the things they buy and use (Kumari et al. 2023).

6. Future Prospects and Innovations

6.1 Emerging Nanomaterials

Several nanomaterials have been introduced with some excellent properties and are gaining popularity with their various applications. Still, the development of nanomaterials is being advanced and modified to be used in agriculture. The carbon-based classes of nanomaterials like carbon dots, carbon nanotubes, and graphene with reduced size but multifunctional properties have emerged. The reduced sizes of carbon-based nanomaterials have a broad surface area for pesticides, and the nutrient delivery system aims to promote the quality of crop production. With the discovery of these nanomaterials, agriculture has become more advanced. The exceptional properties like photoluminescence and semi-conductivity make the carbon nanomaterials suitable to be used in sensing and bio-imaging. The problems related to crops can be diagnosed easily (Balusamy et al. 2023). In the future, nanomaterials may play an important role in every field of agriculture, like crop protection, crop production, nutrient transport systems, soil health maintenance, gene transformation, conservation of biodiversity, and the development of sustainable agriculture. We can say that there could be war between the nations over the resources of nanoparticles by observing their importance in every field of life.

6.2 Integration with Smart Farming Technologies

Two emerging fields, nanotechnology and the Internet of Things (IoT), are being used together to provide various novel applications. The Internet of Things (IoT) is the network of physical gadgets, automobiles, home appliances, and other items equipped with sensors, software, and network connectivity to collect and exchange data. In comparison, nanotechnology is the study of controlling matter at the nanoscale, which ranges in size between 1 to 100 nanometers. Integrating IoT and nanotechnology can develop a range of applications with improved capabilities, including the development of smart sensors that provide precise results in very little time (Malik et al. 2023). Nanotechnology develops ultra-small sensors that can be inserted into everyday objects, allowing them to remain organized, under control, and transmit information about their overall status. By integrating sensors and IoT devices, we may create a network of interconnected devices that can monitor air quality, water levels, and building structures to ensure their integrity and accuracy. Nanosensors enable high sensitivity and selectivity, resulting in error-free and self-controllable data. In this way, we can monitor and sense the production and protection stages of crops in no time, precisely, which may lead to the development of sustainable agriculture (Fu et al. 2024).

6.3 Sustainable Agriculture Practices

Sustainable agriculture has been transformed by synthetic nanotechnology with enormous promise. Their application can improve plant germination, growth, development, and resistance to biotic and abiotic problems, resulting in healthier crops and higher yields. Furthermore, nanotechnology helps to improve soil health, promoting a sustainable and balanced ecology. Furthermore, its uses improve human food safety by reducing the presence of toxic chemicals, opening the path for a more environmentally friendly and sustainable agricultural system overall. Conservation of biodiversity may be achieved by using nanopesticides with green synthesis precursors in place of synthetic pesticides, which impose hazardous impacts on biodiversity (Singh et al. 2023).

Nanomaterials and organic farming both play an important role in the development of sustainable and regenerative agriculture. Both have long-lasting effects in comparison to synthetic inorganic farming, as they work in precise and efficient ways. NMs and organic matter, when combined, represent enormous promise for sustainable agriculture as better alternatives to agrochemicals and are becoming a popular notion in the agricultural sector. Organic farming in conjunction with nanotechnology may be preferred over chemical-based farming due to its higher efficacy in resource utilization, sustained and slow release of nutrients, increased crop productivity with a lower dose of nutrients, and minimal negative impacts on soil. Furthermore, the usage of NMs and organic materials is cost-effective and less hazardous to the environment (Sharma and Kumar 2024).

7. Conclusion

An increase in food production worldwide and in the provision of other essentials provided by soil is a pressing concern given the growing global population. Roughly 25% of the world's soils have deteriorated to the point where it is significantly harder to meet human needs. Therefore, the topic of this review is deteriorated soils and their restoration. The primary focus of this graphic review is on the nano-remediation of polluted soils using nanomaterials, with an emphasis on the function of soil microbes and grown plants through the nanoparticle–plant–microbe nexus and its interactions. On the other hand, excessive use of nanomaterials in agricultural management or nano-remediation could cause nano-toxicity to food plants that are grown or farmed, which could have detrimental effects on human health as a result. Therefore, rules intended to prevent environmental and human health issues should be followed when nano-restoring deteriorated soils. Further research is necessary to have a better knowledge of how plants, microorganisms and nanoparticles interact in degraded or polluted soils to retain soil fertility through nano-restoration. All potential interactions between any soil element other than water, plants that have been cultivated or produced and any agrochemicals present in the soil, such as pollutants, fertilizers, or nanoparticles should be carefully studied to understand their combined effects on soil health, plant growth, and environmental safety. More research is still needed to determine the predicted fate and behavior of agrochemicals in the soil, water, plant, and microbial system under various circumstances, particularly in light of climate change. The importance of soil microbes is growing daily, and many of their negative effects are now being transformed into positive ones, especially in the pharmaceutical industry. Globally, companies and scholars are becoming increasingly concerned about the integrative role of soil bacteria, plants, and nanoparticles in the ecosystem. There is still a wealth of scientific research needed on the interacting systems between soil bacteria, plants, pollutants, agrochemicals, and nanomaterials to maintain soil fertility.

8. Acknowledgement

The authors gratefully acknowledge the support of the Institute of Plant Protection, Muhammad Nawaz Sharif University of Agriculture, Multan for the support in this publication.

References

Irzh, A., N. Perkaz and A. Gedanken. 2007. Microwave-assisted coating of PMMA beads by silver nanoparticles. Langmuir. 23: 9891–9897.

Abdollahi Boraei, S.B., B. Bakhshandeh, F. Mohammadzadeh, D.M. Haghighi and Z. Mohammadpour. 2024. Clay-reinforced PVC composites and nanocomposites. Heliyon. 10: e29196.

Abulyazied, D.E. and A. Ene. 2021. An investigative study on the progress of nanoclay-reinforced polymers: Preparation, properties, and applications: A review. Polym. Basel. 13: 4401.

Amanullah, S. Khalid, A. Muhammad, M. Yar, M. Ahmad, H.M. Akram, I. Imran et al. 2021. Integrated use of biofertilizers with organic and inorganic phosphorus sources improve dry matter partitioning and yield of hybrid maize. Commun. Soil Sci. Plant Anal. 52: 2732–2747.

Ambrosini, A., R. De Souza and L.M.P. Passaglia. 2015. Ecological role of bacterial inoculants and their potential impact on soil microbial diversity. Plant Soil. 400: 193–207.

An, K. and G.A. Somorjai. 2012. Size and shape control of metal nanoparticles for reaction selectivity in catalysis. ChemCatChem. 4: 1512–1524.

Anjum, M.R., J.M.J. Maheswari, K. Anusha, B. Sravya, G. Narasimha and K.A. Abd-Elsalam. 2023. Maximizing crop yield with macro and micro nano enhanced fertilizers. pp. 1–33. *In*: Nanotechnology in the Life Sciences.

Arab, D., A. Kantzas and S.L. Bryant. 2018. Nanoparticle stabilized oil in water emulsions: A critical review. J. Pet. Sci. Eng. 163: 217–242.

Arienzo, M. and L. Ferrara. 2022. Environmental fate of metal nanoparticles in estuarine environments. Water. 14: 1297.

Ashraf, S.A., A.J. Siddiqui, A.E.O. Elkhalifa, M.I. Khan, M. Patel, M. Alreshidi et al. 2021. Innovations in nanoscience for the sustainable development of food and agriculture with implications on health and environment. Sci. Total Environ. 768: 144990.

Bastús, N.G., F. Merkoçi, J. Piella and V. Puntes. 2014. Synthesis of highly monodisperse citrate-stabilized silver nanoparticles of up to 200 nm: Kinetic control and catalytic properties. Chem. Mater. 26: 2836–2846.

Biliuță, G. and S. Coseri. 2019. Cellulose: A ubiquitous platform for eco-friendly metal nanoparticle preparation. Coord. Chem. Rev. 383: 155–173.

Burke, D., N. Pietrasiak, S.F. Situ, E. Abenojar, M. Porche, P. Kraj et al. 2015. Iron oxide and titanium dioxide nanoparticle effects on plant performance and root associated microbes. Int. J. Mol. Sci. 16: 23630–23650.

Buzea, C., I. Pacheco and K. Robbie. 2007. Nanomaterials and nanoparticles: Sources and toxicity. Biointerphases. 2: MR17–MR71.

Chatterjee, B. and V. Ravishankar. 2023. Nanofertilizers: A promising approach to boost plant health and yield. pp. 455–506. *In*: Nanotechnology in the Life Sciences.

Chau, C.F., S.H. Shiuan-Huei and G.C. Yen. 2007. The development of regulations for food nanotechnology. Trends Food Sci. Technol. 18: 269–280.

Chaudhary, P., A. Chaudhary, V. Dhaigude, S.A. Fayssal, P. Kumar, A. Sharma et al. 2023. Environmental-friendly nanoparticles in agriculture. pp. 267–289. *In*: CRC Press eBooks.

Chavez-Hernandez, J.A., A.J. Velarde-Salcedo, G. Navarro-Tovar and C. González. 2024. Safe nanomaterials: From their use, application and disposal to regulations. Nanoscale Adv.

Compant, S., A. Samad, H. Faist and A. Sessitsch. 2019. A review on the plant microbiome: Ecology, functions, and emerging trends in microbial application. J. Adv. Res. 19: 29–37.

Daima, H.K. and P.N. Navya. 2016. Rational engineering of physicochemical properties of nanomaterials for biomedical applications with nanotoxicological perspectives. Nano Converg. 3: 1–14.

Darr, J.A., J. Zhang, N.M. Makwana and X. Weng. 2017. Continuous hydrothermal synthesis of inorganic nanoparticles: Applications and future directions. Chem. Rev. 117: 11125–11238.

Das, I., B. Gogoi, B. Sharma and D. Borah. 2022. Role of metal-nanoparticles in farming practices: An insight. 3 Biotech. 12: 294.

Di Falco, S. and E. Zoupanidou. 2017. Soil fertility, crop biodiversity, and farmers' revenues: Evidence from Italy. Ambio. 46(2): 162–172.

Dimkpa, C.O. and P. Bindraban. 2016. Fortification of micronutrients for efficient agronomic production: A review. Agron. Sustain. Dev. 36: 1.

Dubey, A., M.A. Malla, F. Khan, K. Chowdhary, S. Yadav, A. Kumar et al. 2019. Soil microbiome: A key player for conservation of soil health under changing climate. Biodivers. Conserv. 28: 2405–2429.

Elemike, E.E., I.M. Uzoh, D.C. Onwudiwe and O.O. Babalola. 2019. The role of nanotechnology in the fortification of plant nutrients and improvement of crop production. Appl. Sci. 9: 499.

Farooq, M.A., F. Hannan, F. Islam, A. Ayyaz, N. Zhang, W. Chen et al. 2022. The potential of nanomaterials for sustainable modern agriculture: Present findings and future perspectives. Environ. Sci. Nano. 9: 1926–1951.

Fatima, F., A. Hashim and S. Anees. 2020. Efficacy of nanoparticles as nanofertilizer production: A review. Environ. Sci. Pollut. Res. Int. 28: 1292–1303.

Fei, P., Y. Su, Y. Zhong, C. Fan, S. Lee and Y. He. 2014. Silicon nanomaterials platform for bioimaging, biosensing, and cancer therapy. Acc. Chem. Res. 47: 612–623.

Fen, L.B., A.H.A. Rashid, N.I. Nordin, M.A.M. Hossain, S.M.K. Uddin, M.R. Johan et al. 2022. Applications of nanomaterials in agriculture and their safety aspect. pp. 243–299. *In*: Apple Academic Press [eds.]. Apple Academic Press, Location.

Fischer, H.C. and W.C.W. Chan. 2007. Nanotoxicity: The growing need for *in vivo* study. Curr. Opin. Biotechnol. 18: 565–571.

Ge, Y., J.H. Priester, M. Mortimer, C.H. Chang, Z. Ji, J.P. Schimel et al. 2016. Long-term effects of multiwalled carbon nanotubes and graphene on microbial communities in dry soil. Environ. Sci. Technol. 50: 3965–3974.

Hänsch, R. and R.R. Mendel. 2009. Physiological functions of mineral micronutrients (Cu, Zn, Mn, Fe, Ni, Mo, B, Cl). Curr. Opin. Plant Biol. 12: 259–266.

Holmes, J.D., D.M. Lyons and K.J. Ziegler. 2003. Supercritical fluid synthesis of metal and semiconductor nanomaterials. Chem. Eur. J. 9: 2144–2150.

HuangXing-Feng, M. ChaparroJacqueline, K.F. Reardon, R. ZhangRuifu, SHENQi-Rong and M. VivancoJorge. 2014. Rhizosphere interactions: Root exudates, microbes, and microbial communities. Botany. 92: 267–275.

Iqbal, M., S. Umar and Mahmooduzzafar. 2019a. Nano-fertilization to enhance nutrient use efficiency and productivity of crop plants. pp. 473–505. *In*: Springer eBooks [eds.]. Springer, Location.

Iqbal, M., S. Umar and Mahmooduzzafar. 2019b. Nano-fertilization to enhance nutrient use efficiency and productivity of crop plants. pp. 473–505. *In*: Springer eBooks [eds.]. Springer, Location.

Zhu, J., S. Liu, O. Palchik, Y. Koltypin and A. Gedanken. 2000. Shape-controlled synthesis of silver nanoparticles by pulse sono-electrochemical methods. Langmuir. 16: 6396–6399.

Jiao, K., B. Yang, H. Wang, W. Xu, C. Zhang, Y. Gao et al. 2022. The individual and combined effects of polystyrene and silver nanoparticles on nitrogen transformation and bacterial communities in agricultural soil. Sci. Total Environ. 820: 153358.

Kakkar, R., E.D. Sherly, K. Madgula, D.K. Devi and B. Sreedhar. 2012. Synergetic effect of sodium citrate and starch in the synthesis of silver nanoparticles. J. Appl. Polym. Sci. 126(S1)

Kalia, A. and H. Kaur. 2018. Nanofertilizers. pp. 45–61. *In*: CRC Press eBooks [eds.]. CRC Press, Location.

Kaphle, A., P.N. Navya, A. Umapathi and H.K. Daima. 2017. Nanomaterials for agriculture, food and environment: Applications, toxicity and regulation. Environ. Chem. Lett. 16: 43–58.

Karamesouti, M. and D. Gasparatos. 2017. Sustainable management of soil phosphorus in a changing world. pp. 189–214. *In*: Springer eBooks [eds.]. Springer, Location.

Karimi, E. and E.M. Fard. 2017. Nanomaterial effects on soil microorganisms. pp. 137–200. *In*: Soil Biology [eds.]. Springer, Location.

Khan, I., K. Saeed and I. Khan. 2019. Nanoparticles: Properties, applications and toxicities. Arab. J. Chem. 12: 908–931.

Khan, S., M. Naushad, A. Al-Gheethi and J. Iqbal. 2021. Engineered nanoparticles for removal of pollutants from wastewater: Current status and future prospects of nanotechnology for remediation strategies. J. Environ. Chem. Eng. 9: 106160.

Khan, S., A. Wong, S. Zeb, B. Mortari, J.E. Villa and M. Del Pilar Taboada Sotomayor. 2023. Characterization of polymeric nanoparticles. pp. 141–163. *In*: Elsevier eBooks [eds.]. Elsevier, Location.

Khodakovskaya, M. and M. Marmiroli. 2023. Editorial: Polymeric nanoparticles for sustainable plant agriculture and food industry. Front. Plant Sci. 14: 1183938.

Klanjšček, T., E. Müller, P.A. Holden and R.M. Nisbet. 2017. Host–Symbiont Interaction model explains non-monotonic response of soybean growth and seed production to Nano-CEO2 exposure. Environ. Sci. Technol. 51: 4944–4950.

Krämer, U. 2015. Planting molecular functions in an ecological context with Arabidopsis thaliana. eLife. 4.

Kumari, R., K. Suman, S. Karmakar, V. Mishra, S.G. Lakra, G.K. Saurav et al. 2023. Regulation and safety measures for nanotechnology-based agri-products. Front. Genome Editing. 5.

Kumawat, K.C., N. Razdan and K. Saharan. 2022. Rhizospheric microbiome: Bio-based emerging strategies for sustainable agriculture development and future perspectives. Microbiol. Res. 254: 126901.

Li, Y., J. Liu, Y. Zhong, J. Zhang, Z. Wang, L. Wang et al. 2011. Biocompatibility of Fe3O4@Au composite magnetic nanoparticles *in vitro* and *in vivo*. Int. J. Nanomed. 2805.

Liu, J., P.C. Williams, J. Geisler-Lee, B.M. Goodson, M. Fakharifar, M. Peiravi et al. 2018. Impact of wastewater effluent containing aged nanoparticles and other components on biological activities of the soil microbiome, Arabidopsis plants, and earthworms. Environ. Res. 164: 197–203.

Lv, H. and S. Mu. 2014. Nano-ceramic support materials for low temperature fuel cell catalysts. Nanoscale. 6(10): 5063–5074.

Mahdieh, M., M.R. Sangi, F. Bamdad and A. Ghanem. 2018. Effect of seed and foliar application of nano-zinc oxide, zinc chelate, and zinc sulphate rates on yield and growth of pinto bean (*Phaseolus vulgaris*) cultivars. J. Plant Nutr. 41(18): 2401–2412.

Maurice, N. 2023. Impact of nanoparticles on soil microbes for enhancing soil fertility and productivity. In: Apple Academic Press eBooks, pp. 75–127.

Meisinger, J.J., J.S. Schepers and W.R. Raun. 2015. Crop nitrogen requirement and fertilization. In: Agronomy Monograph/Agronomy, pp. 563–612.

Mengel, K., E.A. Kirkby, H. Kosegarten and T. Appel. 2001. Fertilizer application. *In*: Springer eBooks, pp. 337–396.

Mody, V., R.C. Siwale, A.V. Singh and H.R. Mody. 2010. Introduction to metallic nanoparticles. J. Pharm. Bioallied Sci. 2(4): 282.

Muzammil, S., A. Ashraf, M.H. Siddique, B. Aslam, I. Rasul, R. Abbas et al. 2023. A review on toxicity of nanomaterials in agriculture: Current scenario and future prospects. Sci. Prog. 106(4).

Luong, N.H., N.H. Hai, N.D. Phu and D.A. MacLaren. 2011. Co-Pt nanoparticles encapsulated in carbon cages prepared by sonoelectrodeposition. Nanotechnology. 22: 285603.

Nel, A., T. Xia, L. Madler and N. Li. 2006. Toxic potential of materials at the nanolevel. Science. 311: 622–627.

Pang, Z., J. Chen, T. Wang, C. Gao, Z. Li, L. Guo et al. 2021. Linking plant secondary metabolites and plant microbiomes: A review. Front. Plant Sci. 12.

Patel, A., S. Tiwari, P. Parihar, R. Singh and S.M. Prasad. 2019. Carbon nanotubes as plant growth regulators. *In*: Elsevier eBooks, pp. 23–42.

Perveen, S., A. Parveen, M. Saeed, R. Arshad and S. Zafar. 2022. Interactive effect of glycine, alanine, and calcium nitrate Ca(NO3)2 on wheat (*Triticum aestivum* L.) under lead (Pb) stress. Environ. Sci. Pollut. Res. Int. 29(25): 37954–37968.

Rahme, K. PhD. 2020, October 1. Synthesis and characterization of nanoparticles for potential applications in wound healing. Available at: http://ir.ndu.edu.lb:8080/xmlui/handle/123456789/1259.

Rajput, V.D., A. Kumari, S.K. Upadhyay, T. Minkina, S. Mandzhieva, A. Ranjan et al. 2023. Can nanomaterials improve the soil microbiome and crop productivity? Agric. 13(2): 231.

Ramesh, T., N. Bolan, M. Kirkham, H. Wijesekara, M. Kanchikerimath, C.S. Rao, S. Sandeep et al. 2019. Soil organic carbon dynamics: Impact of land use changes and management practices: A review. Adv. Agron. 1–107.

Ramírez-Rodríguez, G.B., G.D. Sasso, F.J.G. Carmona, C. Miguel-Rojas, A. Pérez-de-Luque, N. Masciocchi et al. 2020. Engineering biomimetic calcium phosphate nanoparticles: A green synthesis of slow-release multinutrient (NPK) nanofertilizers. ACS Appl. Bio Mater. 3(3): 1344–1353.

Ramteke, P., V.V. Gabhane, P.R. Kadu, V.K. Kharche and S. Ghosh. 2023. Perspective chapter: Conservation and enhancement of soil health for sustainable agriculture. *In*: IntechOpen eBooks.

Rashidi, L. and K. Khosravi-Darani. 2011. The applications of nanotechnology in food industry. Crit. Rev. Food Sci. Nutr. 51: 723–730.

Raynolds, L.T. 2004. The globalization of organic agro-food networks. World Dev. 32: 725–743.

Santoyo, G., C.A. Urtis-Flores, P.D.L. Lara and B.R. Glick. 2021. Rhizosphere colonization determinants by plant growth-promoting rhizobacteria (PGPR). Biol. 10(6): 475.

Shrivastava, P. and R. Kumar. 2015. Soil salinity: A serious environmental issue and plant growth promoting bacteria as one of the tools for its alleviation. Saudi J. Biol. Sci. 22(2).

Singh, P. and A.P. Singh. 2021. Nanomaterials in soil health management and crop production: Potentials and limitations. pp. 1221–1245. *In*: Handbook of Nanomaterials and Nanocomposites for Energy and Environmental Applications.

Sojka, R., D.L. Bjorneberg, J.A. Entry, R. Lentz and W.J. Orts. 2007. Polyacrylamide in agriculture and environmental land management. *In*: Adv. Agron. pp. 75–162.

Srivastava, D., R. Maurya, N.A. Khan, M.A. Nayyer, A. Mishra, F. Fatima et al. 2019. General introduction of bio-inputs versus chemical inputs in agriculture and ill effects. In: Apple Academic Press eBooks, pp. 1–22.

Suazo-Hernández, J., N. Arancibia-Miranda, R. Mlih, L. Cáceres-Jensen, N. Bolan and M.L. Mora. 2023. Impact on some soil physical and chemical properties caused by metal and metallic oxide engineered nanoparticles: A review. Nanomaterials (Basel). 13(3): 572.

Wang, A., S.M. Marinakos, A.R. Badireddy, C.M. Powers and K.A. Houck. 2013. Characterization of physicochemical properties of nanomaterials and their immediate environments in high-throughput screening of nanomaterial biological activity. Wiley Interdiscip. Rev. Nanomed. Nanobiotechnol. 5(5): 430–448.

Wang, Z., L. Yue, O.P. Dhankher and B. Xing. 2020. Nano-enabled improvements of growth and nutritional quality in food plants driven by rhizosphere processes. Environ. Int. 142: 105831.

Yadav, G.S., D. Kumar, Y.S. Shivay and H. Singh. 2010. Zinc-enriched urea improves grain yield and quality of aromatic rice. Better Crops. 94(2): 6–7.

Zhang, R., A.F. Khalizov, L. Wang, M. Hu and W.v. (2011). Nucleation and growth of nanoparticles in the atmosphere. Chemical Reviews. 112(3): 1957–2011.

Zhou, W., X. Bai, E. Wang and S. Xie. 2009. Synthesis, structure, and properties of single-walled carbon nanotubes. Adv. Mater. 21(45): 4565–4583. https://doi.org/10.1002/adma.200901071.

Zhu, G., P. Zhang, W. Zhao, N. Shakoor, Y. Sun, Q. Wang et al. 2024. Toxicity of metal-based nanomaterials in different organisms. Toxin Rev. 1–19

CHAPTER-5

Nanosensors for Soil Monitoring and Assessment

Ghulam Mustafa,[1,*] *Nauman Sadiq,*[1,2] *Muhammad Zahid,*[3]
Mubushar Majeed[1] and *Safeer Ahmed*[2]

1. Introduction

Nanosensors are sensitive mechanical or chemical instruments that detect the presence of certain chemical species or nanoparticles or monitor physical characteristics like temperature at the nanoscale. These sensors are perfect for detecting subtle changes in physical and chemical surroundings due to their nanoscale size (John et al. 2022). Nanosensors measure in the nanometer range, allowing them to detect small particles and identify changes in physical and chemical properties. Computer systems built at the nanoscale level possess the capacity to detect minor modifications within physical and chemical measures. Detecting changes leads nanosensors to transform the information into detectable signals for subsequent evaluation. The fabrication of nanosensors requires three steps: material manufacturing, characteristic and property evaluation, and device production. The term nanosensor describes sensing devices with one measurement lower than 100 nm. They can transform chemical, biological, or physical substances into detectable signals. The ability of these devices to detect minute physical and chemical changes has emerged from their extensive use across multiple industries, including agriculture, security, medical diagnostic, and even COVID-19 detection. Their small sizes enable effective detection and measurement of physical and chemical changes, frequently in challenging surroundings. In recent years, nanosensors have attracted attention due to their optical, chemical, magnetic, and mechanical properties. Unlike traditional sensors, such as electrical, optical, and mechanical sensors, nanosensors provide notable advantages such as high sensitivity and the ability to work with extremely low concentrations, which results in high signal-to-noise ratios even in label-free detection applications. Additionally, incorporating different nanomaterials and metals into constructing these sensors can further enhance their performance. These materials possess unique and exceptional properties that improve the sensitivity and functionality of nanosensors, making them even more effective in various fields such as communication, agriculture, transportation, households, medicine, and energy (Adam and Gopinath 2022). In nanotechnology, nanosensors have proven crucial for monitoring chemical and

[1] Department of Chemistry, University of Okara, Okara 56300, Pakistan.
[2] Department of Chemistry, Quaid-i-Azam University, 15320, Islamabad, Pakistan.
[3] Department of Chemistry, University of Agriculture Faisalabad, Pakistan.
* Corresponding author: ghulam.mustafa@uo.edu.pk

DOI: 10.1201/9781003558354-5

physical phenomena. Their adaptability and precision make them critical tools for cutting-edge technologies and solutions that require accurate data collection at the nanoscale.

Plant growth requires soils beyond the simple cultivation medium because they provide essential ecosystem services such as nutrient cycling, water purification, carbon sequestration, and habitat provision for biodiversity. The environment depends on the soil to maintain its quality on regional, local, and worldwide scales. Soil health reacts to soil structure elements and institutional elements resulting from agricultural management (Rinot et al. 2019). Soil quality assessment measures the grading and the state of soil for investigating nutritional viability alongside structural aspects and its plant-growing capacity. Soil health evaluation requires examining chemical attributes with physical attributes alongside biological attributes to determine both the health status and production capability. Assessing soil health is a fundamental requirement for responsible agricultural methods and sustainable land management alongside environmental preservation activities. Multiple soil features need analysis to reveal their fertility levels, nutrient amounts, structure composition, biological function, and organic matter quantities. Soil assessments reveal meaningful data regarding present conditions, which helps discover blocks and possible problems that affect plant expansion and agricultural output potential. Soil assessment explains how soil translates into plant growth support and its functions related to nutrient cycling, water retention, and carbon storage capacity. Soil health understanding gives land managers and farmers the tools to pick suitable land practices while deciding appropriate soil management techniques and determining the need for fertilizers and soil amendments to boost soil fertility and productivity (Maurya et al. 2020).

The analytical power of nanosensors, combined with their ability to detect very low amounts of material, makes them advantageous for sustainable agricultural purposes. Portable devices based on nanotechnology find extensive applications across agriculture because they monitor plant health together with soil conditions (pH and soil fertility), toxic chemical detection, measurement of soil moisture, and disease-causing pests (Sharma et al. 2021, Nadporozhskaya et al. 2022). The main goal of precision farming is to optimize agricultural problems through precise plant control, resource efficiency, and soil and plant analysis while also reducing environmental contamination from waste reduction techniques. Effective agricultural techniques receive substantial improvement through nanosensors in precision farming systems. Soutter says that nanosensors based on the global positioning system have been useful in identifying soil conditions and crop growth. Moreover, he pointed out that smart sensors enabled precision farming to enhance field production. To sum up, precision agriculture, which makes use of nanosensors, provides an eco-friendly solution to the problems that contemporary farming faces (Zain et al. 2023, Soutter 2013).

2. Working Mechanism of Nanosensors

Nanosensors can measure changes at the level of single molecules. They generally consist of four parts: analyte, receptor, transducer, and detector. These sensors measure electrical changes in the receptor materials. When the analyte migrates from the solution to the receptor surface, it responds in a predictable and efficient way. This contact alters the physicochemical properties of the transducer surface, resulting in changes in its optical or electrical characteristics. These changes are subsequently translated into electrical impulses that may be recognized and studied (John et al. 2022). The operational basis of nanosensors relies on identifying electrical changes inside the sensing material. The analyte interacts with the receptor surface, initiating a process that alters the transducer's characteristics. The change of the transducer surface results in a quantifiable electrical signal (Khazaei et al. 2023). Electrochemical transduction in nanosensors indicates the mechanism by which a nanosensor transforms a chemical or biological interaction into a quantifiable electrical signal, generally by observing variations in current or potential at an electrode surface. This is accomplished through the utilization of nanomaterials that enhance the selectivity and sensitivity of the detection process; fundamentally, the nanosensor "translates" a molecular binding event into an

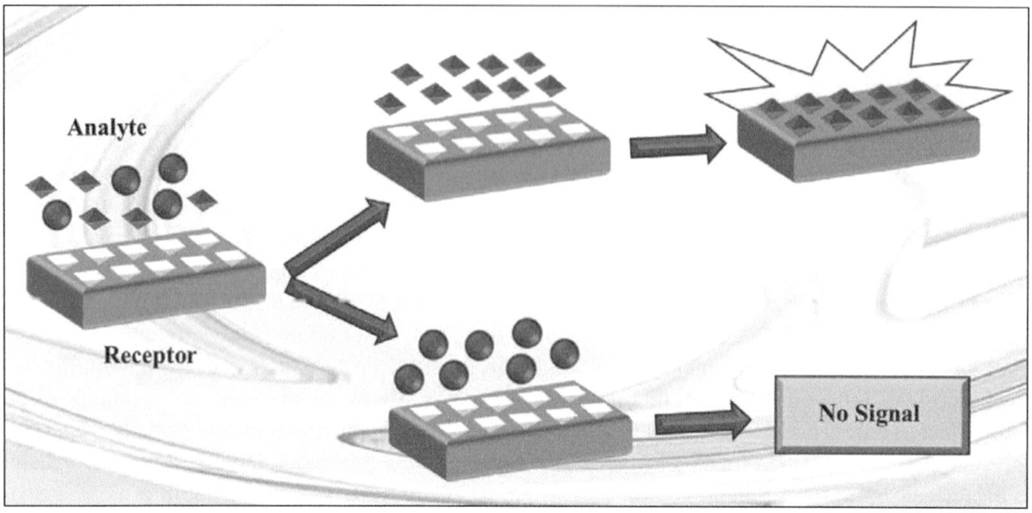

Figure 1. Working mechanism of nanosensor.

electrical signal that can be interpreted by the device (Grieshaber et al. 2008, Imtiaz et al. 2022). The working mechanism of the nanosensor can be seen in Figure 1.

From environmental monitoring to robotics development, artificial intelligence (AI) is a fast-growing technology with many practical applications in today's world. It allows for the production of high-quality items with minimal environmental impact when used in conjunction with automation, especially in industries like agriculture. Real-time monitoring of characteristics including humidity, soil quality, radiation, pollution, and more is driven by AI and IoT. Coupled with IoT and nanotechnology, precision agriculture has been transformed, opening up new possibilities for collecting data in real-time and making intelligent choices. Nanosensors powered by nanomaterials are useful for detecting and measuring a broad variety of agricultural analytes. In this chapter, we'll look at how nanosensors, nanotechnology, and the Internet of Things may all help to enhance precision agriculture. These instruments allow farmers to monitor soil quality, crop health, and other environmental factors in real-time (Pratama and Rahayu 2024, Rajak et al. 2023, Senoo et al. 2024).

3. Types of Nanosensors used in Soil Monitoring

Soil is monitored by various devices. Some of the nanosensors are discussed below, including electrochemical nanosensors, physical nanosensors, optical nanosensors, and nano biosensors. Figure 2 describes the display of these nanosensors.

3.1 Electrochemical Nanosensors

Electrochemical nanosensors generally consist of a working electrode, a reference electrode, and a counter electrode. Amperometric and voltammetric approaches have been shown to be beneficial in the quantitative evaluation of plant sensing, as well as in industrial environmental applications such as air and water monitoring. By incorporating nanomaterials into electrode architecture, these sensors offer a high active surface area, enhancing sensitivity. Recent breakthroughs in electrochemical nanosensors are mostly due to developments in metallic nanoparticles, such as gold, and nanocarbon materials, such as carbon nanotubes and graphene, which are renowned for their extraordinary electrical properties (Tan et al. 2019). These sensors can detect soil composition, which is critical for increasing agricultural yields. Many plants, for example, incur oxidative stress as a consequence of harmful metal exposure, which may harm cells. Electrochemical nanosensors can precisely measure the concentration of toxic metal ions in the soil. They also allow for the detection of nutritional ions such as H+, K+, Na+, which are essential for the growth of plants (Shaw and Honeychurch 2022).

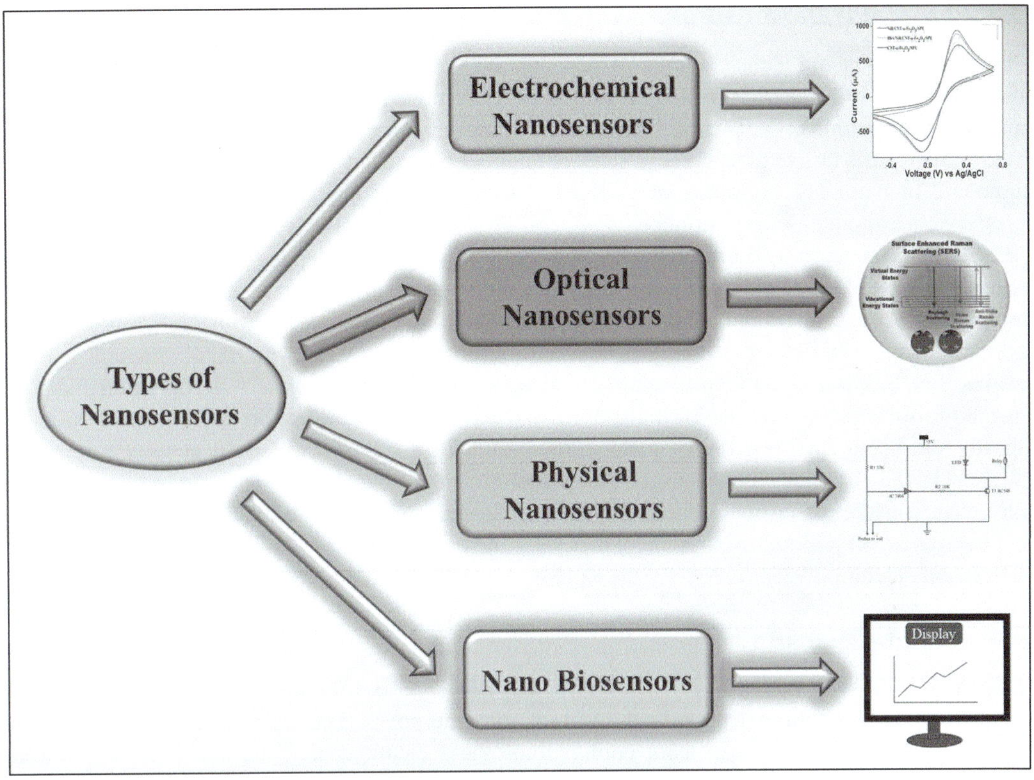

Figure 2. Types of nanosensors and their method of detection for soil monitoring.

Hematite, a stable oxide with excellent biocompatibility, electrochemical properties, large surface area, and chemical stability, has shown potential in biosensing applications. It has been successfully used in detecting formaldehyde in corn. Due to these advantages, hematite-based nanosensors are particularly suitable for agricultural applications (Kundu et al. 2022, 2021).

3.2 Optical Nanosensors

Optical nanosensors function by converting changes in molecular interactions with analytes into measurable data, based on variations in light intensity. These sensors track interactions between radiation and the sample to display optical properties that correlate with analyte concentration. Several optical principles, such as reflection, absorbance, fluorescence, and luminescence, can be applied across different regions of the spectrum (UV, visible, and near-infrared). This permits the measurement of light intensity along with other characteristics like scattering, polarization, lifetime, diffraction, and refractive index. Research-oriented nanosensors frequently include features like high sensitivity, quick detection capacity, high signal-to-noise ratio, and minimal detection thresholds. These devices must operate at low costs while being suitable for measurements at original locations. The mechanisms behind optical nanosensors focus on luminescent sensors, surface plasmon resonance sensors, and quantum dots alongside their preparation with advanced nanosensors such as nano-cellulose silver nanoparticles to develop carbon-based materials. The application area of these sensors continues to expand significantly for bionanotechnology research and personalized health surveillance, as well as environmental and food quality evaluation (Matmin et al. 2024). These soil composition detection capabilities exist in a cost-efficient RGB-based sensor system. The sensor system successfully analyzes salt, sand, and nitrophosphate materials under different RGB lighting conditions (Diaz et al. 2024).

3.3 Physical Nanosensors

Physical nanosensors detect four key aspects: mass, pressure, force, and displacement. The functioning of these sensors relies on nanotube electronic property change, which results from material deformations (Guha and Mishra 2022). Nanoscale devices that detect soil compaction and mechanical stress utilize piezoelectric sensors mainly constructed from zinc oxide, quartz, and lead zirconate titanate. The nanosensors produce electrical output signals directly indicating applied pressure measurements for continuous soil aeration and structural assessment. Soil aeration needs proper maintenance because root health and water permeability depend on it, and piezoelectric nanosensors serve precision agriculture needs (Hornyak et al. 2008). The measurement of soil temperature using thermal nanosensors depends on nanostructured thermistors or metal-oxide semiconductors for detecting direct effects on microbial activity, enzyme reactions, and nutrient availability. Monitoring soil temperature remains vital since it regulates seed germination and plant metabolic processes while determining optimal yield results. Nanosensors exhibit quick reaction speeds and heightened sensitivity, making them suitable for enabling farmer adaptations (Yin et al. 2021). The capacitive nanosensors detect soil moisture changes by measuring alterations of a dielectric constant using nano-based dielectric materials. The acquired data supports scheduling optimal irrigation periods, minimizing water loss, and enhancing farms' ability to withstand drought (Hefft and Adetunji 2023).

3.4 Nano Biosensors

The "nano biosensors" series investigates different biosensors and biochips by analyzing nanostructure applications for medical and biological purposes. The main diagnostic components of electrochemical nano biosensors contain biological elements that work alongside electrode transducers. These systems are based on nanostructures which are connected to the bioreceptor at the nanoscale level (Huang et al. 2021). A research group designed a urine detection sensor and soil enzymatic activity tester that was used for laboratory testing before transferring them to controlled test environments. Nano biosensors help enhance agricultural sustainability by maximizing resource effectiveness and increasing production efficiency (Mandal et al. 2020). The sensors can effectively identify natural and artificial stress that occurs in plants before damage. The use of nano fertilizers results in improved efficiency of agrochemicals, although decreasing their concentration has adverse effects on the environment. Both nano biosensors offer real-time assessment ability which measures nutrient levels and metabolites, pesticides, pathogens, and soil moisture together with temperature. The obtained data proves fundamental for precision farming since it enables efficient resource management as well as enhanced crop defense systems. Two essential criteria ensure the successful implementation of smart technologies into agricultural practices: first, to educate farmers about nanotechnology usage, and second, to perform field tests under farming conditions. Nano biosensors consist of three main elements: bioreceptor, transducer, and detector in their standard design. A bioreceptor functions as the biological component in the nano biosensor to detect analytes while creating chemical signals using biological materials such as antibodies, enzymes, microorganisms, and nucleic acids. This signal is then sent to the transducer, which acts as an interface, converting it into a measured electrical signal. Finally, the detector amplifies and analyzes the signal (Miguel-Rojas and Pérez-de-Luque 2023).

4. Parameters Monitored by Nanosensors

Nanosensors can be used to monitor different parameters such as soil moisture, pH levels, nutrient content (N, P, K), toxic metal contaminants, soil microbial activity, temperature, and salinity. These parameters are discussed in detail, and their pictorial diagram can be viewed in Figure 3.

Figure 3. Different parameters monitored by nanosensors.

4.1 Soil Moisture

Soil moisture has a significant impact on soil's agronomic, geological, ecological, and hydrological qualities. Water in the soil is required for mineral absorption by plant roots, making soil and water crucial to plant development. Ongoing studies focus on the significance of detecting soil moisture, with findings applied to conditions from everyday life. The combination of nanosensors with soil moisture detectors provides novel and precise approaches for monitoring the properties of soil. These sensors allow for the quantification of soil nutrients and moisture, thereby enhancing precision in agricultural management. Nanosensors play an important role in assessing the interactions among soil, roots, and organisms while also ensuring optimum crop health. Nonetheless, it is necessary to improve the speed, accuracy, and scope of these sensors, while addressing ethical issues related to nanotechnology in consumer goods and its ecological consequences (Umasankareswari et al. 2022, Mustafa 2025).

4.2 pH Levels

Nanosensors intended to monitor pH have great sensitivity and can work in the 5.5 to 9.0 range, which includes the extracellular pH (pHe) of most healthy cells. They are also used to analyze abnormal cell pH levels in cancerous cells (Kromer et al. 2024). pH tests are relevant in agriculture, industry, and environmental science fields. The optical fluorescence-based sensing technique stands out as an attractive method for pH measurement because it provides high sensitivity and selectivity and is a non-invasive operation. The fluorescence intensity of mesoporous silica nanoparticles containing rhodol dye increases ninefold between pH 3.0 and pH 9.0 (Lin et al. 2022). Clinical diagnostics, agricultural, and environmental monitoring are just a few of the many industries that may benefit from adopting pH nanosensors since the value gives essential information about the local chemical conditions (Orouji et al. 2022).

4.3 Nutrient Content (N, P, K)

Nanosensors improve plant production by accurately determining specific agricultural nutrients, which include nitrogen (N), phosphorus (P), and potassium (K). The application of animal slurry for nutrient addition should be controlled carefully because overuse leads to environmental degradation. There are two essential ways to reduce the risk to monitor the nutrients in the slurry and then apply appropriate measured amounts to agricultural fields. A mobile low-field nuclear magnetic resonance sensor is an example of real-time monitoring of nitrogen, ammonium, potassium, and phosphorus levels in animal slurry. The technology equipment (nanosensors) used in modern substitutes for conventional off-site laboratory testing, which enables to precise nutrient monitoring (Sørensen et al. 2015). Integration of NPK sensors with internet of things leads to data-based farming practices for agriculture management that improve production efficiency and environmental protection (Al-Mamun et al. 2021, Yohannes et al. 2024, Zhang et al. 2021, Adhikary et al. 2024).

4.4 Toxic Metals Contamination

Industrialization has led to the release of toxic metal ions into the environment, contaminating soil, water, and air. These metals have significant health risks as they can accumulate in the food chain. Nanosensors are being developed to detect harmful metals like mercury, arsenic, cadmium, and lead in the soil. These sensors help to mitigate the risk of heavy metal contamination, which can result from industrial activities such as fuel emissions and improper waste disposal. The development of nanosensors for on-field applications is crucial to monitor and prevent the harmful effects of metal contamination on both the environment and human health (Sadiq et al. 2023, Potadar et al. 2022). Advances in nanotechnology have permitted the creation of multianalyte nanosensor arrays, which can simultaneously detect various contaminants in soil, air, and water (Chakraborty et al. 2021).

4.5 Soil Microbial Activity

Plant productivity and soil health are greatly influenced by soil microbiomes. Soil microorganisms help to keep plants healthy by minimizing stress and sickness in a range of environments. Conventional methods for detecting soil microorganisms, such as isotope probing and DNA analysis, take a significant amount of time and money. Nanotechnology enables miniaturized, cost-effective means of monitoring microbial activity on-site. Nano biosensors, by increasing soil bacterium identification, may provide information on their role in sustainable agriculture and nutrient cycling. Nanomaterials improve sensor performance for real-time detection of microbial life in soil, and recent developments in metagenomics and sensor-based technologies have helped to enhance our understanding of the soil microbiome (Zia et al. 2022, Sashidhar et al. 2019).

4.6 Temperature and Salinity

Temperature sensors convert temperature variations into electrical impulses, and differences in potential and resistance offer precise readings. Zinc oxide nanoparticles employed in temperature sensors have excellent sensitivity owing to their piezoelectric characteristics. For example, flexible zinc oxide nanowire sensors respond to temperature changes in a well-defined manner, with current decreasing as temperature increases from 10 to 110 (Xue et al. 2014). Nanosensors can detect heat even in very hostile environments, including within living cells. The chemical, electronics, and industrial sectors are big users of these sensors, which should keep their market share growing. Environmental monitoring and precision agriculture are two other areas that are investigating the potential use of temperature nanosensors (Behera et al. 2021).

5. Applications in Soil Monitoring and Assessment

Nanosensors have a variety of applications, including monitoring different stressors, soil quality metrics, and detecting heavy metals. They increase data collection, allowing for accurate decision-making and effective resource management in agricultural systems. One notable innovation is the nanoencapsulation of traditional fertilizers (known as nano-fertilizers) and insecticides (known as nano-pesticides). This method enables the steady and continuous release of chemicals into the soil, giving plants exact dosages while limiting environmental effects. Furthermore, nano-based disease detection kits have gained popularity as early and quick diagnostic tools for viral infections. Numerous novel techniques, including biosynthesized nanoparticles, have been tested at different sizes, but actual uses of nanotechnology in soil-plant systems confront significant hurdles. The evaluation of biosynthesized nanoparticles occurs at various scales, but several obstacles remain in the implementation of nanotechnology in soil-plant systems. Nanotechnology faces four major barriers: field-service delivery effectiveness, consumer acceptance, safety matters, and regulatory compliance. Nanotechnology adoption in agriculture requires eliminating these barriers because they limit broader acceptance. Specific research strategies and policy implementations should address current difficulties so sustainable agricultural productivity can be secured (Zain et al. 2024). The detection of NOx, CO, and volatile organic compounds using nanosensors contributes to effective urban smog reduction and industrial emission reduction programs by recognizing trace pollutant concentrations. Real-time air quality monitoring is possible because these sensors provide high sensitivity that enables better studies of environmental health conditions. Nanosensors exhibit high efficiency in measuring toxic compounds and regulating essential nutrients that plants need for development through their role in soil quality assessment. This ability proves essential for sustainable farming because it prevents contamination and creates better soils (Mahmoud and Fawzy 2021, Vikesland 2018). For the last 15 years, researchers have built nanomaterial-based sensors that show potential as advanced environmental contaminant detection tools because of their specificity and high sensitivity. The field with low-cost technology shows a comprehensive ecological condition assessment system before developing, essential for environmental monitoring (Willner and Vikesland 2018). Fabricating sensitive, durable, and selective soil sensors for health assessment requires nanomaterials such as carbon nanotubes, graphene, metal oxide nanoparticles, and quantum dots. Wireless technology integrated with data processing systems allows precise, time-based soil tracking because of nanosensors. Implementing nanosensors in soil requires a comprehensive evaluation of their effects on the environment and ecological systems. Agricultural applications of nanosensors require standardized protocols and regulatory frameworks to protect their safe implementation (Parameswari et al. 2024).

6. Advantages of Nanosensors in Soil Monitoring

Due to the nanosensors' exceptional sensitivity, they can detect minute levels of pathogens, pollutants, soil moisture content, and nutrients, making them highly beneficial for soil monitoring. This precision allows for focused and efficient agricultural practices while lessening the negative effects on the environment. These attributes include highly sensitive real-time monitoring, small size, ability to analyze multiple factors simultaneously, and localized examination within the soil matrix. Ultimately, this will enhance the effectiveness of agriculture and positively improve how fertilizer and water are consumed (El-Chaghaby and Rashad 2024). The pictorial diagram of nanosensors' advantages can be seen from Figure 4.

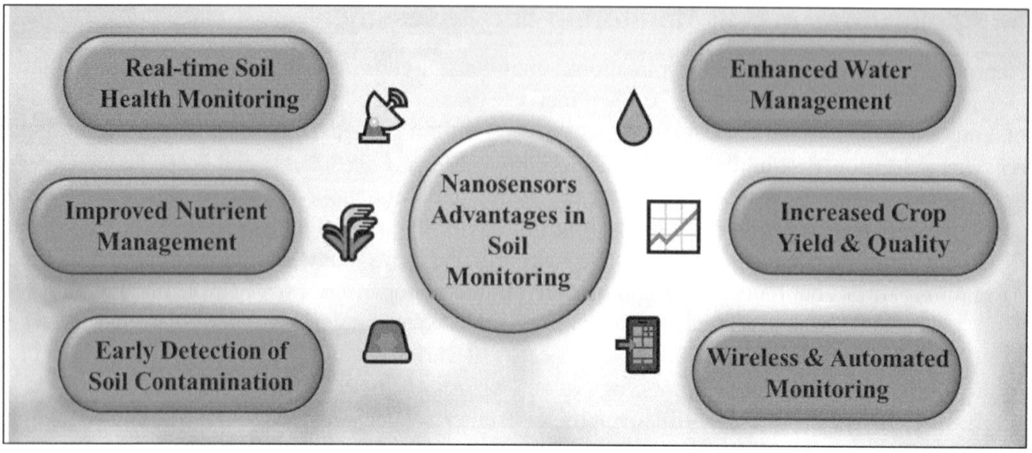

Figure 4. Advantages of nanosensors in soil monitoring.

7. Limitations and Challenges

The principal restraints and issues regarding nanosensors are as follows: complex fabrication processes, high cost, sensitivity to environmental factors, potential toxicity, sensitivity of signal, biocompatibility, absence of standardization in manufacturing processes, and the question of stability and reliability over extended periods of time. All of these issues stem from the fact that nanosensors are, at best, a few nanometers in size, which makes it increasingly complicated to control and manufacture the sensors due to their ever-growing sensitivity to the immediate environment (Han et al. 2023, Ghorbian et al. 2025).

Some challenges are given below.

- Developing new synthesis techniques for controlled fabrication of nanosensors with consistent properties.
- Exploring biocompatible materials and surface modifications to enhance biocompatibility
- Improving signal processing algorithms to accurately interpret nanosensors data.
- Establishing standardized protocols for calibration and testing of nanosensors.
- Investigating methods for long-term stability and reliability of nanosensors in complex environments.

8. Conclusion and Future Prospects

This chapter describes the use of soil nanosensors and how effectively nanoscale sensors can be used to monitor and analyze soil. These nanosensors provide high sensitivity and rapid response time, allowing researchers and farmers to identify and make timely decisions regarding soil health, contamination levels, and nutrient management. Apart from these benefits, nanosensors also provide continuous monitoring, unlike traditional methods, which are highly labor-intensive. Though these nanosensors are highly advanced, several barriers hinder farmers from adopting their technology without regulation. High production costs, environmental problems with soil ecosystem nanomaterial accumulation, and rigid regulatory standards frame concerns. There is also the barrier to advancing the research and development needed to implement AI and predictive analytics with wireless nanosensor networks. Solving these issues is critical in ensuring nanosensors' scalability and practicality for global usage. However, the possibility of nanosensors integrating AI with soil monitoring is impressive. The development and improvement of nanosensor fabrication techniques open the possibility of integrating advanced algorithms into soil monitoring and analysis. Moreover, machine learning assistance and nanosensor networks with big data analytics will allow for improving predictive models. This advancement makes it possible to understand soil degradation proactively and optimize nutrient recommendations. Implementing multi-functional nanosensors that detect multiple soil parameters simultaneously will also increase agricultural productivity and sustainability. Additional research and development are required to connect nanosensors with wireless sensor networks and artificial intelligence for the purpose of statistical forecasting. It will be necessary to find solutions to these challenges in order to undertake their scalability and practical implementation on a global scale. In the field of soil monitoring, the use of nanosensors seems to have a promising future. It is possible that the creation of biodegradable nanomaterials, technological advancements in nanofabrication techniques, and increased sensor miniaturization can contribute to the lowering of costs and the improvement of environmental problems. Furthermore, predictive modeling will be improved by integrating machine learning and big data analytics into nanosensor networks, enabling earlier identification of soil deterioration and more accurate fertilizer recommendations. Utilizing multipurpose nanosensors that can identify many soil factors at once would improve agricultural sustainability and efficiency even further. Additionally, advances in sensor technology will result from multidisciplinary cooperation between environmental specialists, agronomists, and material scientists. To provide fair access to nanosensor technologies, especially for small-scale and resource-constrained farmers, governments and authorities must also support cost-effective manufacturing techniques, standardize laws, and facilitate research. Precision farming and soil monitoring might be completely transformed by nanosensors, which would increase crop yields, lessen their negative effects on the environment, and improve food security worldwide. These gadgets will be a crucial part of smart farming systems as research and technology develop, opening the door to a more resilient and sustainable agricultural future. Unlocking the full potential of nanosensors in soil assessment will require addressing present constraints via sustained research and legislative assistance.

9. Acknowledgement

The Higher Education Commission of Pakistan (HEC) supported this work (Grant No. 17273/NRPU/R&D/HEC).

References

Adam, T. and S.C. Gopinath. 2022. Nanosensors: Recent perspectives on attainments and future promise of downstream applications. Process Biochem. 117: 153–173.

Adhikary, R. S.J. Choudhury and T. Shankar. 2024. Real-time soil nutrient monitoring using NPK sensors: Enhancing precision agriculture. Int. J. Exp. Res. Rev. 45: 197–202.

Al-Mamun, M.R. M.R. Hasan, M.S. Ahommed, M.S. Bacchu, M.R. Ali and M.Z.H. Khan. 2021. Nanofertilizers towards sustainable agriculture and environment. Environ. Technol. Innov. 23: 101658.

Behera, A.J. Pan and A. Behera. 2021. Temperature nanosensors for smart manufacturing. *In*: Nanosensors for Smart Manufacturing. Elsevier.

Chakraborty, U. G. Kaur and G.R. Chaudhary. 2021. Development of environmental nanosensors for detection monitoring and assessment. *In*: New Frontiers of Nanomaterials in Environmental Science. pp. 91–143.

Diaz, F.J., A. Ahmad, L. Parra, S. Sendra and J. Lloret. 2024. Low-cost optical sensors for soil composition monitoring. Sensors. 24: 1140.

El-Chaghaby, G.A. and S. Rashad. 2024. Nanosensors in agriculture: Applications, prospects, and challenges. In: Handbook of Nanosensors: Materials and Technological Applications. pp. 1303–1330.

Ghorbian, M., M. Ghobaei-Arani, M.R. Babaei and S. Ghorbian. 2025. Nanotechnology and nanosensors in personalized healthcare: A comprehensive review. Sensing Bio-Sensing Res. 100740.

Grieshaber, D., R. Mackenzie, J. Vörös and E. Reimhult. 2008. Electrochemical biosensors-sensor principles and architectures. Sensors. 8: 1400–1458.

Guha, P. and S. Mishra. 2022. Wireless nanosensor network for agricultural applications. In: Nanosensors for Smart Agriculture. Elsevier.

Han, D. H. Hosamo, C. Ying and R. Nie. 2023. A comprehensive review and analysis of nanosensors for structural health monitoring in bridge maintenance: innovations, challenges, and future perspectives. Appl. Sci. 13: 11149.

Hefft, D. and C.O. Adetunji. 2023. Sensing and artificial intelligence solutions for food manufacturing. CRC Press.

Hornyak, G.L., H.F. Tibbals, J. Dutta and J.J. Moore. 2008. Environmental nanotechnology. *In*: Introduction to Nanoscience and Nanotechnology. CRC Press.

Huang, X., Y. Zhu and E. Kianfar. 2021. Nano biosensors: properties, applications and electrochemical techniques. J. Mater. Res. Technol. 12: 1649–1672.

Imtiaz, T., A. Shah, N. Ullah, F.J. Iftikhar, I. Shah, S.M. Shah et al. 2022. Electrochemical nanosensor for ultrasensitive detection of malachite green and monitoring of its photocatalytic degradation. npj Clean Water. 5: 69.

John, S.A., A. Chattree, P.W. Ramteke, P. Shanthy, T.A. Nguyen and S. Rajendran. 2022. Nanosensors for plant health monitoring. *In*: Nanosensors for Smart Agriculture. Elsevier.

Khazaei, M., M.S. Hosseini, A.M. Haghighi and M. Misaghi. 2023. Nanosensors and their applications in early diagnosis of cancer. Sensing Bio-Sensing Res. 100569.

Kromer, C., A. Katz, I. Feldmann, P. Laux, A. Luch and H.R. Tschiche. 2024. A targeted fluorescent nanosensor for ratiometric pH sensing at the cell surface. Sci. Rep. 14: 12302.

Kundu, M., P. Krishnan, K.A. Chobhe, K. Manjaiah, R. Pant and G. Chawla. 2022. Fabrication of electrochemical nanosensor for detection of nitrate content in soil extract. J. Soil Sci. Plant Nutr. 22: 2777–2792.

Kundu, M., Rajesh, P. Krishnan and S. Gajjala. 2021. Comparative studies of screen-printed electrode based electrochemical biosensor with the optical biosensor for formaldehyde detection in corn. Food Bioprocess Technol. 14: 726–738.

Lin, Z., F. Hu, G. He, Y. Yang, Y. Liao, X. Luo et al. 2022. Highly photostable and pH− sensitive nanosensors. Chemosensors. 10: 354.

Mahmoud, A.E.D. and M. Fawzy. 2021. Nanosensors and nanobiosensors for monitoring the environmental pollutants. *In*: Waste Recycling Technologies for Nanomaterials Manufacturing, pp. 229–246.

Mandal, N., S. Adhikary and R. Rakshit. 2020. Nanobiosensors: recent developments in soil health assessment. In: Soil Analysis: Recent Trends and Applications, pp. 285–304.

Matmin, J., F. Mohamad, R.A. Wahab and M.H.M. Hatta. 2024. Advanced optical nanosensors. *In*: Handbook of Nanosensors: Materials and Technological Applications. Springer.

Maurya, S., J.S. Abraham, S. Somasundaram, R. Toteja, R. Gupta and S. Makhija. 2020. Indicators for assessment of soil quality: a mini-review. Environ. Monit. Assess. 192: 1–22.

Miguel-Rojas, C. and A. Pérez-de-Luque. 2023. Nanobiosensors and nanoformulations in agriculture: new advances and challenges for sustainable agriculture. Emerg. Top. Life Sci. 7: 229–238.

Mustafa, G. 2025. Nanosensor in soil-water agroecosystem. In: Nanoscience and Soil-Water Interaction in Agroecosystem, p. 1660.

Nadporozhskaya, M., N. Kovsh, R. Paolesse and L. Lvova. 2022. Recent advances in chemical sensors for soil analysis: a review. Chemosensors. 10: 35.

Orouji, A., S. Abbasi-Moayed, F. Ghasemi and M.R. Hormozi-Nezhad. 2022. A wide-range pH indicator based on colorimetric patterns of gold@silver nanorods. Sens. Actuators B Chem. 358: 131479.

Parameswari, P., N. Belagalla, B.V. Singh, G. Abhishek, G. Rajesh, D. Katiyar et al. 2024. Nanotechnology-based sensors for real-time monitoring and assessment of soil health and quality: a review. Asian J. Soil Sci. Plant Nutr. 10: 157–173.

Potadar, R.S., D.N. Moholkar and K.D. Pawar. 2022. Nanosensors for the detection of heavy trace metals in soil. In: Nanosensors for Smart Agriculture. Elsevier.

Pratama, A. and S. Rahayu. 2024. Biosensors enabled by nanotechnology: advancing precision agriculture through IoT monitoring. J. Intell. Connect. Emerg. Technol. 9: 25–39.

Rajak, P., A. Ganguly, S. Adhikary and S. Bhattacharya. 2023. Internet of Things and smart sensors in agriculture: scopes and challenges. J. Agric. Food Res. 14: 100776.

Rinot, O., G.J. Levy, Y. Steinberger, T. Svoray and G. Eshel. 2019. Soil health assessment: a critical review of current methodologies and a proposed new approach. Sci. Total Environ. 648: 1484–1491.

Sadiq, N., M. Shafique, M. Akbar, M. Shakoor, A. Mujahid, T. Hussain et al. 2023. An ion-imprinted polymer-receptor-based electrochemical sensor for the sensitive and selective detection of cadmium. ChemistrySelect. 8: e202204824.

Sashidhar, P., M.K. Dubey and M. Kochar. 2019. Sensing soil microbes and interactions: how can nanomaterials help? pp. 213–236. *In*: Microbial Nanobionics: Volume 2, Basic Research and Applications.

Senoo, E.E.K., L. Anggraini, J.A. Kumi, B.K. Luna, E. Akansah, H.A. Sulyman et al. 2024. IoT solutions with artificial intelligence technologies for precision agriculture: definitions, applications, challenges, and opportunities. Electronics. 13: 1894.

Sharma, P., V. Pandey, M.M.M. Sharma, A. Patra, B. Singh, S. Mehta et al. 2021. A review on biosensors and nanosensors application in agroecosystems. Nanoscale Res. Lett. 16: 1–24.
Shaw, D.S. and K.C. Honeychurch. 2022. Nanosensor applications in plant science. Biosensors. 12: 675.
Sørensen, M.K., O. Jensen, O.N. Bakharev, T. Nyord and N.C. Nielsen. 2015. NPK NMR sensor: online monitoring of nitrogen, phosphorus, and potassium in animal slurry. Anal. Chem. 87: 6446–6450.
Soutter, W. 2013. Nanotechnology in agriculture. Retrieved on 1, 2020.
Tan, X., S. He, X. Liu, G. Zhao, T. Huang and L. Yang. 2019. Ultrasensitive electrochemical sensing of dopamine by using dihydroxylatopillar [5] arene-modified gold nanoparticles and anionic pillar [5] arene-functionalized graphitic carbon nitride. Microchim. Acta. 186: 1–8.
Umasankareswari, T., V. Dharshana, K. Saravanadevi, R. Dorothy, T. Sasilatha, T.A. Nguyen et al. 2022. Soil moisture nanosensors. pp. 185–216. *In*: Nanosensors for Smart Agriculture. Elsevier.
Vikesland, P.J. 2018. Nanosensors for water quality monitoring. Nat. Nanotechnol. 13: 651–660.
Willner, M.R. and P.J. Vikesland. 2018. Nanomaterial enabled sensors for environmental contaminants. J. Nanobiotechnol. 16: 1–16.
Xue, F., L. Zhang, W. Tang, C. Zhang, W. Du and Z.L. Wang. 2014. Piezotronic effect on ZnO nanowire film based temperature sensor. ACS Appl. Mater. Interfaces. 6: 5955–5961.
Yin, H., Y. Cao, B. Marelli, X. Zeng, A.J. Mason and C. Cao. 2021. Soil sensors and plant wearables for smart and precision agriculture. Adv. Mater. 33: 2007764.
Yohannes, Z., T. Yoseph, A. Kiflu, T. Ayalew and A. Haile. 2024. Liquid bio-slurry enhances the productivity of N-fertilized maize under field conditions in Ethiopia. Int. J. Exp. Res. Rev. 43: 13–31.
Zain, M., H. Ma, M. Nuruzzaman, S. Chaudhary, M. Nadeem, N. Shakoor et al. 2023. Nanotechnology based precision agriculture for alleviating biotic and abiotic stress in plants. Plant Stress. 10: 100239.
Zain, M., H. Ma, S.U. Rahman, M. Nuruzzaman, S. Chaudhary, I. Azeem et al. 2024. Nanotechnology in precision agriculture: advancing towards sustainable crop production. Plant Physiol. Biochem. 206: 108244.
Zhang, Z., D. Liu, M. Wu, Y. Xia, F. Zhang and X. Fan. 2021. Long-term straw returning improves soil K balance and potassium supplying ability under rice and wheat cultivation. Sci. Rep. 11: 22260.
Zia, R., M. Iftikhar, A. Rafiq, S. Hakim, M.S. Nawaz, A. Imran et al. 2022. Nanosensors for microbial detection in soil. pp. 367–400. *In*: Nanosensors for Smart Agriculture. Elsevier.

CHAPTER-6

Impact of Nanoparticles on the Biogeochemistry of the Soil

Muhammad Junaid Nazir,[1] *Naila Ghafoor,*[2] *Amina Rashid,*[3,*]
Hamna Bashir,[4] *Muhammad Mahroz Hussain,*[5] *Muhammad Hamza Amin*[4]
and *Muhammad Tahir Shehzad*[4,*]

1. Introduction

The soil system is a very vital and valuable natural resource in the provision of earthly amenities such as food, feed, fiber, and fuel which humans require for survival (Berazneva and Güereña 2019, Cabini et al. 2018). There is a strong dependency between the most important parts of the human ecosystem, which are soil, water, vegetation, microorganisms, people, and animals; therefore, agroecosystems are critical for supporting human life (Hurni et al. 2015). Hence, several researchers have proposed to examine the importance of agroecosystems for maintaining and enriching the fertility of the soils, which implies the soil's ceramic potential to provide important nutrients to crops. Different aspects of soil fertility may be brought about through degradation processes such as pollution in the soil-plant-water system, alkalinity and salinity, and competition from other nutrients or elements in the process of interventions, such as in agriculture (Gavrilescu 2021). The soil bacteria are one of the active fractions of an agroecosystem that play a significant role in sustenance. Especially in the rhizosphere, the bacteria are an essential factor because they elicit the desired interactions in processes involved in affecting soil fertility (Kumawat et al. 2022). In the area of interest, the rhizosphere is characterized by several activities that involve the secretion of root exudates or plant metabolites.

These substances are released to provide support to the soil microbial population, which in turn helps in the transformation of nutrients (Condron et al. 2010). Consequently, it is important to enhance soil organic content and microbial communities to attain soil fertility and promote sustainable agriculture (Tahat et al. 2020, Scotti et al. 2015). Several studies have identified elements that are linked to interactions between various compartments of the agroecosystem. The interconnections between nexuses and their relationship to soil and water have been extensively studied. These include the systems of soil–water–climate change, water–land–energy–food, soil–

[1] Shenzhen Key Laboratory of Ecological Remediation and Carbon Sequestration, Institute of Environment and Ecology, Tsinghua Shenzhen International Graduate School, Tsinghua University, Shenzhen 518055, China.
[2] Department of Zoology, Wildlife and Fisheries. University of Agriculture, Faisalabad 38040, Pakistan.
[3] Department of Agronomy, University of Agriculture, Faisalabad-38040, Pakistan.
[4] Institute of Soil and Environmental Sciences, University of Agriculture, Faisalabad-38040, Pakistan.
[5] College of Environmental Science and Engineering, Yangzhou University, Jiangsu China.
* Corresponding authors: aminauafl@gmail.com; tahir_kumboh@hotmail.com

DOI: 10.1201/9781003558354-6

food–environment–health, sustainable water–energy–environment, water–food–energy–climate, water–energy–waste, water–energy–food, water–food–land–ecosystem, water–energy–carbon, soil health–human health, and soil–water–plant–human (Nadarajah and Abdul Rahman 2021, Bargaz et al. 2018, Jacoby et al. 2017, Moreira-Grez et al. 2019). Photographic or graphical pieces may be particularly successful in conveying these intricate connections. As a result, there have been numerous recent researches that utilize illustrations and/or photographs to emphasize various subjects, including smart agriculture, the relationship between soil and humans, the management of salt-affected soils, the comparison between higher plants and mushrooms, nano-farming, nano-grafting, and the interconnectedness of water, plants, soil, and human beings (Bogardi et al. 2012, Garg et al. 2023).

2. Nano-Farming: Revolutionizing the Future of Agriculture

Nanoparticle engineering is a cutting-edge technical advancement that exhibits distinct targeted properties with increased potency. Professor Norio Taniguchi of Tokyo University of Science introduced the term "nanotechnology" in 1974 (Chaudhari et al. 2024). While the concept of nanotechnology has been widely discussed in several fields, the use of nanoparticles (NPs) in agricultural development is a relatively new and ongoing scientific advancement. Recent developments in the synthesis of nanomaterials with a wide range of sizes and shapes have led to their widespread use in environmental science, medical, agricultural, and food processing industries. Throughout the past, agriculture has consistently derived advantages from these developments (Rahman et al. 2024).

Furthermore, agriculture is currently confronted with various and unparalleled obstacles, such as decreased crop productivity caused by living and non-living factors, including lack of essential nutrients and contamination of the environment. However, nanotechnology has presented encouraging possibilities for precise agricultural practices. Precision agriculture, or precision farming, refers to the wireless networking and miniaturized sensors to monitor, evaluate, and manage agricultural processes. It pertains to the precise handling of crops in particular locations, including many areas of agriculture before and after production. This includes both horticultural and field crops (Shafi et al. 2019). The scientific achievement of using tissue engineering and engineered nanomaterials (ENMs) to deliver CRISPR/Cas mRNA and sgRNA for genetically modifying crops is a recent advancement that has been recognized (Abd-Elsalam and Lim 2021). In addition, nanotechnology offers effective solutions for a growing array of environmental issues. For instance, the advancement of nanosensors has great promise in monitoring environmental stress and improving plants' ability to fight against pathogens (Sharma et al. 2022, Chaudhary et al. 2020). Hence, the ongoing advancements in nanotechnology, particularly in issue identification and the establishment of collaborative strategies for agricultural growth that are sustainable, have significant potential to provide extensive societal and fair advantages.

Nanomaterials are among the most distinctive and widely used materials due to their size of less than 100 nm and vast surface space of 1,000 m^2/g (Jeevanandam et al. 2018). ENMs can easily reach other internal parts of living organisms since they are much smaller in size and have the ability to penetrate the hematoencephalic barrier and plant materials. Even though profitable products incorporating nanomaterials are found in numerous installation projects, the capability of adjusting the properties that result in several physical and chemical characteristics of the finished product is the chief factor behind the wide use of nanomaterials. The unique characteristics of nanomaterials have been used in several industrial sectors, including agriculture, cosmetics, aeronautics, and pharmaceutics (Sampathkumar et al. 2020, Xin et al. 2018). This phenomenon takes place both during the manufacturing process and after the utilization of ENMs. Hence, it is imperative to carefully choose a suitable synthetic approach for the fabrication and manufacturing of certain forms and dimensions of ENMs. In the subsequent section, here is a succinct summary of frequently employed methodologies for the synthesis of nanoparticles.

3. Innovative Techniques for Nanomaterial Synthesis

The procedure of producing ENMs is usually chosen based on the required properties of the final nanomaterials, including their size and other physicochemical factors (Albalawi et al. 2021, Cockburn et al. 2012). The production of nanomaterials may be roughly classified into two major techniques: top-down and bottom-up, depending on the kind of application. The phase of the starting materials distinguishes these techniques (Arole and Munde 2014). Reactants in top-down processes are usually solids; in bottom-up approaches, reactants are mostly gases or liquids.

3.1 Advanced Top-Down Techniques for Nanoparticle Synthesis

The top-down approaches are often called physical processing methods because they involve treating the starting materials in a solid form. These procedures include physically decreasing the size of large bulk material particles using etching, milling, crushing, grinding, and lithographic techniques (Qin and Riggs 2013). The benefits of these techniques include the ease with which a significant amount of ENMs may be produced and applied to a variety of surfaces, including layered and porous materials and metal oxides (silica, alumina, etc.). There is no need to utilize chemicals in these approaches; therefore, it is possible to avoid the time-consuming purification steps. However, these methods often fall short for the accurate manufacturing of ENMs. Moreover, it is usually not feasible to get the necessary form of ENMs in these methods due to the lack of control over the surface features of ENMs (Yadav 2021).

3.2 Bottom-Up Approaches for the Synthesis of Nanoparticles

These techniques involve the fabrication of materials employing individual atoms, molecules, or clusters at the nanoscale level (Chakraborty et al. 2023). The bottom-up strategies can be classified into three main categories: physical methods, chemical methods, and biological approaches. Chemical techniques are frequently employed due to their efficacy in regulating particle growth and their ability to generate intricate nanostructures (Sampathkumar et al. 2020). Physical techniques, such as abrasion, condensation, melting, and evaporation, utilize mechanical pressure, high electrical or thermal energy, and radiation to create a range of ENMs (Ijaz et al. 2020). Top-down physical operations, such as milling, crushing, and etching, utilize solid-state precursors. On the other hand, bottom-up physical methods, including chemical vapor deposition (CVD), flame pyrolysis, and electrolysis, employ gaseous or liquid chemicals as precursors (Silina et al. 2020).

3.2.1 Chemical Methods for the Synthesis of Nanoparticles

Chemical synthetic methods, on the other hand, involve the use of chemical contents in synthesizing the ENMs from a solution via bottom-up techniques (Xin et al. 2018). In these processes, molecules or atoms are brought to a particular site, and they then coalesce to form nanoparticles. Currently, there are several chemical syntheses for raising high-quality and reproducible ENMs through precise control of manufacturing processes in a liquid phase (Backer et al. 2018). On the advantage side of chemical synthesis methods is the ability to synthesize colloidal nanoparticles of a solid material in a liquid medium that can easily be separated to form a powder. The execution of these reactions is simple and can produce a substantial variety of forms and sizes of ENMs (Albalawi et al. 2021). Many procedures are in practice to synthesize ENMs, which are gradually described in detail in the further sections of the text, such as, among others, the sol-gel process, the microemulsion method, and hydrothermal polyol synthesis.

3.2.2 Sol Gel Method

Sol-gel is used in synthesizing metal oxide and mixed metal oxides nanomaterials and is a liquid chemical technique (Parashar et al. 2020). It has a large ability to affect the hardness and interfacial behavior of the nanoscale parts. ENMs synthesis consists of many phases, for instance, hydrolysis, which is a phase of condensation as well as the drying phase. A sol can therefore be defined as particles dispersed in a liquid, whereas a gel refers to a fluid containing polymeric network. The procedure is mostly done in water, even though solvents which are not liquid are used by rule, contingent on the characteristics of the product. It has been used in the synthesis of several metal as well as bimetallic oxides (Parashar et al. 2020).

3.2.3 Microemulsion Method

Microemulsions are stable, transparent systems that do not undergo phase separation on heating or cooling of the system (Mishra et al. 2014). Throughout microemulsions, surfactants are used in a specific action as an interfacial layer. They assist in demulsification and also avoid coalescence of drops in the two phases of the system. Reverse micellar systems, which are also known as water-in-oil minuscule emulsions, are regarded as effective nanoreactors for the synthesis of ENMs that are uniformly sized. Unique characteristics of this procedure include the following: thermodynamic stability, lubricity, interfacial tension, and interfacial area of certain non-miscible liquids. This process has been used often for the synthesis of various metals and their oxide nanoparticles (Yousefpoor et al. 2024).

3.2.4 Hydrothermal Method

This technology means that organic- or water-based reagents, as the reaction medium, are placed in a pressure vessel to achieve high pressure as well as temperature for the synthesis of green ENMs (Li et al. 2024). It is used to dissolve and recrystallize chemicals that are poorly soluble or, on the other hand, are usually insoluble. The literature has demonstrated that this is a one-step process and it produces nanomaterials that are highly crystalline. These nanomaterials also do not entail any other processes such as annealing or calcination to be carried out on them (Liu et al. 2024). Hydrothermal procedures can also be worked in two types: one is a batch hydrothermal system and the other is a continuous hydrothermal process. The batch system is somewhat basic and transforms the oxidation state of any compound and builds an exact ratio of segments, within which an element may possess a variable oxidation state. In contrast, the continuous hydrothermal treatment technique explains a process in which fast reaction rates dominate, and the time frame is less than a minute. It is possible for this technology to be scaled with the aim of producing large quantities of engineering nanomaterials of desirable quality at a lower price (Shandilya et al. 2016).

3.2.5 Polyol Method

It is an easily and effectively applicable and flexible process, which is responsible for synthesizing small-sized nanoparticles from metal salts being precursors. Polyols serve as the reducing reagents in the thermal reduction of these precursors at elevated temperatures. Polyols, including EG and its greater molecular mass different forms, triethylene glycol (TrEG), diethylene glycol (DEG), and polyethylene glycol (PEG), fulfill numerous functions. In the process, they are the general reaction medium, solvent, reducing agent, and also the complexing ligand. Because of their high dielectric constant, polyols are good solvents for many metal precursors. Besides, polyols have relatively high boiling points (about 320 degrees Celsius), and their reducing ability depends on temperature. These characteristics facilitate the effective regulation of the formation and development of particles, resulting in the synthesis of high-quality ENMs (Baričić et al. 2024).

4. Nanoparticle-Enabled Smart Delivery Systems: Unlocking Sustainable Agriculture

Nanotechnology is regarded as a pivotal technology of the 21st century that has the potential to improve conventional agricultural techniques and facilitate sustainable development by boosting conservation and management strategies while reducing wastage of agricultural resources (Das et al. 2017). Efficient methods for delivering agricultural compounds, such as DNA or oligonucleotides, into plant cells are crucial for accurate farming and sustainable harvests. Conventional methods of applying agrochemicals to crops, including broadcasting or spraying, are often used. As a result, agrochemicals only partly reach the desired areas on crops, which is much less than the least effective concentration needed for optimal plant development (Safari et al. 2020). It is crucial to take into account the nutrients' bioavailability while applying fertilizer. This is due to the fact that the soil may have an impact on nutrient accessibility via processes such as chelation, microbial degradation, evaporated water, overuse, hydrolysis, and problems with runoff (Perea Vélez et al. 2021). When using insecticides, the main goal is to optimize effectiveness by controlling spray drift. As a way to encourage environmentally friendly farming practices, the development of nanotechnology-based synthetic herbicides, insecticides, and delayed or controlled-release fertilizers has drawn profound attention from the agricultural community. The use of nanotechnology has evolved from research in labs to real-world uses. The goal of controlled methods of distribution is to minimize loss or adverse effects on the environment or implicated species while releasing exact and adequate volumes of agrochemicals over a predetermined period of time (Otlewska et al. 2020, Rajput et al. 2022). Nanoparticles provide excellent pesticide administration due to their substantial surface area, ability to adhere easily, and rapid mass transfer. Agrochemicals may include micronic or submicronic nanoparticles by many methods, including trapping them inside the nano-matrix of active substances, attaching them to the surface through ionic or weaker bonds, absorbing them, and encapsulating them. An instance of this occurs when potassium nitrate is coated with layers of graphene oxide, which significantly extends the duration of fertilizer release. This formulation seems to have the potential for large-scale production at an affordable cost. Nanomaterials enhance the stability of agrochemicals, preventing their breakdown and subsequent release into the environment. This increased stability improves their efficacy and reduces the required amount (Khan et al. 2022).

5. Nano Fertilizers: A Source of Balanced Crop Nutrition

Typically, it is necessary to replenish critical nutrients (via element fertilization) in order to enhance crop yield and soil fertility (Ahmed et al. 2021). However, the accurate administration of fertilizers is regarded as a crucial need for achieving sustainable agricultural growth. Nevertheless, access to food is an essential entitlement of every human being. The issue of global food availability is now facing significant challenges on a global scale. The scarcity of natural resources contributes to the endangerment of food security. Projections suggest that the present global population of seven billion is expected to grow and reach around nine billion by the year 2050. In order to sustain the growing population, an additional 60–100% of food will be required (Xue et al. 2022). In order to satisfy the growing need for food, farmers are resorting to intensive farming methods, which in turn result in a detrimental loop of depleting soil fertility and reducing agricultural productivity. Since the adoption of intensive farming practices, an estimated forty percent of the global agricultural land has seen major degradation, which has resulted in a large drop in soil fertility. Therefore, a substantial quantity of fertilizers is applied to enhance soil fertility and increase crop output. Furthermore, it has been indisputably noted that fertilizers account for a third of crop production, while the other two-thirds are influenced by the effectiveness of other agricultural inputs. However, the effectiveness of traditional fertilizers in terms of nutrient use efficiency is often no more than 30–40%. As an instance, the nutrient use efficiency of traditional fertilizers, including nitrogen (N) at a rate of 30–35%, phosphorus (P) at a rate of 18–20%, and potassium (K) at a rate of 35–40%, has been consistent throughout the previous several decades. Furthermore, the effectiveness of traditional fertilizers, whether they are applied directly to the soil or sprayed on the leaves, is greatly influenced by the concentration of the fertilizers that ultimately reaches the intended areas (Nadarajah and Abdul Rahman 2021) .

A minuscule amount, much lower than the least required concentration, reaches the desired destination due to the loss of chemicals by processes such as leaching, runoff, drift, hydrolysis, evaporation, photolytic destruction, or microbiological decomposition. Excessive and frequent use of fertilizers disrupts the natural equilibrium of nutrients in the soil, leading to negative consequences. Furthermore, aquatic habitats have undergone substantial contamination due to the release of detrimental compounds into rivers as well as reservoirs of water. This pollution also affects the quality of the water that is consumed. According to accounts, it was discovered in the early 1970s that a mere 27 kilograms of NPK per hectare was sufficient to yield one metric ton of grain. Nevertheless, by 2008, the need for NPK had escalated to 109 kg per hectare in order to achieve the same level of productivity. Based on data from the International Fertilizer Industry Association (IFIA), there has been a substantial rise in worldwide fertilizer consumption. It is projected that the demand would reach 192.8 million metric tons over the 2016–2017 timeframe. Significant quantities of traditional fertilizers can contribute to the accumulation of chemicals in the soil and other environmental components, causing substantial environmental contamination that can negatively impact the development of plants and animals. Utilizing ENMs in environmentally friendly farming has introduced a novel approach to food production that has the potential to address the unpredictability in crop yield due to limited resources (Munir et al. 2021). The advent of green nanotechnology has significantly transformed the worldwide agricultural landscape. Nanomaterials, namely nano fertilizers, have great potential in addressing the projected increase in worldwide hunger and promoting sustainable agriculture. Nano fertilizers may be a very effective solution for addressing both macro- and micronutrient deficiencies and combating the persistent issue of eutrophication by improving nutrient utilization efficiency (Gelaw and Sanan-Mishra 2022).

Nano fertilizers, designed to control the flow of nutrients based on crop needs and reduce losses, have significant promise. Conventional fertilizers containing nitrogen, such as those mentioned in references 41, 85, and 86, result in significant losses from the soil due to leaching, evaporation, and degradation. These losses may reach up to 50–70%, leading to a decrease in fertilizer efficiency and an increase in production costs. However, nitrogenous fertilizer nanoformulations ensure that the release of Fertilizer-N is synchronized with the crop's absorption needs. Nanoformulations effectively minimize nutrient losses by impeding direct crop internalization, hence minimizing nutrient interactions with soil, water, air, and microbes (Lee et al. 2014). For example, the use of permeable nanomaterials like chitosan, clay, or zeolites may effectively decrease nitrogen losses by improving plant absorption and controlling release according to demand. Zeolites charged with ammonium can enhance the solubility of phosphate minerals, hence increasing the availability and absorption of phosphorus by crops. Graphene oxide sheets, which are carbon-based nanomaterials, may delay the release of potassium nitrate. By reducing leaching losses, the usable life is extended (Ahmed et al. 2021, Dibyanshu et al. 2023). A study on the use of nanomaterials in agriculture showed promising results when compared to traditional fertilizers. The researchers found that the use of nano calcite ($CaCO_3$–40%) in combination with nano SiO_2 (4%), Fe_2O_3 (1%), and MgO (1%) not only increased the absorption of calcium (Ca), magnesium (Mg), and iron (Fe), but also significantly improved the uptake of phosphorus (P), zinc (Zn), and manganese (Mn). Nano-fertilizers are available in a diverse range of distinct forms. Nanofertilizers include several sorts, including control or slow-release fertilizers, control loss fertilizers, magnetic chemical fertilizers, and nanocomposite fertilizers. These fertilizers use integrated nanodevices to provide a diverse array of macro- and micronutrients with the necessary characteristics. Nano-fertilizers are mostly manufactured by the method of encapsulating nutrients with nanoparticles. Nanomaterials are originally synthesized using physical (top-down) as well as chemical (bottom-up) approaches. Subsequently, certain nutrients are encapsulated inside materials that have nano-sized pores, coated with a thin film made of polymer, or administered in the form of particles or emulsions with dimensions at the nanoscale. This process is carried out for positively charged nutrients such as NH_4^+, K^+, Ca^{2+}, Mg^{2+}, or after altering the surface for negatively charged nutrients like NO_3^-, PO_4^{3-}, SO_4^{2-}. By using a well-balanced approach to fertilizer management, efficient irrigation techniques, and deploying high-quality seeds, it is possible to achieve a substantial increase of 35–40% in agricultural production (Perea Vélez et al. 2021, Dibyanshu et al. 2023).

The use of nanoformulated fertilizers has been closely observed and shown to possess significant potential for augmenting agricultural productivity. Combining carbon nanoparticles with fertilizer has the potential to greatly increase the yield of several crops. For example, incorporating carbon nanoparticles into rice may increase grain yields by 10.29%, while spring maize yields can be enhanced by 10.93%. Furthermore, it has been shown that soybean yields may be enhanced by 16.74%, winter wheat by 28.81%, and vegetables by a range of 12.34–19.76%. Zhang et al. (2020) showed that the use of chitosan-NPK fertilizer markedly improves the crop index, harvest index, and mobilization index of wheat yield variables, in comparison to the control variables. Nanomaterials improve key elements of plant biology by taking advantage of the high porosity at the nanoscale in the root and leafy sections of plants, which are the main pathways for nutrient absorption (Awwad et al. 2022). Therefore, the use of nanofertilizers may enhance the uptake of nutrients by plants via these minuscule apertures. In addition, nanofertilizers may enhance the attachment of nutrients to molecular transporters or chemicals produced by the roots by generating new openings. In addition, they may use endocytosis or ion channels to enhance the absorption of nutrients. Moreover, a multitude of research has unequivocally shown that diminishing the dimensions of nanomaterials results in an augmentation in the ratio between the surface area and the mass of particles. Consequently, a significant quantity of nutritional ions is assimilated and then released slowly over an extended duration (Otlewska et al. 2020, Rajput et al. 2022). Nanoformulations of nutrients provide a balanced supply of nourishment to crops throughout their growth cycle, hence improving agricultural production. It is important to highlight that the improved effectiveness of a product might motivate farmers to use the commodity in a more lucrative way.

6. Nanomaterials in Soil: Pathways, Impact, and Environmental Fate

Natural nanomaterials found in the soil include clay, organic compounds, and different metal and metal oxide nanoparticles. However, due to the widespread use of products containing nanomaterials, a number of the previously listed ENMs are also disposed of in soil and water. It is expected that several million tons of metal as well as metal oxide nanoparticles, i.e., Ag, Au, SiO_2, ZnO, TiO_2, and CNT (carbon nanotube) fullerenes, are manufactured annually in the world. When ENMs are manufactured, used, and disposed of, they leak into ecosystems. Key sources of exposure to ENMs include coatings, paints, pigments, electronics, cosmetics, and lenses (Ahmed et al. 2021, Dibyanshu et al. 2023, Perea Vélez et al. 2021). Between 0.1 to 2% of ENMs are released into the environment during the manufacturing process. The product determines how many ENMs are released into the environment (Figure 1). Many nanoparticles from plastics, paper, board, electronics, and packaging end up in landfills and soil. It is crucial to emphasize that 10–25% of ENMs end up in soil and 60–90% are disposed of in landfills. This implies that a significant amount of ENMs is deposited into the soil each year. The most frequent ENMs discharged into ecosystems are SiO_2, TiO_2, ZnO, iron, and Al_2O_3. The time-dependent buildup of ENMs in a variety of environments at high concentrations is accurately predicted by Garner et al.'s dynamic multimodal death and transport model, NanoFate (McKee and Filser 2016).

Research shows that utilizing wastewater for water supply, which is likely to include ENMs, dramatically increases soil nanomaterials. Approximately 90 micrograms of TiO_2 per kilogram of soil are found annually. Engineered nanoparticles were found most in sludge-treated soil. The study also found that Europe deposits 0.3–1.3 grams per kilogram of TiO_2 into the soil yearly, whereas the USA and Europe deposit 0.05 and 0.09 grams per kilogram of ZnO. TiO_2, followed by ZnO, is also being put into soil. The impact of ENMs on soil is significantly influenced by soil type and the features of the ENMs (Hurtado et al. 2022). ENMs can engage with soil-charged particles, dissolve in soil water, or be ingested by soil organisms, depending on their bioavailability. ENMs have the potential to disperse into sediments, bodies of water, and soil. The soil behavior of ENMs is influenced by their size, shape, chemical characteristics, and surface features. These properties can substantially impact the electrical, optical, and catalytic activity of particles, hence determining their fate in the

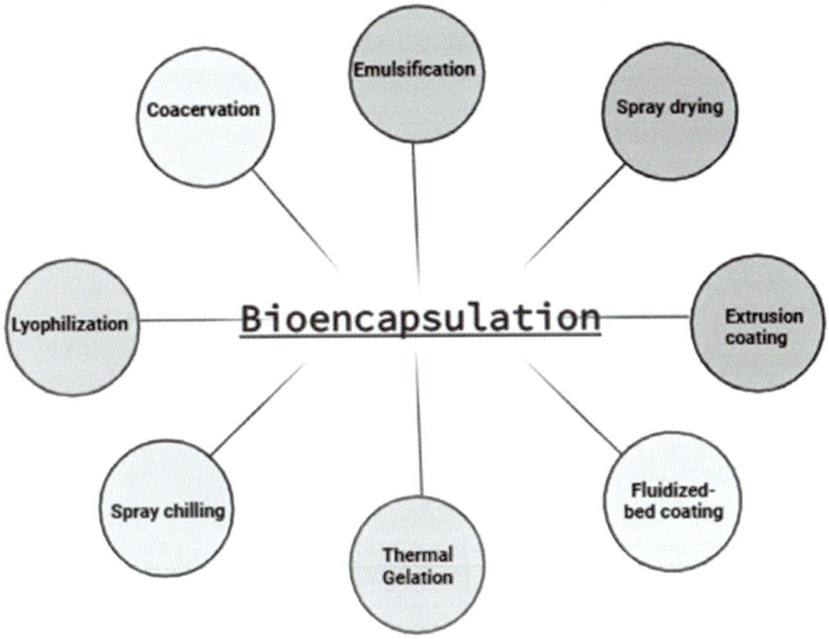

Figure 1. Nanoencapsulation methods for the capsule.

soil (Pereira et al. 2021). In order to enhance the stability of colloids, ENMs are frequently subjected to treatment with inorganic or organic ligands as well as polymeric surfactants. This modification impacts the way they interact with soil particles, which includes the generation of colloidal solutions in soil water and the formation of aggregates. The numbers 151 and 152 are enclosed in square brackets. ENMs undergo chemical conversion, aggregation, as well as disaggregation in soil as they age. Soil ENMs can experience processes such as adsorption, desorption, dissolution, and sulfidation. The surface properties of ENMs may potentially undergo alterations as a result of these processes. The fate and bioavailability of ENMs to plants and microorganisms are influenced by environmental physicochemical and biological processes (Yabur et al. 2007). Agglomeration and deposition/ sedimentation dominate physical transformation. ENMs have high surface energy and seldom exist as discrete particles. ENMs typically homo- or hetero-agglomerate after release. ENMs tend to cluster with soil particles rather than other ENMs since there are fewer ENMs. Besides clumping, additional physical changes like deposition and sedimentation occur in three stages: a slow stage, a quick phase, and a sluggish phase owing to the limited number of particles in suspension. Suspended ENMs deposit on solid surfaces quickly due to fast aggregation, reducing their mobility and bioavailability. ENMs dissolve in soil, releasing metal ions that form compounds. The soil particles absorb these metal ions. Soft metals such as silver (Ag), copper (Cu), and zinc (Zn) react strongly with inorganic sulfur deposits in the soil and bio-macromolecules containing sulfur. Usually, they produce metal sulfides. These metals' strong sulfur affinity and ability to form complexes with sulfur-containing molecules explain this tendency. Therefore, the lifespan of different nanomaterials in diverse settings, such as soil, depends on the intrinsic characteristics of ENMs and other environmental conditions (Muhammadi and Ahmed 2007, Bashan et al. 2014).

6.1 Environmental Impact and Dynamics of NMs

An essential component of environmental risk assessment is understanding the environmental fate and behavior of nanoparticles (NMs) in order to determine their impact on environmental health (Tao et al. 2018). Understanding the origins, routes, changes, and destinations of NMs will be crucial for comprehending the compartments (such as sediment, water, and biota) that interact with the NMs.

Additionally, gathering data on how nanoparticles (NMs) behave in the environment would make it possible to identify certain risks, such as the potential for bioaccumulation in aquatic food chains (Lobo et al. 2019).

6.2 Physico-chemical Characteristics

The significance of NM features in environmental risk is generally acknowledged, while the techniques used to measure these qualities are not always suitable or used correctly. In our previous review (Lee et al. 2016), the significance of transformations was barely emphasized. Nevertheless, the physicochemical characteristics, including dimensions, composition, energy from the surface, and specific surface area of the initial, unaltered (as manufactured or as used) nanomaterials remain significant for two particular reasons. In order to evaluate and manage risks effectively, it is necessary to have a comprehensive grasp of both the potential hazards and advantages associated with NP, as well as a thorough knowledge of the original and modified NMs. Furthermore, the behaviors and rates of transition are highly reliant on these initial features. Nevertheless, it has been shown that the characteristics of the resulting outcomes are far more important than first believed, and they will be thoroughly examined in the subsequent sections (Bernabeu et al. 2018, Lobo et al. 2019).

7. Exposure of Soil Organisms to NMs and their Bioaccumulation Potential

Accumulation and adverse impacts of metal and metal oxide nanoparticles in soil and the capacity of NMs to be assimilated by land-dwelling organisms in soil-based habitats have notably increased during the last 8 to 10 years. However, there is a lack of research specifically examining this problem, unlike the many studies that concentrate on the aquatic ecotoxicology of NMs and the harmfulness of NMs to land-dwelling animals in controlled conditions such as hydroponics. Early studies in this area focused on NMs in their natural form and recorded cases of earthworms accumulating copper oxides and silver NMs (Grasso et al. 2020, Zhang et al. 2023). Additionally, it was found that corn has a limited ability to absorb CeO_2 NMs, pumpkin can absorb Fe_3O_4 NMs, and grain can absorb TiO_2 and ZnO NMs (Du et al. 2011). Recent studies have shown that FeO NMs may pose a risk to clover, CeO_2 NMs might impact the distribution of nutrients in maize, and CeO_2 NMs can accumulate in the root nodules of soybeans. The strong interaction between NMs and soils, especially through hetero-aggregation through various soil minerals and dissolved organic substances, raises concerns about the environmental impact of toxicological edges established using pristine NMs in non-soil (de Jesus et al. 2024).

Likewise, the impact of surface coatings, sizes, and forms of pure NMs on toxicity to soil species, as evaluated in non-soil conditions, will likewise be of minimal environmental significance owing to the aforementioned responses. Additional long-term research is required to provide chronic toxicity facts, since the responses of NMs in soils may need time to completely manifest. While initial studies focused on as-manufactured and uncommon NMs, recent research in soil-based cultivation nanotoxicology has shifted towards studying the NMs that are more likely to be released in large amounts into the environment. This research also investigates the toxicity of the transformed end products of NMs, which occur due to the chemical conditions associated with realistic discharge pathways. As a result, recent research has primarily concentrated on studying TiO_2 and ZnO NMs due to the anticipated presence of relatively high levels of these NMs in sludge with different concentrations (Reches and Gazit 2006). Additionally, there has been a significant focus on investigating Ag NMs due to their well-documented ability to kill microorganisms and their Widespread use in remediating soil and ground water (Rodríguez-Carmona and Villaverde 2010).

Considering the present use patterns, significant quantities of TiO_2, ZnO, and silver NMs, together with their transformed by-products, are anticipated to be introduced into global ecosystems by the application of biosolids on land. As a result of this issue, there is a growing trend of conducting studies on the toxicity associated with these NMs in soil that has been modified with biosolids. There is an increasing number of advanced studies on the toxicity of nanoparticles, but the potential harm they may cause to land ecosystems due to metal and metal oxide nanoparticles is still uncertain. This is because there are similar limitations in understanding their dosage, transformation processes, and long-term effects, just like in aquatic systems. The recent studies conducted by (Rabbani et al. 2020) have shown that nematodes exposed to Ag and Ag_2S NMs exhibit transgenerational sensitivity. Similarly (Sun et al. 2016), found that toxic effects occurred in tomato progeny, but not in parents who were exposed to CeO_2 NMs. These findings emphasize the significant potential of such kinds of studies. Similarly, Li et al. (2024) presented findings from a research project that investigated the impact of adding biosolids to soil. The biosolids included either bulk or dissolved forms of TiO_2, Ag, and ZnO, as well as nanoparticulate forms of TiO_2, Ag, and ZnO. Another group of biosolids had no additional alloys.

Using chemical parameters that simulate those in a genuine WWTP, the biosolids were created by subjecting the waste streams' metal content to prototype WWTP installations. When compared to the bulk/dissolved controls, the model legume *Medicago truncatula* showed less root nodulation after NM treatments. This observation was associated with a notable increase in Zn bioaccumulation. The research also observed significant changes in the microbial communities in soil treated with NMs compared to the bulk/dissolved treatment. Nevertheless, a different study conducted on a less complex system without plants found that the community of microbes in the treatment where the media was in large quantities or dissolved form was significantly different from the treatment without media. However, this change did not lead to a remarkable decline in the functioning of the ecosystem (Baričić et al. 2024). In various WWTP setups, a diverse range of Ag NMs undergo a significant transformation into insoluble Ag_2S NMs.

However, it is also conceivable for Ag-cysteine and AgCl to be formed as end products of this transformation process. There seems to be little danger to terrestrial environments from Ag NMs, since non-labile conversion end products, such as AgCl and Ag_2S, are expected to be environmentally dependable and have limited permeability. However, uncertainties regarding longer-term modifications and dosimetry should also be considered. Soil-based studies have provided additional evidence that Ag_2S NMs pose little risk to plant-microorganism symbioses, crop plants, soil microorganisms, and soil invertebrates (Kappi et al. 2018).

Xiao et al. (2021) investigated the impact of CeO_2 NMs at dosages of 100 and 400 mg/kg on wheat crops that were cultivated in the field for a period of 7 months. No significant impacts were found on the growth of wheat, the amount of grain produced, or the accumulation of Ce. However, there were noticeable changes in the physical and chemical characteristics of the plants, including a decrease in chlorophyll levels, a delay in the blooming process, and an increase in the activities of catalase and superoxide dismutase enzymes. Additional research is required to investigate the interactions and changes that occur between NMs and soil components during extended durations of exposure.

7.1 Accumulation and Harmful Effects of Carbon Nanomaterials in Soil

Consumer items often use carbon-based nanomaterials, such as fullerenes and carbon nanotubes (CNTs). Nevertheless, the quantities of carbon NMs that enter WWTPs are very small. In contrast to silver nanoparticles (Ag NMs), research indicates that carbon NMs are only expected to exhibit toxicity at higher concentrations. While certain studies have indicated toxicity at large quantities in hydroponics and *in vitro*, there has been limited research on the ecotoxicological effects in soil-based media, with little reported toxicity to soil organisms. Research in this field has shown that carbon nanotubes (CNTs) have minimal bioavailability to earthworms in both artificial and natural soils (Waalewijn-Kool et al. 2014). Additionally, further investigations have shown that fullerenes do not have any influence on soil microbial communities. Graphene and other recently developed carbon-based nanomaterials have not been extensively researched yet; however, they seem to have minimal levels of toxicity.

Preliminary studies indicated that gold nanoparticles (Au-NMs) had the ability to undergo biomagnification in a simulated food chain including tobacco plants (*Nicotiana tabacum*) and tobacco hornworm caterpillars (*Manduca sexta*) (Borgatta et al. 2023). This finding raises concerns about the possibility of human exposure to nanoparticles via trophic transfer. Subsequent research has documented the transfer of NMs in various simulated food chains. While the investigations documented the transfer of nutrients, the quantities of transferred NMs were minimal, and the tomato-hornworm research did not find the claimed increase in concentration of these nutrients. Recently, there have been reports of biomagnification of NMs in a terrestrial food chain. Specifically, CeO_2 NM was found to accumulate in kidney bean plants, which then biomagnified in Mexican bean beetles. The beetles were consumed by spined soldier bugs, resulting in the transfer of NMs through the food chain. These results indicate that the movement of NMs across different tiers of a food chain is probable to some degree in the majority of land-based food webs. However, some species and food chains exhibit greater vulnerability to biomagnification, while the underlying causes for this sensitivity remain unidentified (Lead et al. 2018).

8. Soil Environment and Biotransformation

8.1 Soil Organic Matter

Soil organic matter (SOM) refers to any organic substances derived from plants and animals that are added to the soil or deposited on its surface, regardless of whether they are in living form or in different stages of degradation. Soil organic matter (SOM) constitutes just 1–5% of the soil components, yet it plays a crucial role as a primary indicator of soil quality. This is so because SOM is associated with the tasks of maintaining the physical, chemical, and biological characteristics of the soil. Soil Organic matter (SOM) influences some of the critical soil characteristics such as water holding capacity, buffer capacity, cation exchange capacity, aggregation, mineralization, adsorption of heavy metals, availability of micro nutrients, aeration and soil microbial activity (Nowack et al. 2015). Soil organic matter (SOM) may be classified into two separate fractions. The main sources of CFs are hydrophobic and hydrophilic. These fractions are value-added and differ in composition and depend on such factors as geographic location and/or the season. The hydrophobic fraction includes carboxylic and tannic acids, carbohydrate and proteins (El Hadri et al. 2018). Hydrophilic part of the SOM mainly contains humic substances, which are humic acids, fulvic acids, and humin. They are said to account for 60–80% of the total SOM and have a great influence on the chemical nature of soils (Deb and Shukla 2011). The general characteristic of humic compounds is their hydrophilicity, meaning that they are called dissolved organic matter (DOM), if they are in the aqueous solution. The environmental behavior and stability of nanomaterials are significantly influenced by their interactions with soil organic matter (SOM). Key binding mechanisms include hydrogen bonding, electrostatic attraction/repulsion (dependent on surface charge), and hydrophobic effects, changed and improved statement. This interaction leads to the development of a surface coating referred to as the ecocorona. Thus, the differentiation of the nanomaterial occurs, which influences the changes in their behavior and impact on biological systems during exposure through the terrestrial environment (Gelaw and Sanan-Mishra 2022).

The investigation of the interactions between SOM and nanomaterials in actual situations is insufficient owing to the considerable variability of environmental factors (Louie et al. 2016). The adsorption of SOM on nanomaterials is influenced by the intrinsic properties of the material, including its reactive surface area, surface charge, aggregation state, and the presence or absence of a coating or surface modification. Nevertheless, the interaction is also affected by the physical characteristics of SOM, such as its molecular composition, electrical charge, molecular mass, and the arrangement of the coating formed (whether it is a single layer or several layers, and its shape). Considering soil factors such as pH, ionic strength, and valence of ions is crucial. The main modes of interaction between nanomaterials and dissolved organic carbon in the water component of soil are Van der Waals forces, Coulomb forces, hydrophilic forces, H bonding, cation bridging, and surface ion chelation (Deb and Shukla 2011). Several studies have been conducted to understand how nanomaterials travel in water systems, primarily because their colloidal condition is more stable when they are coated with humic acid. Similarly, in the soil, the adsorption of dissolved organic matter (DOM) onto nanoparticles leads to a reduction in the surface charge of the nanomaterials. This, in turn, greatly hinders the speed at which the particles join together and settle down. As a result, the target's ability to move is improved (Park et al. 2016). Furthermore, the study suggests that the SOM may enhance the ability of some carbon nanostructures to regenerate over time in the soil (Li et al. 2024).

Therefore, steric hindrance might cause entrapment of nanoparticles in the upper soil layer, possibly due to their interactions. Nonetheless, it has the potential to be immobilized in the soil permanently in some conditions, but in fact, incorporation of humic acid to the silver nanoparticles promoted the mobility of nanoparticles in the soil, as was proved (Coutris et al. 2012). Similar outcomes were also reported for fullerene (nC60), and this was also established to be due to produced steric repulsive energy. Concerning the stability and transformations of nanomaterials in soil, it was understood that Ag nanoparticles are not immediately available. Nevertheless, the presence of large clusters and large particles was no longer noticeable after 70 days, while the count of accessible Ag^+ ions increased. Ag nanoparticles are capable of providing a steady stream of fresh Ag in this particular context. Furthermore, the study conducted showed that the dissolution of Ag nanoparticles was higher in mineral soil as opposed to organic soil. However, the authors also demonstrated that in the case of zinc oxide (ZnO) nanoparticles, the solubility was higher in soils with higher organic matter levels, as exposed by (Waalewijn-Kool et al. 2014).

8.2 Enzymatic Mechanisms

The different types of nanomaterials pose high toxicity and antibacterial characteristics when they are introduced into the terrestrial environment and therefore reduce the fertility of the soil. Generally, microbial responses to the induced perturbations are determined using enzyme assays (Perea Vélez et al. 2021, Dibyanshu et al. 2023). Thus, the enzymes in the soil are responsible for enhancing and quickening essential soil processes like the breakdown of organic matter, conversion of soil organic matter, releasing the nutrients which are indispensable for plant growth, nitrogen fixation, nitrification, denitrification, and detoxifying foreign substances. In addition, their application affects the carbon, nitrogen, phosphorus, and sulfate cycles. 'Soil enzyme' activity is about the most sensitive bioindicator of changes occurring in the soil concerning biological balance, fertility, and quality. Among the specific enzymes are urease for the nitrogen cycle, acid phosphatase and arylsulfatase for the phosphorus and sulfate cycles, and β-glucosidase for the carbon cycle. Some of the diverse functions of phenol oxidase and peroxidase are in the acquisition of carbon (Acosta Motos et al. 2018, Hernandez-Ruiz et al. 2001). Dehydrogenase and catalase have been associated with the number of aerobic microorganisms and the fertility of the soil, as detected (Kucharski et al. 2009).

Due to the significant role of enzymes in the evaluation of the effects of human activities such as the use or discharge of nanomaterials in soils, the authors find that enzymatic activity should be given consideration in the prognosis of changes in soils. Also, it has been illustrated that enzymes are involved in the biotransformation process of nanomaterials. This is fairly new research, and the majority of studies have been performed on tissues in a petri dish. As for the knowledge we currently possess, to the best of our knowledge, there is no reported literature on the enzymatic biotransformation of nanomaterials in soil. However, there are a few writers who have contended that enzymatic activity is involved in the biotransformation process and the breakdown of nanomaterials. The preliminary study in 2008 explored the capability of the horseradish peroxidase enzyme to degrade single-walled carbon nanotubes (SWCNT). Subsequent investigations have revealed that enzymes have the capacity to degrade carbon nanotubes (CNTs), graphene, as well as fullerene (Allen et al. 2011).

For instance, when Lalwani et al. (2014) observed the effects of the enzyme lignin peroxidase on graphene for 4 h, graphene formed wrinkles and holes. In addition, the graphene structure disappeared after 96 hours and was replaced with carbon agglomerates. Fullerene structure may react with laccase to be biodegraded, in which hydroxyl and epoxy-fullerene are produced (Gitsov et al. 2012). Various studies have explored how enzymes can act to create pathways for the oxidation of nanomaterials, and information is accessible regarding numerous oxidative enzymes. Using H_2O_2 as a type of oxidizing substrate, the peroxidase enzyme has the catalytic cycle described below. The enzyme removes two electrons from hydrogen peroxide (H_2O_2) and resolves into Compound I, a two-electron oxidized enzyme. After that, there are two reduction processes by which Compound I returns to its initial state and forms the intermediate molecule, Compound II—an enzyme with one electron of the heme's iron oxidized (Zhang et al. 2014). The nature of Compound I and of the hydroxyl/hydroperoxyl radicals is highly oxidizing, and these species are thus capable of efficiently disrupting the integrity of carbon nanomaterials. For instance, it is rather difficult to apply solely oxidase enzymes in the process of breaking down the virgin carbon nanotubes (CNT) and graphene (Costello et al. 2009, Lalwani et al. 2014).

Steady reduction in enzyme activity or contact could be attributed to the hydrophobicity of the material, coupled with increased aggregation and sedimentation. Zhou et al. (2019), in their study, established that the process of covalent functionalization of MWCNT could facilitate the rate of enzymatic degradation. However, the functionalization might prevent further biotransformation of nanomaterials. For example, BSA or PEG coatings on GO can reduce the activity of HRPs. This might be explained by the existence of functionalization species on the surface of the nanomaterials, which form a protective carbon core and reduce the action of enzymes. In addition, any structural changes of the nanomaterial in the environment will cause variations in its chemical and physical properties. Therefore, it may engender a number of environmental problems, like toxicity or bioaccumulation (Karaca et al. 2010). The internalization of cells may be facilitated by the insertion of carboxylic groups and the decrease in MWCNT metrics (Raffa et al. 2008). Therefore, the predictability of ecotoxicity should take into account the process of biotransformation.

8.3 Soil pH

The pH of the soil is one of the serious limiting factors in determining the fertility and health status of soils. Regarding the pH levels observed in soils containing engineered nanoparticles (ENPs), it is evident that there is a notable variance in pH depending on the specific type of ENPs, their quantity, and their activity in the soil. Minimal studies have been conducted on different soil types, including silver, phytogenic Fe oxide, ZnO, CuO, Fe_3O_4, and a composite of ZnO and CuO. It was noted that the pH of the soils experienced a modest increase (Fajardo et al. 2015).

Scientists applied 10 mg kg^{-1} of CuO ENPs to sandy soil and observed a pH range of 4. It was indicated that the pH value ranged from 5 to 9, which was quite close to the value of the control soil. Simultaneously, the pH value increased from 5.1 to 5.4 following the administration of a 40 mg kg^{-1} dose of CuO ENPs. The observed increase was smaller than the nominal value, which further confirmed that the soil's ability to sustain a change limited the fluctuation of soil pH. However, certain research studies have indicated that variations in soil pH may be caused by the indirect effects of engineered nanoparticles (ENPs). In Zhang et al.'s study, 100 mg kg^{-1} of silver nanoparticles (Ag ENPs) were introduced into the soil, either with or without the presence of cucumber (*Cucumis sativa*) plants. As to this, Yue et al. explain that the change of pH in the soil is due to the transformation of the metabolites within the presence of Ag ENPs' influence. With concern to this, some research indicated that after the application of ENPs, the pH levels of the soil reduced slightly. Therefore, according to research by Duncan and Owens (Zhang et al. 2020), Australian soils with 500 mg kg^{-1} of CeO and TiO_2 ENPs had a pH drop of 0.1 to 0.3. Moreover, H. Zhang et al. (2020) found that plants' biomass, both fresh and dry, decreased when 50 and 100 mg kg^{-1} of TiO_2 ENPs were administered, while also lowering the rhizosphere solution's pH from 7.3 to 7.1.

The authors did not delve into the underlying cause behind the fall in pH, as it was not the primary objective of the study. The addition of 2, 4, and 6 g per kilogram of Fe, Fe_2O_3, and Fe_3O_4 ENPs respectively to red soil resulted in a reduction in pH by 0.4 to 0. On the seventh day of the week, a total of 8 units were collected. Based on the results obtained from the Wushan soil, it can be inferred that the pH value decreased by 0.60 to 1. On the second day, five samples were measured to have a difference of 10 units higher compared to the control soil. The pH values of the examined soils were analyzed in relation to the hydrolysis of Fe^{3+} ions (Lin et al. 2019). Regarding the capacity to alter the variability of soil pH, it should be noted that the parameters that govern this depend on the features of the matrix and the type of ENPs. According to (Peng et al. 2020), the presence of Fe ENPs in the soil resulted in a significant increase in pH, raising it by nearly 5 units in acidic soil, from 30 to 7.60.

8.4 Redox Potential

The pH of the soil is a crucial determinant of soil fertility and health. The soil pH levels of the ENPs exhibit varying values, ranging from 2.2 to 8.3. Research conducted on several soil types employing ENPs such as phytogenic iron oxide, silver, zinc oxide, copper oxide, a mixture of ZnO and CuO, and magnetite has demonstrated a slight increase in pH levels (Shi et al. 2018). In their study, Zhang et al. (2020) found that sandy soil with a dose of 10–15 mg kg^{-1} of CuO ENPs had pH values ranging from 4.9 to 5, which was quite close to the control soil's pH value. The pH then rose from 5.1 to 5.4 when 40 mg kg^{-1} of CuO ENPs were added. The lower-than-anticipated rise in pH levels indicates that the soil's buffer capacity was responsible for limiting the pH elevation. However, further research has shown that ENPs indirectly cause changes in soil pH. Zhang et al. (2020) subjected soil samples to 100 mg kg^{-1} of silver nanoparticles, either with or without *Cucumis sativa* plants included. The soil pH rose from 5.28 to 5.33 after 60 days of crop growth, and from 5.19 to 5.27 after plant removal. Soil metabolite changes caused by Ag ENP exposure were thought to be responsible for the pH shift. Soil acidity levels were shown to drop slightly when ENPs were added. Soil pH levels in Australia dropped by 0.1 to 0.3 points after being treated with 500 mg kg^{-1} of CeO and TiO_2 ENPs. The rhizosphere pH dropped from 7.3 to 7.1 when 50 and 100 mg kg^{-1} of TiO_2 ENPs were introduced into the soil (Das et al. 2018).

Because it had not been central to their work, the researchers did not go into detail on the process that caused the pH to drop. When 2, 4, and 6 grams per kilogram of iron nanoparticles (FeNPs), iron (III) oxide nanoparticles (Fe_2O_3NPs), and iron (II, III) oxide nanoparticles (Fe_3O_4NPs) were introduced to red soil, the pH of the soil decreased by 0.4 to 0.8 units after 7 days. The pH of the Wushan soil decreased by 0.60 to 1.10 units on day 2, in comparison to the control soils. The acidity of both soils was ascribed to the hydrolysis of Fe^{3+} ions (Li and Zhang 2007). The impact of ENPs on soil pH is contingent upon the characteristics of the soil matrix. When Fe ENPs were added to acidic soil, an increase in pH from approximately 5.30 to 7.60 was reported. Conversely, the pH level of calcareous soil stayed consistently around 8.0, regardless of whether ENPs were present or not. The pH values in the calcareous soil remained constant, likely due to high carbonate concentration and its potential to moderate fluctuations in soil pH ($CaCO_3$ = 5.7% for calcareous soil and 0.15% for acid soil). Moreover, the ENPs released Cu^{2+} ions into the aqueous solution, which promoted the formation of long-lasting compounds including Cu_2S and $Cu(OH)_2$, resulting in an increase in soil pH (Husson 2013).

8.5 Nutrients Availability

Phosphorus, nitrogen, and potassium are crucial macronutrients for the growth and development of all plants. Phosphorus in the environment exists in various chemical forms, such as H_3PO_4, $H_2PO_4^-$, HPO_4^{2-}, and PO_4^{3-}. The dissociation constants for these forms are as follows: $pK_1 = 2.21$, $pK_2 = 7.21$, and $pK_3 = 12.67$, respectively (Peng et al. 2020). Phosphate, in the form of $H_2PO_4^-$ and HPO_4^{2-}, has a vital function in stimulating plant growth and supporting microorganisms in agricultural soil. Nevertheless, the existence of contaminants, such as ENPs, could potentially affect the accessibility of phosphate and thus reduce its advantageous impacts on plants and microorganisms (Zhang et al. 2020). The adsorption of phosphate on several engineered nanoparticles (ENPs), such as CeO_2, magnetic iron oxide, Fe, ZnO, Fe/Cu, and TiO_2, has been confirmed in multiple experiments conducted in water-based systems. Research has demonstrated that phosphorus undergoes chemisorption with active sites of various ENPs, and this interaction is permanent. Furthermore, these experiments demonstrate that cations derived from ENPs have the ability to interact with phosphate, resulting in the formation of complexes and/or precipitates. Although the tests were not conducted on soils, they do provide a potential indication of what could occur within the soil medium. In their work, (Köser et al. 2017) showed that the presence of Al_2O_3 and Fe_3O_4 ENPs enhances the adsorption of phosphate in calcareous soil. Furthermore, the existence of ENPs resulted in the following alterations: a rise in phosphate concentration outside of the hydrochloric acid-phosphate fraction and inside the residual phosphate and sodium hydroxide-phosphate areas. According to this, they hypothesized that the existence of ENPs (Chen et al. 2022, Rajput et al. 2018) leads to a reduction in the availability of phosphate. The hypothesis suggests that the drop in values in these circumstances may be linked to the presence of L-ascorbic acid on the surface of ENPs, as well as the presence of dissolved L-ascorbic acid. The significance of this work lies in the utilization of copper (Cu) or silver (Ag) ENPs as nano-pesticides, which leads to the deposition of considerable amounts of ENPs in soils. The presence of zinc oxide (ZnO) in agricultural soils is quite probable, as it serves as a vital nutrient for plants. ZnO can enter the soil through the application of agricultural pesticides. The association between phosphate ions (PO_4^{3-}) and zinc oxide nanoparticles (ZnO ENPs) involves the release of ions from the nanoparticles. Subsequently, they have the capability to generate a micrometer-scale crystalline zinc phosphate and a nanoscale shell composed of amorphous $Zn_3(PO_4)_2$. Furthermore, Zn is an essential component of the enzymes phytase and phosphatase. Research demonstrated that ZnO ENPs have the ability to stimulate the synthesis of P-mobilizing enzymes, leading to an augmentation in the quantity of accessible phosphorus in the soil. Thus, both direct and indirect factors can potentially affect the influence of ENPs on the bioavailability of P. Soil phytoavailability of rhizosphere-attached P was improved after TiO_2 and Fe_3O_4 ENPs were introduced to sandy loam soil. The results were thought to be caused by the acidity produced by the organic acids excreted from the roots of *Lactuca sativa*. To enhance the concentration of accessible phosphate when iron ENPs and cerium oxide ENPs are present, a comparable method was suggested. A substantial decrease of 27.33% in the content of total phosphorus (TP) in surface water was observed when composites made of CeO_2-functionalized maize straw biochar (CeO_2-MSB) were used. The TP concentration also increased by 7.22 percent in the uppermost layer of soil. Though it's true that this increased P adsorption, it would be beneficial to study the soil-to-plant development and leaf area relationship for rice plants. For this reason, CeO_2-MSB can prevent the surface layer of rice fields from losing phosphorus (Wang and Jiang 2019).

8.6 Enhancing Plant Adaptation with Nanomaterials Amid Climate Change

The increasing global population, coupled with the adverse effects of climate change, has made food security a formidable challenge owing to the scarcity of available resources. Progressive climate change encompasses alterations in the fundamental characteristics of climate over extended periods, including variations in temperature, water scarcity, cold conditions, salinity, alkalinity, and the presence of hazardous metals causing environmental contamination. Hence, the primary objective is to facilitate the rapid adjustment of plants without endangering vulnerable ecosystems when they endeavor to withstand environmental pressures (Stephen Inbaraj and Chen 2016).

To achieve this objective, a comprehensive approach is necessary, including the activation of plant enzymes, hormonal control, expression of stress genes, management of the absorption of harmful metals, and prevention of water scarcity or sudden flooding by reducing the plant's life cycle. Researchers have made many efforts to create technology and methods that promote sustainable agriculture systems while minimizing negative impacts on the environment (Zulfiqar et al. 2019). Nanomaterial engineering advancements indicate that the use of nanofertilizers has the potential to enhance agricultural yield under challenging environmental conditions. Salinity stress significantly restricts agricultural yield in around 23% of the farmed areas globally. On the other hand, studies have shown that the use of nano-SiO_2 enhances seed germination and promotes growth in tomato and squash plants subjected to NaCl stress. This improvement is shown in terms of increased plant fresh weight, dry weight, chlorophyll content, and the accumulation of proline (Boutasknit et al. 2021).

In addition, applying nanoparticles, namely iron sulfate ($FeSO_4$), by foliar spray resulted in improved salt stress tolerance in sunflower cultivars. The researchers found that applying nano $FeSO_4$ had multiple positive effects, including increasing leaf area, shoot dry weight, net carbon dioxide (CO_2) assimilation rate, sub-stomatal CO_2 concentration (Ci), chlorophyll content, maximum photochemical efficiency of photosystem II (Fv/Fm), and iron (Fe) content. Additionally, it significantly reduced the amount of sodium (Na) content in the leaves. Further studies have also looked at the ability of SiNPs to lessen the impacts of stress that result from UV-B in wheat (Kalia et al. 2019).

A study revealed that wheat plants treated with nanofertilizer had a much shorter life cycle, with a reduction of 23.5% (130 days instead of 170 days) compared to plants treated with traditional fertilizer. This resulted in an accelerated yield production starting from the date of seeding. The usage of nanofertilizers has shown significant potential in accelerating plant growth and productivity. This makes them valuable tools in agricultural practices, particularly in areas prone to drought or sudden flash floods, where early crop maturity is crucial for sustainable crop production. Furthermore, nanomaterials have been shown to be very efficient in the process of detoxifying or remediating dangerous contaminants such as heavy metals. Chakraborty et al. (2023) demonstrated that applying nano-Si to rice plants at a dose of 2.5 mM on the leaves effectively enhances their tolerance to Cd stress by controlling the accumulation of Cd. Under a separate investigation, the same group demonstrated that nano-Si is likewise efficient in combating Pb, Cu, and Zn in the presence of Cd. Nano-Si fertilizers could offer an advantage over standard fertilizers in decreasing the buildup of heavy metals (Das et al. 2022).

9. Role of Nanomaterial in Plant Health and Microbiome

Despite the availability of various commercial products utilizing ENMs such as fertilizers, pesticides, and sensors for plant health and soil quality monitoring, the use of ENMs in agriculture is still in its early stages (De Corato 2021). Nano fertilizers based on ENMs have been shown to improve the efficiency of nutrient utilization. Similarly, nano pesticides have demonstrated great effectiveness, enhanced solubility, and longevity, all achieved with small amounts of active ingredients (Zhang et al. 2014). Furthermore, several advantages of agricultural products based on ENMs have been thoroughly documented. The increased use of ENMs for agricultural purposes is also significantly increasing the risk of ENM contamination in the soil. Large quantities of ENMs in the soil may have a significant impact on the plant rhizosphere. This is because ENMs have increased interactions with the soil and plant microbiome, as stated in reference. Specifically, when ENMs are introduced into the underground environment with the intention of gaining immediate advantages, they have the ability to be absorbed by various plants and may have adverse effects over extended periods of time. Plants release a variety of substances that promote the development of microorganisms in the area around their roots, known as the rhizosphere. At the same time, these microorganisms help with the cycling of nutrients, which in turn supports the growth of plants. This interaction between plants and microbes is mutually beneficial and occurs in conjunction with the plant roots (Table 1). Nevertheless, the presence of ENMs has a substantial impact on the microbial populations in the rhizosphere, as well as on the plants and the extracellular chemicals generated by the bacteria (Simonin and Richaume 2015).

Table 1. Microbes for Plant Growth and Metabolite Production.

Formulation type	Examples	Function	References
Microbial bioformulations	*Rhizobium (e.g., Rhizobium etli, Rhizobium leguminosarum)*	• Forms a symbiotic relation with pulses and leguminous plants. • Fixes nitrogen and encourages plant growth and health. • Enhances soil fertility. • Minimizes the use of synthetic fertilizers.	(Arora et al. 2017, Comite et al. 2021, Hashem et al. 2019)
	Arbuscular Mycorrhizal Fungi (AMF)	• Greatly improves the absorption of nutrients. • Offers protection against abiotic environmental stress.	(Begum et al. 2019)
	Azotobacter (e.g., Azotobacter chroococcum)	• Produces phytohormones that enhance plant vitality and increase crop productivity. • Functions as biofertilizers by converting atmospheric nitrogen into a usable form, decreasing the need for synthetic fertilizers and supporting sustainable agricultural practices.	(Baslam et al. 2011, Lopes et al. 2021)
	Pseudomonas (e.g., Pseudomonas putida, Pseudomonas fluorescens)	• Develops metabolites and chemicals that promote plant development. • Enhances plant health by secreting antimicrobial chemicals that suppress soil-borne diseases. • Increases the absorption of nutrients and enhances the ability to tolerate stress.	(Hashem et al. 2019, Kecskés et al. 2016, Rizvi and Khan 2018, Santos et al. 2019)
	Trichoderma (e.g., Trichoderma harzianum)	• Enhances the growth and protects pathogenic organisms via symbiotic connection. • Synthesizes a diverse range of secondary metabolites that exhibit bio stimulant and antifungal activities. • Enhances nutrient uptake efficiency, promotes root development, and prevents plants from stress tolerance, which collectively contribute to increased yields and crop resilience.	(Ahmad et al. 2018, Lopes et al. 2021)
Metabolite-based free culture supernatant bioformulations	*Trichoderma* spp. *(metabolites)*	• Synthesizes a diverse range of secondary metabolites that exert both bio stimulant and protective effects on plants.	(Ahmad et al. 2018, Backer et al. 2018)
	Burkholderia spp.	• Produces ACC deaminase, which has a function in promoting plant development under stressful situations.	(Khan et al. 2023, Onofre-Lemus et al. 2009)
	Pseudomonads	• Synthesizes siderophores, high-affinity iron-chelating compounds that facilitate iron acquisition from the environment.	(Khan et al. 2023, Řezanka et al. 2018)
	Pseudomonads	• Produces Phenazines that possess antibiotic characteristics that hinder the formation of plant pathogens, such as fungi and nematodes in cell membranes, disrupt oxidative phosphorylation, and generate reactive oxygen species (ROS).	(Biessy and Filion 2018, Yu et al. 2018)
	Fluorescent Pseudomonads	• Production of 2,4-diacetyl phloroglucinol, a compound that triggers plant defense mechanisms by inducing systemic resistance.	(Almario et al. 2017)
	Streptomyces sp. MR14	• Production of a metabolite that has biocontrol and plant growth-boosting characteristics.	(Kaur et al. 2019)
	Bacillus amyloliquefaciens (LZN01)	• Synthesis of myriocin, methyl-2-oxovaleric acid, sphingofungin, gabapentin, and sphingofungin with antifungal properties.	(Xu et al. 2019)

Soil ENMs affect plant health in two ways: first, by acting directly on plant cells, and second, by altering the plant and soil microbiome. It is well known that plants and other organisms absorb them, and that they subsequently interact with various biomolecules (Gelaw and Sanan-Mishra 2022). It is possible that ENMs in the soil pose a risk to food security due to the potential effects on plant physiology. The phenomenon known as nano-phytotoxicity is to blame for this. Nanoparticle interactions with crops have both beneficial and negative effects on biochemistry, physiology, and genetics, according to certain researchers. Several factors influence the results. They include the dose, size, surface charge, and chemical characteristics of ENMs. Crucial aspects that determine the effect of these nanoparticles on plant growth include the mobility of ENMs within the soil, their potential to transfer from soil to plants, and their distribution across distinct plant tissues. Therefore, to correctly predict how ENMs will affect plant growth, it is crucial to understand how ENMs migrate in soil and how they travel internally in plants. The chemical-physical features of ENMs and their surrounding environment impact their mobility. As a general rule, ENMs are found in the soil around a plant's roots, which may then transport them to the plant's above-ground components, such as its leaves. The direct transfer has a significant impact on the development of these plant elements, as pointed out (Das et al. 2022). After being taken up by the roots, the question of whether these nanoparticles may reach the plant's aerial tissues, such as leaves, has been explored in several studies (Daly et al. 2021). The cuticle and cell walls of plants are only two of the numerous barriers that ENMs must traverse before they can reach plant tissues. They may enter plant tissues via epidermal pores or wounds. Even if the ENMs' dimensions are larger than the cell wall pores, research suggests that certain ENMs may still permeate plant tissues by expanding the cell wall pores (Zhang et al. 2023).

The results showed that the concentrations of glucan and pectin in the *Arabidopsis thaliana* root cell wall were reduced when CuO was added. The foliar spray technique of ENM delivery also exposes the plant's above-ground parts directly. Spraying ENMs onto leaves improves their absorption via the leaf's outer layer or the tiny holes called stomata. In the cuticular pathway, polar solutes diffuse and permeate via polar aqueous pores, whereas nonpolar solutes use these same mechanisms (Giller et al. 2021, Schreefel et al. 2020). Once ENMs have passed through the stomata and reached the leaf apoplast, they are transported via the vascular system. The aerosol process by which watermelon plants absorb ENMs was shown by Wang et al. According to the study, ENMs made their way to the plant's roots after entering the leaf via the stomatal channel and traveling down the stem. Nanoparticles have the potential to apoplastically or simplistically move into plants. Simplistic migration takes place via plasmodesmata, while apo plastic migration takes place via extracellular spaces and xylem vessels. Research on the movement of multi-walled carbon nanotubes (MWNCTs) and titanium dioxide nanoparticles (TiO_2 NPs) from soil to plants (wheat and red clover) revealed that the former two substances exhibited little migration from soil to effluents. Several theories have been advanced to explain the chemical basis of nanomaterial toxicity (Ganie et al. 2021, Kumar et al. 2021).

10. Conclusions

Nanomaterials refer to organic and inorganic compounds that exhibit unique and desirable features at the nanoscale level. Additionally, it is possible to modify or create nanomaterials using physical, chemical, or biological methods, specifically for various businesses and sectors including agriculture. Due to the large-scale production and widespread release of these ENMs into several habitats, including soil, whether deliberate or inadvertent, no risk assessments have been conducted for them. Recent research has revealed several types of nanomaterials that are being released in significant amounts into the soil. These include silver, SiO_2, titania, iron, iron oxides, zinc oxide, alumina, and others. As a result of the unregulated and ongoing discharge of these significant amounts of ENMs into the environment, they have built up in the soil. Research is being conducted not only on the accumulation of these compounds, but also on the annual rate of fluctuation in their concentrations per kilogram of soil samples. The variation in concentration is contingent upon the stability of ENMs in the soil and the manner in which they are released. Upon entering the soil environment, these can undergo various transformations and modifications inside this medium. These ENMs have an impact on the number and activity of all microbial populations in the soil and plants, which in turn affects the health of the plants. The efficiency of ENMs is influenced by several factors, including the dimensions and shapes of the particles, their chemical composition, surface charges, and hydrophobicity. The relationship between nanomaterials and stabilized soil ecology is variable due to differences in soil composition, microbiome content, and plant microbiomes across diverse plant/soil systems.

Several research studies have evaluated the impacts of ENMs on various soil species, including bacteria and plants. Microorganisms play a crucial role in the movement of nutrients within the ecosystem. They are responsible for converting nutrients into forms that may be easily used by plants, hence influencing soil fertility and plant health. Nevertheless, it is worth mentioning that several of these tests used excessively high quantities of ENMs, although the concentrations present in the soil are significantly lower. Significantly, the research revealed that the concentrations that were relatively lower exhibited a nearly inactive response. Previous study has shown that low doses of ENMs can have positive and growth-promoting effects. Hence, it is essential to examine the toxicity or growth-promoting effects of these active compounds and take into consideration the actual dosages used.

Therefore, it is imperative to regulate the emission of ENMs across the entire lifecycle, including production, usage, and disposal stages, in a consistent manner. Furthermore, there has been a lack of thorough investigation and documentation regarding the effects of ENMs on field development. Additionally, more studies are required to identify potential risks to the microbiomes. For example, it has been discovered that Ag possesses the ability to enhance plant development. However, when Ag is present in plants and their consumable products, it can have adverse effects on human health. This is because, when the frequencies persist and intensify over time, the presence of Ag can contaminate the soil beyond a specific threshold that is considered acceptable for feeding, hence resulting in negative consequences for the well-being of organisms. Hence, a thorough assessment of all components and their corresponding hazards is needed in order to ensure the responsible and secure implementation of nanotechnology in the agricultural sector. Biocompatible ENMs are highly suitable for this purpose due to their ability to undergo biodegradation, resulting in a shorter half-life while present in the soil. Furthermore, the utilization of these materials has minimal impact on soil microbiomes and may be recycled. Therefore, the use of ENMs in soil can be beneficial, provided that their concentration is kept at a low level and they are regularly monitored and tailored to specific requirements. The unregulated release of ENMs into the environment increases their levels in the soil, posing a significant risk to food production and agriculture. Furthermore, this practice has detrimental effects on human health and continuously degrades the soil, which is a vital and irreplaceable resource.

11. Acknowledgement

Authors want to thank the university of Agriculture Faisalabad and Yangzhou University for the support in this publication. Authors also want to thank Professor Dr Abdul Wakeel from Soil Fertility and Plant Nutrition Laboratory, University of Agriculture Faisalabad, Pakistan.

References

Abd-Elsalam, K.A. and K.T. Lim. 2021. CRISPR and RNAi Systems: Nanobiotechnology approaches to plant breeding and protection. Elsevier, Amsterdam.

Acosta-Motos, J. R. and M. PedroDiaz-Vivancos, 2018. Effect of Biostimulants on Plant Responses to Salt Stress.

Ahmad, P., C.A. Jaleel, M.A. Salem, G. Nabi and S. Sharma. 2010. Roles of enzymatic and nonenzymatic antioxidants in plants during abiotic stress. Crit. Rev. Biotechnol. 30(3): 161–175.

Ahmad, R. and A. Mirza. 2018. Facile one pot green synthesis of Chitosan-Iron oxide (CS-Fe2O3) nanocomposite: Removal of Pb (II) and Cd (II) from synthetic and industrial wastewater. Journal of Cleaner Production 186: 342–352.

Ahmed, B., A. Rizvi, K. Ali, J. Lee, A. Zaidi, M.S. Khan et al. 2021. Nanoparticles in the soil–plant system: a review. Environ. Chem. Lett. 19(2): 541–560. https://doi.org/10.1007/s10311-020-01138-y.

Albalawi, F., M.Z. Hussein, S. Fakurazi and M.J. Masarudin. 2021. Engineered nanomaterials: The challenges and opportunities for nanomedicines. Int. J. Nanomedicine. 16: 161–184.

Allen, C.S., M.D. Elkin, G. Burnell, B.J. Hickey, C. Zhang, S. Hofmann et al. 2011. Transport measurements on carbon nanotubes structurally characterized by electron diffraction. Phys. Rev. B Condens. Matter Mater. Phys. 84(11): 115444.

Arole, V.M. and S.V. Munde. 2014. Fabrication of nanomaterials by top-down and bottom-up approaches—an overview. J. Mater. Sci. 1: 89–93.

Awwad, E., M.I. Mohamed, A. Abd El-Hameedb and E. Zaghloul. 2022. The co-addition of soil organic amendments and natural bio-stimulants improves the production and defenses of the wheat plant grown under the dual stress of salinity and alkalinity. Egypt. J. Soil Sci. 62(2). https://doi.org/10.21608/ejss.2022.148406.1513.

Backer, R., J.S. Rokem, G. Ilangumaran, J. Lamont, D. Praslickova, E. Ricci et al. 2018. Plant growth-promoting rhizobacteria: Context, mechanisms of action, and roadmap to commercialization of biostimulants for sustainable agriculture. Front. Plant Sci. 9, 1473.

Bargaz, A., K. Lyamlouli, M. Chtouki, Y. Zeroual and D. Dhiba. 2018. Soil microbial resources for improving fertilizers efficiency in an integrated plant nutrient management system. Front. Microbiol. 9.

Baričić, M., J.M. Nuñez, M.H. Aguirre, D. Hrabovsky, M. Seydou, C. Meneghini et al. 2024. Advancements in polyol synthesis: Expanding chemical horizons and Neel temperature tuning of CoO nanoparticles. Sci. Rep. 14(1): 12529.

Bashan, Y., L.E. de-Bashan, S.R. Prabhu and J.P. Hernandez. 2014. Advances in plant growth-promoting bacterial inoculant technology: Formulations and practical perspectives (1998–2013). Plant Soil. 378(1–2): 1–33.

Berazneva, J. and D. Güereña. 2019. Soil management for smallholders: Lessons from Kenya and Nepal. Choices. The Mag. Food Farm Resour. Issues. 34(2).

Bogardi, J.J., D. Dudgeon, R. Lawford, E. Flinkerbusch, A. Meyn, C. Pahl-Wostl et al. 2012. Water security for a planet under pressure: Interconnected challenges of a changing world call for sustainable solutions. Curr. Opin. Environ. Sustain. 4(1): 35–43.

Borgatta, J., J.C. White and N. Youssef. 2023. Food chain transfer of nanomaterials in agriculture. pp. 493–502. *In*: M.R.S.D. [eds.]. Nano-Enabled Sustainable and Precision Agriculture. Elsevier, Amsterdam.

Boutasknit, A., M. Baslam, M. Ait-El-Mokhtar, M. Anli, R. Ben-Laouane, Y. Ait-Rahou et al. 2021. Assemblage of indigenous arbuscular mycorrhizal fungi and green waste compost enhance drought stress tolerance in carob (*Ceratonia siliqua* L.) trees. Sci. Rep. 11(1): 22835.

Butt, N.M. (n.d.). Nanotechnology and explosion in nano-medicine development.

Cabini, E., L. Fontana, P. Malavasi and I. Iavicoli. 2018. Land use: The perception of risk by the citizens and local administrators in the North of Italy. Land Use Policy. 76. https://doi.org/10.1016/j.landusepol.2018.02.036.

Chakraborty, R., A. Mukhopadhyay, S. Paul, S. Sarkar and R. Mukhopadhyay. 2023. Nanocomposite-based smart fertilizers: A boon to agricultural and environmental sustainability. Sci. Total Environ. 863. https://doi.org/10.1016/j.scitotenv.2022.160859.

Chaudhari, R., V. Patel and A. Kumar. 2024. Cutting-edge approaches for targeted drug delivery in breast cancer: Beyond conventional therapies. Nanoscale Adv.

Chaudhary, R., K. Nawaz, A.K. Khan, C. Hano, B.H. Abbasi and S. Anjum. 2020. An overview of the algae-mediated biosynthesis of nanoparticles and their biomedical applications. Biomolecules. 10(11): 1498.

Chen, Y., X. Zhang and W. Liu. 2022. Effect of metal and metal oxide engineered nanoparticles on nitrogen bio-conversion and its mechanism: A review. Chemosphere. 287: 132097.

Cockburn, A., R. Bradford, N. Buck, A. Constable, G. Edwards, B. Haber et al. 2012. Approaches to the safety assessment of engineered nanomaterials (ENM) in food. Food Chem. Toxicol. 50(6): 2224–2242.

Condron, L., C. Stark, M. O'Callaghan, P. Clinton and Z. Huang. 2010. The role of microbial communities in the formation and decomposition of soil organic matter. Soil Microbiol. Sustainable Crop Prod. 81–118.

Costello, A., M. Abbas, A. Allen, S. Ball, S. Bell, R. Bellamy et al. 2009. Managing the health effects of climate change. Lancet and University College London Institute for Global Health Commission. Lancet. 373(9676). https://doi.org/10.1016/S0140-6736(09)60935-1.

Coutris, C., E.J. Joner and D.H. Oughton. 2012. Aging and soil organic matter content affect the fate of silver nanoparticles in soil. Sci. Total Environ. 420: 327–333.

Das, P., S. Barua, S. Sarkar, N. Karak, P. Bhattacharyya, N. Raza et al. 2018. Plant extract–mediated green silver nanoparticles: Efficacy as soil conditioner and plant growth promoter. J. Hazard. Mater. 346: 62–72.

Das, P.P., K.R. Singh, G. Nagpure, A. Mansoori, R.P. Singh, I.A. Ghazi et al. 2022. Plant-soil-microbes: A tripartite interaction for nutrient acquisition and better plant growth for sustainable agricultural practices. Environ. Res. 214. https://doi.org/10.1016/j.envres.2022.113821.

Das, R.K., V.L. Pachapur, L. Lonappan, M. Naghdi, R. Pulicharla, S. Maiti, et al. 2017. Biological synthesis of metallic nanoparticles: plants, animals and microbial aspects. In Nanotechnology for Environmental Engineering. 2(1). https://doi.org/10.1007/s41204-017-0029-4.

De Corato, U. 2021. Effect of value-added organic co-products from four industrial chains on functioning of plant disease suppressive soil and their potentiality to enhance soil quality: A review from the perspective of a circular economy. Appl. Soil Ecol. 168. https://doi.org/10.1016/j.apsoil.2021.104221.

de Jesus, R.A., G.C. de Assis, R.J. de Oliveira, J.A.S. Costa, C.M.P. da Silva, H.M.N. Iqbal et al. 2024. Metal/metal oxide nanoparticles: A revolution in the biosynthesis and medical applications. Nano-Structures Nano-Objects. 37. https://doi.org/10.1016/j.nanoso.2023.101071.

Deb, S.K. and M.K. Shukla. 2011. A review of dissolved organic matter transport processes affecting soil and environmental quality. J. Environ. Anal. Toxicol. 1(02).

Dibyanshu, K., C. Chhaya and T. Raychoudhury. 2023. A review on the fate and transport behavior of engineered nanoparticles: possibility of becoming an emerging contaminant in the groundwater. Int. J. Environ. Sci. Technol. 20(4). https://doi.org/10.1007/s13762-021-03835-9.

El Hadri, H., S.M. Louie and V.A. Hackley. 2018. Assessing the interactions of metal nanoparticles in soil and sediment matrices–a quantitative analytical multi-technique approach. Environ. Sci. Nano. 5(1): 203–214.

Fajardo, C., M. Gil-Díaz, G. Costa, J. Alonso, A.M. Guerrero, M. Nande et al. 2015. Residual impact of aged nZVI on heavy metal-polluted soils. Sci. Total Environ. 535: 79–84.

Ganie, A.S., S. Bano, N. Khan, S. Sultana, Z. Rehman, M.M. Rahman et al. 2021. Nanoremediation technologies for sustainable remediation of contaminated environments: Recent advances and challenges. Chemosphere. 275. https://doi.org/10.1016/j.chemosphere.2021.130065.

Garg, S., N.P. Rumjit and S. Roy. 2023. Smart agriculture and nanotechnology: Technology, challenges, and new perspective. Adv. Agrochem.

Gavrilescu, M. 2021. Water, soil, and plants interactions in a threatened environment. Water. 13(19): 2746.

Gelaw, T.A. and N. Sanan-Mishra. 2022. Nanomaterials coupled with microRNAs for alleviating plant stress: a new opening towards sustainable agriculture. Physiol. Mol. Biol. Plants. 28(4). https://doi.org/10.1007/s12298-022-01163-x.

Giller, K.E., R. Hijbeek, J.A. Andersson and J. Sumberg. 2021. Regenerative Agriculture: An agronomic perspective. Outlook Agric. 50(1). https://doi.org/10.1177/0030727021998063.

Gitsov, I., A. Simonyan, L. Wang, A. Krastanov, S.W. Tanenbaum and D. Kiemle. 2012. Polymer-assisted biocatalysis: unprecedented enzymatic oxidation of fullerene in aqueous medium. J. Polym. Sci. Part A: Polym. Chem. 50(1): 119–126.

Grasso, G., D. Zane and R. Dragone. 2020. Microbial nanotechnology: Challenges and prospects for green biocatalytic synthesis of nanoscale materials for sensoristic and biomedical applications. Nanomaterials. 10(1). https://doi.org/10.3390/nano10010011.

Hernandez-Ruiz, J., M.B. Arnao, A.N.P. Hiner, F. García-Cánovas and M. Acosta. 2001. Catalase-like activity of horseradish peroxidase: relationship to enzyme inactivation by H2O2. Biochem. J. 354(1): 107–114.

Hurni, H., M. Giger, H. Liniger, R.M. Studer, P. Messerli, B. Portner et al. 2015. Soils, agriculture and food security: the interplay between ecosystem functioning and human well-being. Curr. Opin. Environ. Sustain. 15: 25–34.

Hurtado, A., A.A.A. Aljabali, V. Mishra, M.M. Tambuwala and Á. Serrano-Aroca. 2022. Alginate: Enhancement Strategies for Advanced Applications. Int. J. Mol. Sci. 23(9). https://doi.org/10.3390/ijms23094486.

Husson, O. 2013. Redox potential (Eh) and pH as drivers of soil/plant/microorganism systems: a transdisciplinary overview pointing to integrative opportunities for agronomy. Plant Soil. 362: 389–417.

Ijaz, I., E. Gilani, A. Nazir and A. Bukhari. 2020. Detail review on chemical, physical and green synthesis, classification, characterizations and applications of nanoparticles. Green Chem. Lett. Rev. 13(3). https://doi.org/10.1080/17518253.2020.1802517.

Jacoby, R., M. Peukert, A. Succurro, A. Koprivova and S. Kopriva. 2017. The role of soil microorganisms in plant mineral nutrition—current knowledge and future directions. Front. Plant Sci. 8. https://doi.org/10.3389/fpls.2017.01617.

Jeevanandam, J., A. Barhoum, Y.S. Chan, A. Dufresne and M.K. Danquah. 2018. Review on nanoparticles and nanostructured materials: history, sources, toxicity and regulations. Beilstein J. Nanotechnol. 9(1): 1050–1074.

Kalia, A., S.P. Sharma and H. Kaur. 2019. Nanoscale fertilizers: harnessing boons for enhanced nutrient use efficiency and crop productivity. In: Nanobiotechnology Applications in Plant Protection. 2: 191–208.

Kappi, F.A., G.Z. Tsogas, A.-M. Routsi, D.C. Christodouleas and D.L. Giokas. 2018. Based devices for biothiols sensing using the photochemical reduction of silver halides. Anal. Chim. Acta. 1036: 89–96.

Karaca, A., S.C. Cetin, O.C. Turgay and R. Kizilkaya. 2010. Effects of heavy metals on soil enzyme activities. In: Soil Heavy Metals. pp. 237–262.

Khan, A., A. V. Singh, S. S. Gautam, A. Agarwal, A. Punetha, V. K. Upadhayay et al. 2023. Microbial bioformulation: a microbial assisted biostimulating fertilization technique for sustainable agriculture.

Khan, S.T., S.F. Adil, M.R. Shaik, H.Z. Alkhathlan, M. Khan and M. Khan. 2022. Engineered nanomaterials in soil: Their impact on soil microbiome and plant health. Plants. 11(1). https://doi.org/10.3390/plants11010109.

Köser, J., M. Engelke, M. Hoppe, A. Nogowski, J. Filser and J. Thöming. 2017. Predictability of silver nanoparticle speciation and toxicity in ecotoxicological media. Environ. Sci. Nano. 4(7): 1470–1483.

Kucharski, J., E. Boros and J. Wyszkowska. 2009. Biochemical activity of nickel-contaminated soil. Pol. J. Environ. Stud. 18(6).

Kumar, L., V. Ragunathan, M. Chugh and N. Bharadvaja. 2021. Nanomaterials for remediation of contaminants: a review. Environ. Chem. Lett. 19(4). https://doi.org/10.1007/s10311-021-01212-z.

Kumawat, K.C., N. Razdan and K. Saharan. 2022. Rhizospheric microbiome: Bio-based emerging strategies for sustainable agriculture development and future perspectives. Microbiol. Res. 254: 126901.

Lalwani, G., W. Xing and B. Sitharaman. 2014. Enzymatic degradation of oxidized and reduced graphene nanoribbons by lignin peroxidase. J. Mater. Chem. B 2(37): 6354–6362.

Lead, J.R., G.E. Batley, P.J.J. Alvarez, M. Croteau, R.D. Handy, M.J. McLaughlin et al. 2018. Nanomaterials in the environment: behavior, fate, bioavailability, and effects—an updated review. Environ. Toxicol. Chem. 37(8): 2029–2063.

Lee, S.K., H.S. Lur, K.J. Lo, K.C. Cheng, C.C. Chuang, S.J. Tang et al. 2016. Evaluation of the effects of different liquid inoculant formulations on the survival and plant-growth-promoting efficiency of *Rhodopseudomonas palustris* strain PS3. Appl. Microbiol. Biotechnol. 100(18). https://doi.org/10.1007/s00253-016-7582-9.

Lee, W.H., C.Y. Loo, P.M. Young, D. Traini, R.S. Mason and R. Rohanizadeh. 2014. Recent advances in curcumin nanoformulation for cancer therapy. Expert Opin. Drug Deliv. 11(8): 1183–1201.

Li, J., X. Cai, P. Jiang, H. Wang, S. Zhang, T. Sun et al. 2024. Co-based Nanozymatic Profiling: Advances Spanning Chemistry, Biomedical, and Environmental Sciences. Adv. Mater. 36(8): 2307337.

Li, X. and W. Zhang. 2007. Sequestration of metal cations with zerovalent iron nanoparticles: a study with high resolution X-ray photoelectron spectroscopy (HR-XPS). J. Phys. Chem. C.111(19): 6939–6946.

Lin, J., M. Sun, B. Su, G. Owens and Z. Chen. 2019. Immobilization of cadmium in polluted soils by phytogenic iron oxide nanoparticles. Sci. Total Environ. 659: 491–498.

Liu, H., X. Zhong, Q. Pan, Y. Zhang, W. Deng, G. Zou et al. 2024. A review of carbon dots in synthesis strategy. Coord. Chem. Rev. 498: 215468.

Lobo, C.B., M.S. Juárez Tomás, E. Viruel, M.A. Ferrero and M.E. Lucca. 2019. Development of low-cost formulations of plant growth-promoting bacteria to be used as inoculants in beneficial agricultural technologies. Microbiol. Res. 219. https://doi.org/10.1016/j.micres.2018.10.012.

Lopes, M. J. d. S., M. B. Dias-Filho and E. S. C. Gurgel. 2021. Successful Plant Growth-Promoting Microbes: Inoculation Methods and Abiotic Factors.

Louie, S.M., R.D. Tilton and G.V. Lowry. 2016. Critical review: impacts of macromolecular coatings on critical physicochemical processes controlling environmental fate of nanomaterials. Environ. Sci. Nano. 3(2): 283–310.

McKee, M.S. and J. Filser. 2016. Impacts of metal-based engineered nanomaterials on soil communities. Environ. Sci. Nano. 3(3): 506–533.

Mishra, A., R. Panola and A.C. Rana. 2014. Microemulsions: As drug delivery system. J. Sci. Innov. Res. 3(4): 467–474.

Moreira-Grez, B., M. Muñoz-Rojas, K. Kariman, P. Storer, A.G. O'Donnell, D. Kumaresan et al. 2019. Reconditioning degraded mine site soils with exogenous soil microbes: Plant fitness and soil microbiome outcomes. Front. Microbiol. 10(July). https://doi.org/10.3389/fmicb.2019.01617.

Muhammadi and N. Ahmed. 2007. Genetics of bacterial alginate: Alginate genes distribution, organization and biosynthesis in bacteria. Curr. Genomics. 8(3). https://doi.org/10.2174/138920207780833810.

Munir, N., M. Hanif, D.A. Dias and Z. Abideen. 2021. The role of halophytic nanoparticles towards the remediation of degraded and saline agricultural lands. Environ. Sci. Pollut. Res. 28(43). https://doi.org/10.1007/s11356-021-16139-9.

Nadarajah, K. and N.S.N. Abdul Rahman. 2021. Plant–microbe interaction: Aboveground to belowground, from the good to the bad. Int. J. Mol. Sci. 22(19). https://doi.org/10.3390/ijms221910388.

Nowack, B., M. Baalousha, N. Bornhöft, Q. Chaudhry, G. Cornelis, J. Cotterill et al. 2015. Progress towards the validation of modeled environmental concentrations of engineered nanomaterials by analytical measurements. Environ. Sci. Nano. 2(5): 421–428.

Onofre-Lemus, J., I. Hernández-Lucas, L. Girard and J. Caballero-Mellado. 2009. ACC (1-aminocyclopropane-1-carboxylate) deaminase activity, a widespread trait in Burkholderia species, and its growth-promoting effect on tomato plants. Applied and Environmental Microbiology 75.

Otlewska, A., M. Migliore, K. Dybka-Stępień, A. Manfredini, K. Struszczyk-Świta, R. Napoli et al. 2020. When salt meddles between plant, soil, and microorganisms. Front. Plant Sci. 11. https://doi.org/10.3389/fpls.2020.553087.

Parashar, M., V.K. Shukla and R. Singh. 2020. Metal oxides nanoparticles via sol–gel method: A review on synthesis, characterization and applications. J. Mater. Sci. Mater. Electron. 31(5): 3729–3749.

Park, C.M., J. Heo, N. Her, K.H. Chu, M. Jang and Y. Yoon. 2016. Modeling the effects of surfactant, hardness, and natural organic matter on deposition and mobility of silver nanoparticles in saturated porous media. Water Res. 103: 38–47.

Peng, C., H. Tong, C. Shen, L. Sun, P. Yuan, M. He et al. 2020. Bioavailability and translocation of metal oxide nanoparticles in the soil-rice plant system. Sci. Total Environ. 713: 136662.

Perea Vélez, Y.S., R. Carrillo-González and M.d.C.A. González-Chávez. 2021. Interaction of metal nanoparticles–plants–microorganisms in agriculture and soil remediation. J. Nanopart. Res. 23(9). https://doi.org/10.1007/s11051-021-05269-3.

Pereira, A.d.S., C.P.L. Souza, L. Moraes, G.C. Fontes-Sant'ana and P.F.F. Amaral. 2021. Polymers as encapsulating agents and delivery vehicles of enzymes. Polymers. 13(23). https://doi.org/10.3390/polym13234061.

Qin, D. and B.A. Riggs. 2013. Nanotechnology: A top–down approach. pp. 1–9. *In*: Murthy, D.S.S.R. and R.S.S. Murthy [eds.]. Encyclopedia of Supramolecular Chemistry-Two-Volume Set (Print). CRC Press.

Rabbani, M., M.E. Hoque and Z. Bin Mahbub. 2020. Nanosensors in biomedical and environmental applications: Perspectives and prospects. *In*: Nanofabrication for Smart Nanosensor Applications. https://doi.org/10.1016/B978-0-12-820702-4.00007-6.

Rahman, M.M., J. Uddin, A.M. Asiri and M.R. Rahman. 2024. Toxicity of nanoparticles: Recent advances and new perspectives.

Rajput, V.D., T. Minkina, S. Sushkova, V. Tsitsuashvili, S. Mandzhieva, A. Gorovtsov et al. 2018. Effect of nanoparticles on crops and soil microbial communities. J. Soils Sediments. 18: 2179–2187.

Rajput, V.D., T. Minkina, S.K. Upadhyay, A. Kumari, A. Ranjan, S. Mandzhieva et al. 2022. Nanotechnology in the restoration of polluted soil. Nanomaterials. 12(5). https://doi.org/10.3390/nano12050769.

Reches, M. and E. Gazit. 2006. Molecular self-assembly of peptide nanostructures: Mechanism of association and potential uses. Curr. Nanosci. 2(2). https://doi.org/10.2174/157341306776875802.

Řezanka, T., A. Palyzová, H. Faltýsková and K. Sigler. 2018: Siderophores: Amazing Metabolites of Microorganisms.

Rodríguez-Carmona, E. and A. Villaverde. 2010. Nanostructured bacterial materials for innovative medicines. Trends Microbiol. 18(9). https://doi.org/10.1016/j.tim.2010.06.007.

Safari, M., E. Motamedi, H.K. Dolatabad and S.A.M.M. Sanavy. 2020. Nano-carriers effects on the viability and efficiency of Pseudomonas strains as phosphate solubilizing bacteria. Heliyon. 6(10).

Sampathkumar, K., K.X. Tan and S.C.J. Loo. 2020. Developing nano-delivery systems for agriculture and food applications with nature-derived polymers. iScience. 23(5). https://doi.org/10.1016/j.isci.2020.101055.

Schreefel, L., R.P.O. Schulte, I.J.M. de Boer, A.P. Schrijver and H.H.E. van Zanten. 2020. Regenerative agriculture – the soil is the base. Global Food Security. 26. https://doi.org/10.1016/j.gfs.2020.100404.

Scotti, R., G. Bonanomi, R. Scelza, A. Zoina and M.A. Rao. 2015. Organic amendments as sustainable tools to recover fertility in intensive agricultural systems. J. Soil Sci. Plant Nutr. 15(2): 333–352.

Shafi, U., R. Mumtaz, J. García-Nieto, S.A. Hassan, S.A.R. Zaidi and N. Iqbal. 2019. Precision agriculture techniques and practices: From considerations to applications. Sensors. 19(17): 3796.

Shandilya, M., R. Rai and J. Singh. 2016. Hydrothermal technology for smart materials. Adv. Appl. Ceram. 115(6): 354–376.

Sharma, S., S. Singh, A. Bahuguna, B. Yadav, A. Barthwal, R.N.S. Khatana et al. 2022. Nanotechnology: An efficient tool in plant nutrition management. In Ecosystem Services: Types, Management and Benefits.

Shi, J., J. Ye, H. Fang, S. Zhang and C. Xu. 2018. Effects of copper oxide nanoparticles on paddy soil properties and components. Nanomaterials. 8(10): 839.

Silina, Y.E., K.V. Gernaey, D. Semenova and I. Iatsunskyi. 2020. Application of organic-inorganic hybrids in chemical analysis, bio- and environmental monitoring. Appl. Sci. (Switzerland) 10(4). https://doi.org/10.3390/app10041458.

Stephen Inbaraj, B. and B.H. Chen. 2016. Nanomaterial-based sensors for detection of foodborne bacterial pathogens and toxins as well as pork adulteration in meat products. J. Food Drug Anal. 24(1). https://doi.org/10.1016/j.jfda.2015.05.001.

Sun, J., H. Peng, J. Chen, X. Wang, M. Wei, W. Li et al. 2016. An estimation of CO2 emission via agricultural crop residue open field burning in China from 1996 to 2013. J. Clean. Prod. 112. https://doi.org/10.1016/j.jclepro.2015.09.112.

Tahat, M., K. Alananbeh, Y. Othman and D. Leskovar. 2020. Soil health and sustainable agriculture. Sustainability. 12(12): 4859.

Tao, S., Z. Wu, X. He, B.C. Ye and C. Li. 2018. Characterization of biochar prepared from cotton stalks as efficient inoculum carriers for Bacillus subtilis SL-13. BioResources. 13(1): 1773–1786. https://doi.org/10.15376/biores.13.1.1773–1786.

Waalewijn-Kool, P.L., S. Rupp, S. Lofts, C. Svendsen and C.A.M. van Gestel. 2014. Effect of soil organic matter content and pH on the toxicity of ZnO nanoparticles to Folsomia candida. Ecotoxicol. Environ. Saf. 108: 9–15.

Wang, C. and H. Jiang. 2019. Real-time monitoring of sediment bulking through a multi-anode sediment microbial fuel cell as reliable biosensor. Sci. Total Environ. 697: 134009.

Xiao, Z., N. Fan, L. Yue, F. Chen, H. Ji, Y. Shu et al. 2021. Dose-dependent effects of CeO2 nanomaterials on tomato plant chemistry and insect herbivore resistance. Environ. Sci. Nano. 8(12): 3577–3589. https://doi.org/10.1039/d1en00837d.

Xin, X., Z. He, M.R. Hill, R.P. Niedz, X. Jiang and B.S. Sumerlin. 2018. Efficiency of biodegradable and pH-responsive polysuccinimide nanoparticles (PSI-NPs) as smart nanodelivery systems in grapefruit: *In vitro* cellular investigation. Macromol. Biosci. 18(7). https://doi.org/10.1002/mabi.201800159.

Xue, Y., H. Kang, Y. Cui, S. Lu, H. Yang, J. Zhu et al. 2022. Consistent plant and microbe nutrient limitation patterns during natural vegetation restoration. Front. Plant Sci. 13. https://doi.org/10.3389/fpls.2022.885984.

Yabur, R., Y. Bashan and G. Hernández-Carmona. 2007. Alginate from the macroalgae Sargassum sinicola as a novel source for microbial immobilization material in wastewater treatment and plant growth promotion. J. Appl. Phycol. 19(1). https://doi.org/10.1007/s10811-006-9109-8.

Yadav, R.H. 2021. Soil-plant-microbial interactions for soil fertility management and sustainable agriculture. In: Microbes in Land Use Change Management. pp. 341–362. Elsevier.

Yousefpoor, Y., H. Baharifar, S.S. Esnaashari, F. Gheybi, M. Mehrabi, M. Osanloo et al. 2024. A review of topical micro-and nanoemulsions for common skin diseases. Nanomed. J. 11(3).

Zhang, H., M. Huang, W. Zhang, J.L. Gardea-Torresdey, J.C. White, R. Ji et al. 2020. Silver nanoparticles alter soil microbial community compositions and metabolite profiles in unplanted and cucumber-planted soils. Environ. Sci. Technol. 54(6): 3334–3342.

Zhang, J., F. Li, D. Liu, Q. Liu and H. Song. 2023. Engineering extracellular electron transfer pathways of electroactive microorganisms by synthetic biology for energy and chemicals production. Chem. Soc. Rev. 53(3). https://doi.org/10.1039/d3cs00537b.

Zhang, Q., J. Ma, A. Gonzalez-Ollauri, Y. Yang and F. Chen. 2023. Soil microbes-mediated enzymes promoted the secondary succession in post-mining plantations on the Loess Plateau, China. Soil Ecol. Lett. 5(1). https://doi.org/10.1007/s42832-022-0148-0.

Zhang, T., Y. Lu and G. Luo. 2014. Synthesis of hierarchical iron hydrogen phosphate crystal as a robust peroxidase mimic for stable H2O2 detection. ACS Appl. Mater. Interfaces. 6(16): 14433–14438.

Zhang, W., J. Long, J. Li, M. Zhang, X. Ye, W. Chang et al. 2020. Effect of metal oxide nanoparticles on the chemical speciation of heavy metals and micronutrient bioavailability in paddy soil. Int. J. Environ. Res. Public Health. 17(7): 2482.

Zhang, J., Li, F., Liu, D., Liu, Q., and Song, H. (2023). Engineering extracellular electron transfer pathways of electroactive microorganisms by synthetic biology for energy and chemicals production. In Chemical Society Reviews (Vol. 53, Issue 3). https://doi.org/10.1039/d3cs00537b.

Zhang, Q., Ma, J., Gonzalez-Ollauri, A., Yang, Y. and Chen, F. (2023). Soil microbes-mediated enzymes promoted the secondary succession in post-mining plantations on the Loess Plateau, China. Soil Ecology Letters, 5(1). https://doi.org/10.1007/s42832-022-0148-0.

Zhou, Y., Y. Fang and R.P. Ramasamy. 2019. Non-covalent functionalization of carbon nanotubes for electrochemical biosensor development. Sensors. 19(2): 392.

Zulfiqar, F., M. Navarro, M. Ashraf, N.A. Akram and S. Munné-Bosch. 2019. Nanofertilizer use for sustainable agriculture: Advantages and limitations. Plant Sci. 289: 110270.

CHAPTER-7

Nanoparticles and Plant-Soil Interactions

Waleed Asghar,[1,*] *Mehran Yasir,*[2] *Kelly D. Craven,*[1] *Ahmad Mahmood,*[3] *Ryota Kataoka*[4] and *Md. Tareq Bin Salam*[5]

1. Introduction

Agricultural systems have recently used soil amendments to boost plant growth and development, especially by adding organic and inorganic nutrients. Introducing particular compounds into the soil improves its ability to support plant growth (Asghar et al. 2024, Asghar and Kataoka 2024, Rubin et al. 2023). Multiple research investigations have shown that soil organic amendments offer various advantages, such as improved soil texture, restored soil health, increased soil fertility, and ultimately, enhanced crop yield (Khan et al. 2024). Organic-based fertilizers, on the other hand, have become a viable option to address nutrient deficits and ensure proper nutrient equilibrium in agroecosystems (Asghar et al. 2022b, Qaswar et al. 2019). However, overuse and improper use of organic fertilizers can have negative ecological consequences. Plant growth-promoting microorganisms (PGPM) and nanomaterials (NMs) or nanoparticles (NPs) are being considered as a new method to enhance crop yield and soil fertility (Kumari et al. 2023, Rajput et al. 2023). As a result, it is increasingly important to investigate novel solutions that improve soil fertility using organic and natural approaches while also promoting sustainable soil and plant health.

The presence of antibiotic-resistant microbes, antibiotic residues, and antibiotic-resistant genes in agricultural organic amendments poses significant risks to human health (Upadhayay et al. 2023). Consequently, these contaminants are considered major emerging pollutants in agroecosystems (Garbowski et al. 2023, Upadhayay et al. 2023). Soil supplements should include characteristics that ensure ecological protection and do not have adverse effects on soil fertility, soil composition, or the natural ecosystem. Due to their distinctive characteristics, PGPMs and NPs have gained significant attention in recent years as prospective soil supplements that can reduce the potential hazards associated with other soil additions in both normal and unfavorable situations (Kumari et al. 2023, Rajput et al. 2023).

[1] Department of Biochemistry and Molecular Biology, Oklahoma State University, Stillwater, OK, USA.
[2] Department of Physics, Oklahoma State University, Stillwater, OK, USA.
[3] Departments of Climate Change, and Soil and Environmental Sciences, MNS-University of Agriculture, Multan, Pakistan.
[4] Department of Environmental Sciences, Faculty of Life and Environmental Sciences, University of Yamanashi, Japan.
[5] School of Interdisciplinary Engineering and Sciences (SINES), National University of Sciences and Technology (NUST), Islamabad, Pakistan.
* Corresponding author: Waleedasghar978@gmail.com

DOI: 10.1201/9781003558354-7

A nanoparticle, the fundamental unit of nanotechnology, is a particle with a diameter smaller than 100 nm. Nanoparticles exhibit physicochemical features that vary with their size, which typically differ from those of larger or smaller particles in the micron or sub-micron range (Dasgupta et al. 2017). One of the most distinctive characteristics of nanoparticles is their relatively high surface area-to-volume ratio, which gives them a high level of reactivity and physicochemical dynamicity (Mauter et al. 2018). Nanoparticles are classified into various groups based on their structure and chemical compositions. These groups include zero-valent metals, nanopolymers, metal oxides, lipids, quantum dots, dendrimers, semiconductors, and carbonaceous materials (Sudha et al. 2018). However, multiple publications and meta-analyses have found that NPs have positive effects on soil health and agronomic productivity (Urra et al. 2019, Verma et al. 2022). They have also identified the factors that contribute to the NPs' ability to improve these conditions. During plant growth, the NPs have also mitigated a variety of environmental pressures (Dilnawaz et al. 2023, Pramanik et al. 2023). They find applications in several sectors, ranging from industry to agriculture, because of their numerous qualities.

Plant growth-promoting microorganisms (PGPMs), especially plant growth-promoting rhizobacteria (PGPRs), are useful for plant and soil health (Salam et al. 2024). Researchers have documented and verified that the majority of PGPRs enhance plant productivity by mitigating environmental stresses (Asghar et al. 2020). By employing both direct and indirect tactics, they can improve overall soil quality and increase plant productivity (Asghar et al. 2020). Direct mechanisms, such as nitrogen fixation, phosphate and potassium solubilization, the synthesis of growth-promoting phytohormones like indole-acetic acid (IAA), and the production of siderophores, carry out PGPR functions (Mahmood and Kataoka 2020). However, indirect pathways connect to the synthesis of lytic enzymes and antibiotics, leading to a reduction in soil pH and the formation of exopolysaccharides (Mahmood et al. 2022). Multiple studies have evaluated the effectiveness of PGPR in preserving a sustainable agroecosystem under both normal and challenging circumstances (Magnabosco et al. 2023, Timofeeva et al. 2023). Several studies currently use PGPRs and NPs to enhance agronomic production and soil quality in plant-soil systems. The underlying premise in these investigations has been that the NPs would enhance the availability of nutrients and create a favorable environment for the development of PGPRs. The purpose of the book chapter is to fill a knowledge gap in understanding nanoparticles and their interactions with plant-soil systems. Furthermore, this text underscores the forthcoming research paths aimed at enhancing the combined benefits of Plant Growth-Promoting Rhizobacteria (PGPR) and Nanoparticles (NPs) in sustainable plant-soil systems.

2. Synergistic Impacts of NPs and PGPRs on Plants-Soil Ecosystems

Nanoparticles (NPs) have the ability to enhance both the process of seed germination and the subsequent lateral development stages (Aslani et al. 2014). The use of SiO_2 NPs and TiO_2 NPs improved the functionality of nitrate reductase, thereby promoting seed germination (Aslani et al. 2014). Furthermore, the combined application of these nanoparticles yielded even more significant results (Changmei et al. 2002). Adding titanium dioxide nanoparticles (TiO_2 NPs) to spinach seeds made them better at taking in nitrogen and making food through photosynthesis, which led to faster plant growth. Another study found that the use of low concentrations of SiO_2 NPs improved the germination of tomato seeds (Siddiqui and Al-Whaibi 2014). Zinc oxide nanoparticles (ZnO NPs) also had a beneficial impact on the germination of tobacco seeds (*Nicotiana tabacum* L.). In terms of physiological characteristics, the presence of ZnO nanoparticles increased the levels of chlorophyll and protein in cluster beans (*Cyamopsis tetragonoloba* L.) (Mazaheri-Tirani and Dayani 2020). A recent experiment found that applying SiO_2 nanoparticles through foliar spray enhances the quantum yield of electron transport during photosynthesis or photoprotection (Elsheery et al. 2020). A new study discovered that Mn_3O_4 and Fe NPs greatly increased the highest possible quantum yield and the production of mesophyll ATP. The potential mechanism involves the enhancement of electron transport rate and photocurrent by these nanoparticles, as evidenced by a photochemical study. Consequently, this modification affects the metabolites of mesophyll cells in spinach (Wang et al. 2020). Furthermore, the simultaneous use of 24-epibrassinolide and ZnO NPs from an external source improved the growth characteristics of tomato plants (Faizan et al. 2021b). This was achieved by minimizing the harmful effects of copper and enhancing the levels of chlorophyll and the photosynthetic machinery. Additionally, the presence of ZnO nanoparticles reduced the negative effects of salt stress on tomato plants by enhancing the antioxidant system and photosynthetic machinery (Faizan et al. 2021a). Another research investigation found that treating seeds with silver nanoparticles (Ag NPs) enhanced the plant's physiology, increased relative water content, elevated proline levels, and promoted the activity of antioxidant enzymes. This treatment also reduced oxidative stress in pearl millet (*Pennisetum glaucum* L.) (Khan et al. 2020). Abiotic stress factors have a significant negative impact on plants' physiological processes, mostly by causing an overproduction of reactive oxygen species, which are extremely toxic (Hussain et al. 2021, Sheteiwy et al. 2021). While several studies have highlighted the beneficial effects of NPs on plants, there is also much documentation on their role in mitigating abiotic stress factors. Recent results indicate that TiO_2 nanoparticles (NPs) have been successful in lowering the toxic effects of arsenic on *Vigna radiata* L. (Katiyar et al. 2020). Enhancing the activities of antioxidant enzymes and reducing the excessive formation of reactive oxygen species prevents oxidative damage (Zahedi et al. 2021).

On the other hand, ecological roles characterize soils, as they carry out a wide range of these functions. From the standpoint of both agriculture and the environment, soil health is defined as the ability of soil to operate within the limits of ecosystems and land use in order to support biological productivity, preserve environmental quality, and enhance the overall health of plants and animals (Anikwe and Ife 2023, Sharma et al. 2023). The soil organic matter, pH, and readily available macronutrients (nitrogen, phosphorus, and potassium) are crucial and important for soil health. The most commonly used biological indicators include soil respiration, microbial biomass, nitrogen mineralization, and extracellular enzymatic activities. On the other hand, the most frequently employed physical indicators are bulk density, structural stability, and water retention (Bünemann et al. 2018). This book chapter will assess the impact of simultaneously applying plant growth-promoting rhizobacteria (PGPR) on improving soil quality, focusing on these specific criteria. Researchers have proposed the simultaneous application of different PGPRs as a highly successful method for enhancing soil quality (Akhtar et al. 2021, Nayana et al. 2020). Since PGPR improves their nutrition for survival, their presence could potentially enhance the effectiveness of soil microbial communities. By employing both direct and indirect tactics, it can improve the overall soil quality and increase plant productivity (Asghar et al. 2020). Direct mechanisms, such as nitrogen fixation, phosphate and potassium solubilization, the synthesis of growth-promoting phytohormones like indole-acetic acid (IAA), and the production of siderophores, carry out PGPR functions (Mahmood and Kataoka 2020). However, indirect pathways connect to the synthesis of lytic enzymes and antibiotics, leading to a reduction in soil pH and the formation of exopolysaccharides (Mahmood et al. 2022). Multiple studies have evaluated the effectiveness of PGPR in preserving a sustainable agroecosystem under both normal and challenging circumstances (Magnabosco et al. 2023, Timofeeva et al. 2023). Figure 1 has examined the effects of co-applying different plant growth-promoting rhizobacteria (PGPRs) and nanoparticles (NPs) on crop productivity and soil quality (Alharbi et al. 2023, Rajput et al. 2023).

2.1 *Enhancing Nutrients in Soil: The Influence of PGPR and Nanoparticles (NPs)*

Several studies assessed the effects of simultaneously applying PGPR and NPs on soil quality by analyzing the soil's physical, biological, and chemical properties (Akhtar et al. 2021, Alharbi et al. 2023, Tang et al. 2023). When NPs (nanoparticles) and PGPRs (plant growth-promoting

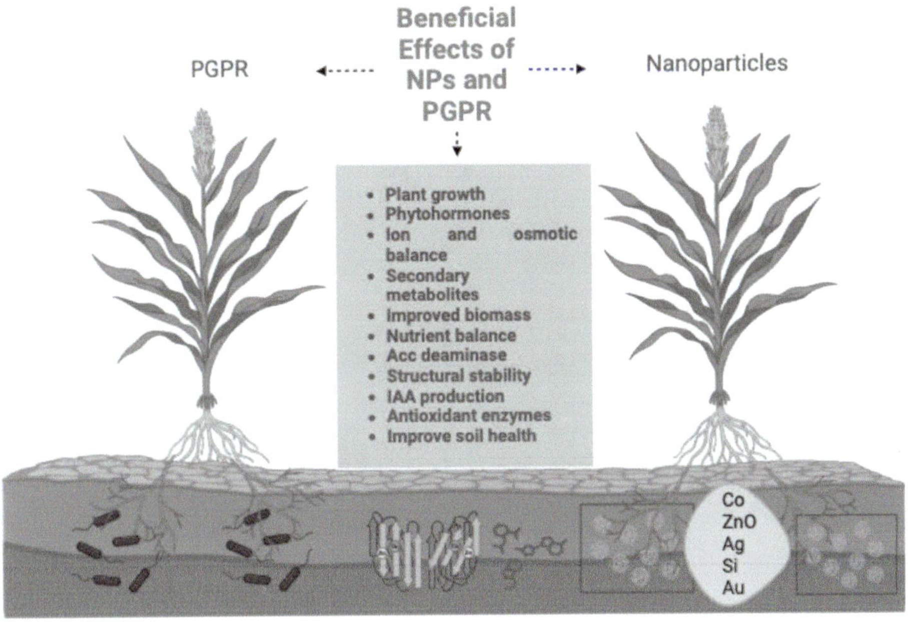

Figure 1. Beneficial Impacts of NPs and PGPRs on Plant-Soil Ecosystems.

rhizobacteria) are applied together (Table 1), they tend to improve the mineral content of the soil around the roots (rhizospheric soils) more than when NPs or PGPRs are applied separately. Compared to adding nitrogen alone, adding plant growth-promoting rhizobacteria (PGPRs) and nitrogenous fertilizers significantly increased the amount of nitrate in the soil (Akhtar et al. 2021, Asghar et al. 2022a). Nanoparticles are applied by spray to plant leaves at low concentrations due to their cost-effectiveness and environmental advantages (Ameen et al. 2021, Liu et al. 2022). Nanoparticles (NPs) have gained attention from plant physiologists due to their ability to function as nano-growth regulators, enhancing plant growth and development. Plant physiologists primarily attribute this to their efficient supply, accumulation, and transfer of essential minerals (Awan et al. 2021). When combined with soil, they undergo dissolution and decomposition as a result of soil characteristics and microbial activity, leading to the release of nutrients (Küpe et al. 2023). Plant growth-promoting rhizobacteria (PGPR), particularly those that enhance the solubilization of organic phosphate, expedite the accumulation of mineral nutrients from nanoparticles (NPs). Combining PGPR and NPs relieves soil microbes of the need to actively search for certain nutrients, enabling them to focus on absorbing other essential components (Ameen et al. 2021). This alteration enhances the efficiency of enzymes and facilitates the release of additional nutrients. As a result, the process of acquiring nutrients becomes more efficient, leading to an overall improvement in soil fertility and sustaining soil health in the long term (Liu et al. 2022, Rajput et al. 2023).

2.2 Improving Soil Water Retention and Modulating Soil Microbial Communities by Nanoparticles (NPs) and PGPRs

Nanoparticles (NPs) have a greater ratio of surface area to volume, which means they have the potential to enhance the soil's ability to store more water, particularly in soils with coarse surfaces. The beneficial effect of nanoparticles (NPs) on water retention capacity is additionally seen in various experimental studies that assess the combined use of NPs and plant growth-promoting rhizobacteria (PGPR). When comparing the application of PGPR and NPs alone, the combined application of PGPR and NPs increased the soil's water-holding capacity (WHC) (Nayana et al. 2020, Tang et al. 2023). Although the use of PGPR has not been proven to enhance soil water-holding capacity (WHC) and water content, it can enhance the ability of crop plants to withstand

Table 1. Plants under stress can benefit from the simultaneous application of plant growth-promoting rhizobacteria (PGPRs) and nanoparticles (NPs).

Crops	Stress	NPS and PGPRs strains	Their effects on soil and crop attributes	References
Wheat (*Triticum aestivum* L.)	Salinity	*Bacillus coagulans, Azospirillum lipferum, Bacillus subtilies* with foliar spray of ZnO-NPs	In treated plants, the application of PGPR+ZnO-NPs resulted in a 57% increase in nitrogen absorption. Applying a combination of PGPR and ZnO-NPs enhanced the activity of soil urease (80%) and dehydrogenase (232%) during saline stress.	(Alharbi et al. 2023)
Wheat (*Triticum aestivum* L.)	Water deficiency and hot weather	Combined NPs and PGPR (ZnO+*Pseudomonas* sp.), and single PGPR	Improved plant productivity and enhanced ability for resistance to stress. The interaction between nanoparticles (NPs) and *Pseudomonas* sp. provides protection against stress circumstances by increasing the production of proline, superoxide dismutase (SOD), peroxidase (POD), catalase (CAT).	(Azmat et al. 2022)
Sugar beet (*Beta vulgaris* L.)	Salinity	*Bacillus coagulans* and *Azospirillum koreensis* with Si-NPs	The study investigates the effects of applying PGPR (*Bacillus coagulans* and *Azospirillum koreensis*) and Si-NPs (Silicon Nanoparticles) on the agronomic responses and productivity of sugar beet plants. The experiment involves exposing the plants to both normal and saline water irrigation in soil with high salt content.	(Alharbi et al. 2022)
Soybean (*Glycine max* L.)	Galaxolide-contaminated soil	Se-NPs with *Actinobacterium* sp.	Chl *a* levels dropped by 58%, stomatal conductance (gs) dropped by 57%, and the maximum quantum yield of photosystem II (Fv/Fm) dropped by 36% when plant growth promoting bacteria (*Actinobacterium* sp.) and selenium (Se) were used together. In plants exposed to galaxolide contamination, the rate of photosynthetic CO_2 assimilation increases by 36% and biomass production increases by 74% compared to unaffected plants.	(Halawani and Aloufi 2023)
Rice (*Oryza sativa* L.)	Chromium	Fe-NPs with *Staphylococcus aureus* (Chromium resistant bacterium)	Having chromium-tolerance bacteria (*Staphylococcus aureus*) around made Fe-NPs work much better by changing chromium ions (Cr^{6+}) into a less harmful form of chromium (Cr^{3+}). The addition of bacteria reduced the buildup of Cr in plant roots through the process of adsorbing Cr ions.	(Ma et al. 2022)
White clover (*Trifolium repens* L.)	Soil polluted with Cadmium	TiO_2-NPs with *Pseudomonas fluorescens*	Titanium dioxide nanoparticles (TiO_2-NPs) and plant growth-promoting rhizobacteria (*Pseudomonas fluorescens*) have the potential to reduce the concentration of TiO_2-NPs in phytoremediation to remove harmful ions. The combined application of *Trifolium repens* effectively sustained development in polluted soil and significantly enhanced the accumulation of Cd in plants.	(Zand et al. 2020)

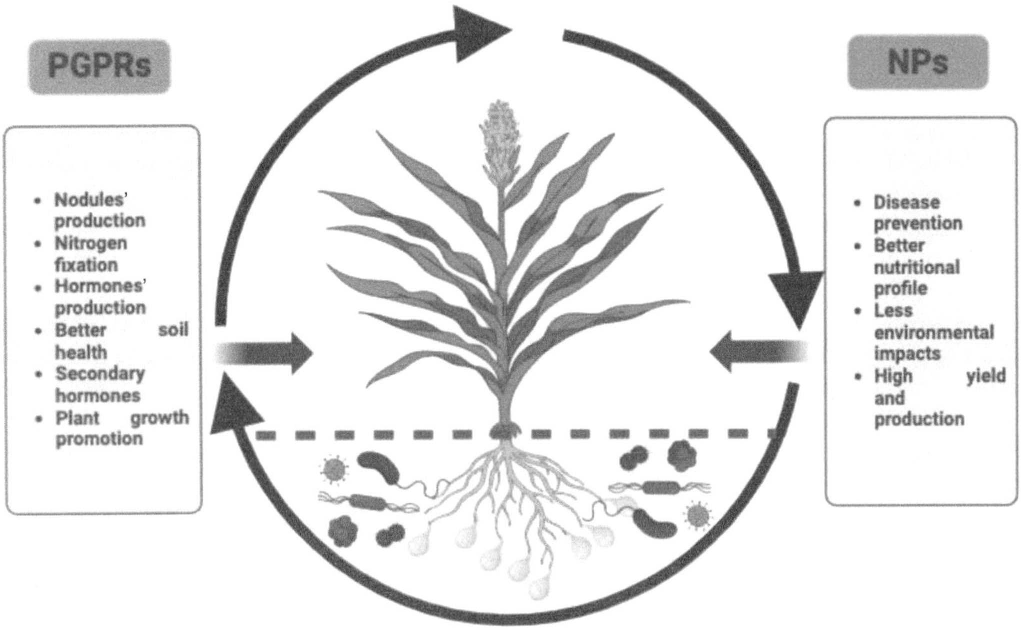

Figure 2. Advantages of NPs and PGPRs applications on plant growth and soil rhizosphere.

water deficits. Nevertheless, the utilization of NPs in conjunction with PGPR can improve water-holding capacity (WHC) by promoting nutrient cycling, decomposition of soil organic matter, and microbial communication, particularly in conditions of increased soil moisture (Figure 2) (Ahmad et al. 2022, Chieb and Gachomo 2023). It is important to note that the indirect importance of PGPRs with NPs application has not been studied yet.

Nanoparticles improve the soil's biological and physicochemical properties, which supports the activity of native soil microbial communities (Upadhayay et al. 2023). Nanoparticles (NPs) have the ability to increase water-holding capacity (WHC), substrate content, soil pH, and nutrient availability, which in turn enhances the growth and diversity of microbial biomass. Moreover, earlier studies have demonstrated that NPs enhance the nodulation process between legume plants and natural rhizobia. NPs help increase airflow, which gives nodule bacteria better access to air. These bacteria can stay on the porous outer layer of NPs for a long time before moving into the roots (Flores-Duarte et al. 2022, Swarnalakshmi et al. 2020).

Adding nanoparticles (NPs) can further increase the mutualistic link between existing microbes and plants (Ameen et al. 2021). The microorganisms that solubilize phosphate (*Pseudomonas* sp.) enhanced phosphorus availability in soil by solubilizing it from the NPs, leading to higher root colonization and enhanced plant growth (De Souza-Torres et al. 2021). The synergistic interaction between nanoparticles (NPs) and plant growth-promoting rhizobacteria (PGPR) may enhance the overall prevalence of certain microbial groups in soil, leading to an overall improvement in soil health. The researchers attributed the increase in beneficial bacteria to the enhanced soil organic matter content and its decomposition resulting from the combined effect of nanoparticles (NPs) and the inoculant (Ahmad et al. 2022, De Souza-Torres et al. 2021).

3. Adverse Effect of Nanoparticles (NPS) on Plants

In addition to their positive impacts on plants, nanoparticles (NPs) can also have phytotoxic consequences, depending on the dosage and characteristics of the NPs used (Rastogi et al. 2017). The primary detrimental impacts on plant physiological characteristics include reduced biomass accumulation, decreased germination rate, diminished leaf count, and inhibited root elongation (Dev et al. 2018, Jośko et al. 2017). Nanoparticles (NPs) resulted in a decrease in plant elongation, the initiation of oxidative damage, a reduction in photosynthesis, impairments to biological membranes, disruptions to the water transport system and water balance, and, in some cases, even plant mortality (Azhar et al. 2020, Khan et al. 2019, Noori et al. 2020). The presence of nanoparticles (NPs) within cellular organs led to changes in gene expression, cellular processes, and biological pathways such as metabolism and proliferation. Additionally, NPs disrupted the transport activity of electrons and ions across the cell membrane, ultimately impacting the overall growth and development of plants (de la Rosa et al. 2021, Moreno-Olivas et al. 2014). Nanoparticles (NPs) have the ability to disrupt the flow of energy and the transportation routes of electrons by suppressing the degree of gene expression. NPs induced the upregulation of specific genes associated with water channels and stress responses (Hossain et al. 2016, Tripathi et al. 2017). Another study found that tiny pieces of aluminum oxide (Al_2O_3), titanium dioxide (TiO_2), and zinc oxide (ZnO) seriously damage the DNA structure of *Allium cepa* (Debnath et al. 2020). Furthermore, studies have revealed that ZnO nanoparticles cause genotypic and cytotoxic effects on the root cells of many crops (Faizan et al. 2021a). When used on *R. sativus* sprouts, Ag NPs made it much harder for the plants to absorb mineral nutrients and disrupted the integrity of cell components like lignin, pectin, and cellulose (Zuverza-Mena et al. 2016). Several recent studies have reported the role of ethylene signaling as a biomarker for the harmful effects of ZnO NPs and CuO NPs in *Arabidopsis thaliana*. Their research indicated that ethylene enhanced the oxidative damage caused by ZnO NPs and CuO NPs in hydroponic situations. However, additional field studies are necessary to confirm these findings. Many recent research efforts have examined the harmful impacts of nanoparticles (NPs) on several aspects of plant growth (Azhar et al. 2020, Khan et al. 2019). Table 2 explains the methods by which plants resist these toxic effects.

Table 2. Toxic effects of nanoparticles NPs on plant growth.

Plant species	Detoxification strategies	NPs stress	Alterations	References
Rice (*Oryza sativa* L.)	Humic acid used for seed priming	ZnO-NPs	Seedlings growth and seed germination up but caused ROS-accumulation and cell-ultrastructural damages	Sheteiwy et al. (2017)
Wheat (*Triticum aestivum* L.)	Nitric oxide	ZnO-NPs	Zn-uptake, photosynthetic efficiency and biomass up but changed the H_2O_2 and ascorbate-glutathione cycle enzymes	Tripathi et al. (2017)
Rice (*Oryza sativa* L.)	Nitric oxide	ZnO-NPs	Antioxidants, Zn-accumulation and growth traits up but reduced the lipid production	Chen et al. (2015)
Arabidopsis thaliana	Glucosinolate	ZnO-NPs	Increased the chlorophyll, biomass and lateral root length but caused the oxidative damage and Zn-accumulation	Tao et al. (2021)
Wheat (*Triticum aestivum* L.) and maize (*Zea mays* L.)	Bioactive actinomycete	As-NPs	Increased the soluble sugar, photosynthesis and growth but also increased the uptake of As and caused oxidative damage	Selim et al. (2021)

4. Conclusions

In conclusion, researchers are interested in finding ways to make plant-soil system more resilient by using microbes or their derived products made from microbes, nanoparticles (NPs), and green manuring. These methods aim to increase rhizosphere activities, boost crop productivity, and keep soil quality high. Because PGPR (Plant Growth Promoting Rhizobacteria) and NPs (Nanoparticles) can greatly increase crop yield and soil quality in a range of environmental conditions, our conclusion emphasizes how important it is to use these technologies in environmentally friendly farming methods. In addition, these methods have the potential to improve the resilience of plant-soil system, reduce the amount of pollution in the soil, and improve the processes that occur in the rhizosphere.

5. Acknowledgment

Figures for the book chapter were created by BioRender (https://biorender.com).

The authors would like to acknowledge the support received from the Soil and Environmental Microbiology Laboratory, University of Yamanashi.

References

Ahmad, H.M., S. Fiaz, S. Hafeez, S. Zahra, A.N. Shah, B. Gul et al. 2022. Plant growth-promoting rhizobacteria eliminate the effect of drought stress in plants: a review. Front. Plant Sci. 13: 875774.

Akhtar, N., N. Ilyas, R. Hayat, H. Yasmin, A. Noureldeen and P. Ahmad. 2021. Synergistic effects of plant growth promoting rhizobacteria and silicon dioxide nano-particles for amelioration of drought stress in wheat. Plant Physiol. Biochem. 166: 160–176.

Alharbi, K., E. Hafez, A.E.-D. Omara, A. Awadalla and Y. Nehela. 2022. Plant growth promoting rhizobacteria and silica nanoparticles stimulate sugar beet resilience to irrigation with saline water in salt-affected soils. Plants. 11: 3117.

Alharbi, K., E.M. Hafez, A.E.-D. Omara, E. Rashwan and T. Alshaal. 2023. Zinc oxide nanoparticles and PGPR strengthen salinity tolerance and productivity of wheat irrigated with saline water in sodic-saline soil. Plant Soil. 493: 475-495.

Ameen, F., K. Alsamhary, J.A. Alabdullatif and S. ALNadhari. 2021. A review on metal-based nanoparticles and their toxicity to beneficial soil bacteria and fungi. Ecotoxicol. Environ. Saf. 213: 112027.

Anikwe, M. and K. Ife. 2023. The role of soil ecosystem services in the circular bioeconomy. Front. Soil Sci. 3: 1209100.

Asghar, W., M.O. Akça, H. Akça, O.J. Tarf, R. Kataoka and O.C. Turgay. 2022a. Alternative strategies to synthetic chemical fertilizers: revitalization of soil quality for sustainable agriculture using organic-based approaches. pp. 1–30. *In*: New and Future Developments in Microbial Biotechnology and Bioengineering. Elsevier.

Asghar, W., K.D. Craven, R. Kataoka, A. Mahmood, N. Asghar, T. Raza et al. 2024. The application of *Trichoderma* spp., an old but new useful fungus, in sustainable soil health intensification: A comprehensive strategy for addressing challenges. Plant Stress. 100455.

Asghar, W. and R. Kataoka. 2024. Co-application of green manure and *Trichoderma* spp. induced plant growth promotion by nutrient improvement and increased fungal biomass in soil. Agric. Res. 1–10.

Asghar, W., S. Kondo, R. Iguchi, A. Mahmood and R. Kataoka. 2020. Agricultural utilization of unused resources: liquid food waste material as a new source of plant growth-promoting microbes. Agronomy. 10: 954.

Asghar, W., A. Mahmood, F. Iftikhar, B. Ahmad, R. Ullah, M. Bilal etal. 2022b. Efficient utilization of organic wastes effluent for nitrogen mineralization and plant growth promotion in mono-cropping soil of China. Sarhad J. Agric. 38.

Aslani, F., S. Bagheri, N. Muhd Julkapli, A.S. Juraimi, F.S.G. Hashemi and A. Baghdadi. 2014. Effects of engineered nanomaterials on plants growth: an overview. Sci. World J. 2014: 641759.

Awan, S., K. Shahzadi, S. Javad, A. Tariq, A. Ahmad and S. Ilyas. 2021. A preliminary study of influence of zinc oxide nanoparticles on growth parameters of *Brassica oleracea* var. italic. J. Saudi Soc. Agric. Sci. 20: 18–24.

Azhar, W., A.R. Khan, N. Muhammad, B. Liu, G. Song, A. Hussain et al. 2020. Ethylene mediates CuO NP-induced ultrastructural changes and oxidative stress in *Arabidopsis thaliana* leaves. Environ. Sci. Nano. 7: 938–953.

Azmat, A., Y. Tanveer, H. Yasmin, M.N. Hassan, A. Shahzad, M. Reddy et al. 2022. Coactive role of zinc oxide nanoparticles and plant growth promoting rhizobacteria for mitigation of synchronized effects of heat and drought stress in wheat plants. Chemosphere. 297: 133982.

Bünemann, E.K., G. Bongiorno, Z. Bai, R.E. Creamer, G. De Deyn, R. De Goede et al. 2018. Soil quality-A critical review. Soil Biol. Biochem. 120: 105–125.

Chen, J., X. Liu, C. Wang, S.-S. Yin, X.-L. Li, W.-J. Hu et al. 2015. Nitric oxide ameliorates zinc oxide nanoparticles-induced phytotoxicity in rice seedlings. J. Hazard. Mater. 297: 173–182.

Chieb, M. and E.W. Gachomo. 2023. The role of plant growth promoting rhizobacteria in plant drought stress responses. BMC Plant Biol. 23: 407.

Dasgupta, N., S. Ranjan and C. Ramalingam. 2017. Applications of nanotechnology in agriculture and water quality management. Environ. Chem. Lett. 15: 591–605.

de la Rosa, G., E. Vázquez-Núñez, C. Molina-Guerrero, A.H. Serafín-Muñoz and I. Vera-Reyes. 2021. Interactions of nanomaterials and plants at the cellular level: current knowledge and relevant gaps. Nanotechnol. Environ. Eng. 6: 1–19.

De Souza-Torres, A., E. Govea-Alcaide, E. Gomez-Padilla, S.H. Masunaga, F.B. Effenberger, L.M. Rossi et al. 2021. Fe_3O_4 nanoparticles and *Rhizobium* inoculation enhance nodulation, nitrogen fixation and growth of common bean plants grown in soil. Rhizosphere. 17: 100275.

Debnath, P., A. Mondal, K. Sen, D. Mishra and N.K. Mondal. 2020. Genotoxicity study of nano Al_2O_3, TiO_2 and ZnO along with UV-B exposure: An *Allium cepa* root tip assay. Sci. Total Environ. 713: 136592.

Dev, A., A.K. Srivastava and S. Karmakar. 2018. Nanomaterial toxicity for plants. Environ. Chem. Lett. 16: 85–100.

Dilnawaz, F., A.N. Misra and E. Apostolova. 2023. Involvement of nanoparticles in mitigating plant's abiotic stress. Plant Stress. 100280.

Elsheery, N.I., V. Sunoj, Y. Wen, J. Zhu, G. Muralidharan and K. Cao. 2020. Foliar application of nanoparticles mitigates the chilling effect on photosynthesis and photoprotection in sugarcane. Plant Physiol. Biochem. 149: 50–60.

Faizan, M., J.A. Bhat, C. Chen, M.N. Alyemeni, L. Wijaya, P. Ahmad et al. 2021a. Zinc oxide nanoparticles (ZnO-NPs) induce salt tolerance by improving the antioxidant system and photosynthetic machinery in tomato. Plant Physiol. Biochem. 161: 122–130.

Faizan, M., J.A. Bhat, A. Noureldeen, P. Ahmad and F. Yu. 2021b. Zinc oxide nanoparticles and 24-epibrassinolide alleviate Cu toxicity in tomato by regulating ROS scavenging, stomatal movement and photosynthesis. Ecotoxicol. Environ. Saf. 218: 112293.

Flores-Duarte, N.J., E. Mateos-Naranjo, S. Redondo-Gómez, E. Pajuelo, I.D. Rodriguez-Llorente and S. Navarro-Torre. 2022. Role of nodulation-enhancing rhizobacteria in the promotion of *Medicago sativa* development in nutrient-poor soils. Plants. 11: 1164.

Garbowski, T., D. Bar-Michalczyk, S. Charazińska, B. Grabowska-Polanowska, A. Kowalczyk and P. Lochyński. 2023. An overview of natural soil amendments in agriculture. Soil Tillage Res. 225: 105462.

Halawani, R.F. and F.A. Aloufi. 2023. Galaxolide-contaminated soil and tolerance strategies in soybean plants using biofertilization and selenium nanoparticle supplementation. Front. Plant Sci. 14: 1221780.

Hossain, Z., G. Mustafa, K. Sakata and S. Komatsu. 2016. Insights into the proteomic response of soybean towards Al_2O_3, ZnO, and Ag nanoparticles stress. J. Hazard. Mater. 304: 291–305.

Hussain, S., M. Mumtaz, S. Manzoor, L. Shuxian, I. Ahmed, M. Skalicky et al. 2021. Foliar application of silicon improves growth of soybean by enhancing carbon metabolism under shading conditions. Plant Physiol. Biochem. 159: 43–52.

Jośko, I., P. Oleszczuk and E. Skwarek. 2017. Toxicity of combined mixtures of nanoparticles to plants. J. Hazard. Mater. 331: 200–209.

Katiyar, P., B. Yadu, J. Korram, M.L. Satnami, M. Kumar and S. Keshavkant. 2020. Titanium nanoparticles attenuate arsenic toxicity by up-regulating expressions of defensive genes in *Vigna radiata* L. J. Environ. Sci. 92: 18–27.

Khan, A.R., A. Wakeel, N. Muhammad, B. Liu, M. Wu, Y. Liu et al. 2019. Involvement of ethylene signaling in zinc oxide nanoparticle-mediated biochemical changes in *Arabidopsis thaliana* leaves. Environ. Sci.: Nano 6: 341–355.

Khan, I., M.A. Raza, S.A. Awan, G.A. Shah, M. Rizwan, B. Ali et al. 2020. Amelioration of salt induced toxicity in pearl millet by seed priming with silver nanoparticles (AgNPs): The oxidative damage, antioxidant enzymes and ions uptake are major determinants of salt tolerant capacity. Plant Physiol. Biochem. 156: 221–232.

Khan, N., T. Raza, N.S. Eash, M. Mechri and W. Asghar. 2024. Effect of combined FYM and P fertilizer application on soil health and crop performance. Commun. Soil Sci. Plant Anal. 55: 579–595.

Kumari, A., S.K. Singh, B. Mathpal, K.K. Verma, V.K. Garg, M. Bhattacharyya et al. 2023. The biosynthesis, mechanism of action, and physiological functions of melatonin in horticultural plants: A review. Horticulturae. 9: 913.

Küpe, M., F. Hacımüftüoğlu and E. Yağanoğlu. 2023. Effects of PGPR bacteria applications on soil properties, plant growth and yield values in Karaerik and Narince grape varieties. J. Agric. Prod. 4: 128–137.

Liu, L., H. Nian and T. Lian. 2022. Plants and rhizospheric environment: Affected by zinc oxide nanoparticles (ZnO NPs). A review. Plant Physiol. Biochem. 185: 91–100.

Lu ChangMei, L.C., Zhang ChaoYing, Z.C., Wen JunQiang, W.J., Wu GuoRong, W.G. and Tao MingXuan, T.M. 2002. Research of the effect of nanometer materials on germination and growth enhancement of *Glycine max* and its mechanism.

Ma, J., H. Alshaya, M.K. Okla, Y.A. Alwasel, F. Chen, M. Adrees et al. 2022. Application of cerium dioxide nanoparticles and chromium-resistant bacteria reduced chromium toxicity in sunflower plants. Front. Plant Sci. 13: 876119.

Pramanik, B., P. Sar, R. Bharti, R.K. Gupta, S. Purkayastha, S. Sinha et al. 2023. Multifactorial role of nanoparticles in alleviating environmental stresses for sustainable crop production and protection. Plant Physiol. Biochem. 201: 107831.

Qaswar, M., W. Ahmed, H. Jing, F. Hongzhu, S. Xiaojun, J. Xianjun et al. 2019. Soil carbon (C), nitrogen (N) and phosphorus (P) stoichiometry drives phosphorus lability in paddy soil under long-term fertilization: A fractionation and path analysis study. PLoS One. 14: e0218195.

Rajput, V.D., A. Kumari, S.K. Upadhyay, T. Minkina, S. Mandzhieva, A. Ranjan et al. 2023. Can nanomaterials improve the soil microbiome and crop productivity? Agric. 13: 231.

Rastogi, A., M. Zivcak, O. Sytar, H.M. Kalaji, X. He, S. Mbarki et al. 2017. Impact of metal and metal oxide nanoparticles on plant: a critical review. Front. Chem. 5: 78.

Rubin, R., E. Oldfield, J. Lavallee, T. Griffin, B. Mayers and J. Sanderman. 2023. Climate mitigation through soil amendments: quantification, evidence, and uncertainty. Carbon Manag. 14: 2217785.

Salam, M.T.B., A. Mahmood, W. Asghar, K. Ito and R. Kataoka. 2024. Phytomicrobiomes: A Potential Approach for Sustainable Pesticide Biodegradation. Appl. Sci. 14: 2740.

Selim, S., H. AbdElgawad, S.S. Alsharari, M. Atif, M. Warrad, N. Hagagy et al. 2021. Soil enrichment with actinomycete mitigates the toxicity of arsenic oxide nanoparticles on wheat and maize growth and metabolism. Physiologia Plant. 173: 978–992.

Sharma, A., R.N. Singh, X.-P. Song, R.K. Singh, D.-J. Guo, P. Singh et al. 2023. Genome analysis of a halophilic Virgibacillus halodenitrificans ASH15 revealed salt adaptation, plant growth promotion, and isoprenoid biosynthetic machinery. Front. Microbiol. 14: 1229955.

Sheteiwy, M.S., D.F.I. Ali, Y.-C. Xiong, M. Brestic, M. Skalicky, Y.A. Hamoud et al. 2021. Physiological and biochemical responses of soybean plants inoculated with Arbuscular mycorrhizal fungi and Bradyrhizobium under drought stress. BMC Plant Biol. 21: 1–21.

Sheteiwy, M.S., Q. Dong, J. An, W. Song, Y. Guan, F. He et al. 2017. Regulation of ZnO nanoparticles-induced physiological and molecular changes by seed priming with humic acid in *Oryza sativa* seedlings. Plant Growth Regul. 83: 27–41.

Siddiqui, M.H., and M.H. Al-Whaibi. 2014. Role of nano-SiO_2 in germination of tomato (*Lycopersicum esculentum* seeds Mill.). Saudi J. Biol. Sci. 21: 13–17.

Sudha, P.N., K. Sangeetha, K. Vijayalakshmi and A. Barhoum. 2018. Nanomaterials history, classification, unique properties, production and market. *In*: Emerging Applications of Nanoparticles and Architecture Nanostructures, pp. 341–384. Elsevier.

Swarnalakshmi, K., V. Yadav, D. Tyagi, D.W. Dhar, A. Kannepalli and S. Kumar. 2020. Significance of plant growth promoting rhizobacteria in grain legumes: Growth promotion and crop production. Plants. 9: 1596.

Tang, L., Y. Shi, Y. Zhang, D. Yang and C. Guo. 2023. Effects of plant-growth-promoting rhizobacteria on soil bacterial community, soil physicochemical properties, and soil enzyme activities in the rhizosphere of alfalfa under field conditions. Diversity. 15: 537.

Tao, H., S. Hu, C. Xia, M. Wang, T. Wang, W. Zeng et al. 2021. Involvement of glucosinolates in the resistance to zinc oxide nanoparticle-induced toxicity and growth inhibition in Arabidopsis. Environ. Sci. Process. Impacts. 23: 1040–1049.

Timofeeva, A.M., M.R. Galyamova and S.E. Sedykh. 2023. Plant growth-promoting soil bacteria: nitrogen fixation, phosphate solubilization, siderophore production, and other biological activities. Plants. 12: 4074.

Tripathi, D.K., S. Singh, S. Singh, R. Pandey, V.P. Singh, N.C. Sharma et al. 2017. An overview on manufactured nanoparticles in plants: uptake, translocation, accumulation and phytotoxicity. Plant Physiol. Biochem. 110: 2–12.

Upadhayay, V.K., Chitara, M.K., Mishra, D., Jha, M.N., Jaiswal, A., Kumari, G. et al. 2023. Synergistic impact of nanomaterials and plant probiotics in agriculture: A tale of two-way strategy for long-term sustainability. Front. Microbiol. 14: 1133968.

Urra, J., Alkorta, I. and Garbisu, C. 2019. Potential benefits and risks for soil health derived from the use of organic amendments in agriculture. Agron. 9: 542.

Verma, K.K., Song, X.-P., Joshi, A., Tian, D.-D., Rajput, V.D., Singh, M. et al. 2022. Recent trends in nano-fertilizers for sustainable agriculture under climate change for global food security. Nanomaterials. 12: 173.

Wang, A., Jin, Q., Xu, X., Miao, A., White, J.C., Gardea-Torresdey, J.L. et al. 2020. High-throughput screening for engineered nanoparticles that enhance photosynthesis using mesophyll protoplasts. J. Agric. Food Chem. 68: 3382–3389.

Zahedi, S.M., Hosseini, M.S., Daneshvar Hakimi Meybodi, N. and Peijnenburg, W. 2021. Mitigation of the effect of drought on growth and yield of pomegranates by foliar spraying of different sizes of selenium nanoparticles. J. Sci. Food Agric. 101: 5202–5213.

Zand, A.D., Mikaeili Tabrizi, A. and Vaezi Heir, A. 2020. Application of titanium dioxide nanoparticles to promote phytoremediation of Cd-polluted soil: contribution of PGPR inoculation. Bioremediation J. 24: 171–189.

Zuverza-Mena, N., Armendariz, R., Peralta-Videa, J.R. and Gardea-Torresdey, J.L. 2016. Effects of silver nanoparticles on radish sprouts: root growth reduction and modifications in the nutritional value. Front. Plant Sci. 7: 90.

CHAPTER-8

Nanotechnology for Sustainable Soil Management Practices

Subhan Ali, Akhtar Hameed, Rana Binyamin, Hasan Riaz, Muhammad Arslan Khan* and *Hira Akhtar*

1. Introduction

Soil management is an essential activity in agriculture that aims at sustaining soil quality to support continued production for years. In the past, the management of soil was done by the application of crop rotation, cover cropping, organic manipulation, and use of chemical fertilizers and chemical pesticides (Lal 2015). New techniques of farming affect the environment in many ways, though they are very crucial in feeding the world's increasing population. The application of agricultural techniques has resulted in alarming levels of soil pollution from numerous toxic substances such as heavy metals and pesticides (Jakhar et al. 2022, Alengebawy et al. 2021). These pollutants threaten humanity and the environment, and hinder the productivity and fertility of agricultural land due to their effects on the food chain (Chu et al. 2023, Dhanapal et al. 2024).

This situation thus calls for the understanding of better and environmentally friendly methods that can be used in management to reduce the impacts of soil pollution, and at the same time, facilitate the quick regain of soil vitality. The incorporation of nanotechnology in agriculture is highly useful in solving the complex problems faced in food production systems (He et al. 2019). The advancements in nanomaterials (NMs) in improving soil fertility, nutrient delivery, precision farming, and pest control offer potentially unexplored potential for efficient farming with minimal impacts on the soil and environment (Masireh et al. 2021).

Nanotechnology has recently come to the forefront of discussions as a potential solution to the multifaceted issue of pollution in agricultural soils (Singh 2021, Dhanapal et al. 2024). Key attributes of NMs include large surface area, high reactivity, and the ability to alter physicochemical factors, and due to these reasons, NMs have been preferred for soil remediation work (Singh et al. 2021). Praised for being able to provide more precise, effective, and environmentally friendly solutions regarding the contamination of soil with chemicals used in farming, nanotechnology can transform the approaches toward soil management in modern agriculture (Ahmad et al. 2022). For instance, nanofertilizers enable the slow release of nutrients to the plant, which enhances the nutrient use rate and reduces the loss of nutrients in the environment. Like nano-pesticides, they offer a slow

Institute of Plant Protection, MNS University of Agriculture, Multan 66000, Pakistan.
* Corresponding author: akhtar.hameed@mnsuam.edu.pk

DOI: 10.1201/9781003558354-8

release of active ingredients, which allows the use of smaller amounts of chemicals and reduces the impact on other species (Servin and White 2016).

The application of nanotechnology in soil has the capability to enhance soil fertility and productivity, while also reducing the effects of soil degradation and other agricultural impacts. It is proven that NMs can be engineered to bond effectively with soil particles, improve swelling properties, and encourage desired microbial species (Chen and Yada 2011). In addition, nanotechnology offers novel solutions for the treatment of soil contamination, including the elimination of numerous pollutants present in soils, such as heavy metals and organic compounds (Zhang 2003). However, the effectiveness of nanotechnology in soil management also raises concerns, particularly regarding the ecological implications of nanoparticles (NPs) in the soil environment (Kah et al. 2013).

1.1 Importance of Soil Health in Agriculture and Ecosystem Services

Healthy soils are essential for farming and for meeting the ecological function demands in food production. Healthy soils ensure the growth and development of plants through the provision of physical, chemical, and biological requirements for plant growth. They also help control water cycles, remove pollutants, and sequester carbon, which is important for combating climate change (Shukla et al. 2024). Soils are the basis of crop production, water and nutrient balance, and are home to various microbial communities that are vital to soil productivity (Lehmann and Kleber 2015). Furthermore, fertile soils facilitate carbon storage, playing an important role in reducing climate change impacts due to their effectiveness in storing carbon dioxide in the form of organic carbon (Lal 2004). The importance of soil health is illustrated in Figure 1.

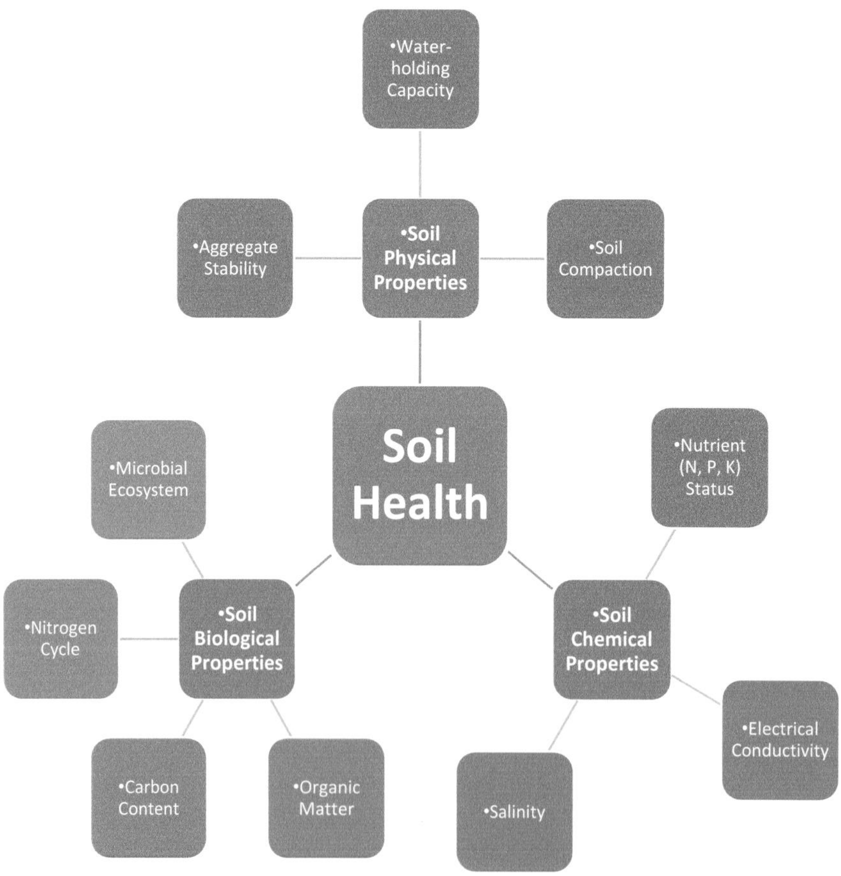

Figure 1. A general conceptual framework for defining soil health.

However, soil health is not only important for agricultural purposes since it supports ecosystem services such as maintaining biodiversity, water purification, and preventing soil erosion (Dominati et al. 2010). Soil health also determines the vulnerability of an agricultural system to biophysical stresses, including water deficits, pests, and diseases (Kibblewhite et al. 2008). Hence, practices such as conservation tillage and crop rotation are crucial in sustaining the health of the soil and thus the security of agricultural production. However, the erodibility of soil through unsustainable agriculture, deforestation, and the expansion of urban centers is on the rise and is a global threat to the sustainability of the soil (Montgomery 2007). Soil compaction and nutrient depletion result in poor water and nutrient holding capacities, leading to poor crop production and increased vulnerability to land degradation through erosion (Lal 2015).

1.2 Objectives of the Chapter

This chapter focuses on the application of nanotechnology to sustainable soil management practices. It gives an overview of the challenges that are facing soil today regarding its degradation and the shortcomings of the conventional methods, and then identifies how nanotechnology can be utilized to improve the practice of soil management, including nanofertilizers, nanopesticides, and remediation techniques. New developments in nanotechnology are also discussed in the chapter, including new NMs for water management and soil microbial function. Some potential applications of nanotechnology in soil management are the promotion of soil fertility, better water-holding capacity, and removal of contaminated soils. The chapter also covers some of the environmental issues, such as toxicity, bioaccumulation, and life cycle assessment. Finally, the chapter urges researchers, policymakers, and farmers to be more innovative.

2. Soil Degradation Issues

2.1 Soil Degradation Sources and Types

Soil degradation is a major environmental issue worldwide that affects ecosystem quality and crop yields. Soil is a vital and diverse category and performs multiple ecosystem services, including water provision, timber, food, and fiber, climate regulation, flood protection and erosion prevention, aesthetics and additives, cultural and spiritual significance, and framework support for plants, animals, human beings, and structures (Li et al. 2023, Dhanapal et al. 2024). Soil health refers to the ability of soil to function as a complex, self-sustaining living ecosystem and its capacity to support plant and animal productivity within the specified constraints of an ecosystem and the land use practices (Li et al. 2023). Gases, salts, minerals, organic and inorganic debris, and living things are just a few of the many constituents that make up soil, a dynamic natural resource. Its physical, chemical, and biological properties are malleable to change. These changes can come from natural occurrences like ore weathering, volcanic eruptions, forest fires, and more commonly, from a variety of human activities like disposing of industrial and household waste and applying chemical pesticides and fertilizers to increase crop yields (Tuti et al. 2022).

Pollution of the soil is mostly caused by naturally occurring changes in the environment and chemicals that humans manufacture. The disintegration of underground storage connections, the application of pesticides, the seepage of contaminated surface water into subsurface strata, the leaching of wastes from landfills, the disposal of fuel and oil, and the direct release of industrial wastes into the soil are the common causes of soil contamination (Gurusamy and Thangam 2023). Many toxic chemical elements from both manmade and natural sources are always present in the environment. This is only one of the numerous elements causing pollution in the environment (Shetty et al. 2023). Because dangerous compounds are emitted from man-made sources as a result of industrialization and urbanization, many portions of the world have been poisoned. Both agricultural and non-agricultural sources, such as mining, smelting, and industrial pollutants, can pollute soil (Shetty et al. 2023).

A number of factors affect production, filtering, buffering, and other soil activities, including the skeletal nature of the soil, depth, nutrient availability, humus content, structure, reactivity, presence of foreign materials, and edaphon (Bretzel et al. 2016). Analyzing the qualities of the soil is necessary to determine its quality, as it cannot be determined instantly (Bretzel et al. 2016). It is more accurate to take into account the chemical, physical, biological, and biochemical traits that are impacted by changes in the environment and land management practices. Reaction, carbon and nitrogen content, and nutrient content are examples of chemical parameters; bulk density, porosity, temperature, and water-holding capacity are examples of physical qualities (Bretzel et al. 2016, Dhanapal et al. 2024). Although soil enzymes, respiration, C and N mineralization, biological N_2 fixation, and the total biomass of soil microorganisms may all be quantified, microbial components appear to be especially useful in tracking heavy-metal pollutants of soil (Zornoza et al. 2015). However, the recent literature describes the relationship between soil, enhanced technology, microbial ecology, climate change, sustainable conduct, and previous studies, which laid the foundation for research on conventional causes and forms of soil contamination. These understandings are pivotal for formulating integrated strategies for addressing the dynamic issues concerning soil degradation and creating a more resilient and sustaining environment (Dhanapal et al. 2024).

2.2 Impact of Soil Degradation on Crop Productivity and Food Security

The impacts of soil degradation are enormous as far as crop yield and food security are concerned, especially in the developing world where agriculture is the main source of livelihood. Soil degradation reduces the fertility of the soil, leading to low crop yield and susceptibility to environmental factors (Montgomery 2007, Tahat et al. 2020).

2.3 Reduced Crop Productivity

Soil degradation has many effects, but possibly the most obvious one is the decline in crop yield. It also results in poor water and nutrient retention in the soil, so that crops do not grow well, and yields decline (Lal 2015). For example, soil erosion—one of the most gradual types of degradation—removes the nutrient-rich topsoil necessary for plant growth and development and replaces it with less fertile soil suitable for minimal agricultural production (Pimentel and Burgess 2013). In the same way, soil compaction due to factors such as use of machinery or livestock leads to a reduction in porosity and water infiltration into the soil, causing waterlogging and poor root development (Hamza and Anderson 2005).

Soil degradation also impacts the availability of nutrients, including nitrogen, phosphorus, and potassium—all of which are vital in plant production. Degraded soils have a lower ability to hold nutrients, which makes them wash away and become deficient, greatly affecting crop productivity (Barber 1995). At times, farmers attempt to replenish lost nutrients by adding more fertilizers, which in turn accelerates soil degradation due to nutrient imbalances and increased soil acidity (Guo et al. 2010). However, degraded soils typically have limited nutrient availability, low levels of organic matter, and poor structure, all of which lead to low crop yields (Jie et al. 2002). In this case, the level of soil degradation is estimated to have cost about 11.9–13.4% of the world's agricultural commodity supply (AbdelRahman 2023)

2.4 Impact on Food Security

Soil degradation affects not only particular farms or areas but the food security of the entire world. As the global population rises, the demand for food increases, placing pressure on agricultural production (D'Amour et al. 2017). However, soil degradation reduces the capacity of these systems to feed the population in the future. Since crop yields decline, the ability of farmers to feed people both locally and globally is affected, leading to food scarcity and higher food prices (Godfray et al. 2010).

These problems are especially alarming in areas where the risk of food deficiency is rising due to climate change, population growth, and economic crises (Beddington et al. 2012).

Climate change is also worsened by soil degradation, which in turn intensifies its effects on agriculture (Corwin 2021). Soil degradation makes soils more sensitive to harsh conditions such as droughts and floods, which are occurring more frequently and intensely as a result of climate change (Reynolds et al. 2015, Corwin 2021). Soil deterioration and climate change are dual threats in areas where crop failure and food insecurity are imminent possibilities (Vermeulen et al. 2012).

Additionally, soil degradation results in erosion, prompting farmers to shift to less fertile lands for cultivation. Such a change is referred to as land degradation and leads to the overall reduction of arable and food-producing lands. It can also lead to the forced displacement of rural dwellers, further escalating food insecurity (Gibbs and Salmon 2015). The gradual decline in soil productivity through continued cultivation on degrading land has emphasized the need for sound and sustainable soil management practices that can restore and maintain soil health for long-term crop production.

3. NMs Synthesis

NPs have been synthesized using mechanical pressure, high-energy radiation, and heat energy, where material is either evaporated, condensed, abraded, or melted physically (Escorcia-Díaz et al. 2023). Physical techniques are preferred over chemical ones because they allow for solvent-free thin films and optimal NP distribution. These techniques are top-down approaches and therefore eliminate the need for solvents, while affording monodisperse NPs consistently. Some of the physical methods used in manufacturing NPs include laser ablation, high-energy ball milling, physical vapor deposition, laser pyrolysis, and inert gas condensation (Abid et al. 2022). There are two widely used approaches for producing NPs: the top-down approach and the bottom-up method, as used by Chung et al. (2016). Some of the approaches that have been used to synthesize NMs include mechanical milling, sputtering, laser ablation, electron beam evaporation, laser pyrolysis, and nanolithography, among others (Khan et al. 2022b). Figure 2 shows the formation of NPs using physical, chemical, and biological methods.

3.1 Physical Methods

3.1.1 Mechanical Milling

Mechanical milling is one of the most efficient and affordable ways to turn large-block materials into NMs (El-Eskandarany et al. 2021). This process is effective in phase mixing and is crucial for the formation of nanocomposites (El-Eskandarany et al. 2021, Dhanapal et al. 2024). This procedure uses the ball-milling technique as outlined above. Mechanical milling has been employed in the development of wear-resistant spray coatings, aluminum alloys with oxides and carbides, Cu-Mg-Ni-Al nanoalloys, and other nanocomposite materials (Priyadarshini 2021). A new class of NPs has emerged, including ball-milled carbon NMs, offering solutions in many fields related to conversion, storage, and purification of energy (El-Eskandarany et al. 2021).

3.1.2 Sputtering

Sputtering is the process of depositing materials through the interaction of high-energy particles such as plasma or gas with solid surfaces (Sergievskaya et al. 2022). During the sputtering deposition process, gaseous ions directly strike the target surface, and depending on the incident gaseous-ion energy, tiny clusters of atoms are ejected (Sergievskaya et al. 2022, Dhanapal et al. 2024). This process commonly occurs in a vacuum environment with the inclusion of sputtering gas, where magnetron radio frequency diode and DC diode sputtering are used. When a high voltage is applied to the cathode target, the gas accelerates and collides with free electrons, producing gas ions. These positively charged ions are accelerated at relatively high speed and bombard the cathode target

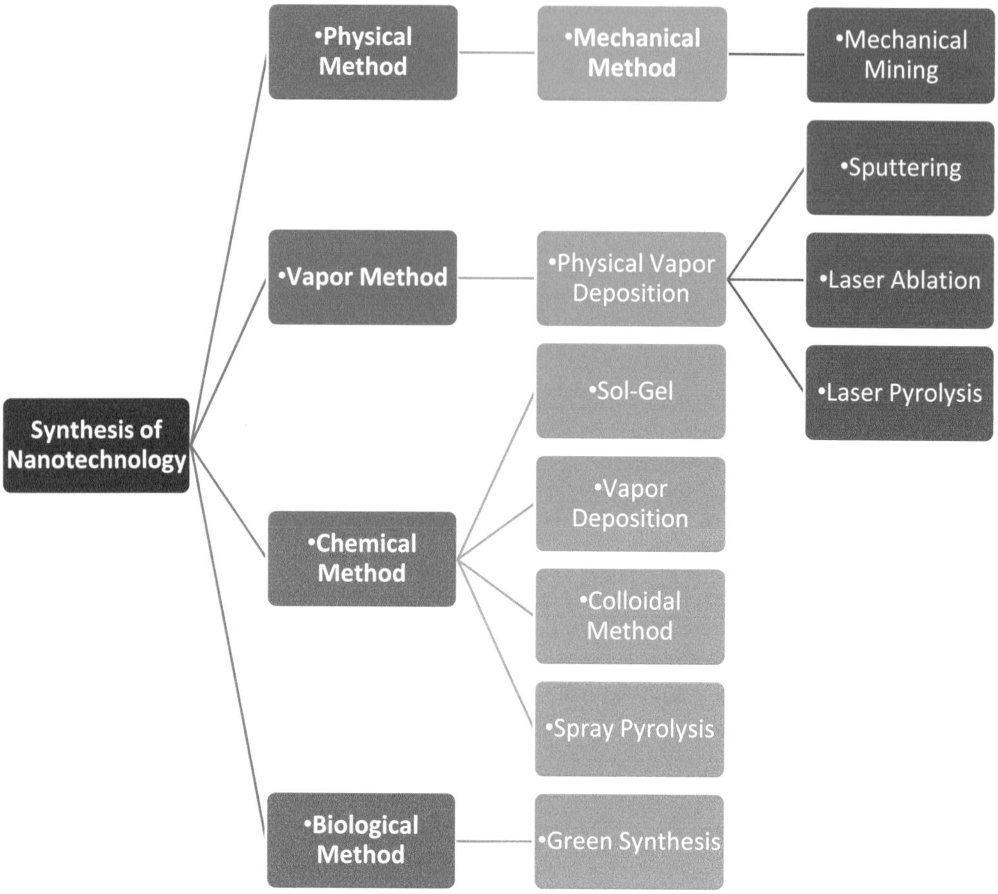

Figure 2. Nanotechnology synthesis by numerous methods.

in succession, knocking out atoms from its surface. For instance, WSe_2-layered nanofilms have been deposited on SiO_2 and carbon paper substrates via magnetron sputtering (Baig et al. 2021, Dhanapal et al. 2024). This method is comparatively cheaper than electron-beam lithography since the material sputtered is nearly pure, comprising mainly the target material (Sergievskaya et al. 2022, Baig et al. 2021).

3.1.3 Laser Ablation

Laser ablation synthesis produces NPs by focusing a high-energy laser light on the material, and the source material sprays due to laser radiation pressure ((Semaltianos 2016). Semaltianos et al. pointed out that laser ablation is a green technology that does not require chemical solvents or stabilizers. Thus, it can be used to build many kinds of NMs, including oxide composites, metallic NPs, carbon NMs, and ceramics. A rather unique technique that allows synthesizing colloidal nanoparticle solutions containing completely homogeneous particles that do not require ligands or surfactants is pulsed laser ablation in liquids (Subhan et al. 2022). In this way, it is possible to adjust the characteristics of NPs—such as size and dispersity—by varying the fluence level and wavelength, and by using the laser salt. It has also been observed that both the fluence and wavelength of the pulsed laser can affect the size of the Pd NPs generated in the process (Kim et al. 2014).

3.2 Chemical Methods

3.2.1 Sol-Gel

The sol-gel process consists of five major steps: hydrolysis, polycondensation, aging, drying, and heat breakdown (Danks et al. 2016). The metal precursors are treated through organic (non-aqueous) solvents and aqueous (water) solvents so that they undergo hydrolysis. Adjacent molecules come closer; water or alcohol is eliminated, and metal oxides are formed through polycondensation. Due to the continuous occurrence of condensation, the aging process alters the structure (Danks et al. 2016). When materials are rendered gel and then dried through methods such as freeze-drying, thermal drying, or supercritical fluid drying, various structures like aerogels and cryogels are formed (Koçyigit 2023). The final process is heat treatment, during which any remaining residuals, water, or alcohol molecules are extracted. This step is important to achieve the final density of the material, making the heat-treatment temperature one of the key control parameters (Danks et al. 2016).

3.2.2 Chemical Vapor Deposition

Chemical vapor deposition (CVD) is an essential process that directly aids in the synthesis of carbon-based NMs (Manawi et al. 2018). This is the procedure through which vapor-phase precursors undergo a chemical reaction and form a thin layer on the surface of the substrate. Manawi et al. (2018) noted that a precursor is suitable for CVD if it has high volatility, high purity, stability during evaporation, and is inexpensive, non-toxic, and has a long shelf life. Their decomposition also did not retain any residual contamination that could be traced. Consider, for example, the CVD technique for producing carbon nanotubes, where a substrate is placed in an oven and heated, and a carbon source in the form of hydrocarbons is slowly introduced as gas. Manawi et al. (2018) reported that the gas is thermally decomposed and the atmosphere is transformed into carbon nanotubes. The type of catalyst used is one of the most critical factors that determine the kind and shape of the nanomaterial synthesized. CVD-based graphene synthesis with Ni and Co catalysts yields multilayer graphene, while Cu as the catalyst yields monolayer graphene. Overall, CVD is a successful approach for high-quality NM fabrication and is widely accepted as effective for the fabrication of two-dimensional nanomaterials (Manawi et al. 2018).

3.3 Biological Methods

3.3.1 Green Synthesis

The nanotech industry seeks to support nanotechnology as a sustainable option for the development and enhancement of existing industries while minimizing environmental impact (Rajakumar et al. 2023). It aims to achieve sustainable economic growth through reduced energy and resource consumption. Because plant extract-mediated nanoparticle production requires less time for cell upkeep than microbial methods, it is an environmentally friendly approach (Abuzeid et al. 2023). Precursor preparation for nanoparticle production, characterization, phytochemical screening, and leaf extract extraction are crucial stages. Synthesis is influenced by variables such as pH, temperature, and duration. Using plant extracts, the green method breaks down organic compounds, primarily polyphenols. Despite a rise in the production of NPs, their limited application in wastewater treatment is evolving, potentially offering an alternative water supply (Kamarudin et al. 2021).

4. Nanotechnology and Various Nanomaterial Applications

NMs may be divided into four main categories: metallic, carbonaceous, dendritic, and composite (Khan et al. 2019). Carbon-based materials have been essential throughout human history, widely used in homes and for daily necessities from prehistoric times (Soffian et al. 2022). Numerous carbon

Table 1. NMs for improving soil structure and associated properties.

Nanomaterial	Associated Soil Physical Properties	References
Nano-silica, nano-zeolite	Soil moisture content, shear strength, swelling characteristics, dry density, Atterberg limits	(Hareesh and Vinothkumar 2016)
Nano-copper, nanoclay, nano-magnesium	Soil compaction characteristics and strength, Atterberg limits	(Majeed et al. 2014)
Nano-lime	Soil strength	(Govindasamy et al. 2017)
Carbon nanotube, carbon nanofibers, nanocarbon made of coconut shell	Soil moisture content, dry density, hydraulic conductivity, Atterberg limits, interparticle spacing, available soil water content	(Alsharef et al. 2016)
Nano-titanium dioxide and nano fly ash	Load-bearing capacity of soil, shear strength, dry density	(Pramanik et al. 2020)
Nano-Fe_3O_4, nano-MgO	Bulk density of soil, tensile strength of soil aggregates	(Bayat et al. 2018)
Nano-alumina (Al_2O_3)	Swell–shrink behavior of soil, hydraulic conductivity	(Taha and Taha 2012)
Graphene oxide nanosheet	Tensile and shear strength of soil	(Naseri et al. 2016)

compounds in nanoform that function at the molecular and submolecular levels have also been made possible by the development of nanotechnology. Compared to their bulk-size counterparts, these carbon nanostructures exhibit unique capabilities that have piqued the interest of many researchers exploring their electrical, physical, mechanical, sensory, and chemical properties (Soffian et al. 2022). Research on the various carbon nanomaterial allotropes and their intrinsic characteristics is now underway. The functionalization of carbon-based materials is a critical requirement for various applications. These materials are increasingly being used commercially in technological, environmental, and agricultural fields (Fritea et al. 2021). Here is the summary of NMs for improving soil structure and associated properties.

4.1 Dendrimers

Dendrimers have a highly branched molecular structure with complex three-dimensional architecture (Fernandes et al. 2022). These structures have led to significant interest in appending various molecules for enhanced properties such as solubility and bioavailability. As pointed out by (Palmerston Mendes et al. 2017), dendrimers have found applications in drug delivery, including gene delivery, triggered drug release, and antimicrobial and anticancer therapies. Dendrimer-based biosensors are typically fabricated using a layer-by-layer approach and are used as electrochemical, fluorescence, glucose, and QCM detectors. In their recent work, Ramesh et al. described an example of such a system (Ramesh et al. 2022).

4.2 Composite Nanomaterial

Natural polymer nanocomposites have gained interest in scientific and industrial fields because they help overcome the environmental challenges of petroleum-based polymers (Wypij et al. 2023). The packaging and food industries utilize sustainable nanocomposites based on biopolymers derived from chitosan, starch, cellulose, and alginate (Baranwal et al. 2022, Dhanapal et al. 2024). In essence, they act as slow-release nanocarriers of agrochemicals and enhance crop production in the process. Protection from pathogens, seed germination, shelf-life extension, and coating are other functionalities of hydrogels and nanofilms based on biopolymers. The incorporation of biopolymers and nanofillers improves the mechanical strength and barrier properties of food packaging. This article contains information on the use of some of these coatings in agricultural products, hydrogels, and nanocarriers. It also examines the challenges, risks, opportunities, and public perception of nanotechnology in packaging, food, and farming (Baranwal et al. 2022).

5. Nanotechnology Applications in Soil Management

Nanotechnology is a relatively new aspect in agriculture, especially in enhancing the phytoremediation capabilities of plants regarding the soil and water, which defines its agricultural relevance (Romeh and Saber 2020). For these reasons, NPs show several advantages over conventional techniques for remediating soil (Sathish et al. 2024). Due to their very small particle size, these products offer very high efficacy when used in the application of nanotechnology. This looks at the application of nanotechnology in phytoremediation, as well as the possible uses of nanotechnology. As described in detail below, there are several types of NPs that help in cleaning and detoxifying numerous pollutants (Dhanapal et al. 2024). Further investigation into this fascinating field is still expected to be provided by the current research projects and the cooperation between various academic disciplines. However, planetary reclamation is still a formidable challenge, and nanophytoremediation has emerged as a feasible approach to address it. In their view, Gomiero (2016) indicate that soil degradation and pollution are significant global environmental issues that affect food security, agricultural productivity, and human health. One of the major causes of soil exploitation and degradation is the exhaustion of soil resources that comes with the compounded demand for food production under the growing world population (Gomiero 2016). Excessive use of pesticides and heavy metals, among other persistent organic pollutants (POPs), has also contributed to the increase in soil pollution. Sites with such compounded, contaminated soil are at great risk of infiltrating the food chain since these chemicals mostly tend to bioaccumulate (Alengebawy et al. 2021).

Agricultural output has been drastically lost, as there is limited capacity to meet two divergent needs: the need for more food production and the need to prevent the loss of soil health. Among them, Nano-Enabled Soil Remediation appears to be one of the optimal and eco-friendly solutions to soil reclamation (Prakash et al. 2023). The application of nanotechnology is considered one of the efficient techniques in the management of environmental recoveries and treatment of soils from contaminants, as it is practiced at lower costs and with ease of application (Ibrahim et al. 2016). For this, a study based on the results obtained through the analysis of pesticides, heavy metals, their residues, and POPs considered the potential of nanotechnology involving soil remediation and discussed how it could further improve phytoremediation and bioremediation (Tufail et al. 2022). Consequently, the usage of nanotechnology has steadily risen in the field of soil management. It evaluates the benefits and drawbacks of NMs for soil, human health, and terrestrial creatures. The use of nanotechnology for soil remediation has presented a number of challenges, one of which is the possible inhibition of soil enzyme activities by the negative effects of NPs on pathogens (Figure 3).

There is no shortage of data on the novel applications of nanotechnology and nanoremediation, especially in relation to soil pollution. Hexavalent chromium is remarkably 100% removed by iron NPs (Prema et al. 2022). Using NMs in soil remediation eliminates soil disposal and reduces contaminants, cleaning time, and costs (Alazaiza et al. 2021). According to Liang et al. (2022), carbon nanotubes have a high adsorption capability for both organic and inorganic cleaning, whereas nZVI NPs immobilize heavy metals. Studies on the biological synthesis of nZVI soil remediation materials have also been carried out recently, and the results are encouraging (Liang et al. 2022). While NPs serve in soil treatment, ecosystems and soil health are seriously threatened by the growing concentration of metal and metallic oxide engineered NPs (ENPs) in agricultural soils (Suazo-Hernandez et al. 2023). They also contribute to improved hydraulic conductivity and affect nutrient access and availability. Chemical and physical properties play a significant role in the quality and health of the soil, and ENPs' import every year may lead to the degradation of such qualities (Suazo-Hernandez et al. 2023). Furthermore, the interactions between the nutrients, ENPs, or cations released in the soil could lead to the formation of precipitates and complexes that affect the concentration of nutrients soluble in the soil (Suazo-Hernandez et al. 2023). In the past, nanotechnology was applied in the management of soil with increased success, and the newer focus has been on individualized applications, preservation, intelligent delivery systems, disciplinary

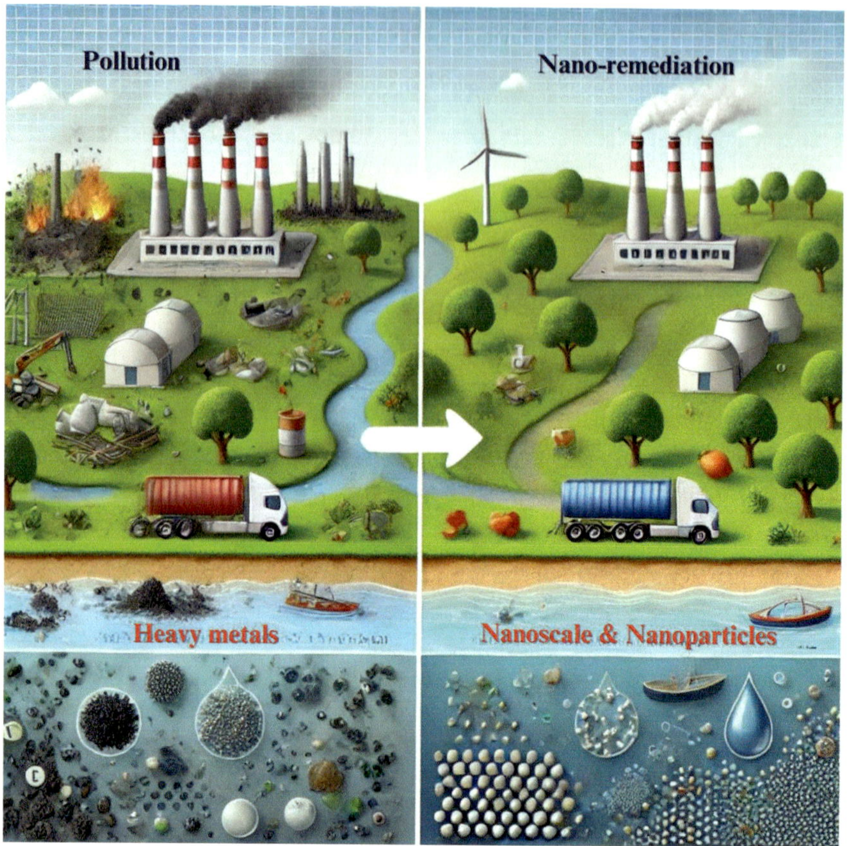

Figure 3. Ecological pollution management using nanotechnological methods.

cooperation, and a complete understanding of the risks. These discoveries contribute to the progress of nanotechnology in preventing and treating this form of land degradation in an effective and sustainable manner.

5.1 Mechanism of Nanotechnology in Soil Management

Soil management commonly uses in-situ methods. For the treatment of soil pollutants, a range of *in situ* techniques predominate in the management of soil. Different approaches have been employed to decontaminate soil pollutants, such as adsorption, Fenton and Fenton-like oxidation, immobilization, reduction processes, and nanotechnology combined with bioremediation (Usman et al. 2022). Controversy has recently emerged regarding the possible link between nanotechnology and bioremediation. Table 2 also provides a summary of the NMs and nanotechnology used for in-situ remediation of soil contaminated with heavy metals, organic compounds, or metalloids. Normally, adsorption with the help of NPs is effective for eliminating inorganic contaminants consisting of metalloids and heavy metals (Dhanapal et al. 2024). It is also possible to eliminate organic pollutants in parallel with the action of catalysts via degradation and reduction. NMs enhance adsorption and oxidation mechanisms to facilitate the degradation and elimination of micropollutants that persist in the soil (Nzilu et al. 2023). Carbon NMs, iron(III) oxide (Fe_3O_4), titanium oxide (TiO_2), zinc oxide (ZnO), nZVI, and nanocomposites are among the most frequently employed nanotechnological applications in soil management for pollutant elimination. However,

Table 2. List of Heavy Metals and Environmental Pollution Sources.

Heavy metals	Sources	References
As	Ceramics, coals, geochemistry, animal supplements, metallurgy, electrical production, and pesticides.	(Cooper et al. 2020)
Ag	Electroplating, refining of copper, gold, nickel, and jewelry industries.	(Sudarman et al. 2023)
Cd^{2+}	Coal burning, metal-coated batteries and pigments.	(Bhuyan et al. 2019)
Cr(VI)	Ferroalloys, metallurgy, mining, leather tanning, stainless steel production.	(Tchounwou et al. 2012)
Cu^{2+}	Metals, pharmaceutical and chemical sectors	(Baby et al. 2019, Hong et al. 2020)
Fe	Industrial emissions, steel manufacturing, and natural geological sources	(Jayakumar et al. 2021)
Hg^{1+}	Industries of catalyst, metallurgy, paper and pulp, and agriculture and pharmaceuticals	(Alina et al. 2012)
Ni^{2+}	Catalyst, ceramics and glass batteries.	(Baby et al. 2019, Hong et al. 2020)
Pb^{2+}	Pesticides, fertilizers, metal plating, batteries and leaded gasoline.	(Wei et al. 2019, Gebeyehu and Bayissa, 2020)
Zn^{2+}	PVC stabilizers, paint, rubber, zinc alloys and stabilizers	(Schutte et al. 2008)

nZVI is the most commonly applied nanoparticle for eradicating heavy metal pollution due to its effectiveness against key pollutants such as hazardous chlorinated organic compounds, metals, and inorganic chemicals (Ken and Sinha 2020).

5.1.1 Reduction Reaction

Reduction processes have great potential to remediate organic compounds and heavy metals in contaminated soil, water, and groundwater when in the presence of nZVI NPs (Galdames et al. 2020). The extensive application of nZVI particles in numerous fields is attributed to their large surface area and nanoscale size, which improve remediation efficacy by enabling direct contact with contaminants. After nano zero-valent iron particles are introduced into polluted soil, they exhibit strong reductive capabilities and high adsorption capacity. For example, chromium(VI) is reduced to chromium(III) or forms ferrous chromite and other compounds (Liang et al. 2022). Biochar enhances iron molecule dispersion and reduces the mobility of the mixture in the soil by increasing reduction reaction activity and enhancing the effectiveness of nZVI NPs (Sharma et al. 2022, Dhanapal et al. 2024). For instance, 66% of the Cr(VI) concentration in the soil was eliminated when biochar and nZVI were used together. In one study, 1 g of nZVI applied to polluted soil reduced 28% of 1 kg of chromium(VI). Within a 24-hour period, 98% of the chromium(VI) was eliminated at a pH of 5 (Sharma et al. 2022).

5.1.2 Immobilization

Low-cost and eco friendly management of polluted soil through in-situ immobilization has proven feasible and has attracted significant global interest (Mak et al. 2022). In general, the characteristics of the pollutant and soil have a significant influence on the choice of NPs for immobilization treatment. Functionalized metal oxide and carbon NPs are commonly applied in immobilization remediation processes (Baragaño et al. 2021). In immobilization remediation, the carbon nanomaterials such as graphene, fullerene and carbon nanotubes function as adsorbent through π–π interactions and van der Waals forces for the removal of organic pollutants (Zhang et al. 2022). Carbon NPs' hydrophobic surface and strong adsorption enhance their ability to remove organic pollutants from soils. Of all the compounds present in the soil, carbon nanotubes have remarkable adsorption properties regarding the organic molecules. They prefer π–π and cation–π interactions and show selectivity in adsorbing ionizable organic compounds such as insecticides. Carbon nanotubes form hydrogen bonds with electron charges and have low-barrier surfaces. Various studies have explored the use of carbon nanotubes in the electrochemical reduction of organic species, especially polycyclic aromatic hydrocarbons (PAHs) (Lv et al. 2024). For example, the availability of C-nanotubes in soil limits the ability of PAHs to come in contact with crops and soil-dwelling microbes by hindering the PAH transport. The capacity for adsorption of carbon nanotubes depends on the concentration of oxygen, while the functional group of -OH enhances the adsorption by enhancing the interaction between π and π and the -OH functional group (Qiu et al. 2023).

5.1.3 Nanobiosorbents

Current pollution levels are high due to the evolution of the industrial sector, which is increasingly modernized and urbanized. Environmental assessments also show that various pollutants present in the environment have detrimental effects on humans, animals, and plants, and they also degrade the beauty of natural surroundings. In response, scientists and researchers have developed environmentally sustainable, efficient, and innovative techniques for toxin elimination. One such approach is the development of nanobiosorbents for the extraction of contaminants using various natural and renewable resources (Tofan and Suteu 2023). Due to their applications in water and wastewater purification and the availability of diverse adsorbent materials, nanosorbents are generally regarded as the most effective method across all level-1 techniques. According to Hassan et al. (2020), numerous biosorbents and nanoadsorbents can potentially be used for the removal of contaminants from fluids. Some of these include agricultural wastes, microbial biomass, nano-MgO, Fe_3O_4, CaO/Fe_3O_4, and activated carbon/Fe_3O_4 nanocomposites (Osman et al. 2023).

5.1.4 Nanobiosurfactants

Serious threats to human and environmental health have been associated with the massive global pollution of coastal zones and marine sediments by various pollutants, including heavy metals, halogenated compounds, crude oil and Polycyclic Aromatic Hydrocarbons. The quality of human life, stocks of fish, the structure of marine life, and the overall economy suffer from these chemicals. The methods used in cleaning up contaminated deposal sediments are urgently required to address this issue in an environmentally friendly approach. Despite difficulties in solubilizing certain contaminants, microbial and plant-based biological remediation techniques are becoming more popular despite the strength of physiochemical treatments. As a result, more research is being done on combined biotechnologies that use biosurfactant replenishment in management systems. Amphipathic biomolecules that make up biosurfactants provide special qualities like decreased surface tension, wettability, low toxicity, enhanced solubility, high emulsification, low critical micelle concentration, and chemical stability in harsh environments (Bjerk et al. 2021). This section explores how biosurfactants can be used to remediate organically polluted marine sediments, with an emphasis on how different coastal locations can maintain their environmental sustainability. The biosynthesis of biosurfactants under both aerobic and anaerobic conditions, environmental appropriateness characteristics (EACs), methods of applying the product, and the behavior of biosurfactants and contaminants during the clean-up process are all considered in the discussion. Also given are the recent advancements and future trends in fabrication of efficient, environmentally friendly biosurfactants for bioremediation of marine silt (Gayathiri et al. 2022).

6. Recent Advances in Nanotechnology

6.1 Innovative NMs

6.1.1 Types of NMs Used in Soil Management

The emergence of nanotechnology in recent years has led to the synthesis of new NMs for soil management. These include metal oxide nanoparticles, carbon nanoparticles, and polymeric nanoparticles, each with properties that make them soil-compatible. Some important metal oxide nanoparticles used in crop cultivation are zinc oxide (ZnO), titanium dioxide (TiO_2), and iron oxide (Fe_3O_4), which are vital in soil management for enhanced nutrient uptake and efficient soil remediation (Kah et al. 2013). These nanoparticles can be used to enhance agricultural practices while minimizing the adverse impacts of chemical fertilizers and pollutants. For example, a relevant application of ZnO NPs includes the addition of zinc to marginal soils with low zinc content for crop growth and development (Pramudita et al. 2018).

Additionally, polymer-based nanomaterials such as polymeric nanogels and polymeric nanocapsules have been evaluated due to their ability to release nutrients and agrochemicals either rapidly or slowly, as needed. Such materials can manipulate target performance by encapsulating active ingredients or preventing their degradation, then dispersing them into the soil in a controlled manner (Solomon et al. 2015). This targeted and efficient delivery technique reduces the quantity of chemicals required, thus lowering environmental risks while improving soil health.

6.1.2 Properties Enhancing Soil Health and Productivity

NMs have some special features that apply to the general enhanced silicon improved soil and general productivity. Some of these properties are high surface area to volume ratio, which make the particles more reactive and good in adsorption. It also increases the NMs' ability to interact with soil particles, nutrients, and contaminants, thereby improving the soil structure and fertility (Raliya et al. 2015). Based on the functions, the NMs bear more enhanced catalytic and photocatalytic efficiency for soil purification and nutrient recycles because of size and surface area level. For example, ZnO NPs derived from the plant extract can remove the organic contaminants in the soil as well to improve the growth of healthy soil (Gogos et al. 2012). Similarly, iron oxide NPs can mediate redox processes and result in the sedimentation of heavy metals to reduce their bioaccessibility and uptake by plants through water (Zhang 2003).

The other main feature of NMs is the improvement they have on water absorption and nutrient transport in soils. Graphene oxide and other forms of amorphous carbon-based NMs enhance the ability of the soil to retain moisture, which aids plants in resisting dryness (Liu et al. 2018). In addition, NMs have the potential to enhance microbial motility and abundance in a soil ecosystem by creating suitable habitats for beneficial microorganisms. For instance, some NPs like silver and copper NPs have the ability to inhibit microbial growth of soil pathogens, while other NPs are capable of promoting the growth of beneficial microbes required for soil nutrient cycling (Simonin and Richaume 2015).

6.2 Nanotechnology in Remediation of Heavy Metals

Around the world, heavy metal pollution is a serious environmental problem. One of the main reasons for concern is the steady rise in soil pollution from heavy metals brought on by human activities like mining and urbanization. Rubbish accumulation sites are locations where large amounts of waste are generated during mining operations (Briffa et al. 2020). Some areas are negatively impacted by these growing rubbish heaps, and some agricultural areas may become wastelands (Rashid et al. 2023). Heavy metals poison aquatic creatures, plants, and people when they are exposed to levels higher than those advised. The toxicity of heavy metals is high. While most people believe that heavy metals are detrimental to living things, lightweight metals like lithium and beryllium may also be dangerous. Some heavy metals are harmless in moderation and are required for human health, such as Cr3+, Fe3+, Fe2+, and so on. Acute or chronic toxicity can result from exposure to metals, and the degree of toxicity is contingent upon the exposure duration, route, and dose/quantity. The most stable oxidation states of chromium (Cr) are +3 and +6, while it can exist in other states as well. Chromium in its +3 form is necessary for humans because of its special nutritional and biological characteristics (Baby et al. 2022). Though naturally occurring, the pace at which heavy metals are created and discharged into the environment is concerning as a result of growing urbanization and industry (Table 1). These days, one effective way to eliminate pollutants from the atmosphere is through the application of biosynthetic NPs in nanotechnology. Adsorption is a popular method for removing heavy metals. Because of their small size and large surface area, NMs are strong sorbents with excellent adsorption capabilities that may remove heavy metal ions from contaminated water (Kolluru et al. 2021).

NMs are tiny enough to reduce some of the issues with conventional site rehabilitation at a cost that makes sense. They might also be retained for long enough durations to initiate the development of an *in situ* target (Kumar 2023). The *in situ* and *ex situ* capabilities of nanoremediation are equivalent to those of traditional methods (Table 2). Contaminants are removed at their source when using the *in situ* approach. They are moved to a different area for cleanup under the *ex situ* technique (Hussain et al. 2022). According to data from many studies, using NPs might enhance Pb, Cr, Cd, Zn, and Ni phytoremediation (Prekash et al. 2023). Recent studies indicate that metal oxide NPs have promise for removing dangerous metal ions from wastewater (Naseem and Durrani 2021). Only a small number of metallic NPs have been investigated for sorption because they are unstable when they agglomerate or separate. Furthermore, it is a difficult process to extract individual metallic NPs from the effluent. Although earlier studies have shown that nanotechnology may be used to remediate heavy-metal contamination, more recent research has focused on customized nanomaterial qualities, environmentally friendly synthesis methods, multifunctional materials, *in situ* applications, and sophisticated monitoring tools. These revelations advance the further development of nanotechnology for the environmentally safe and efficient cleanup of heavy metal-contaminated areas. Multiple NPs used for bioremediation are illustrated in Table 3.

Table 3. Different NPs were used for bioremediation.

NPs	Application/advantage	References
Cobalt and cobalt oxide NPs	A large surface area and sunlight	(Adekunle et al. 2020)
Electrospun cyclodextrin fibers	Bacterial bioremediation	(San Keskin et al. 2018)
Electrospun nanofibrous webs	Color loss due to biological processes	(Sarioglu et al. 2017)
Enzyme immobilized NPs	Immobilized laccase oxidation	(Ji et al. 2017)
Graphene oxide and carbon nanotubes	Numerous morphological features of mixed nanocomposites, including high surface area, π–π interactions and/or hydrophobicity, as well as holes between MOFs and platforms	(Ahsan et al. 2020)
Mesoporous organosilica NPs (MONs)	Increased conjugation and surface area as a result of noncovalent interaction mediated by ferrocene	(Yang et al. 2019)
NiO and MgO NPs	Zn^{2+} absorption is chemical and exothermic, whereas Cu^{2+} and $Cr3^{+}$ adsorption occurs spontaneously, endothermically, and physically	(Abuhatab et al. 2020)
Silica NPs	Photocatalytic degradation	(Mohanraj et al. 2022)
Zirconia NPs	Zwitter ions exhibit robust electrostatic interactions and chemisorptions	(Debnath et al. 2020)

6.2.1 Nanobioremediation of Organic Contaminants in Soil

NPs containing plants and microorganisms like bacteria or fungi, extracted using nanotechnology from polluted areas or regions containing heavy metals and various organic and inorganic pollutants, can be eradicated from the environment (Hidangmayum et al. 2023). This procedure is commonly referred to as nanobioremediation. As highlighted by Hidangmayum et al. (2023), nanobioremediation is gradually gaining recognition as a versatile technique for sustainable bioremediation. According to recent findings, bioremediation appears to be a cost-effective and environmentally sustainable solution for eradicating contaminants from the environment (Ayilara and Babalola 2023). Bioremediation can be performed through microbial, plant-based, or enzyme-mediated methods. For instance, nanobioremediation employs biological and physiochemical systems and is currently being developed in several polluted areas. It involves reducing pollutants to forms amenable to biodegradation using NPs, after which the pollutants undergo biodegradation (Hidangmayum et al. 2023). Nanobioremediation involves the removal of pollutants from land and water using biogenic NPs from microorganisms or plant extracts. Due to their high effectiveness, low price, and resistance to pollution, NMs are considered to be potential contenders as replacement options for standard medical practices over the past twenty years (Ayilara and Babalola 2023).

Iron-based treatments are known to be highly effective for treating polluted soil and groundwater, and several methods are available (Rather et al. 2023). Zerovalent iron NPs are particularly beneficial and important for nanoremediation, having been shown to efficiently eliminate acidic water contaminated with heavy metals, due to pollutants' solubility at the NP interface (Plessl et al. 2023). Nanobioremediation is a novel and dynamically developing method for decontaminating environments using biologically produced NPs (Chaudhary et al. 2023). The use of nanotechnology products such as nano-pesticides, nano-fertilizers, and sensors in agriculture is also gaining increased attention, with the aim of improving crop production. Many researchers have examined the prospects of nanotechnology, particularly its potential to enable pesticide dispersion through nanoencapsulation (Machado et al. 2023).

Nanoencapsulated pesticides represent a naturally evolved method to protect crops from pests, and their reduced usage also minimizes human exposure. One of the primary effects occurs during degradation at the plant surface, and the distinct characteristics of nanopastes significantly influence pesticide breakdown (Rani et al. 2023). Although earlier studies clearly demonstrated the capability of nanobioremediation in addressing organic pollutants in soil, recent advancements focus on precision, microbial consortium compatibility, smart delivery, biodeposition, field trials, and efficient monitoring techniques.

6.2.2 Nano-phytoremediation of Soil Pollutants

Soil pollution is a severe global issue because heavy metals are potentially toxic. Soil contamination with heavy metals is one of the major threats to ecosystem and human health (Ekeoma et al. 2023). Due to the rapid growth of the agricultural sector and allied industries, environmental and food safety concerns have intensified, with heavy metals becoming a top priority (Dhanapal et al. 2024). Moreover, disruption of the Earth's ecosystem due to global population expansion has raised the levels of heavy metals in our environment. Managing and preventing heavy-metal pollution remains one of the most prominent areas of research in environmental science (Rashid et al. 2023). The methods for eliminating heavy metals from the environment include chemical, physical, and biological processes. As noted by Ayilara et al. (2023), the decline in soil microbial populations, coupled with additional costs and side effects on soil texture, are significant drawbacks of physical and chemical restoration techniques.

Bioremediation is an innovative method involving the elimination of pollutants from the environment with minimal ecological damage using microbes, plants, animals, and other biological agents (Dhanapal et al. 2024). Phytoremediation, a subcategory of bioremediation, is an effective, inexpensive, and eco-friendly approach. The cleanup of heavy metals and hazardous chemical compounds through this method is becoming increasingly common (Dhanapal et al. 2024). It can also be employed to treat contaminated water and farmlands affected by radioactive materials. Phytoremediation is one of the cheapest technologies which do very well when pollutants are localized in the root zone of plants. Flax (Linum usitatissimum) is similarly suitable for Cu-phytoremediation, since the plant can be grown to produce flax seeds and tolerate and uptake medium to high Cu concentrations as well (Saleem et al. 2020).

There are various types of phytoremediation, namely phytostabilization, rhizofiltration, phytoextraction and phytovolatilization, through which heavy metal pollutants can be removed (Nedjimi 2021). In the context of depolluting the contaminated environment, this procedure implicates not only plants but also the rhizosphere microorganisms. Using the same scientific practice as nature does, microbes and plants employ phytoremediation as a way to minimize organic and inorganic pollutants (Joshi et al. 2023). While earlier studies revealed the efficiency of nanophytoremediation of soil pollutants, its latest aspects include understanding the uptake process, friendly plant–microbe interactions, tailored NPs, large-scale implementation, eco-friendly synthesis, multiple pollutants, risk analysis, and community engagement (Dhanapal et al. 2024).

6.3 Enhancements in Soil Microbial Activity

6.3.1 Interaction of NPs with Soil Biota

Soil biota interaction with NPs is one of the significant issues in present-day soil management. The functional properties of NPs can affect microbial function and, hence, microbial processes such as nutrient cycling, decomposition, and nutrient availability in the soil. Antimicrobial NPs present in the soil may affect the growth of both beneficial and harmful microorganisms. For example, copper and silver NPs possess antibacterial and antifungal properties that hinder the growth of harmful pathogens (Peralta-Videa et al. 2011). On the other hand, carbon-based NMs like graphene oxide enhance the growth of beneficial fungi and bacteria, which help in the disintegration of organic matter and the liberation of nutrients (Liu et al. 2018). Thus, the potential consequences of NPs on long-term nutrient cycling processes are still unknown, and the impacts of NPs on soil microbial communities are debatable. Further investigations into their risks and benefits are required.

6.3.2 Implications for Soil Fertility

Soil is a dynamic system in which NPs exert direct influence over microbial activity. Increased microbial activity enhances soil fertility. It is likely that NPs contribute to microbial growth and metabolic processes, accelerating nutrient cycling and the decomposition of organic matter, thus improving soil fertility and plant growth (Raliya et al. 2015). For example, iron oxide NPs have been shown to increase the soil's ability to absorb phosphorus—an essential plant nutrient (Kah et al. 2013).

However, it is clear that research conducted in the literature highlights that the NPs alter the soil fertility in a different manner based on the type of the nanoparticle that is used, concentration, and other factors. For instance, according to Ippolito et al. (2012), biochar is among the NPs that enhance the fertility of the soil while Simonin and Richaume (2015) established that the wrong application of NPs is risky for the soil. Therefore, more research on the pros and cons of NPs in its application to the soil is needed to understand how it enhances or harms the fertility and sustainability of the soil.

7. Benefits of Nanotechnology in Soil Management

7.1 Enhancing Soil Fertility and Productivity

7.1.1 Controlled Release of Nutrient

There are several benefits to applying nanotechnology in soil management, including the ability to address the limitations of conventional fertilizers. These nutrients are easily leached and volatile; rather than benefiting the plant, they often end up polluting the environment (DeRosa et al. 2010). A controlled release formulation at the nanoscale was developed with the nutrient-releasing properties to meet the requirement of a plant for an extended period. This allows less leaching of nutrients to reduce pollution and increases nutrient uptake, hence increased production (Naderi and Danesh-Shahraki 2013). For example, if the nutrient is encapsulated within nanoclay or a polymer-based NP, then the nutrient will only be used at a time when it is needed by the plant. This selective placement has been useful in increasing the fertility of the soils and reducing the cycles of application of fertilizers which enhances sustainable agriculture (Solanki et al. 2015).

7.1.2 Improved Nutrient Uptake by Plants

Nanotechnology also improves nutrient uptake by plants and enhances soil quality, thereby increasing fertility and productivity. Nanofertilizers can be formulated to have higher bioavailability than commercial fertilizers, making it easier for plants to absorb nutrients (Raliya et al. 2015). Since NPs are small in size, they can easily penetrate the plant cells, which ensures delivery of important nutrients such as nitrogen, phosphorus, and potassium.

Therefore, besides helping in improving nutrient uptake efficiency, nanofertilizers can stimulate formation of more roots, implying that the area of roots will be larger and therefore the plant's ability to uptake water and nutrients from the soil is enhanced (Pramudita et al. 2018). This in turn improves nutrient uptake by plant tissues leading to healthier plants, increased crop yield and better soil health. In addition, the use of nanofertilizers may reduce the use of chemical fertilizers to avoid cases of soil erosion and ecological pollution (Naderi and Danesh-Shahraki 2013).

8. Soil Remediation

8.1 Removal of Contaminants Using NMs

Soil degradation is probably one of the most significant threats facing agriculture due to industrial pollution and overuse of chemical fertilizers. The use of nanotechnology in the soil remediation processes involves the use of NMs in cleaning up the material from the soil, as highlighted in the following: heavy metals, organic pollutants, and pesticides (Zhang 2003). Carbon-based NMs and nZVI are highly reactive and have large surface area that allows them to adsorb and reduce the contaminants in the soil matrix (He and Zhao 2005). For instance, nZVI particles can reduce the toxicity of heavy metals such as arsenic and lead and encapsulate and immobilize them in the soil, thus they are not able to leach into groundwater (Zhang 2003). Similarly, it has been demonstrated that graphene oxide NPs are capable of adsorbing organic pollutants, thus reducing their bioavailability in the soil and their toxicity (Liu et al. 2018). They are subsequent NMs that not only help in cleaning the soil but make it workable once more for farming.

8.2 Restoring Degraded Soils

Besides the aspect of detoxification, nanotechnology can also be applied in the reclamation of degraded soils. Findings evidence that NMs have the potential for better influence on both physical and chemical properties of the soil to rehabilitate degraded soils (Rajput et al. 2022). For instance, application of the biochar NPs on farmland has shown beneficial influences on soil fertility and healthy development of microorganisms, with the consequence of improved soil health and better production (Liu et al. 2018).

9. Environmental Impacts and Sustainability Considerations

9.1 Advanced and Sustainable Practices

The research has specified focus on the bioremediation of contaminated agricultural soil, which is an area where nanotechnology opens new opportunities for an emerging environmental problem. This dual challenge of environmental science and nanotechnology is well captured by the following list of NMs that have been explored for soil remediation within the last few years, as identified by Dhanapal et al. (2024). One of them is on using nZVI particles, especially since the particles have great potential in the degradation of various contaminants (Shi et al. 2018). These particles might be some means through which the quality of the soil can be enhanced because they can be made to specifically capture specific pollutants. Further, it was observed that the NMs with the most efficient targeted content and adsorption were carbon-based NPs and TiO2 in modifying the negative impacts of the agricultural practices on the soil health (Sharma and Kumar 2024). In a similar manner, nanoscale delivery systems, including nanocarriers and nanosensors, have also been deployed in precision agriculture for pollution control (Yadav et al. 2023).

Thus, via the utilization of these nanodevices for the targeted distribution of repair chemicals as well as real-time identification of soil conditions in need of repair for agricultural purposes, soil remediation may be effectively achieved in the future in a proactive and sustainable manner. However, it is essential to analyze the risks that could be associated with the application of NMs in the agricultural sector. Based on the above-discussed revelations, implications for acquiring knowledge of the nanotechnology-based soil remediation techniques, their trend and longevity, destiny of NPs, and their toxicity are discussed. Nanotechnology, when applied to the remediation of agricultural soil, opened a lot of opportunities in coming up with innovative solutions to the complex issues that arise from pollution of the soil.

9.2 Potential Environmental Risks of NMs

9.2.1 Toxicity to Soil Microorganisms

NMs, despite their potential benefits in agriculture, pose significant risks to soil microorganisms, which are crucial for maintaining soil health and fertility. Microbe, including bacteria, fungi, and other microbes, are important in nutrient cycling, decomposition of organic matter, and enhancing plant growth (Simonin and Richaume 2015). When NPs enter the soil solution, they are able to interfere with microbial populations, which can become detrimental to the soil.

First, one of the main issues relating to NP use is their toxicity. For instance, Ag and Cu NPs possess antimicrobial properties. At the same time as these NPs may be useful in the containment of dangerous pathogens, they are equally detrimental to the useful bacteria that enrich the soil (Kah et al. 2013). For instance, copper NPs can inhibit the functionality of soil enzymes, which play crucial roles in the cycling of nutrients and degradation of organic matter in the soil (Peralta-Videa et al. 2011).

The impacts of nanoparticle exposure on the long-term persistence of MS populations in soil have not been ascertained. More recent evidence suggests that long-term exposure to NPs will alter microbial community structures, which decreases the stability of soil communities against environmental stress (Simonin and Richaume 2015). Therefore, it becomes necessary to evaluate the effects of NPs used in agriculture and to identify ways to minimize adverse effects on the soil microbial communities.

9.2.2 Bioaccumulation Concerns

Another environmental factor that is linked to NMs is the issue of bioaccumulation. Bioaccumulation is the gradual increase in the concentration of a particular chemical, including NPs, in the body tissues of living organisms. This process happens in plant and animal circumstances, which means the NPs pass through the food network and create risk to human life and the environment (Servin and White 2016).

The particles can get into the food chain in several ways, including through plant uptake from the soil. It has been established that some NPs like ZnO and TiO_2 are bioavailable and could be translocated through the root system into the aerial parts of the plant including leaves and fruits (Raliya et al. 2015). These NPs, once present in the plant tissues, can also accumulate and be ingested by herbivores, and hence transferred to higher trophic levels.

Bioaccumulation indicates that incorporation of NMs in agriculture poses long-term dangers to health and the environment. While some NPs may be relatively non-hazardous and safe for the environment and human health, others, especially with toxic characteristics, may be detrimental to the health of wildlife and humans if they enter the food chain (Servin and White 2016). Moreover, NPs can remain stable in the environment and hence can accumulate in the soil as well as water sources, which will further worsen their effects on ecosystems. More investigations, therefore, need to be done on the bioaccumulation of different NMs and measures put in place that can be used to ensure their safe applications in agricultural practices.

9.3 Life Cycle Assessment of Nanotechnology in Soil Management

Life cycle assessment (LCA) is one of the most important techniques for estimating the possible environmental implications of nanotechnology in relation to the management of soil. The estimation by LCA includes not only the environmental burden connected with the manufacture of products but also from raw material extraction to manufacturing, usage, and final disposal of the product (Cucurachi et al. 2013). It is with this that the LCA puts into focus areas where nanotechnology can become highly threatening to the environment, ensuring that best practices are in place for environmental stewardship.

Thus, thirdly, LCA also ensures better information with a centralized approach towards the impacts of NMs on the environment. For instance, even though the main aim of using nanofertilizers (NFs) is to cut down on the use of chemical fertilizers and nutrient leaching, these NFs can basically involve the process of energy consumption (Hischier and Walser 2012). The strength of the LCA is in providing a holistic outlook on the impact of NMs and possible ways to mitigate the undesirable effects.

Besides, LCA can also be used for decision-making by comparing the environmental impact of nanotechnology-based techniques for soil management with traditional practices. For example, a comparative LCA could show that although nanotechnology can yield numerous benefits including lower nutrient losses and optimum yields, it can also result in higher energy intensities and GHG emissions during production (Khanna and Bakshi 2009). With such pros and cons in mind, the stakeholders of the soil would be in a better position to decide whether they should embrace or reject the use of nanotechnology and what measures should be taken to reduce the negative effects of its application.

9.4 Ensuring Sustainable Practices and Safety Regulations

Due to its potential threat to both the environment and the well-being of mankind, it is crucial that a standard framework and guidelines for nanotechnology in soil management be developed in order to avoid the adverse effects. NMs should be appropriately regulated based on properties such as toxicity, bioaccumulation, and persistency in the environment (Kah et al. 2013). These regulations should be informed by sound scientific evidence and should make it mandatory for manufacturers to present information on the safety of their NMs for use in agriculture (Gogos et al. 2012).

Along with regulations, the most important factor for the development of norms of appropriate usage of the nanotechnology in agriculture is vital. Such practices should aim at reducing the emissions of the NMs—for instance, power usage in production, application of nanofertilizers and pesticides (which should be applied in moderation), and elimination of improper disposal of NMs, which can pollute the soil and water systems (Raliya et al. 2015). Moreover, accreditations, farmer education, or training in the safe use and handling of NMs themselves can effectively support the sustainable adoption of those materials.

10. Challenges and Considerations

10.1 Technical Challenges in the Application of Nanotechnology

Soil pH is one of the important characteristics dependent on NP existence. This heavily influences nutrient recycling within the soil, health, seed germination, and plant growth. A lot of emphasis has been placed on the fact that NPs facilitate changes in pH in soils, and since pH is one of the most influential factors that determine nutrient availability in soils, microbial activity, soil health, and plant growth and development (Rajput et al. 2022). In addition, NPs of Ag, Au, Ti, and Zn can alter the pH of soil and reduce the population of beneficial bacteria and nematodes in the soil (Khan et al. 2021). The fate of NPs in the soil depends on the type and concentration of NPs, the enzymatic activity of the soil, and the type of soil. Furthermore, lower dehydrogenase activity correlates with higher NP concentration, which creates an unfavorable balance of nutrients and soil fertility. Additionally, the uptake and incorporation of these NPs affects microbes, negatively interfering with mycelial processes such as cellular activities that occur periodically (Dhanapal et al. 2024).

The opportunity for nanotechnology in the field of agriculture is promising but not free of technical complexities. One of the major issues is the fact that there are limited data on the performance of NPs in various soil conditions. Soils are also complex systems in terms of their physical, chemical, and biological characteristics, which are indicated to influence the NMs' performance and stability (Gogos et al. 2012). For example, the adsorption of NPs with soil particles, organic matter, and microorganisms can affect their mobility as well as their bioavailability and efficacy in the context of soil application. Due to these factors, it becomes highly challenging to predict the future behavior or effects of NMs in the soil environment.

Some technical considerations are associated with the proper choice of NMs' delivery systems. Nevertheless, nanofertilizers and nano-pesticides are effective because they have controlled release and targeted delivery, but the challenge lies in how to deliver these materials to the target site in the soil or plant tissues (Kah et al. 2013). The dispersion and penetration of NPs in the soil may be restricted by certain factors, such as the size of NPs, surface charge, and environmental conditions.

Maintaining NMs for agricultural production also brings more technical challenges that need to be addressed to enhance their production. The issue with NMs is that it is quite hard to produce them in large numbers while maintaining their properties and, simultaneously, managing to produce them as cheaply as possible. Further, there is a need to ensure that NMs in large-scale production are stable and exhibit uniformity, which is fundamental to applying them in agriculture (Dimkpa and Bindraban 2017).

10.2 Economic and Regulatory Barriers

However, there are still some economic and legal issues pertaining to the use of nanotechnology in agriculture, which might also pose some technical difficulties. Among the primary concerns raised in the economic sphere is the great expense of creating and manufacturing NMs. Nanotechnology investment involves significant capital expenditure on structures, tools, and personnel with high skill sets in research and innovation. Thus, the initial cost of NMs can pose a significant barrier to entry for farmers, especially smallholder and resource-poor farmers in the developing world (Roco 2011). Due to high costs, nanotechnology application could be expensive and out of reach for many farmers, especially those living in the developing world.

Another economic factor is the absence of reasonable market pull, since nanotechnology products—especially in agriculture—may not be readily accepted by the market. However, the value of the products derived through the application of nanotechnology is not highly noticeable by farmers and consumers. This ignorance and consequently the poor perception of investing in nanotechnology can limit its market acceptance and commercialization (Kuzma and VerHage 2006). Furthermore, the application of nanotechnology requires substantiation regarding the costs associated with implementation and the benefits expected to be received from this technology in the future.

Regulatory issues are yet another concern that have major implications for the utilization of nanotechnology within the agriculture business. At the moment, a significant number of countries worldwide have no explicit policies and standards with regard to NM application in the agricultural sector (Kah et al. 2013). Since the NM has not been fully legally defined, it remains problematic in terms of its regulatory status and the ability of firms to gain approvals for new products. Further, international regulations remain uncoordinated, making the export and import of such agricultural products containing nanoparticles a difficult affair due to existing variations in safety standards across countries.

10.3 Potential Risks and Ethical Considerations of Using Nanotechnology in Agriculture

Despite their perceived benefits, nanotechnology applications in agriculture entail the following risks and ethical issues that need to be considered. Environmental toxicity and impact on organisms other than pests is one of the biggest dangers per the argument outlined. NMs are potentially hazardous to soil quality, water, and biodiversity if they are allowed to build up or if they are released in large amounts (Simonin and Richaume 2015). For instance, utilization of nano-pesticides could prove highly detrimental to useful insects, soil organisms, and aquatic life and hence cause an interruption of balance in ecosystems.

Another significant concern is the toxicity of NMs to the human body. Despite its advantages in agriculture, there is evidence that shows the possible dangers of being exposed to NPs in the long run. There are reasons to consider NPs as a potential hazard to food safety when they are taken up by plants or contaminate water used in plant irrigation (Servin and White 2016). Concerning the bioaccumulation and toxicity of the NPs to humans and animals, there are still more questions that need answering and precautions that need to be taken to avoid their adverse effects.

There are also obvious ethical issues involved in the availability and distribution of nanotechnology in agriculture. These constraints might result in inequality between commercial large-scale farmers and smallholders or those farmers who just sustain themselves. Mearns (2007) notes that this raises concerns about the possibilities of growing inequality in the agricultural sector whereby the wealthy farmers derive more benefits from the advancement in technologies and the poor farmers are left farther behind, as pointed out by Roco (2011). Because precise and accurate information on nanotechnology development can be expensive, it falls under the important ethical consideration to ensure farmers across all categories benefit from the technology.

Besides the issues of effectiveness and efficiency, there is the issue of transparency and public participation in the application of nanotechnology in agriculture. The covert nature of nanotechnology and resultant products can cause consumer skepticism and hostility. This should be facilitated through stakeholder engagement that encompasses farmers, consumers, and environmental groups to ensure appropriate and responsible deployment and application of nanotechnology (Kuzma and VerHage 2006). Creating an environment of dialogue with the public about the potential of nanotechnology and the potential threats it might pose to society, on a positive note, is important in order to create a positive attitude toward the use of this technology as well as its proper utilization.

11. Conclusion

Agriculture is one of the sectors that can benefit a lot from nanotechnology in sustainable soil management such as nanofertilizers, nano-pesticides, and NMs for soil remediation. These innovations can help in increasing soil fertility, increasing nutrient uptake by plants, and reducing the negative effects of conventional practices on the environment. Environmental health in the future is shown by nanotechnology besides giving directions on solving the immediate issue of soil pollution. For this reason, the expansion of research work and development on nanotechnology should be given impetus when used in the treatment of soil ailments. The future management and prevention of soil pollution will largely rely on ongoing advancements in nanotechnology. To make NMs more effective and safer for use, an efficient look at their life span consequences on the environment should be done. Furthermore, it is important to investigate new research directions including the synthesis of new NMs and enhancement of nanophytoremediation processes. Nanotechnology can only be beneficent if farmers are informed and encouraged to apply the techniques, with programs and tutorials for them to grasp the utility and drawbacks of the innovation. The integration of nanotechnology with various industries can help bring improvements in agricultural productivity, environmental protection, and overall social welfare. Site-specific remediation measures will be facilitated by the routine monitoring of NPs' characteristics under different types of soil conditions. From an evolutionary standpoint, sustainable practices and nanotechnology must coexist in order to protect and improve soil health for coming generations..

12. Acknowledgement

The authors gratefully acknowledge the support of the Institute of Plant Protection, Muhammad Nawaz Sharif University of Agriculture, Multan for the support in this publication.

References

AbdelRahman, M.A. 2023. An overview of land degradation, desertification and sustainable land management using GIS and remote sensing applications. Rend. Lincei. Sci. Fis. Nat. 34: 767–808.

Abid, N., A.M. Khan, S. Shujait, K. Chaudhary, M. Ikram, M. Imran et al. 2022. Synthesis of nanomaterials using various top-down and bottom-up approaches, influencing factors, advantages, and disadvantages: A review. Adv. Colloid Interface Sci. 300: 102597.

Abuhatab, S., A. El-Qanni, H. Al-Qalaq, M. Hmoudah and W. Al-Zerei. 2020. Effective adsorptive removal of Zn^{2+}, Cu^{2+}, and Cr^{3+} heavy metals from aqueous solutions using silica-based embedded with NiO and MgO nanoparticles. J. Environ. Manage. 268: 110713.

Abuzeid, H.M., C.M. Julien, L. Zhu and A.M. Hashem. 2023. Green synthesis of nanoparticles and their energy storage, environmental, and biomedical applications. Crystals. 13: 1576.

Adekunle, A.S., J.A. Oyekunle, L.M. Durosinmi, O.S. Oluwafemi, D.S. Olayanju, A.S. Akinola et al. 2020. Potential of cobalt and cobalt oxide nanoparticles as nanocatalyst towards dyes degradation in wastewater. Nano-Struct. Nano-Objects. 21: 100405.

Adeyemi, J.O., A.O. Oriola, D.C. Onwudiwe and A.O. Oyedeji. 2022. Plant extracts mediated metal-based nanoparticles: synthesis and biological applications. Biomolecules. 12: 627.

Ahmad, Z., S. Tahseen, A. Wasi, I.B. Ganie, A. Shahzad, A. Emamverdian et al. 2022. Nanotechnological interventions in agriculture. Nanomaterials. 12: 2667.

Ahsan, M.A., V. Jabbari, M.A. Imam, E. Castro, H. Kim, M.L. Curry et al. 2020. Nanoscale nickel metal organic framework decorated over graphene oxide and carbon nanotubes for water remediation. Sci. Total Environ. 698: 134214.

Alazaiza, M.Y., A. Albahnasawi, G.A. Ali, M.J. Bashir, N.K. Copty, S.S.A. Amr et al. 2021. Recent advances of nanoremediation technologies for soil and groundwater remediation: A review. Water. 13: 2186.

Alengebawy, A., S.T. Abdelkhalek, S.R. Qureshi and M.Q. Wang. 2021. Heavy metals and pesticides toxicity in agricultural soil and plants: Ecological risks and human health implications. Toxics. 9: 42.

Alina, M., A. Azrina, A.S. Mohd Yunus, S. Mohd Zakiuddin, H. Mohd Izuan Effendi and R. Muhammad Rizal. 2012. Heavy metals (mercury, arsenic, cadmium, plumbum) in selected marine fish and shellfish along the Straits of Malacca. Int. Food Res. J. 19: 1.

Aliyari Rad, S., K. Nobaharan, N. Pashapoor, J. Pandey, Z. Dehghanian, V. Senapathi et al. 2023. Nano-microbial remediation of polluted soil: a brief insight. Sustainability. 15: 876.

Alsharef, J.M., M.R. Taha, A.A. Firoozi and P. Govindasamy. 2016. Potential of using nanocarbons to stabilize weak soils. Appl. Environ. Soil Sci. 2016: 5060531.

Asghar, N., A. Hussain, D.A. Nguyen, S. Ali, I. Hussain, A. Junejo et al. 2024. Advancement in nanomaterials for environmental pollutants remediation: a systematic review on bibliometrics analysis, material types, synthesis pathways, and related mechanisms. J. Nanobiotechnol. 22: 26.

Ayilara, M.S. and O.O. Babalola. 2023. Bioremediation of environmental wastes: the role of microorganisms. Front. Agron. 5: 1183691.

Baby, R., B. Saifullah and M.Z. Hussein. 2019. Carbon nanomaterials for the treatment of heavy metal-contaminated water and environmental remediation. Nanoscale Res. Lett. 14: 341.

Baby, R., M.Z. Hussein, A.H. Abdullah and Z. Zainal. 2022. Nanomaterials for the treatment of heavy metal contaminated water. Polymers. 14: 583.

Baig, N., I. Kammakakam and W. Falath. 2021. Nanomaterials: A review of synthesis methods, properties, recent progress, and challenges. Mater. Adv. 2: 1821–1871.

Baragaño, D., R. Forján, C. Sierra and J.L.R. Gallego. 2021. Nanomaterials for soil remediation: Pollutant immobilization and opportunities for hybrid technologies. pp. 701–723. *In*: Sorbents Materials for Controlling Environmental Pollution. Elsevier.

Baranwal, J., B. Barse, A. Fais, G.L. Delogu and A. Kumar. 2022. Biopolymer: A sustainable material for food and medical applications. Polymers. 14: 983.

Bayat, H., Z. Kolahchi, S. Valaey, M. Rastgou and S. Mahdavi. 2018. Novel impacts of nanoparticles on soil properties: tensile strength of aggregates and compression characteristics of soil. Arch. Agron. Soil Sci. 64: 776–789.

Beddington, J.R., M. Asaduzzaman, M.E. Clark, A. Fernández Bremauntz, M.D. Guillou, D.J.B. Howlett et al. 2012. What next for agriculture after Durban? Science. 335: 289–290.

Bhuyan, M.S., M.A. Bakar, M. Rashed-Un-Nabi, V. Senapathi, S.Y. Chung and M.S. Islam. 2019. Monitoring and assessment of heavy metal contamination in surface water and sediment of the Old Brahmaputra River, Bangladesh. Appl. Water Sci. 9: 1–13.

Bjerk, T.R., P. Severino, S. Jain, C. Marques, A.M. Silva, T. Pashirova et al. 2021. Biosurfactants: properties and applications in drug delivery, biotechnology and ecotoxicology. Bioengineering. 8: 115.

Bretzel, F., M. Calderisi, M. Scatena and R. Pini. 2016. Soil quality is key for planning and managing urban allotments intended for the sustainable production of home-consumption vegetables. Environ. Sci. Pollut. Res. 23: 17753–17760.

Briffa, J., E. Sinagra and R. Blundell. 2020. Heavy metal pollution in the environment and their toxicological effects on humans. Heliyon. 6: e05059.

Chaudhary, P., L. Ahamad, A. Chaudhary, G. Kumar, W.J. Chen and S. Chen. 2023. Nanoparticle-mediated bioremediation as a powerful weapon in the removal of environmental pollutants. J. Environ. Chem. Eng. 11: 109591.

Chen, H. and R. Yada. 2011. Nanotechnologies in agriculture: new tools for sustainable development. Trends Food Sci. Technol. 22: 585–594.

Chu, S., L. Xian, C. Lai, W. Yang, J. Wang, M. Long et al. 2023. Migration and risks of potentially toxic elements from sewage sludge applied to acid forest soil. J. For. Res. 34: 2011–2026.

Chung, I.M., I. Park, S.H. Kim, M. Thiruvengadam and G. Rajakumar. 2016. Plant-mediated synthesis of silver nanoparticles: their characteristic properties and therapeutic applications. Nanoscale Res. Lett. 11: 1–14.

Cooper, A.M., D. Felix, F. Alcantara, I. Zaslavsky, A. Work, P.L. Watson et al. 2020. Monitoring and mitigation of toxic heavy metals and arsenic accumulation in food crops: A case study of an urban community garden. Plant Direct. 4: e00198.

Corwin, D.L. 2021. Climate change impacts on soil salinity in agricultural areas. Eur. J. Soil Sci. 72: 842-862.

Cucurachi, S., C. van der Giesen and J. Guinée. 2018. Ex-ante LCA of emerging technologies. Procedia CIRP. 69: 463–468.

Danks, A.E., S.R. Hall and Z.J.M.H. Schnepp. 2016. The evolution of 'sol–gel' chemistry as a technique for materials synthesis. Mater. Horiz. 3: 91–112.

Debnath, B., M. Majumdar, M. Bhowmik, K.L. Bhowmik, A. Debnath and D.N. Roy. 2020. The effective adsorption of tetracycline onto zirconia nanoparticles synthesized by novel microbial green technology. J. Environ. Manag. 261: 110235.

DeRosa, M.C., C. Monreal, M. Schnitzer, R. Walsh and Y. Sultan. 2010. Nanotechnology in fertilizers. Nat. Nanotechnol. 5: 91.

Dhanapal, A.R., M. Thiruvengadam, J. Vairavanathan, B. Venkidasamy, M. Easwaran and M. Ghorbanpour. 2024. Nanotechnology approaches for the remediation of agricultural polluted soils. ACS Omega. 9: 13522–13533.

Dimkpa, C.O. and P.S. Bindraban. 2017. Nanofertilizers: new products for the industry? J. Agric. Food Chem. 66: 6462–6473.

Dominati, E., M. Patterson and A. Mackay. 2010. A framework for classifying and quantifying the natural capital and ecosystem services of soils. Ecol. Econ. 69: 1858–1868.

Ekeoma, B.C., L.N. Ekeoma, M. Yusuf, A. Haruna, C.K. Ikeogu, Z.M.A. Merican et al. 2023. Recent advances in the biocatalytic mitigation of emerging pollutants: a comprehensive review. J. Biotechnol. 369: 14–34.

El-Eskandarany, M.S., A. Al-Hazza, L.A. Al-Hajji, N. Ali, A.A. Al-Duweesh, M. Banyan et al. 2021. Mechanical milling: a superior nanotechnological tool for fabrication of nanocrystalline and nanocomposite materials. Nanomaterials. 11: 2484.

Escorcia-Díaz, D., S. García-Mora, L. Rendón-Castrillón, M. Ramírez-Carmona and C. Ocampo-López. 2023. Advancements in nanoparticle deposition techniques for diverse substrates: a review. Nanomaterials. 13: 2586.

Fernandes, T., A.L. Daniel-da-Silva and T. Trindade. 2022. Metal-dendrimer hybrid nanomaterials for sensing applications. Coord. Chem. Rev. 460: 214483.

Fritea, L., F. Banica, T.O. Costea, L. Moldovan, L. Dobjanschi, M. Muresan and S. Cavalu. 2021. Metal nanoparticles and carbon-based nanomaterials for improved performances of electrochemical (Bio) sensors with biomedical applications. Mater. 14: 6319.

Galdames, A., L. Ruiz-Rubio, M. Orueta, M. Sánchez-Arzalluz and J.L. Vilas-Vilela. 2020. Zero-valent iron nanoparticles for soil and groundwater remediation. Int. J. Environ. Res. Public Health. 17: 5817.

Gayathiri, E., P. Prakash, N. Karmegam, S. Varjani, M.K. Awasthi and B. Ravindran. 2022. Biosurfactants: potential and eco-friendly material for sustainable agriculture and environmental safety-a review. Agron. 12: 662.
Gebeyehu, H.R. and L.D. Bayissa. 2020. Levels of heavy metals in soil and vegetables and associated health risks in Mojo area, Ethiopia. PLoS One. 15: e0227883.
Gibbs, H.K. and J.M. Salmon. 2015. Mapping the world's degraded lands. Appl. Geogr. 57: 12–21.
Godfray, H.C.J., J.R. Beddington, I.R. Crute, L. Haddad, D. Lawrence, J.F. Muir et al. 2010. Food security: the challenge of feeding 9 billion people. Science. 327: 812–818.
Gogos, A., K. Knauer and T.D. Bucheli. 2012. Nanomaterials in plant protection and fertilization: current state, foreseen applications, and research priorities. J. Agric. Food Chem. 60: 9781–9792.
Gomiero, T. 2016. Soil degradation, land scarcity and food security: Reviewing a complex challenge. Sustainability. 8: 281.
Govindasamy, P., M.R. Taha, J. Alsharef and K. Ramalingam. 2017. Influence of nanolime and curing period on unconfined compressive strength of soil. Appl. Environ. Soil Sci. 2017: 8307493.
Guo, J.H., X.J. Liu, Y. Zhang, J.L. Shen, W.X. Han, W.F. Zhang et al. 2010. Significant acidification in major Chinese croplands. Science. 327: 1008–1010.
Gurusamy, S. and R.S.P. Thangam. 2023. Potential health risk assessment of contaminants in soil-like material recovered from landfill mining. Environ. Monit. Assess. 195: 330.
Hamza, M.A. and W.K. Anderson. 2005. Soil compaction in cropping systems: A review of the nature, causes and possible solutions. Soil Till. Res. 82: 121–145.
Hareesh, P. and R. Vinoth Kumar. 2016. Assessment of nano-materials on geotechnical properties of Clayey soils. In: International Conference on Engineering Innovations and Solutions. pp. 66–70.
Hassan, M., R. Naidu, J. Du, Y. Liu and F. Qi. 2020. Critical review of magnetic biosorbents: their preparation, application, and regeneration for wastewater treatment. Sci. Total Environ. 702: 134893.
He, F. and D. Zhao. 2005. Preparation and characterization of a new class of starch-stabilized bimetallic nanoparticles for degradation of chlorinated hydrocarbons in water. Environ. Sci. Technol. 39: 3314–3320.
He, X., H. Deng and H.M. Hwang. 2019. The current application of nanotechnology in food and agriculture. J. Food Drug Anal. 27: 1–21.
Hidangmayum, A., A. Debnath, A. Guru, B.N. Singh, S.K. Upadhyay and P. Dwivedi. 2023. Mechanistic and recent updates in nano-bioremediation for developing green technology to alleviate agricultural contaminants. Int. J. Environ. Sci. Technol. 20: 11693–11718.
Hischier, R. and T. Walser. 2012. Life cycle assessment of engineered nanomaterials: State of the art and strategies to overcome existing gaps. Sci. Total Environ. 425: 271–282.
Hong, J., J. Xie, S. Mirshahghassemi and J. Lead. 2020. Metal (Cd, Cr, Ni, Pb) removal from environmentally relevant waters using polyvinylpyrrolidone-coated magnetite nanoparticles. RSC Adv. 10: 3266–3276.
Hussain, A., F. Rehman, H. Rafeeq, M. Waqas, A. Asghar, N. Afsheen, et al. 2022. *In situ, Ex situ*, and nano-remediation strategies to treat polluted soil, water, and air – A review. Chemosphere. 289: 133252.
Ibrahim, R.K., M. Hayyan, M.A. AlSaadi, A. Hayyan and S. Ibrahim. 2016. Environmental application of nanotechnology: air, soil, and water. Environ. Sci. Pollut. Res. 23: 13754–13788.
Jakhar, A.M., I. Aziz, A.R. Kaleri, M. Hasnain, G. Haider, J. Ma et al. 2022. Nano-fertilizers: A sustainable technology for improving crop nutrition and food security. NanoImpact. 27: 100411.
Jayakumar, M., U. Surendran, P. Raja, A. Kumar and V. Senapathi. 2021. A review of heavy metals accumulation pathways, sources and management in soils. Arab. J. Geosci. 14: 2156.
Ji, C., L.N. Nguyen, J. Hou, F.I. Hai and V. Chen. 2017. Direct immobilization of laccase on titania nanoparticles from crude enzyme extracts of *P. ostreatus* culture for micro-pollutant degradation. Sep. Purif. Technol. 178: 215-223.
Joshi, S., S. Gangola, G. Bhandari, N.S. Bhandari, D. Nainwal, A. Rani. 2023. Rhizospheric bacteria: the key to sustainable heavy metal detoxification strategies. Front. Microbiol. 14: 1229828.
Kah, M., S. Beulke, K. Tiede and T. Hofmann. 2013. Nanopesticides: state of knowledge, environmental fate, and exposure modeling. Crit. Rev. Environ. Sci. Technol. 43: 1823–1867.
Kah, M. and T. Hofmann. 2014. Nanopesticide research: current trends and future priorities. Environ. Int. 63: 224–235.
Kamarudin, N.S., R. Jusoh, H.D. Setiabudi, N.F. Sukor and J.H. Shariffuddin. 2021. Potential nanomaterials application in wastewater treatment: Physical, chemical and biological approaches. Mater. Today Proc. 42: 107–114.
Ken, D.S. and A. Sinha. 2020. Recent developments in surface modification of nano zero-valent iron (nZVI): Remediation, toxicity and environmental impacts. Environ. Nanotechnol. Monit. Manag. 14: 100344.
Khan, A.H., N.A. Khan, M. Zubair, M.A. Shaida, M.S. Manzar, A. Abutaleb, et al. 2022. Sustainable green nanoadsorbents for remediation of pharmaceuticals from water and wastewater: A critical review. Environ. Res. 204: 112243.

Khan, I., K. Saeed and I. Khan. 2019. Nanoparticles: Properties, applications and toxicities. Arab. J. Chem. 12: 908–931.

Khan, M., M.S.A. Khan, K.K. Borah, Y. Goswami, K.R. Hakeem and I. Chakrabartty. 2021. The potential exposure and hazards of metal-based nanoparticles on plants and environment, with special emphasis on ZnO NPs, TiO2 NPs, and AgNPs: A review. Environ. Adv. 6: 100128.

Khan, Y., H. Sadia, S.Z. Ali Shah, M.N. Khan, A.A. Shah, N. Ullah et al. 2022. Classification, synthetic, and characterization approaches to nanoparticles, and their applications in various fields of nanotechnology: A review. Catalysts 12: 1386.

Khanna, V. and B.R. Bakshi. 2009. Carbon nanofiber polymer composites: Evaluation of life cycle energy use. Environ. Sci. Technol. 43: 2078–2084.

Kibblewhite, M.G., K. Ritz and M.J. Swift. 2008. Soil health in agricultural systems. Philos. Trans. R. Soc. B Biol. Sci. 363: 685–701.

Kim, J., D.A. Reddy, R. Ma and T.K. Kim. 2014. The influence of laser wavelength and fluence on palladium nanoparticles produced by pulsed laser ablation in deionized water. Solid State Sci. 37: 96–102.

Koçyigit, N. 2023. Aerogels: Characteristics, classification, structure, properties, preparation, types and applications. In: Important Research in the Field of Chemistry in a Globalizing World. pp. 3–42.

Kolluru, S.S., S. Agarwal, S. Sireesha, I. Sreedhar and S.R. Kale. 2021. Heavy metal removal from wastewater using nanomaterials-process and engineering aspects. Process Saf. Environ. Prot. 150: 323–355.

Kumar, S. 2023. Smart and innovative nanotechnology applications for water purification. Hybrid Adv. 3: 100044.

Kuzma, J. and P. VerHage. 2006. Nanotechnology in agriculture and food production: anticipated applications. Project on Emerging Nanotechnologies.

Lal, R. 2004. Soil carbon sequestration impacts on global climate change and food security. Science. 304: 1623–1627.

Lal, R. 2015. Restoring soil quality to mitigate soil degradation. Sustainability. 7: 5875–5895.

Lehmann, J. and M. Kleber. 2015. The contentious nature of soil organic matter. Nature. 528: 60–68.

Li, T., L. Cui, Z. Xu, H. Liu, X. Cui and P. Fantke. 2023. Micro-and nanoplastics in soil: Linking sources to damage on soil ecosystem services in life cycle assessment. Sci. Total Environ. 166925.

Liang, W., G. Wang, C. Peng, J. Tan, J. Wan, P. Sun et al. 2022. Recent advances of carbon-based nano zero valent iron for heavy metals remediation in soil and water: A critical review. J. Hazard. Mater. 426: 127993.

Lv, N., B. Wang, H. Wang, T. Xiao, B. Dong and Z. Xu. 2024. The occurrence characteristics, removal efficiency, and risk assessment of polycyclic aromatic hydrocarbons in sewage sludges from across China. Chemosphere. 351: 141033.

Lyu, R., J. Zhang, M. Xu and J. Li. 2018. Impacts of urbanization on ecosystem services and their temporal relations: A case study in Northern Ningxia, China. Land Use Policy 77: 163–173.

Ma, W., T. Sun, Y. Xu, S. Zheng and Y. Sun. 2022. In-situ immobilization remediation, soil aggregate distribution, and microbial community composition in weakly alkaline Cd-contaminated soils: A field study. Environ. Pollut. 292: 118327.

Machado, S., R. Pereira and R.M.O. Sousa. 2023. Nanobiopesticides: Are they the future of phytosanitary treatments in modern agriculture? Sci. Total Environ. 166401.

Majeed, Z.H., M.R. Taha and I.T. Jawad. 2014. Stabilization of soft soil using nanomaterials. Res. J. Appl. Sci. Eng. Technol. 8: 503-509.

Manawi, Y.M., I. Samara, A. Al-Ansari and M.A. Atieh. 2018. A review of carbon nanomaterials' synthesis via the chemical vapor deposition (CVD) method. Materials 11: 822.

Mohanraj, R., B.M. Gnanamangai, S. Poornima, V. Oviyaa, K. Ramesh, G. Vijayalakshmi et al. 2022. Decolourisation efficiency of immobilized silica nanoparticles synthesized by actinomycetes. Mater. Today Proc. 48: 129–135.

Montgomery, D.R. 2007. Soil erosion and agricultural sustainability. Proc. Natl. Acad. Sci. U.S.A. 104: 13268–13272.

Naderi, M.R. and A. Danesh-Shahraki. 2013. Nanofertilizers and their roles in sustainable agriculture.

Naseem, T. and T. Durrani. 2021. The role of some important metal oxide nanoparticles for wastewater and antibacterial applications: A review. Environ. Chem. Ecotoxicol. 3: 59–75.

Naseri, F., M. Irani and M. Dehkhodarajabi. 2016. Effect of graphene oxide nanosheets on the geotechnical properties of cemented silty soil. Arch. Civil Mech. Eng. 16: 695–701.

Nedjimi, B. 2021. Phytoremediation: A sustainable environmental technology for heavy metals decontamination. SN Appl. Sci. 3: 286.

Nzilu, D.M., E.S. Madivoli, D. Sujee Makhanu, B.V. Otenda, P.G. Kareru, P. Kinoti Kairigo et al. 2023. Environmental remediation using nanomaterial as adsorbents for emerging micropollutants. Environ. Nanotechnol. Monit. Manag. 20: 100789.

Osman, A.I., E.M.A. El-Monaem, A.M. Elgarahy, C.O. Aniagor, M. Hosny, M. Farghali et al. 2023. Methods to prepare biosorbents and magnetic sorbents for water treatment: A review. Environ. Chem. Lett. 21: 2337–2398.

Palmerston Mendes, L., J. Pan and V.P. Torchilin. 2017. Dendrimers as nanocarriers for nucleic acid and drug delivery in cancer therapy. Molecules 22: 1401.

Peralta-Videa, J.R., L. Zhao, M.L. Lopez-Moreno, G. de la Rosa, J. Hong and J.L. Gardea-Torresdey. 2011. Nanomaterials and the environment: A review for the biennium 2008–2010. J. Hazard. Mater. 186: 1–15.

Pimentel, D. and M. Burgess. 2013. Soil erosion threatens food production. Agric. 3: 443–463.

Plessl, K., A. Russ and D. Vollprecht. 2023. Application and development of zero-valent iron (ZVI) for groundwater and wastewater treatment. Int. J. Environ. Sci. Technol. 20(6): 6913–6928.

Prakash, P. 2023. Nano-phytoremediation of heavy metals from soil: A critical review. Pollutants 3(3): 360–380.

Pramanik, P., P. Ray, A. Maity, S. Das, S. Ramakrishnan and P. Dixit. 2020. Nanotechnology for improved carbon management in soil. In: Carbon management in tropical and sub-tropical terrestrial systems, pp. 403–415.

Pramudita, D., I. Iskandar and A. Indarto. 2018. Nano-enhanced materials for reclamation of mine spoils. *In*: Bio-geotechnologies for mine site rehabilitation, Elsevier, pp. 201–214.

Prema, P., V.H. Nguyen, K. Venkatachalam, J.M. Murugan, H.M. Ali, M.Z. Salem et al. 2022. Hexavalent chromium removal from aqueous solutions using biogenic iron nanoparticles: Kinetics and equilibrium study. Environ. Res. 205: 112477.

Pretty, J. 2018. Intensification for redesigned and sustainable agricultural systems. Science. 362(6417): eaav0294.

Priyadarshini, B.G. 2021. Mechanical Milling of Copper Oxide Nanoparticles. *In*: ICCAP 2021: Proceedings of the First International Conference on Combinatorial and Optimization, ICCAP 2021, December 7–8, 2021, Chennai, India, pp. 236. European Alliance for Innovation.

Qiu, C., L. Jiang, Y. Gao and L. Sheng. 2023. Effects of oxygen-containing functional groups on carbon materials in supercapacitors: A review. Mater. Des. 230: 111952.

Rajakumar, G., P.N. Sudha and M. Thiruvengadam. 2023. Synthesis, bioactivity evaluation and application of plant-based nanoparticles. Molecules. 28(12): 4783.

Rajput, V.D., T. Minkina, S.K. Upadhyay, A. Kumari, A. Ranjan, S. Mandzhieva et al. 2022. Nanotechnology in the restoration of polluted soil. Nanomaterials. 12(5): 769.

Raliya, R., R. Nair, S. Chavalmane, W.N. Wang and P. Biswas. 2015. Mechanistic evaluation of translocation and physiological impact of titanium dioxide and zinc oxide nanoparticles on the tomato (*Solanum lycopersicum* L.) plant. Metallomics. 7(12): 1584–1594.

Ramesh, M., R. Janani, C. Deepa and L. Rajeshkumar. 2022. Nanotechnology-enabled biosensors: A review of fundamentals, design principles, materials, and applications. Biosensors. 13(1): 40.

Rani, N., A. Duhan, A. Pal, P. Kumari, R.K. Beniwal, D. Verma et al. 2023. Are nano-pesticides really meant for cleaner production? An overview on recent developments, benefits, environmental hazards and future prospectives. J. Clean. Prod. 411: 137232.

Rashid, A., B.J. Schutte, A. Ulery, M.K. Deyholos, S. Sanogo, E.A. Lehnhoff et al. 2023. Heavy metal contamination in agricultural soil: environmental pollutants affecting crop health. Agronomy. 13(6): 1521.

Rather, M.A., S. Bhuyan, R. Chowdhury, R. Sarma, S. Roy, and P.R. Neog. 2023. Nanoremediation strategies to address environmental problems. Sci. Total Environ. 886: 163998.

Reynolds, T.W., S.R. Waddington, C.L. Anderson, A. Chew, Z. True and A. Cullen. 2015. Environmental impacts and constraints associated with the production of major food crops in Sub-Saharan Africa and South Asia. Food Sec. 7: 795–822.

Roco, M.C. 2011. The long view of nanotechnology development: the National Nanotechnology Initiative at 10 years. J. Nanopart. Res. 13: 427–445.

Romeh, A.A. and R.A.I. Saber. 2020. Green nano-phytoremediation and solubility improving agents for the remediation of chlorfenapyr contaminated soil and water. J. Environ. Manag. 260: 110104.

Saleem, M.H., S. Ali, S. Hussain, M. Kamran, M.S. Chattha, S. Ahmad et al. 2020. Flax (*Linum usitatissimum* L.): A potential candidate for phytoremediation? Biol. Econ. Plants. 9(4): 496.

San Keskin, N.O., A. Celebioglu, O.F. Sarioglu, T. Uyar and T. Tekinay. 2018. Encapsulation of living bacteria in electrospun cyclodextrin ultrathin fibers for bioremediation of heavy metals and reactive dye from wastewater. Colloids Surf. B Biointerfaces. 161: 169–176.

Sarioglu, O.F., N.O. San Keskin, A. Celebioglu, T. Tekinay and T. Uyar. 2017. Bacteria encapsulated electrospun nanofibrous webs for remediation of methylene blue dye in water. Colloids Surf. B Biointerfaces. 152: 245–251.

Sathish, T., N. Ahalya, M. Thirunavukkarasu, T.S. Senthil, Z. Hussain, M.I.H. Siddiqui et al. 2024. A comprehensive review on the novel approaches using nanomaterials for the remediation of soil and water pollution. Alex. Eng. J. 86: 373–385.

Schutte, R., T.S. Nawrot, T. Richart, L. Thijs, D. Vanderschueren, T. Kuznetsova et al. 2008. Bone resorption and environmental exposure to cadmium in women: a population study. Environ. Health Perspect. 116(6): 777–783.

Semaltianos, N.G. 2016. Nanoparticles by laser ablation of bulk target materials in liquids. Handb. Nanopart. 67–92.

Sergievskaya, A., A. Chauvin and S. Konstantinidis. 2022. Sputtering onto liquids: a critical review. Beilstein J. Nanotechnol. 13(1): 10–53.

Servin, A.D. and J.C. White. 2016. Nanotechnology in agriculture: next steps for understanding engineered nanoparticle exposure and risk. NanoImpact. 1: 9–12.

Sharma, N., G. Singh, M. Sharma, S. Mandzhieva, T. Minkina and V.D. Rajput. 2022. Sustainable use of nano-assisted remediation for mitigation of heavy metals and mine spills. Water. 14(23): 3972.

Sharma, R. and V. Kumar. 2024. Nano enabled agriculture for sustainable soil. Waste Manag. Bull.

Shetty, S.S., D. Deepthi, S. Harshitha, S. Sonkusare, P.B. Naik and H. Madhyastha. 2023. Environmental pollutants and their effects on human health. Heliyon. 9(9).

Shi, D., X. Zhang, J. Wang and J. Fan. 2018. Highly reactive and stable nanoscale zero-valent iron prepared within vesicles and its high-performance removal of water pollutants. Appl. Catal. B Environ. 221: 610–617.

Shukla, K., V. Mishra, J. Singh, V. Varshney, R. Verma and S. Srivastava. 2024. Nanotechnology in sustainable agriculture: A double-edged sword. J. Sci. Food Agric. 104(10): 5675–5688.

Simonin, M. and A. Richaume. 2015. Impact of engineered nanoparticles on the activity, abundance, and diversity of soil microbial communities: a review. Environ. Sci. Pollut. Res. 22: 13710–13723.

Singh, K.P. 2021. A Special Section on Applications of Nanotechnology in Agriculture. J. Nanoscience Nanotechnol. 21(6): 3335–3336.

Soffian, M.S., F.Z.A. Halim, F. Aziz, M.A. Rahman, M.A.M. Amin and D.N.A. Chee. 2022. Carbon-based material derived from biomass waste for wastewater treatment. Environ. Adv. 9: 100259.

Solanki, P., A. Bhargava, H. Chhipa, N. Jain and J. Panwar. 2015. Nano-fertilizers and their smart delivery system. In: Nanotechnologies in food and agriculture, pp. 81–101.

Suazo-Hernández, J., N. Arancibia-Miranda, R. Mlih, L. Cáceres-Jensen, N. Bolan and M.D.L.L. Mora. 2023. Impact on some soil physical and chemical properties caused by metal and metallic oxide engineered nanoparticles: A review. Nanomaterials. 13(3): 572.

Subhan, A., A.H.I. Mourad and Y. Al-Douri. 2022. Influence of laser process parameters, liquid medium, and external field on the synthesis of colloidal metal nanoparticles using pulsed laser ablation in liquid: A review. Nanomaterials. 12(13): 2144.

Sudarman, F., M. Shiddiq, B. Armynah and D. Tahir. 2023. Silver nanoparticles (AgNPs) synthesis methods as heavy-metal sensors: A review. Int. J. Environ. Sci. Technol. 20(8): 9351–9368.

Taha, M.R. and O.M.E. Taha. 2012. Influence of nano-material on the expansive and shrinkage soil behavior. J. Nanoparticle Res. 14: 1–13.

Tahat, M., M. Alananbeh, K.A. Othman, Y. and D.I. Leskovar. 2020. Soil health and sustainable agriculture. Sustainability. 12(12): 4859.

Tchounwou, P.B., C.G. Yedjou, A.K. Patlolla and D.J. Sutton. 2012. Heavy metal toxicity and the environment. In: Mol. Clin. Environ. Toxicol. Vol. 3: Environ. Toxicol., pp. 133–164.

Tofan, L. and D. Suteu. 2023. Renewable resource biosorbents for pollutant removal from aqueous effluents in column mode. Separations. 10(2): 143.

Tufail, M.A., J. Iltaf, T. Zaheer, L. Tariq, M.B. Amir, R. Fatima et al. 2022. Recent advances in bioremediation of heavy metals and persistent organic pollutants: A review. Sci. Total Environ. 850: 157961.

Tuti, M.D., M.K. Rapolu, D. Sreedevi, N. Bandumula, S. Kuchi, S. Bandeppa et al. 2022. Sustainable intensification of a rice–maize system through Conservation Agriculture to enhance system productivity in Southern India. Plants. 11(9): 1229.

Usman, M., S. Jellali, I. Anastopoulos, Y. Charabi, B.H. Hameed and K. Hanna. 2022. Fenton oxidation for soil remediation: A critical review of observations in historically contaminated soils. J. Hazard. Mater. 424: 127670.

Vermeulen, S.J., B.M. Campbell and J.S. Ingram. 2012. Climate change and food systems. Annu. Rev. Environ. Resour. 37(1): 195–222.

Wei, J., M. Duan, Y. Li, A.S. Nwankwegu, Y. Ji and J. Zhang. 2019. Concentration and pollution assessment of heavy metals within surface sediments of the Raohe Basin, China. Sci. Rep. 9(1): 13100.

Wypij, M., J. Trzcińska-Wencel, P. Golińska, G.D. Avila-Quezada, A.P. Ingle and M. Rai. 2023. The strategic applications of natural polymer nanocomposites in food packaging and agriculture: Chances, challenges, and consumers' perception. Front. Chem. 10: 1106230.

Xia, X., X. Chen, R. Liu and H. Liu. 2011. Heavy metals in urban soils with various types of land use in Beijing, China. J. Hazard. Mater. 186(2-3): 2043–2050.

Yadav, A., K. Yadav, R. Ahmad and K.A. Abd-Elsalam. 2023. Emerging frontiers in nanotechnology for precision agriculture: advancements, hurdles and prospects. Agrochemicals. 2(2): 220–256.

Yang, S., S. Chen, J. Fan, T. Shang, D. Huang, and G. Li. 2019. Novel mesoporous organosilica nanoparticles with ferrocene group for efficient removal of contaminants from wastewater. J. Colloid Interface Sci. 554: 565–571.

Zhang, W.X. 2003. Nanoscale iron particles for environmental remediation: an overview. J. Nanopart. Res. 5: 323–332.

Zhang, Y., J. Luo, H. Zhang, T. Li, H. Xu, Y. Sun et al. 2022. Synthesis and adsorption performance of three-dimensional gels assembled by carbon nanomaterials for heavy metal removal from water: A review. Sci. Total Environ. 852: 158201.

Zornoza, R., J.A. Acosta, S. Martínez-Martínez, A. Faz and E. Bååth. 2015. Main factors controlling microbial community structure and function after reclamation of a tailing pond with aided phytostabilization. Geoderma. 245: 1–10.

CHAPTER-9

Engineered Nanomaterials and their Impact on Soil Health

Muhammad Ashir Hameed,[1] *Mehreen Malik,*[1] *Amina Rashid,*[2] *Muhammad Junaid Nazir,*[3] *Hamna Bashir,*[1] *Hamza Amin,*[1] *Muhammad Tahir Shahzad,*[1] *Muhammad Mahroz Hussain*[4,5,*] and *Shengsen Wang*[4,*]

1. Introduction

Nanotechnology is becoming an essential component of the developing world due to its wide application and usage in different industries, particularly agriculture, medicine, and electronics. Materials having one or more dimensions of 100 nanometers or less are referred to as nanomaterials (Kreyling et al. 2010). In comparison with bulk materials, these materials frequently display exceptional and singular physicochemical and biological features because of their distinct architectures and unique functionality. The growing use of sophisticated nanomaterials for their capacity to tackle present and future challenges is the main driver of the expansion of enterprises centered around nanotechnology (Bhardwaj and Kaushik 2017, Shang et al. 2019). Currently, there is a huge number of nanomaterials-based commercial products available in the market, which is expected to further increase (Hansen et al. 2020). Engineered nanomaterials (ENMs) is a term specifically used to describe intentionally prepared nanomaterials, as opposed to naturally occurring nanomaterials like montmorillonite (MMT), kaolinite, saponite, etc. Despite their unprecedented prospects and applications, nanomaterials significantly threaten human health and the environment (Ji et al. 2021).

At various stages of their synthesis, production, usage, and eventual disposal, environmental nanomaterials (ENMs) can discharge into the environment. However, the synthesis and production have a minimal risk of releasing ENMs into the environment because they are often prepared and incorporated into various products under controlled environments. Nonetheless, there is a

[1] Institute of Soil and Environmental Sciences, University of Agriculture, Faisalabad-38040, Pakistan.
[2] Department of Agronomy, University of Agriculture Faisalabad-38040, Pakistan.
[3] School of Emergency Management, School of Environment and Safety Engineering, Jiangsu Province Engineering Research Center of Green Technology and Contingency. Management for Emerging Pollutants, Jiangsu University, Zhenjiang, 212013, People's Republic of China.
[4] College of Environmental Science and Engineering, Yangzhou University, Yangzhou 225127, China.
[5] HAM Organics (PVT) Limited, Pakistan.
* Corresponding authors: wangss@yzu.edu.cn, hmahroz@gmail.com

DOI: 10.1201/9781003558354-9

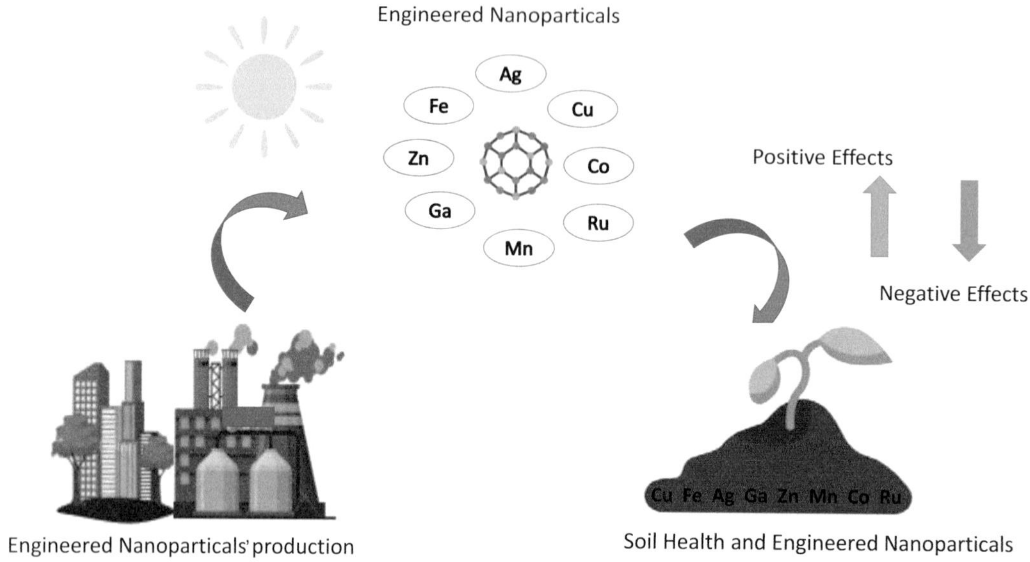

Figure 1. A schematic diagram of engineered nanoparticles' fate and impacts on the soil health.

significant chance that ENMs will find their way into the soil throughout the use and disposal phases, endangering the immediate ecosystem (Loureiro et al. 2018). For instance, the breakdown of damaged nanoproducts during their usage continuously releases ENMs into the soil. Furthermore, the soil is seriously endangered by the widespread use of nanopesticides, fertilizers, and other nanoadsorbent materials in agricultural and environmental remediation. As a usual practice, these products are directly applied to plants and soil, thus greatly accelerating the rate at which ENMs are released into the soil. Furthermore, the wastewater could potentially carry ENMs into the soil (de Oliveira Pereira et al. 2020, Sun et al. 2019). Here is the illustration of the fate of engineered nanoparticles along with their impacts on soil health in Figure 1.

It is vital to comprehend the impacts of these ENMs on the soil ecosystem as well as the routes by which they can enter the soil system because, in general, a huge number of ENMs, during their life cycle, either directly or indirectly, enter the soil ecosystem. Understanding the ENMs' behavior and their persistence in the soil is crucial. It is also recommended to gain a better understanding of the role of nanomaterials in soil pollution by evaluating the life cycle of ENMs in soil and how features of the soil influence their activity. For instance, the chemical form, mobility, bioavailability, and toxicity of ENMs may be influenced by the physicochemical characteristics of the soil, including its pH, temperature, hydrostatic pressure, and ionic composition (Schlich and Hund-Rinke 2015). Researching how ENMs affect plant growth and the microbiota in plants and soil is also useful. In this study, we provide a fundamental overview of nanomaterials, their synthesis, and a detailed overview of how the ENMs could alter the soil and plant microbiomes, which have a major impact on soil fertility and plant health (Sergaki et al. 2018).

2. Relevance and Importance of Studying ENMs and Nanomaterials in Soil

Because of their miniature size, ENMs are able to deeply permeate a variety of biological tissues, including plant tissues and the blood-brain barrier. The vast array of commercial uses of nanoparticles can be attributed to their highly tunable physicochemical features. These unique qualities of nanoparticles have been used in a variety of industrial applications, including pharmaceuticals, cosmetics, aircraft, and agriculture (Khan 2020). Throughout their life cycle, millions of tons of ENMs are dumped in various habitats, such as landfills, sediments, soils, water, etc. Therefore, the selection of an appropriate synthesis process is essential to designing ENMs with the appropriate sizes and shapes. In this regard, the following section elaborates on the widely used techniques for synthesizing nanoparticles (Jutz and Böker 2011, Khan et al. 2021).

Concern over nanoparticles being released into the environment is also growing. There are two ways that ENMs can enter the terrestrial environment: directly and indirectly. The main sources of direct exposure to ENMs are the use of fertilizers and plant protection products that contain ENMs, the use of ENMs in soil remediation procedures, nanowastes from consumer and manufacturing products, and unintentional discharges of ENMs. The soil matrix is indirectly exposed to ENMs through the disposal of sediments and biosolids from wastewater treatment and agricultural fertilizer runoff in landfills. Since many ENMs wind up in landfills, the OECD Working Group on Waste and Resource Productivity reported that landfills represent a conduit for ENM exposure that requires attention (Kwak and An 2016).

3. Innovative, Scope and Objectives of the Chapter

This chapter examines the significant effects of engineered nanomaterials (ENMs) on soil health, which is vital to ecosystem health and agricultural output, the stages by which ENMs enter soil, how different soil features affect their behavior and fate, and potential toxicological effects on soil biota. To reduce the negative effects of ENMs on soil ecosystems, it is crucial to comprehend these factors to design appropriate risk assessment and management techniques. To guarantee that the advantages of nanotechnology are realized without jeopardizing soil sustainability and health, the chapter also highlights the necessity of interdisciplinary research and creative methods. Understanding the intricate relationships between soil health and nanomaterials is crucial for developing best practices in environmental management and agricultural development.

This chapter aims to provide a thorough analysis of the effects of engineered nanomaterials (ENMs) on soil health, highlighting the potential hazards to human health and the environment, as well as the ENMs' distinctive features and broad variety of uses. Its objectives are to describe in detail the processes of ENM increment into the soil, investigate how soil parameters affect their behavior and fate, and assess the toxicological consequences of these elements on the soil biota, which includes microbial and plant communities. The chapter highlights how crucial it is to comprehend how ENMs are transformed and go through their life cycle in soil in order to provide thorough risk assessment and management plans.

4. Types of Engineered Nanomaterials

4.1 Metal-based Nanomaterials (e.g., Silver, Gold, Zinc Oxide)

Nanoparticles (NPs), which disperse throughout the environment, are released from growing manufacturing of nanoproducts. The characteristics of NPs dictate their use in many industrial domains, but they also endanger the well-being of the ecosystem (Auffan et al. 2012, Burns et al. 2013). The NPs are manufactured at such a huge level that it is inevitable to obstruct them from finding their way into the soil and other environmental components. Nanoparticles can be intentionally added to soil for remediation (nZVI, for example), or they can be introduced through the use of mineral fertilizers (TiO2) or organic fertilizers (biosolids, particularly sewage sludge). Pesticides (Cu, CuO) can also cause nanoparticles to move into the soil (Ma et al. 2014). The use of goods containing NPs, such as gasoline or tires for cars, may also result in NP contamination of the soil. The deposition of nanoparticles in the air or groundwater is making the soil indirectly contaminated (Pan and Xing 2012). Figure 2 illustrates the types of nanoparticles and their impacts on different living mediums.

Jośko et al. (2014) obtained results indicating that the nanoparticles with the strongest detrimental effect on the examined enzymes can be identified under certain conditions. Thus, nano-CuO was found to have the greatest detrimental effect on dehydrogenase. For urease, nano-ZnO and

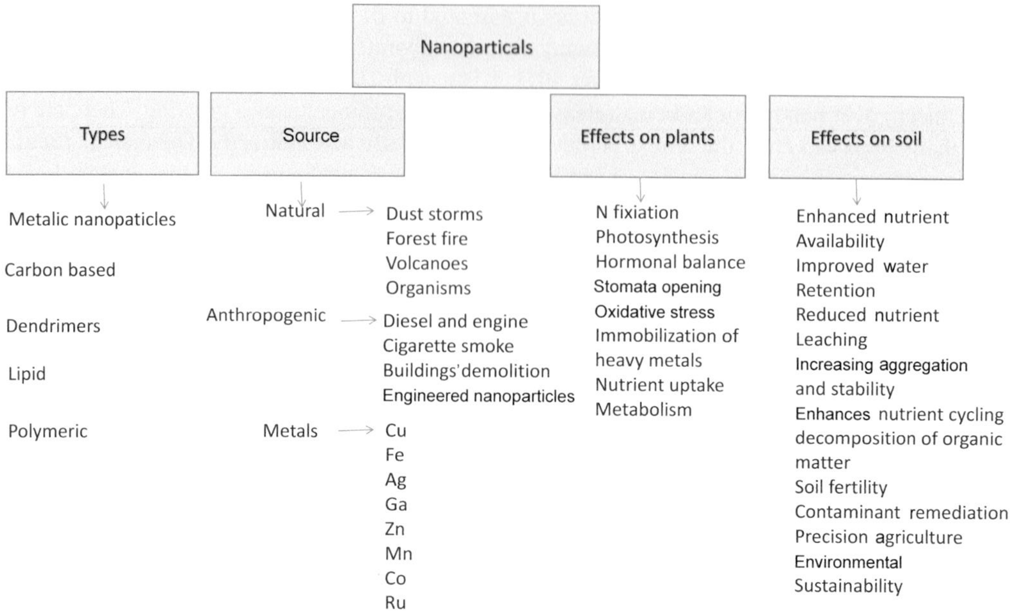

Figure 2. Types of nanoparticles their impacts and toxicities on different living mediums.

nano-CuO were the most hazardous. On the other hand, the most sensitive enzymes to nano-CuO were acid phosphatase and alkaline phosphatase, with nano-Ni and nano-Cr2O3 also affecting the latter. Among the investigated NPs, nano-CuO was the most often reported substance that was found to impede enzyme activity.

4.2 Carbon-based Nanomaterials (e.g., Carbon Nanotubes, Graphene)

Research into carbon-based nanomaterials, including fullerenes, carbon nanotubes, graphene and its derivatives, nanodiamonds, and other nanocarbon allotropes, has increased dramatically in the last several years. Because of their small size—they are comparable to the size of many fundamental biomolecules—their high specific surface area, strong electrical and thermal conductivity, special optical qualities, and superior mechanical capabilities, carbon nanostructures provide countless opportunities for modification and customization. The potential biomedical uses of carbon nanotubes (CNTs), including drug delivery, tissue engineering, diagnostics, and biosensing, have been the focus of research over the last 20 years (Simon et al. 2019). Carbon nanotubes in particular can function as contrast agents in several imaging techniques. They can provide a strong spatial resolution for indicating the presence and localization of target cells once they have been functionalized and conjugated with different biomarkers. One promising option for addressing the therapeutic issues related to neurological disorders like ischemic stroke is carbon nanotubes (Gong et al. 2013).

In regenerative medicine, electrical cell stimulation is a critical technique for nerve regeneration, with applications in body reconstruction, artificial organs, and neurological prostheses. To achieve accurate and efficient indirect electrical stimulation, substrate materials must be both highly conductive and biocompatible.

Graphene nanoplatelets (GNPs) are an effective means to achieve this. However, their strong agglomeration tendency degrades the quality of the coatings produced. For this reason, choosing the appropriate surfactant concentration is essential to achieve excellent conductivity and coating quality. Highly promising nanomaterials for the creation of scaffolds for tissue repair are graphene and its derivatives. It appears that immune cells' reactions to these graphene-based materials are essential for promoting regeneration. Therefore, before using them to create any kind of structure, it is imperative to analyze this response. The diversity of graphene-based materials—particularly

Table 1. Engineered carbon-based nanoparticles' (NP) impacts on the soil plant health.

NP Type	Plant	Size (nm)	Quantity	Soil type	Duration	Impacts	Impacts on Soil	References
Carbon Nanotubes	*Arabidopsis thaliana*	20–30	100 mg/kg	Sandy loam	8 weeks	Enhanced root growth, increased biomass.	No significant impact on microbial activity.	(Elbakyan and Zaporotskova 2024, Hughes et al. 2024, Li et al. 2022, Liu et al. 2024, Wang et al. 2022, Wildgoose et al. 2006)
Graphene Oxide	*Zea mays* (Maize)	10–20	50 mg/kg	Clay loam	10 weeks	Improved germination rate, better stress tolerance.	Slight increase in soil pH.	
Carbon Quantum Dots	*Triticum aestivum* (Wheat)	5–10	75 mg/kg	Loamy sand	6 weeks	Enhanced photosynthetic efficiency.	No noticeable changes.	
Fullerenes (C60)	*Solanum lycopersicum* (Tomato)	30–50	25 mg/kg	Silt loam	12 weeks	Increased fruit yield, better nutrient uptake.	Minor alterations in soil enzymatic activities.	
Carbon Nanofibers	*Glycine max* (Soybean)	50-100	150 mg/kg	Peat soil	14 weeks	Reduced growth, signs of phytotoxicity.	Decrease in nitrogen fixation.	
Carbon Nanotubes (CNTs)	Tomato (*Solanum lycopersicum*)	1–100 nm in diameter	0.1–1 mg/L in nutrient solution	Sandy loam	60 days	Enhanced growth, increased antioxidant activity.	Increased microbial diversity, carbon sequestration.	
Graphene oxide (GO)	Rice (*Oryza sativa*)	10–500 nm	5–50 mg/kg soil	Clayey	90 days	Reduced root growth, altered nutrient uptake.	Increased soil pH, decreased enzyme activity.	
Fullerenes (C60)	*Arabidopsis thaliana*	1–10 nm	0.01–0.1 mg/L in water	Sandy	30 days	Decreased photosynthesis, oxidative stress.	Increased water retention, enhanced microbial activity.	
Graphene (G)	Wheat (*Triticum aestivum*)	1–5 µm lateral size	0.5–2 mg/kg soil	Sandy loam	28 days	Increased growth, photosynthesis, and antioxidant activity.	Increased microbial activity and nutrient cycling.	
Fullerene (C60)	Soybean	0.7–0.9 nm diameter	1–10 mg/kg soil	Silty clay loam	28 days	Improved seed germination and root development in soybean.	Increased soil respiration and microbial biomass.	
Graphene, Graphene Oxide	Maize	1–10 µm lateral size	50–200 mg/kg soil	Sandy loam	21 days	Increased photosynthetic pigments, plant height and biomass in maize.	Increased soil organic carbon, microbial biomass carbon and enzyme activities.	
Biochar		1–10 mm particle size	10–50 t/ha	Sandy loam	180 days	Not reported.	Improved soil structure, increased soil organic carbon and nutrient retention.	
Carbon Nanotubes (CNTs)	Tomato (*Solanum lycopersicum*)	10–30 nm diameter, 1–2 µm length	10–100 mg/kg soil	Loamy soil	21 days	Increased plant growth and yield.	Increased soil aggregation and water holding capacity.	

their surface chemistry, that is, the kind and quantity of oxygen functional groups—is another important consideration. Using graphene's chemically inert qualities to coat medical equipment is an additional strategy. To prepare the substrate surface for coating and optimize the chemical vapor deposition process for improved bonding—for example, by using the imperfect interface side of the graphene layer for bonding—effective coating is required (Cicuéndez et al. 2021).

Another area of carbon nanomaterials research that is a hotspot is clean energy technology. Although hydrogen fuel is a useful substitute for fossil fuels, there are certain drawbacks to it as well. For example, contaminants in the metal catalysts that regulate the reaction of hydrogen fuel can result in hydrogen fuel poisoning. Separating H_2 from other gases like CO, CO_2, CH_4, N_2, and H_2O is therefore essential. Three different rotation velocities were applied in molecular dynamics simulations to assess the performance of double-walled rotating carbon nanotube membranes intended for hydrogen separation. By spinning the nanotubes, they demonstrated how crucial angular velocity is for the process of separating hydrogen (Muraru and Ionita 2020).

4.3 Other Types of Nanomaterials

The possible advantages of polymer nanomaterials and nanoclays for soil health are also making them more and more well-liked. Hydrogels and fertilizers embedded in nanoparticles are examples of polymer nanomaterials that are intended to enhance soil water retention and regulate nutrient release. Particularly in arid areas, hydrogels can absorb vast volumes of water and release it gradually, thus ensuring that plants always have access to water. Conversely, fertilizers contained in nanoparticles can regulate the delivery of nutrients, reducing leaching and increasing plant uptake. Layered double hydroxides (LDHs) and nanoclays enhance the structure, water-holding ability, and nutrient-holding capacity of soil. Additionally, they can enhance the gradual release of nutrients, giving plants a steady supply. These developments in clay- and polymer-based nanomaterials present encouraging paths towards improved agricultural production and sustainable soil management (Guilherme et al. 2015, Kalia et al. 2020).

Typically, hydrocarbon-based molecules, or carbon and hydrogen atoms, make up organic nanomaterials. Through the use of organic molecular structures' weak intermolecular interactions, organic nanomaterials can be engineered to self-assemble into nanostructures with distinct dimensions or targeted characteristics (Yao et al. 2013). Lipid-based, polymer-based, and carbon-based organic nanomaterials are the several types of organic nanomaterials that can be made to function as antimicrobial agents or as nanocarriers in antimicrobial applications. Different classes of lipid nanoparticles, such as liposomes, emulsions, solid lipid nanoparticles, and nanostructured lipid carriers, can be created by nanoformulation of a range of lipid candidates, such as free fatty acids, phospholipids, glycolipids, sphingolipids, fatty alcohols, glycerides, and waxes (Gordillo-Galeano and Mora-Huertas 2018). Table 2 illustrates the impact of engineered nanoparticles on soil health under different experimental conditions.

5. Sources and Pathways of ENMs in Soil

The extensive use of engineered nanomaterials (ENMs) has accelerated the development of numerous industries and technologies, including those related to energy, petroleum, construction, medical, agriculture, consumer goods (like textiles and cosmetics), electrical and electronic products, and transportation. However, when a variety of industrial and home goods are made, used, and disposed of, ENMs are released into the environment. Ecotoxicological hazards are associated with the growing amount of ENMs accumulating in the environment. Ecotoxicity can result from plant systems being exposed to metallic ENMs in soil and aquatic environments. It is crucial to have a complete awareness of the complete life cycle of environmental nanomaterials (ENMs), particularly those that could potentially find their way into plant systems. In addition to this, aquatic life is also threatened by the harms of ENMs, as all sources of ENMs would ultimately lead to disposal in the aquatic system (Zhu and Huang 2023).

Table 2. Impact of engineered nanoparticles on the soil health under different experimental conditions.

NMs	Test species	Primary exposure concentration (mg/kg)	Test soil and amount	Test duration	Container	Endpoint	Effects	References
Silver nanoparticles (AgNPs)	Soil microbial community	0.015 and 1.5 µg kg^{-1}	Sandy clay loam Mediterranean soil	7 and 30 days	Microcosm	Soil microbial biomass, soil enzyme activities, microbial community composition.	Decreased bacterial biomass in SA at 7 days, decreased Gram+ biomass in SA at 30 days, increased bacterial biomass in SU at 30 days, fungal biomass more resistant, changes in bacterial community composition and fungal richness.	(de Oca-Vásquez et al. 2020, Chen et al. 2020, Nie et al. 2020)
SiO2, TiO2, Fe3O4 ENMs	Maize plants	100 mg kg^{-1}	Agricultural soil from Hailun (site N, 126° 38′E and 47° 26′N) mixed with potting soil (Miracle-Gro, Beijing, China) 100 g growth media (40 g agricultural soil and 60 g potting soil)	4 weeks	Plastic containers (9 cm × 9 cm × 7 cm)	Soil and plant metabolomics.	Altered leaf, root, and soil metabolite profiles in an ENM-dependent manner, no negative impacts on plant growth, significant increase in soil metabolites (levoglucosan, linolenic acid, 4-hydroxycinnamic acid, and allo-inositol), changes in SOC pool.	
MgO nanoparticles (nMgO)	*Phytophthora nicotianae*, *Thielaviopsis basicola* (fungi), Tobacco plants	500 µg/ml	Not specified	Not specified	Not specified	Fungicidal activity, fungal growth inhibition, spore germination inhibition, sporangium development inhibition, disease control efficiency.	nMgO inhibited fungal growth, spore germination, and sporangium development more efficiently than mMgO, caused cell morphological changes, disturbed zeta potential, and induced oxidative stress in fungal cells, reduced tobacco black shank by 36.58% and black root rot by 42.35% in greenhouse, final control efficiency reached 50.20% and 62.10%, respectively, with root irrigation	

Table 2 contd. ...

...Table 2 contd.

NMs	Test species	Primary exposure concentration (mg/kg)	Test soil and amount	Test duration	Container	Endpoint	Effects	References
CuO nanoparticles (CuO NPs)	*Arabidopsis thaliana* (root tips)	10 mg/L	Not specified	Not specified	Not specified	Growth inhibition, deposition on root surface, intercellular penetration, cell wall damage, cell arrangement disorder, changes in cell wall components.	CuO NPs inhibited root growth, deposited on root surface, penetrated intercellular spaces, caused severe damage to root tips including cell wall splitting and disordered cell arrangement. Cu2+ release also contributed to effects.	
TiO2	Soil microorganisms	10–500	Sandy loam soil, 1 kg	30–90 days	Pots	Soil microbial community, enzyme activities, nutrient cycling.	Decreased microbial biomass and diversity, impaired soil enzyme activities, disrupted nutrient cycling.	
ZnO	Earth worms, soil bacteria	50–1000	Clay loam soil, 2 kg	28–120 days	Microcosms	Mortality, growth, reproduction, soil respiration.	Increased mortality, reduced growth and reproduction, decreased soil respiration.	
CuO	Soil plants, microarthropods	100–1000	Loamy sand, 1.5 kg	60–180 days	Pots	Root growth, biomass, community structure.	Inhibited root growth, reduced biomass, altered community structure.	
Ag	Soil fauna, microbial communities	1–100	Loamy soil, 2 kg	45–120 days	Microcosms	Abundance, diversity, enzyme activities.	Decreased abundance and diversity, impaired enzyme activities.	

5.1 Agricultural Inputs (Fertilizers, Pesticides, Biosolids)

Fertilizers, insecticides, and biosolids are examples of agricultural inputs that are necessary to maintain soil fertility and boost crop output. On the other hand, new concerns about soil health have emerged due to the addition of ENMs in these inputs. The goal of nano-fertilizers is to increase the effectiveness of nutrient delivery to plants. The requirement for frequent application can be decreased by using ENMs, as their vast surface area combined with small size allows for a targeted and gradual release of nutrients. This can lessen runoff and nutrient leaching, which can check the pollution in the ecosystem. The buildup of nanoparticles in soil, which may change the chemistry of the soil and have an impact on microbial populations that are essential to the cycling of nutrients, has sparked worries regarding the continued use of nano-fertilizers (Bottero et al. 2017).

The use of less active chemicals in nanopesticides results in improved pest control and a decrease in pollution and toxicity to non-target organisms. The efficacy of pest management can be enhanced by the regulated release of nanopesticides and their enhanced adherence to plant surfaces. Their long-term impacts on the health of the soil and their fate in the environment are still unknown. It is possible for nanoparticles to stay in the soil for a long time, which could disrupt the microbial balance and possibly result in the emergence of resistant pest species (Pathak et al. 2024).

By adding organic matter and necessary nutrients to the soil, biosolids improve the fertility and structure of the soil. But biosolids can also carry ENMs into the soil, particularly if they come from nanomaterial-based wastewater treatment systems. The mobility, bioavailability, and toxicity of ENMs can be impacted by their interactions with soil constituents. Unintended effects include soil contamination, changes in the microbiological activity of the soil, and the possibility of plants that enter the food chain absorbing nanoparticles (Jesmer et al. 2017).

5.2 Industrial Emissions and Waste

Measurements taken at facilities manufacturing various ENM products show that worker exposures often occur, for instance, during the handling and processing of dry powders. However, worker exposure measures by themselves are not helpful for assessing ENM emissions to the environment (Mazari et al. 2021). As long as data regarding the amount or flow of ENMs in at least one environmental compartment is not taken into consideration, these occupational exposure measurements cannot be employed. This means that the mass flow per unit time of the pollutant reaching the air (for example, from exhaust gas) is required for the assessment of ENM environmental emissions in addition to the concentration of the pollutant in the indoor air. Merely knowing the amount of particulate matter present in the wastewater is insufficient; further information is needed to determine the mass flow and pollutant concentration of the tributary in question. When ENMs are produced, emissions into the environment also result from the manufacturing of items that include ENMs. It is assumed that the ENM release has a probability distribution that ranges from 0% to 2%. The following release coefficients for the mixture formulation (not contained in a matrix) are assumed by the default parameters listed in ERC25: 2.5% in air, 2% in water prior to the end of STP, and 0.01% in soil. The material formulations (contained in/on the matrix) have equivalent values of 30%, 0.2%, and 0.1% (Samaddar et al. 2018).

The usage, recycling, and disposal of products containing ENMs provide the greatest risk for the unintentional release of ENMs into the environment. Intentional releases from products (like wearing nanotextiles) or unintended releases from products (like applying sunscreen containing ENMs to the skin) can both result in ENM releases during the ENM life cycle. While the origins and magnitude of purposeful emissions of greenhouse gases (ENMs) are known, undesired emissions resulting from product life cycle events like degradation, reformulation, and recycling (Gottschalk and Nowack 2011). The amount of ENM incorporated into a product largely determines the amount and pattern of ENM release from it. While ENMs embedded in solid matrices are released gradually and only partially throughout the course of the product's life, ENMs in liquids are released quickly and, in most cases, entirely during the use phase. While ENMs in paints and textiles are assumed to stay in the matrix for years, ENMs in suspensions, such as sunscreen and cosmetics, are released into the environment in a matter of hours (Kunhikrishnan et al. 2015).

An SMPS (Scanning Mobility Particle Count) was used in a simulation chamber to measure the nano-TiO_2 emissions from wood, tiles, and polymer coatings. The tiles treated with TiO_2 nanopowder had the highest particle emissions out of the three substrates (wood, polymer, and tiles) that were examined (Hsu and Chein 2007). Due to natural climatic conditions, outside wall paints—both old and new—can emit titanium dioxide in the form of nanoparticles into exterior wall runoff, which can then be discharged into natural waters. Several individual particles were also liberated from the aged exterior walls; however, electron microscope analysis of the freed particles revealed that some TiO_2 ENMs remained lodged in the organic binder. The paint's organic binder was still covering the particles, as was observed (Kaegi et al. 2021, 2013).

Investigations into the Ag-ENMs leaching from nano socks in distilled water revealed the release of both particulate and ionic silver. The presence of silver material with a diameter ranging from one to several hundred nanometers was shown based on TEM and EDX examination of colloids in the wash water. Consequently, rather than just being dissolved ionic silver, some of the nanosilver was also liberated as nanoparticles into the wash water. Particulate silver accounted for 14% to 28% or 30% to 95% of the silver released, depending on the substance. Ben et al. (2013) examined the size and form of silver liberated from different goods using filtration techniques and electron microscopy. There were noticed particles or agglomerates emitted in size fractions larger and smaller than 100 nm. Most of the silver radiated from masks, shirts, and medical cloths was blocked by a 20 nm filter. On the other hand, the majority of the silver released from toothpaste, shampoo, and detergents was eliminated by filters having pore sizes of 100 nm (Gottschalk and Nowack 2011).

5.3 Environmental Pathways (Airborn e Deposition, Water Runoff)

ENPs go through a number of transformation processes after being released, which alter their physicochemical characteristics. The primary processes in the terrestrial, marine, and atmospheric environments are those of biological, chemical, and physical changes (Abbas et al. 2020). Condensation, aggregation/deagglomeration, and wet/dry deposition are important atmospheric processes that are influenced by precursor gases, physicochemical properties, ENP concentrations, and local meteorological circumstances (Kumar and Al-Dabbous 2016). Nevertheless, ENPs linked to atmospheric PM can travel great distances and have detrimental effects on both human health and the climate. Examples of such long-distance transports are from northern China to the Pacific Ocean and from the Sahara Desert to Europe. Essential changes in both marine and land environments include dissolution/deposition through homogeneous or heterogeneous aggregation, adsorption, redox reactions, and sulfidation, which can alter the trajectory, transportation, and ultimately the bioavailability and toxicity of ENPs (Amde et al. 2017, Invernizzi et al. 2015). These processes are mainly regulated by environmental conditions like pH, ionic strength, O_2-limiting conditions, natural organic matter (NOM), as well as biological macromolecules (Espinasse et al. 2018).

The wearing and washing of paints, textiles and personal care products, and cleaning agents are examples of non-point sources of engineered nanoparticles (ENPs) that are released into indoor air and the surrounding atmosphere. Point sources include manufacturing factories, transportation processes, landfills, wastewater treatment plants (WWTPs), and waste incineration (Peng et al. 2020). High amounts of ENPs can momentarily be released into the atmosphere in localized emission scenarios such as pesticide aerosolization and unintentional releases. The addition of ENPs in the atmosphere is significant even though their yearly worldwide emissions into the atmosphere are quite low (about 8100 tons) when compared to other environmental elements like soil and water (John et al. 2017, Keller and Lazareva 2014).

After being released into the environment, ENPs experience chemical and physical changes that have an impact on their behavior and ultimate destination in different environmental zones, such as aquatic and terrestrial settings. The Aitken mode (~ 10 to 100 nm) and the less common nucleation mode (1 to 10 nm) are the two categories into which atmospheric scientists divide ultrafine particles (≤ 100 nm). These particles are quickly mixed and moved by Brownian motion, which causes condensation and the creation of bigger particles. When particle surface area and Brownian diffusivity work together, the condensation rate in polydisperse size distributions rises dramatically at large particle concentrations. In the atmospheric phase, heterogeneous condensation of engineered nanoparticles (ENPs) and natural particles predominates over homogeneous condensation, giving rise to novel forms of particles that combine natural and designed nanoparticles (Baalousha et al. 2016, Tiwari and Marr 2010).

Engineered nanoparticles (ENPs) may purposefully enter soil, water, and sediment systems through wastewater and sewage sludge (including solid waste) discharges. The most common ENMs are produced annually at a rate of about 270,000 tons. According to projections, the biggest sources of ENPs are soils (8–28%) and landfills (63–91%), with releases to aquatic ecosystems coming in second (about 7% of total ENP production) (Bundschuh et al. 2018, Grillo et al. 2018). ENPs are subjected to a number of transformation processes immediately upon release, such as homo- or hetero-polymerization, adsorption, dissolution/sedimentation, oxidation, reduction, sulfidation, and reactions mediated by photochemical and biological processes. These modifications have a profound impact on the bioavailability and mobility of ENPs. The fate and bioavailability of ENPs for many organisms are determined by these dynamic modifications, which make an understanding of them essential. Important physicochemical characteristics of ENPs, like particle size, surface area, zeta potential or surface charge, colloidal stability, and core-shell composition, are critical in controlling these transformation processes (Amde et al. 2017, Bhatt and Tripathi 2011, Dale et al. 2015).

6. Behavior and Fate of ENMs

There are three categories of transformations: chemical, biological, and physical processes. Agglomeration, deposition (in porous media), sedimentation, and aggregation are examples of physical processes. Redox reactions, corona creation, dissolution, and the ensuing morphological changes are examples of chemical processes. Biodegradation and biomodification are examples of biologically mediated processes, which are most likely carried out by microorganisms. It is obvious that the characteristics of the micro network and the surrounding environment will influence changes, but comprehension and forecasting are quite challenging due to the intricacy and unpredictability of these elements (Lead et al. 2018). Solubility plays a significant role in determining the destiny and toxicity of certain nanomaterials, such as ZnO, where the ions are primarily responsible for the toxicity. Intermediate solubility and solubility for other NMs, such as Ag, CuO, and certain quantum dots, imply that both particles and ions are involved. Ions most likely have direct biological effects, while NMs increase transport and concentration inside cells (Hartmann et al. 2014, Leclerc and Wilkinson 2014).

Solubility and dissolution become less significant for carbon-based nanomaterials and many inorganic nanomaterials with lesser solubility, such as titania and ceria. However, dissolution occurs in low-pH microenvironments like the cellular vacuoles. The relative behavior and effects of ions and particles over pertinent time scales must be considered when evaluating the significance of ions in relation to particles for nanoscience in the environment. In the field of toxicology, for instance, the significance of dissolving can be assessed during exposure, and this may vary from equilibrium solubility. Nano-CeO_2 is a valuable material for directly examining nanomaterial effects without having to take into account the complexity of ionic behavior because of its exceptionally low solubility. When compared to materials that are micron-sized, nano-CeO_2 can be used to investigate how it affects behavior and toxicity (Dresselhaus and Terrones 2013, Ghosh 2023, Zarzycki 2022). Possible pathways of engineered nanomaterials, toxicity, and contamination are illustrated in Figure 3.

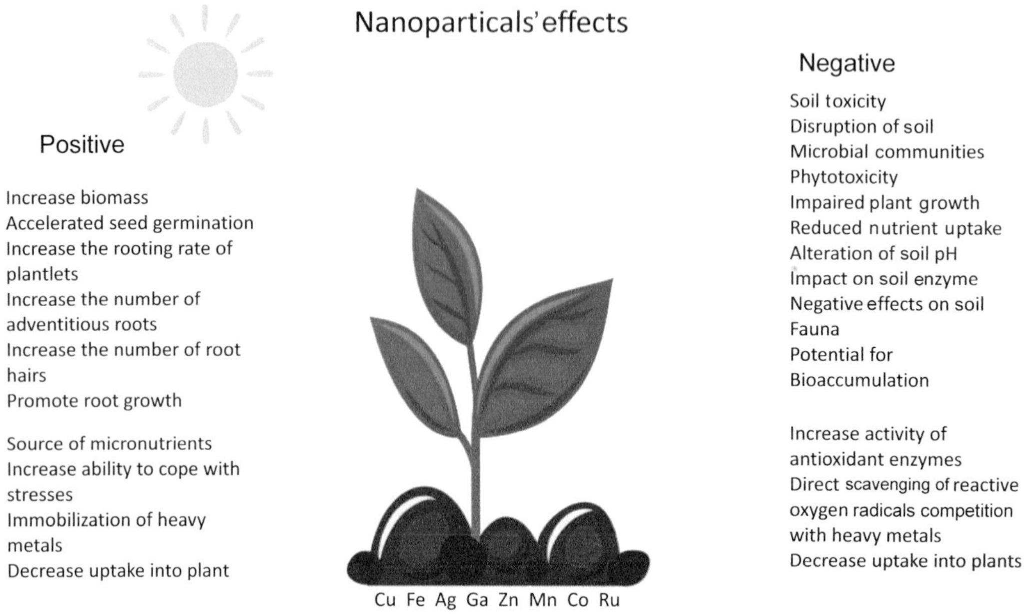

Figure 3. Illustration defining the possible pathways of toxicity and contamination of engineered nanoparticles.

Numerous studies have demonstrated that NMs' inherent characteristics, such as size, coating, and doping, can influence their solubility and dissolution in addition to composition. Although this is being done more successfully now than in 2008, a key issue in this subject, as with other disciplines that aim to correlate features and behavior, is the difficulty of altering the unique qualities of NMs without altering other confounding factors (Adeleye et al. 2018, Toncelli et al. 2017, Tsiola et al. 2017). Furthermore, research has examined how outside variables, including NOM, affect the behavior of NMs and found that when particles develop and new NM precipitates, dissolution may be accelerated. Additionally, organic compounds have a major impact on NM dissolution. However, the effects are complicated and have not yet been fully explained by various research findings. Studies on natural waters from a range of systems have revealed that these intricate systems have a major impact on dissolution (Merrifield et al. 2017, Odzak et al. 2017, Xiao et al. 2018).

Because the effects of these variables are charge-mediated in charge-stabilized systems, variations in pH lead to distinct behaviors, while changes in ionic strength cause aggregation. When NMs are added to a test solution, homoaggregation/agglomeration—where agglomeration is sometimes operationally defined as a weaker contact than aggregation—can happen very instantly, producing aggregate/agglomerate sizes as high as 100 nm. This phenomenon is frequently researched in NMs. Numerous studies have shown how NOMs, such as fulvic and humic acids, stabilize NMs by steric and charge repulsion, preventing aggregation (Yang et al. 2019). Using 1 mg/L nano-CeO_2 added to filtered and unfiltered river water, the significance of heteroaggregation (aggregation between various particles) at high NM concentrations was demonstrated. Following first-order kinetics, heteroaggregation with natural colloids removed 80% to 85% of the ceria in unfiltered water via sedimentation in 4 days. On the other hand, NM was stabilized for up to 12 days by NOM in filtered water. Over 99% of the ceria precipitated at higher ceria concentrations (10 and 100 mg/L in unfiltered water) mostly through homoaggregation (Quik et al. 2012).

The steady increase in steric stability of the zeta potential itself indicates that increasing the amount of NOM in the filtered algal media increased the stability of nano-CeO_2 via adsorption and decreased homoaggregation by increasing electrostatic (i.e., combined electrostatic and steric) repulsion. Keeping in mind that freshwater can have CeO_2 concentrations in the mg/L range, which is significantly lower than the experimental amounts found in most research studies. Although the rate constants are difficult to measure, the illustration of first-order kinetics of sedimentation and dissolution is a valuable addition to upcoming modeling tasks (Abbas et al. 2020, Diegoli et al. 2008).

For a variety of Dutch river flows, the heterogeneous aggregation rates of natural colloids and the sedimentation rates of C60, nano-CeO_2, PVP-coated nano-silver, and SiO_2-coated nano-silver were evaluated. For site-specific modeling, parameters specific to the system are crucial. Further research on the heteroaggregation of clay particles utilizing humic acid colloids and natural clay at high concentrations (0.1 and 0.8 mg/L) of nano-TiO_2 was conducted as a result of these experiments (Loureiro et al. 2018, Quik et al. 2014, Scheringer et al. 2014). Although these investigations offer valuable preliminary data, it's possible that they overstated the significance of homogeneous aggregation. This is because the aggregation kinetics heavily rely on the initial concentration—specifically the number concentration—of the distributed NMs. Aggregation and dispersion can lead to significant differences in the characteristics of NMs in various pools and receptors. Whereas benthic species are exposed to aggregated NMs, which are bigger in bulk but may be less hazardous per unit mass, the water column is likely to include smaller, more dispersed NMs. While aggregation usually lowers the amount of NMs that are bioavailable when exposed to water (Khan et al. 2016, Romero-Franco et al. 2017), aggregation may sometimes enhance bioaccumulation by enhancing absorption rates or particle accessibility. Recently, the number and mass concentration of NMs in core-shell NMs were measured using SP-ICP-MS in order to study the concentration dependency of aggregation (and dissolution). The results of the investigation demonstrated that, at realistic environmental concentrations, homogeneous aggregation has no quantitative significance (Merrifield et al. 2017).

For many metal nanomaterials, sulfidation is a crucial chemical reaction, particularly in situations where sulfide concentrations are high, like in wastewater treatment plant (WWTP) sections or in anoxic sediments. Particle size, surface charge, and solubility are all altered by these reactions, which are commonly attributed to core-shell ($Ag@Ag_2S$) synthesis, in which the Ag_2S shell progressively thickens. In the end, these modifications will impact the NM's bioavailability and effectiveness. Microscopic studies show that sulfidation is taking place, and the detection of silver sulfide (Ag_2S) nanoparticles in sewage sludge provides *in situ* evidence for the sulfidation of a frequently used silver nanomaterial (Auvinen et al. 2017, Spurgeon et al. 2020).

Through a cycle of surface dissolution and reprecipitation, nano-ZnO can undergo gradual sulfidation. A partial sulfide coating, analogous to that observed on partially sulfidized silver nanoparticles, does not necessarily reduce the solubility of ZnO. However, it plays a crucial role in regulating the dissolution rate. Although Zn^{2+} ions are released throughout the conversion process, the overall solubility of the particle decreases with increasing sulfidation. Furthermore, despite the newly formed zinc sulfide (ZnS) domains being smaller, their lower surface charge generally increases their tendency to aggregate. Within a few days, nano-CuO likewise undergoes sulfidation, first generating copper (I) sulfide (Cu_2S) and perhaps converting to CuS (Ma et al. 2013). It has been demonstrated that the sulfided form of pristine nano-CuO is more harmful to aquatic species than the pristine nano-Ag and nano-ZnO due to its higher solubility and greater release of Cu^{2+} than pristine NM (Li et al. 2015, Ma et al. 2014).

More current fate and behavior models incorporate more intricate processes including dissolution, sedimentation, and aggregation, which are frequently connected to river flow and other physical processes. Bottom-up techniques are fully based on the life cycle (Ellis et al. 2020, Liu and Cohen 2014, Mueller and Nowack 2008). The majority of models presuppose steady-state concentrations in several compartments. However, transformation kinetic variability is significant yet poorly understood. Laboratory studies on nanoparticle destiny are frequently carried out under simplified settings, utilizing concentrations much greater than those predicted in the environment. This is in addition to the absence of means to analyze ambient concentrations. Ag NM concentration has been found to affect dissolution, aggregation, and more complex transformations. Below 1 mg/L, a dissolution-dominated regime occurs, and beyond 10 mg/L, an aggregation-dominated regime. These effects have been measured lately. Such alterations have an impact on concentration-dependent bioavailability. Changes in aggregation behavior are probably the mechanism by which these effects work. Similar to this, aggregation may be responsible for the peculiar behavior of TiO_2 NMs' diffusion coefficient with concentration (Baalousha 2017, Merrifield et al. 2017).

7. Impact of ENMs on Soil Properties

Agricultural soils are frequently exposed to engineered nanomaterials (ENMs) when biosolids are used for fertilization or wastewater is treated for irrigation. The use of nanopesticides, nanofertilizers, nanosensors, and soil and groundwater cleanup are examples of intentional exposures. ENM aerosols have the potential to be dispersed in non-agricultural soils, which could result in inadvertent exposure from diffuse sources. Point contamination can also result from uncontrolled releases of industrial nano wastes into the environment. Other possible routes for soil exposure include leaching from landfills and accidental chemical spills (Suppan 2013). The toxicological effects of ENMs on soil biota and their propensity to contaminate groundwater and affect aquatic ecosystems make them an important topic of discussion. Since ENMs typically spend more time in soil than in aquatic environments, a comprehensive study of their life cycle in the soil environment is crucial. ENMs can be generically categorized into two groups: inorganic and carbon-based. To find out what influences the ENMs' interaction with soil constituents, several batch and column studies were carried out in lab settings that mimicked natural settings. The changes, fate, and transit of ENMs in soils have been the subject of current research, which is briefly reviewed below (Ingle 2021).

7.1 Physical Properties

The percentages of sand, silt, and clay in addition to porosity and organic matter define the physicochemical qualities of soil, which are influenced by the soil's geological origin, geographical topography, and climatic conditions. When compared to sandy soils, ENPs often exhibit superior retention (lower migration) in clay soils with a higher clay concentration. The clay-rich mineral structure, which has more binding sites and a larger specific surface area, is primarily responsible for this (Braun et al. 2015).

According to recent research, substrate roughness and surface charge have a big impact on ENP transmission. To facilitate deposition or transport, these variables interact with the system pH and the ENPs' zeta potential. The amount of clay in the soil influences the transport of ENPs. Negatively charged CeO_2 nanoparticles stick poorly to sand, while positively charged CeO_2 nanoparticles stick well to kaolin. In a similar vein, TiO_2 nanoparticles are only destabilized by soil mineral colloids when the pH falls below the zero charge point of TiO_2. Furthermore, through cationic valence states, the presence of Ca^{2+} ions improves the interaction between Ag nanoparticles and clay (Tang and Cheng 2018, Zhang et al. 2018).

Because it offers negatively charged sites for ENPs or derived ions, organic matter (OM) is essential in controlling the fate and transport of ENPs in soil. This can result in heterogeneous aggregation and improved cation exchange capacity. Adsorption of OM, particularly humic acid, onto ENPs or the ions produced from them can cause flocculation, translocation, or suspension. For instance, it has been demonstrated that rice soil rich in OM stabilizes and transports biochar ENPs.

In OM-rich natural soils, TiO_2 ENPs also showed good transportability (51.8–90.6%) (McNew and LeBoeuf 2015, Wang et al. 2016).

7.2 Chemical Properties (pH, Nutrient Availability, Cation Exchange Capacity)

The transport of engineered nanoparticles (ENPs) in soil is influenced by important solution chemistry parameters such as cation valence, pH of the soil, and pore water ionic strength (IS). To replicate the natural ENP transport pathways in soil environments, these components were included in column tests. In electrolyte solutions, ENP transport usually declines as IS increases, and higher IS causes aggregation-induced total inhibition. This happens because, according to the Derjaguin–Landau–Verwey–Overbeek (DLVO) theory, increasing IS increases van der Waals attractions and decreases electrostatic repulsions. Aggregation and sedimentation result from the electrostatic double layer around the ENPs being essentially compressed with rising IS (French et al. 2009, Rawat et al. 2018).

Divalent cations, like Ca^{2+}, can hold ENPs in the soil matrix more successfully than monovalent cations, such as Na^+ and K^+. This is because Na^+/K^+ deposition takes place at the secondary (reversible) minimum of the bilayer colloid/ENP energy barrier, whereas Ca^{2+} deposition occurs at the primary (irreversible) minimum, according to the DLVO theory. The concentration of electrolytes has little bearing on this occurrence. French et al., for instance, found that the micron-scale aggregation rate of TiO_2 ENPs (4–5 nm) was slower in a pH 4.5 solution of 0.0165 M NaCl than in a pH 4.8 solution of 0.0128 M $CaCl_2$, suggesting that the effects of cation valence are concentration-independent (Abbas et al. 2020, Chekli et al. 2016).

The pH of the soil has a significant direct and indirect impact on ENP–soil interactions. Because of homogeneous and heterogeneous aggregation, the retention of sterically and electrostatically stabilized silver nanoparticles rises with lowering pH. Zeta potential (ZP), which is a function of solution pH and is dependent on the surface charge of the ENPs, is significant in this process. The ENPs' zero charge point, or isoelectric point, is reached by the pH of the soil, at which point the particles become neutralized and aggregate. The ENPs stay suspended if the pH of the soil is higher or lower than 30 mV from the zero charge point (Denisse et al. 2021, Hoppe et al. 2014).

Adsorption causes ZP to increase with concentration at lower concentrations. The ZP decreases to zero with increasing concentration, though, as the double-layer colloidal effect takes over the ion–ENP interaction at higher concentrations. In order to gain a better understanding of the fate and transport of engineered nanoparticles (ENPs) in natural environments, certain research on soil columns has taken into account environmental parameters such as temperature, light, UV light, and saturated conditions. Deeper soils close to aquifers are in the saturated (phreatic) zone, whereas natural soils near the surface are in the unsaturated (vadose) zone. In contrast to the predominant soil–water and soil–soil interfaces in the phreatic zone, ENP–media interactions in the vadose zone include air–water and air–water–soil interfaces (Prédélus et al. 2015, Rawat et al. 2018). In contrast to the irreversible energetic interactions at solid–water interfaces suggested by the extended DLVO theory, air–water contacts show hydrophobic interactions that are unsuitable for ENP deposition. TiO_2, CeO_2, and $Cu(OH)_2$ ENPs are preserved in the top 3 cm of natural soil columns under unsaturated conditions as a result of homo- and hetero polymorization with organic matter. Raising the temperature (4–25°C) often improves retention because it enhances Brownian motion and decreases the viscosity of the soil solution, which lowers ENP movement in soils. The fate and transport of ENPs can also be influenced by UV radiation and natural light, particularly in wet surface soils (Conway and Keller 2016, Yecheskel et al. 2016).

7.3 Biological Properties (Microbial Activity, Enzyme Activities, Soil Fauna)

Plant roots release exudates, enhance soil structure, and stop erosion. A wide variety of soil organisms that have a significant impact on plant growth find a home in their rhizosphere. All other trophic groups are impacted if this one is harmed. It is a newly discovered class of chemicals which vehemently finds its applications as insecticides and fertilizers for agriculture. Many plants have been used to test the effects of MENM. Both beneficial and detrimental impacts have been discovered, depending on the plant type, particle properties, soil, and test settings (Gogos et al. 2012, Kah and Hofmann 2014).

Concentrations ranging from 20 to 60 mg kg^{-1} enhanced the yield of borage seeds, while concentrations between 500 and 3000 mg AgNP kg^{-1} boosted the yield of cucumber fruit. On the other hand, AgNPs (100, 1000 mg kg^{-1}) enhanced superoxide dismutase activity and decreased chlorophyll concentration in tomatoes. Higher mustard seed production (25 mg kg^{-1}) was achieved using AuNPs, while oxidative stress ($\geq$10 mg kg^{-1}) was amplified. Only one examined study demonstrated ZnONPs' beneficial impact on crops. ZnONPs reduced wheat biomass (45 mg kg^{-1}), decreased chlorophyll, and caused oxidative stress in pea (125–500 mg kg^{-1}) in addition to increasing peanut pod production (133 mg kg^{-1}). Cowpea yielded no outcomes either. Depending on the application type and particle size, ZnONPs (500 mg kg^{-1}) either decreased growth or showed no physiological effect. TiO_2NPs (Gardea-Torresdey et al. 2014). In natural soils, maize and soybean growth and nutrient content were unaffected by TiO_2-NPs (40–60 nm, primarily anatase) at fresh soil concentrations of less than 200 mg kg^{-1}. When CeO_2NPs were given, tomato fruit weight increased (1.3–130 mg kg^{-1}), whereas bean sprout development reduced (100–1000 mg kg^{-1}). PIONs led to a 250–750 mg kg^{-1} increase in soybean seed production. AgNPs and ZnONPs in particular were responsible for the majority of the detrimental impacts on crop development and physiology in the examined papers (Burke et al. 2014, Gardea-Torresdey et al. 2014).

In the topsoil layer, all plant species typically show robust root growth. 0.14 mg Ag kg^{-1} was added to the soil as a biosolid suspension in the form of AgNPs. For permanent crops like fruit, wine, or asparagus, however, exposure times will be longer. Also, exposure from nearby unmanaged land by wind or surface runoff needs to be taken into account (Colman et al. 2013). The archaea, bacteria, fungi, and microbes form the foundation of the soil food web. They obtain their energy from both organic and inorganic sources, primarily plant roots, their exudates, and leaf litter. By breaking down organic materials, supplying minerals, fixing atmospheric nitrogen, and concurrently fighting living things (pathogens and fungi that trap nematodes), they play a significant role in any soil, but notably in supporting plant growth. Soil animals consume the microbial biomass. Fungi are crucial for the first colonization of leaf litter and mineralization. Due to the antibacterial capabilities of some metal-based MENMs, particularly those based on silver and copper, bacteria are frequently seen as a trophic group at risk for MENMs (Hartemann et al. 2015, Ingle et al. 2014).

These effects can be specific to a particular nanoparticle, vary from metal to metal, and are not always bad. For instance, compared to Ag^+ ions, AgNPs were twice as harmful to microbial growth. The parent material, particle size, and shape can be associated with bacterial toxicity in their review of ENM microbial toxicity. However, no generalizations about the corresponding coatings can be made. For Ag, Al_2O_3, TiO_2, CeO_2, CuO, CdSe, CdTe, FePt, and ZnO NPs, they report harmful effects. Nevertheless, research on Si, Fe, Au, Pd, Ag_2S, and Pt NPs frequently report no effects at all or very minor ones. However, depending on the particle form and coating, these effects are not always constant. For instance, protein-coated Ag_2S nanoparticles (2–20 nm) and silver oleate (4 nm) did not exhibit any inhibitory impact. On the other hand, compared to chemically prepared AgNPs, biologically prepared AgNPs with protein/peptide coatings were more hazardous. The toxicity of ZnONPs (20 nm) and CuONPs (0–200 mmol g^{-1}) to dissolved ions and their bulk and ionic equivalents are compared with respect to soil microbial populations. Agglomerated CuONPs (primary size < 100 nm, 500 mg L^{-1}) had an initial inhibitory impact (Choi and Hu 2008, Dimkpa et al. 2012, Rousk et al. 2012, Suresh et al. 2013).

Gram-negative bacteria are often less sensitive to ENMs than Gram-positive bacteria because of their distinct cell wall structures, a fact that has been extensively researched in medical science. But this isn't always the case. Cu NPs, for instance, adversely affected both population groups, and Gram-negative bacteria were more vulnerable in a study of time-kill studies. The shift in the ratio of fungi to bacteria is more significant from an ecological perspective because the latter are able to break down more complex organic molecules, such as pollutants caused by humans. Since we currently know so little about how ENMs affect soil fungi and archaea, it is unclear if this happens when ENMs are present. A study on silver mineral compounds, including metal-accumulating fungi, supports the idea that bacteria, being prokaryotic, are more sensitive to metals. However, ENMs are also very successful fungicides (Burke et al. 2014, Colman et al. 2013).

Earthworms are regarded as "ecosystem engineers" and they are essential for loosening the soil and moving organic plant detritus to deeper soil layers (Tripathi et al. 2007). The most prevalent detritivores in soils, springtails and mites are, after protozoa and nematodes, and they are essential to the breakdown of litter, which releases nutrients for subsequent breakdown and mineralization by microorganisms. By grazing, they also promote the growth of fungi. Microorganisms involved in the cycle of nutrients are stimulated and regulated by mycophagous invertebrates. Primary consumers are mainly exposed to toxins through skin contact with contaminated soil and food absorption since they live underground or in close proximity to soil (McKee and Filser 2016). In risk evaluations of metal nanoparticles, litter-decomposing invertebrates warrant special consideration because of their significant involvement in soil communities. Loss of function in this trophic group could have detrimental effects, both direct and indirect, on numerous soil processes. However, only a small number of studies have looked at the impacts of MENMs on invertebrates that break down leaf litter in comparison to the quantity of plant and microbial studies that have been published to date (Kuehr et al. 2021).

In addition to larger animals like moles, shrews, or birds that eat litter decomposers, secondary consumers in soil ecosystems include ants, spiders, beetles, predatory mites, and nematodes. By doing this, they not only govern populations of other invertebrates in the soil but also exert significant top-down control over them. This control might be lost if MENMs hurt predators, which might have an indirect impact on all the creatures they interact with in addition to primary consumers. To the best of our knowledge, no information has been released regarding how ENMs affect soil predator growth, survival, or reproduction. To better understand whether this trophic group is possibly more or less at risk than others, this knowledge gap needs to be filled. When assessing the impact of ENMs on ecosystem functions, it is important to consider the possibility that they could pose a risk not only to trophic groups in the soil but also to functional ones (McKee and Filser 2016).

Awet et al. (2018) demonstrated that PS-NPs might considerably lower the biomass and enzyme activity of soil microorganisms. Growing PS-NP application resulted in a drop in soil microbial biomass and a decrease in most enzyme activity, suggesting that PS-NPs have a broad antibacterial effect on ground microbiota and enzymes. Because of their small size and high surface area-to-volume ratio, PS-NPs have inherent nano properties that may allow them to interact directly with microbial cells and subcellular structures, resulting in antibacterial action (Brown et al. 2001). Additionally, PS-NP toxicity to soil microorganisms may result from NPs' great adsorption affinity for chemicals that are harmful to soil. Furthermore, research has demonstrated that, despite the fact that unaltered PS-NPs are typically thought to be non-toxic, their surface charge, size, agglomeration, and aggregation may change in environmental conditions. Strong nanoparticle aggregation decreases the absorption and acute toxicity of nanoparticles while producing bigger particle sizes (micrometers) (Liu et al. 2011, Velzeboer et al. 2014).

However, PS-NPs appear as single particles and can actively affect microorganisms because of the production of stable colloids and low polydispersity index in biological fluids. According to our findings, on day 28, all enzyme activity dramatically dropped. The decrease in dehydrogenase activity was accompanied by a decrease in soil dehydrogenases, which are parts of whole cells and are only found in living cells. The microbial biomass on the 28th day demonstrated that PS-NPs directly impacted the soil microorganism's metabolic processes. Antimicrobial activity's precise mechanism can differ and frequently has a multifaceted impact. It has been suggested that oxidative stress, physical perforation, and direct touch can all disrupt the cell membrane and cause antibacterial activity (Awet et al. 2018, Wang et al. 2013).

8. Regulatory and Policy Considerations

Strong policy and regulatory frameworks are necessary to govern the use of ENMs in agriculture and protect soil health: Provide thorough regulations that particularly address the creation, application, and removal of ENMs from agricultural inputs. The safety guidelines and allowable limits for ENMs in soil and water should be outlined in these regulations. Mandate that life cycle analyses be included for all agricultural products derived from ENMs. Environmental effects from manufacturing to disposal must be evaluated as part of life cycle evaluations to make sure that every step is sustainable and ecologically sound. The labeling for all nano-agricultural goods must comprise comprehensive details about the ENM's composition, application requirements, and any hazards.

It is necessary to evaluate the environmental impact of new nano-agricultural products. Environmental impact studies must evaluate any effects on biodiversity, water quality, and soil health before being approved and commercialized. In order to guarantee a uniform approach to agricultural ENM management, the promotion of cooperation between different regulatory bodies, including those in the environment, agriculture, and health sectors, is crucial. The stakeholders and the general public are actively involved in decision-making about the use of ENMs in agriculture. In addition to ensuring that policies reflect society's values, transparency and public participation can strengthen confidence. The synchronization of national laws on nanotechnology and environmental protection with global norms and guidelines is also important. This will encourage worldwide international cooperation and guarantee the safety and sustainability of agricultural practices used by multinational corporations.

9. Future Directions and Research Needs

Future studies should include storage of ENMs in products over time or resuspension of sedimentary materials in rivers, which may significantly affect the amount of ENMs released to different environmental compartments per unit time. In addition, it is necessary to consider the time-dependent dynamics of ENM production quantities and nanoproduct consumption.

Improving our understanding of agricultural ENMs requires comprehensive and forward-looking research strategies. Emphasis on long-term field studies is essential to understand the ongoing effects of ENMs on microbial communities and ecosystem functioning. These studies are essential to understand the ecological consequences and sustainability of ENM use, especially under natural conditions. To facilitate comparative studies and promote a comprehensive understanding of ENMs in environmental contexts, it is imperative to standardize research methods, especially in the characterization of ENMs and the assessment of their microbial impacts.

For example, soil is an important component that has a significant impact on the behavior, fate, and exposure factors of agrochemicals in agriculture. Therefore, the texture and properties of the soil matrix (pH, porosity, organic/inorganic content, compactness, humus, hydraulic conductivity, and microbial community) play a crucial role in determining the ACV of agrochemicals and their bioavailability to plants. Water, light, and temperature are other environmental parameters that can influence the mode of action of agrochemicals, while affecting biodiversity, crop quality, and yields in conventional agriculture. Global warming and climate change also significantly affect the fate and behavior of agrochemicals by controlling their partitioning among soil matrices.

Therefore, these agricultural and environmental parameters can negatively impact the life cycle assessment of agrochemicals and their bioavailability, which in turn has a significant impact on crop safety and productivity. To overcome these limitations, conventional agrochemicals, such as pesticides, must be sprayed at higher doses and/or high frequencies during warm seasons. As a result, the residual concentrations of some agrochemicals in plants and the environment may double and have dangerous effects on human health and agricultural sustainability. In this sense, the combined effects of climate change and conventional agricultural practices may worsen the situation due to continuous exposure to agrochemical residues. In addition, the overuse of agrochemicals is not conducive to maintaining a sustainable agricultural environment in multiple aspects, especially due to its high operating costs and adverse side effects on plants, the environment, and human health along the food chain. The trends of these impacts have not yet been fully evaluated.

The characteristics and modes of action of some agro-NPs are summarized, along with their positive and negative effects in comparison with conventional agrochemicals. The modes of action and reported effects depend on the nature and surface chemistry of the agro-NPs and the plant species. In most cases, nanospheres and nanogels in the form of nanocapsules have been proposed and favorably developed as protective reservoirs that allow a slow or well-controlled release of active agrochemical ingredients to the target species. Encapsulation of active agrochemical ingredients in nanosphere capsules (liposomes, polymers, and silica) can extend their shelf life and reduce soil contamination, thereby improving environmental safety.

10. Conclusion

Furthermore, proper use of ENM-based nanocapsules can reduce the number of agrochemicals used without sacrificing efficiency. These approaches, known as "precision agriculture," have benefits for agriculture, especially in terms of increasing crop yields without degrading soil and water systems. In some cases, agricultural nanoparticles, such as metal nanoparticles (Ag, Au, Al, and Cu) and metal oxide nanoparticles (ZnO, CuO, SiO_2, and TiO_2), are also conjugated or entrapped in polymers, lipids, lignocellulose, and organic matter. Other ionic capsules/surface carriers are used to produce nanoemulsion formulations such as hydrogels, emulsions, immobilized conjugates, and liposome-based agricultural products. These conjugated nanoformulations/capsules can also be applied to other agricultural purposes, such as protecting plants from infectious fungal/bacterial diseases, precise delivery of nutrients, and bioremediation or reclamation of degraded soils.

In summary, functionally designed agricultural nanoparticles and their nanoemulsion formulations are expected to have significant advantages in terms of cost, biocompatibility, and enhanced activity (minimal doses can significantly increase crop productivity). However, despite the many advantages of ENM-based agrochemicals in agriculture, they also have some disadvantages, such as phytotoxicity and bioaccumulation, depending on the plant species and environmental conditions. Taking exposure time as a factor, agro-NPs undergo various chemical transformation (aging) processes, such as dissolution, deactivation, morphogenesis, aggregation, adsorption, precipitation, sedimentation, and degradation, when released into real agricultural environments. They can also interact with various particulate, inorganic or organic compounds dissolved in natural aquatic and terrestrial ecosystems. These aging and chemical transformation processes may significantly change the surface chemistry of agro-NPs, which in turn affects their fate, life cycle, exposure dynamics, and physiological interaction mechanisms with plants (ultimately affecting food security). All environmental factors that affect the mobility of traditional agrochemicals are likely to also play an important role in controlling the fate/exposure of agro-NPs after practical application. These factors include ionic strength, pH, dissolved and suspended organic matter (DOM), and biodiversity in aquatic environments. Other variables include mineral composition, porosity, microbial communities, and the type of natural organic matter (NOM), such as humic acid, pH, electrolytes, and water flux in soil compartments. In addition, climatic conditions such as temperature, sunlight, and humidity may modify the surface properties of agricultural nanoparticles and their dynamic interactions with the surrounding environmental media in actual farm fields.

According to FAO, the pursuit of sustainable agricultural development is expected to face many difficulties in the production and distribution of food resources by 2050 due to the rapid intensification of climate and environmental changes and their impact on water and energy supply. From a 1987 perspective, the United Nations began to propose the concept of a "circular economy," which is defined as "meeting the needs of the present without compromising the ability of future generations to meet their own needs." In this context, the United Nations adopted 17 Sustainable Development Goals in its 2030 priority agenda in 2015 to help promote innovation and development of safe and more sustainable alternatives to achieve environmental protection, social development, economic growth, and climate neutrality. Based on these Sustainable Development Goals (SDGs), the application of nanotechnology in the agricultural sector (agri-NP) continues to grow to achieve multifaceted smart and sustainable development. In particular, the use of agricultural NPs offers great potential for achieving nutrition-related SDGs, aiming to end poverty (Goal 1), achieve zero hunger (Goal 2), support sustainable management of affordable and sustainable water (Goal 6), clean energy (Goal 7), sustainable economic growth (Goal 8), sustainable consumption and production patterns (Goal 12), combat climate change and its impacts (Goal 13), and halt biodiversity loss by conserving, restoring, or promoting sustainable use of terrestrial ecosystems (Goal 15). agricultural nanoparticles (NPs) are expected to play a vital role in reducing the volume of agrochemicals needed, due to their high efficiency.

11. Acknowledgement

Authors want to thank the university of Agriculture Faisalabad and Yangzhou University for the support in this publication. Authors also want to thank Professor Dr Abdul Wakeel from Soil Fertility and Plant Nutrition Laboratory, University of Agriculture Faisalabad, Pakistan.

References

Abbas, Q., B. Yousaf, M.U. Ali, M.A.M. Munir, A. El-Naggar, J. Rinklebe et al. 2020. Transformation pathways and fate of engineered nanoparticles (ENPs) in distinct interactive environmental compartments: A review. Environ. Int. 138: 105646.

Adeleye, A.S., S. Pokhrel, L. Madler and A.A. Keller. 2018. Influence of nanoparticle doping on the colloidal stability and toxicity of copper oxide nanoparticles in synthetic and natural waters. Water Res. 132: 12–22.

Amde, M., J.F. Liu, Z.Q. Tan and D. Bekana. 2017. Transformation and bioavailability of metal oxide nanoparticles in aquatic and terrestrial environments: A review. Environ. Pollut. 230: 250–267.

Auffan, M., C. Santaella, A. Thiéry, C. Pailles, J. Rose, W. Achouak et al. 2012. Ecotoxicity of inorganic nanoparticles: From unicellular organisms to invertebrates. pp. 623–636. *In*: Bhushan, B. [ed.]. Encyclopedia of Nanotechnology. Springer, Dordrecht.

Auvinen, H., V. Gagnon, D.P. Rousseau and G. Du Laing. 2017. Fate of metallic engineered nanomaterials in constructed wetlands: Prospection and future research perspectives. Rev. Environ. Sci. Biotechnol. 16: 207–222.

Awet, T., Y. Kohl, F. Meier, S. Straskraba, A.L. Grun, T. Ruf et al. 2018. Effects of polystyrene nanoparticles on the microbiota and functional diversity of enzymes in soil. Environ. Sci. Eur. 30: 1–10.

Baalousha, M. 2017. Effect of nanomaterial and media physicochemical properties on nanomaterial aggregation kinetics. NanoImpact. 6: 55–68.

Baalousha, M., G. Cornelis, T. Kuhlbusch, I. Lynch, C. Nickel, W. Peijnenburg et al. 2016. Modeling nanomaterial fate and uptake in the environment: Current knowledge and future trends. Environ. Sci. Nano. 3(2): 323–345.

Ben-Moshe, T., S. Frenk, I. Dror, D. Minz and B. Berkowitz. 2013. Effects of metal oxide nanoparticles on soil properties. Chemosphere 90: 640–3646.

Bhardwaj, V and A. Kaushik. 2017. Biomedical applications of nanotechnology and nanomaterials. Biomed. Appl. Nanotechnol. 8: 298. MDPI.

Bhatt, I. and B.N. Tripathi. 2011. Interaction of engineered nanoparticles with various components of the environment and possible strategies for their risk assessment. Chemosphere. 82(3): 308–317.

Bottero, J.Y., M.R. Wiesner, J. Labille, M. Auffan, V. Vidal and C. Santaella. 2017. Life cycle of nanoparticles in the environment: Nanomaterials and agriculture application—Life cycle through their direct application and indirect application in biosludge-amended soils. pp. 333–346. *In*: Yadav, D., G. Ghosh and D.K. Tiwari [eds.]. Nanotechnology in Agriculture and Food Science. Springer, Cham.

Braun, A., E. Klumpp, R. Azzam and C. Neukum. 2015. Transport and deposition of stabilized engineered silver nanoparticles in water saturated loamy sand and silty loam. Sci. Total Environ. 535: 102–112.

Brown, D.M., M.R. Wilson, W. MacNee, V. Stone and K. Donaldson. 2001. Size-dependent proinflammatory effects of ultrafine polystyrene particles: A role for surface area and oxidative stress in the enhanced activity of ultrafines. Toxicol. Appl. Pharmacol. 175(3): 191–199.

Bundschuh, M., J. Filser, S. Lüderwald, M.S. McKee, G. Metreveli, G.E. Schaumann et al. 2018. Nanoparticles in the environment: Where do we come from, where do we go to? Environ. Sci. Eur. 30: 1–17.

Burke, D.J., S. Zhu, M.P. Pablico-Lansigan, C.R. Hewins and A.C.S. Samia. 2014. Titanium oxide nanoparticle effects on composition of soil microbial communities and plant performance. Biol. Fertil. Soils. 50: 1169–1173.

Burns, R.G., J.L. DeForest, J. Marxsen, R.L. Sinsabaugh, M.E. Stromberger, M.D. Wallenstein et al. 2013. Soil enzymes in a changing environment: Current knowledge and future directions. Soil Biol. Biochem. 58: 216–234.

Chekli, L., G. Brunetti, E. Marzouk, A. Maoz-Shen, E. Smith, R. Naidu et al. 2016. Evaluating the mobility of polymer-stabilised zero-valent iron nanoparticles and their potential to co-transport contaminants in intact soil cores. Environ. Pollut. 216: 636–645.

Chen, J., L. Wu, M. Lu, S. Lu, Z. Li and W. Ding. 2020. Comparative study on the fungicidal activity of metallic MgO nanoparticles and macroscale MgO against soilborne fungal phytopathogens. Front. Microbiol. 11: 511894.

Choi, O. and Z. Hu. 2008. Size dependent and reactive oxygen species related nanosilver toxicity to nitrifying bacteria. Environ. Sci. Technol. 42(12): 4583–4588.

Cicuéndez, M., L. Casarrubios, N. Barroca, D. Silva, M.J. Feito, R. Diez-Orejas et al. 2021. Benefits in the macrophage response due to graphene oxide reduction by thermal treatment. Int. J. Mol. Sci. 22(13): 6701.

Colman, B.P., C.L. Arnaout, S. Anciaux, C.K. Gunsch, M.F. Hochella Jr., B. Kim et al. 2013. Low concentrations of silver nanoparticles in biosolids cause adverse ecosystem responses under realistic field scenario. PLoS One. 8(2): e57189.

Conway, J.R. and A.A. Keller. 2016. Gravity-driven transport of three engineered nanomaterials in unsaturated soils and their effects on soil pH and nutrient release. Water Res. 98: 250–260.

Dale, A.L., G.V. Lowry and E.A. Casman. 2015. Stream dynamics and chemical transformations control the environmental fate of silver and zinc oxide nanoparticles in a watershed-scale model. Environ. Sci. Technol. 49(12): 7285–7293.

de Oca-Vásquez, G.M., F. Solano-Campos, J.R. Vega-Baudrit, R. López-Mondéjar, A. Vera, J.L. Moreno et al. 2020. Organic amendments exacerbate the effects of silver nanoparticles on microbial biomass and community composition of a semiarid soil. Sci. Total Environ. 744: 140919.

de Oliveira Pereira, F.S., D.A. Agarrayua, C.B. Quines and D. Ávila. 2020. Risk assessment of nanofertilizers and nanopesticides. pp. 299–316. *In*: Choudhury, A.S.S. and P.M.R.A. Misra [eds.]. Nanopesticides: From Research and Development to Mechanisms of Action and Sustainable Use in Agriculture. Springer, Cham.

Denisse, V.G., S.C. Roberto, P.H. Hermes, T.G. Patricio, Andrea and F.L. Fabian. 2021. Influence of nanoparticles on the physical, chemical, and biological properties of soils. pp. 151–182. *In*: Nanomaterials for Soil Remediation. Elsevier.

Diegoli, S., A.L. Manciulea, S. Begum, I.P. Jones, J.R. Lead and J.A. Preece. 2008. Interaction between manufactured gold nanoparticles and naturally occurring organic macromolecules. Sci. Total Environ. 402(1): 51–61.

Dimkpa, C.O., J.E. Mclean, D.W. Britt and A.J. Anderson. 2012. CuO and ZnO nanoparticles differently affect the secretion of fluorescent siderophores in the beneficial root colonizer, *Pseudomonas chlororaphis* O6. Nanotoxicology. 6(6): 633–642.

Dresselhaus, M.S. and M. Terrones. 2013. Carbon-based nanomaterials from a historical perspective. Proc. IEEE. 101(7): 1522–1535.

Elbakyan, L. and I. Zaporotskova. 2024. Composite Nanomaterials Based on Polymethylmethacrylate Doped with Carbon Nanotubes and Nanoparticles: A Review. Polymers. 16(9): 1242.

Ellis, L.J.A., S. Kissane and I. Lynch. 2020. Maternal responses and adaptive changes to environmental stress via chronic nanomaterial exposure: Differences in inter and transgenerational interclonal broods of *Daphnia magna*. Int. J. Mol. Sci. 22(1): 15.

Espinasse, D.P., N.K. Geitner, A. Schierz, M. Therezien, C.J. Richardson, G.V. Lowry et al. 2018. Comparative persistence of engineered nanoparticles in a complex aquatic ecosystem. Environ. Sci. Technol. 52(7): 4072–4078.

French, R.A., A.R. Jacobson, B. Kim, S.L. Isley, R.L. Penn, pH and cation and P.C. Baveye. 2009. Influence of ionic strength, valence on aggregation kinetics of titanium dioxide nanoparticles. Environ. Sci. Technol. 43(5): 1354–1359.

Gardea-Torresdey, J.L., C.M. Rico and J.C. White. 2014. Trophic transfer, transformation, and impact of engineered nanomaterials in terrestrial environments. Environ. Sci. Technol. 48(5): 2526–2540.

Ghosh, S. 2023. Promising inorganic nanomaterials for future generation. pp. 247–263. *In*: Applications of Multifunctional Nanomaterials. Elsevier, Amsterdam.

Gogos, A., K. Knauer and T.D. Bucheli. 2012. Nanomaterials in plant protection and fertilization: current state, foreseen applications, and research priorities. J. Agric. Food Chem. 60: 9781–9792.
Gong, H., R. Peng and Z. Liu. 2013. Carbon nanotubes for biomedical imaging: the recent advances. Adv. Drug Deliv. Rev. 65: 1951–1963.
Gordillo-Galeano, A. and C.E. Mora-Huertas. 2018. Solid lipid nanoparticles and nanostructured lipid carriers: A review emphasizing on particle structure and drug release. Eur. J. Pharm. Biopharm. 133: 285–308.
Gottschalk, F. and B. Nowack. 2011. The release of engineered nanomaterials to the environment. J. Environ. Monit. 13(5): 1145–1155.
Grillo, R., M.B. De Jesus and L.F. Fraceto. 2018. Environmental impact of nanotechnology: analyzing the present for building the future. Front. Media SA. 6: 34.
Guilherme, M.R., F.A. Aouada, A.R. Fajardo, A.F. Martins, A.T. Paulino, M.F. Davi et al. 2015. Superabsorbent hydrogels based on polysaccharides for application in agriculture as soil conditioner and nutrient carrier: A review. Eur. Polym. J. 72: 365–385.
Hansen, S.F., O.F.H. Hansen and M.B. Nielsen. 2020. Advances and challenges towards consumerization of nanomaterials. Nat. Nanotechnol. 15: 964–965.
Hartemann, P., P. Hoet, A. Proykova, T. Fernandes, A. Baun, W. De Jong et al. 2015. Nanosilver: Safety, health and environmental effects and role in antimicrobial resistance. Mater. Today. 18: 122–123.
Hartmann, R.R., J. Kono and M.E. Portnoi. 2014. Terahertz science and technology of carbon nanomaterials. Nanotechnology. 25: 322001.
Hoppe, M., R. Mikutta, J. Utermann, W. Duijnisveld and G. Guggenberger. 2014. Retention of sterically and electrosterically stabilized silver nanoparticles in soils. Environ. Sci. Technol. 48: 12628–12635.
Hsu, L.Y. and H.M. Chein. 2007. Evaluation of nanoparticle emission for TiO2 nanopowder coating materials. J. Nanopart. Res. 9: 157–163.
Hughes, K.J., K.A. Iyer, R.E. Bird, J. Ivanov, S. Banerjee, G. Georges et al. 2024. Review of carbon nanotube research and development: materials and emerging applications. ACS Appl. Nano Mater. 7(16): 18695–18713.
Ingle, A.P. 2021. Nanotechnology in plant growth promotion and protection. Wiley Online Library.
Ingle, A.P., N. Duran and M. Rai. 2014. Bioactivity, mechanism of action, and cytotoxicity of copper-based nanoparticles: a review. Appl. Microbiol. Biotechnol. 98: 1001–1009.
Invernizzi, N., G. Foladori, E. Robles-Belmont, E. Záyago Lau, E.A. Figueroa, C. Bagattolli et al. 2015. Nanotechnology for social needs: contributions from Latin American research in the areas of health, energy and water. J. Nanopart. Res. 17: 1–19.
Jesmer, A.H., J.R. Velicogna, D.M. Schwertfeger, R.P. Scroggins and J.I. Princz. 2017. The toxicity of silver to soil organisms exposed to silver nanoparticles and silver nitrate in biosolids-amended field soil. Environ. Toxicol. Chem. 36: 2756–2765.
Ji, Z., W. Guo, S. Sakkiah, J. Liu, T.A. Patterson and H. Hong. 2021. Nanomaterial databases: data sources for promoting design and risk assessment of nanomaterials. Nanomaterials. 11: 1599.
John, A.C., M. Küpper, A.M. Manders Groot, B. Debray, J.M. Lacome and T.A. Kuhlbusch. 2017. Emissions and possible environmental implications of engineered nanomaterials (ENMs) in the atmosphere. Atmosphere. 8: 84.
Jośko, I., P. Oleszczuk and B. Futa. 2014. The effect of inorganic nanoparticles (ZnO, Cr2O3, CuO and Ni) and their bulk counterparts on enzyme activities in different soils. Geoderma. 232: 528–537.
Jutz, G. and A. Böker. 2011. Bionanoparticles as functional macromolecular building blocks-a new class of nanomaterials. Polymer. 52(2): 211–232.
Kaegi, R., A. Voegelin, C. Ort, B. Sinnet, B. Thalmann, J. Krismer et al. 2013. Fate and transformation of silver nanoparticles in urban wastewater systems. Water Res. 47(12): 3866–3877.
Kaegi, R., M. Fierz and B. Hattendorf. 2021. Quantification of nanoparticles in dispersions using transmission electron microscopy. Microsc. Microanal. 27(3): 557–565.
Kah, M. and T. Hofmann. 2014. Nanopesticide research: current trends and future priorities. Environ. Int. 63: 224–235.
Kalia, A., S.P. Sharma, H. Kaur and H. Kaur. 2020. Novel nanocomposite-based controlled-release fertilizer and pesticide formulations: Prospects and challenges. pp. 99–134. *In*: Kalia, A., S.P. Sharma and H. Kaur [eds.]. Multifunctional Hybrid Nanomaterials for Sustainable Agri-Food and Ecosystems. Elsevier, Amsterdam.
Keller, A.A. and A. Lazareva. 2014. Predicted releases of engineered nanomaterials: from global to regional to local. Environ. Sci. Technol. Lett. 1(1): 65–70.
Khan, S.T. 2020. Interaction of engineered nanomaterials with soil microbiome and plants: their impact on plant and soil health. pp. 181–199. *In*: Rai, M. and A.P. Ingle [eds.]. Sustainable agriculture reviews 41: nanotechnology for plant growth and development. Springer, Cham, Switzerland.
Khan, S.T., S.F. Adil, M.R. Shaik, H.Z. Alkhathlan, M. Khan and M. Khan. 2021. Engineered nanomaterials in soil: Their impact on soil microbiome and plant health. Plants. 11(1): 109.

Khan, Z.H., A. Kumar, S. Husain and M. Husain. 2016. Introduction to nanomaterials. pp. 1–23. *In*: Advances in Nanomaterials.

Kreyling, W.G., M. Semmler-Behnke and Q. Chaudhry. 2010. A complementary definition of nanomaterial. Nano Today. 5(3): 165–168.

Kuehr, S., V. Kosfeld and C. Schlechtriem. 2021. Bioaccumulation assessment of nanomaterials using freshwater invertebrate species. Environ. Sci. Eur. 33: 1–36.

Kumar, P., and A.N. Al-Dabbous. 2016. Emission, transformation and fate of nanoparticles in the atmosphere. pp. 4. *In*: Senesi, N. [ed.]. Engineered Nanoparticles and the Environment: Biophysicochemical Processes and Biotoxicity. Springer, Cham, Switzerland.

Kunhikrishnan, A., H.K. Shon, N.S. Bolan, I. El Saliby and S. Vigneswaran. 2015. Sources, distribution, environmental fate, and ecological effects of nanomaterials in wastewater streams. Crit. Rev. Environ. Sci. Technol. 45(4): 277–318.

Kwak, J.I., and Y.J. An. 2016. The current state of the art in research on engineered nanomaterials and terrestrial environments: Different-scale approaches. Environ. Res. 151: 368–382.

Lead, J.R., G.E. Batley, P.J. Alvarez, M.N. Croteau, R.D. Handy and M.J. McLaughlin. 2018. Nanomaterials in the environment: behavior, fate, bioavailability, and effects—an updated review. Environ. Toxicol. Chem. 37(8): 2029–2063.

Leclerc, S. and K.J. Wilkinson. 2014. Bioaccumulation of Nanosilver by *Chlamydomonas reinhardtii*—Nanoparticle or the Free Ion? Environ. Sci. Technol. 48(1): 358–364.

Li, Q., D. Zhang, Z. Song, L. Ren, X. Jin, W. Fang etal. 2022. Organic fertilizer activates soil beneficial microorganisms to promote strawberry growth and soil health after fumigation. Environ. Pollut. 295: 118653.

Li, X., W. Liu, L. Sun, K.E. Aifantis, B. Yu, Y. Fan et al. 2015. Effects of physicochemical properties of nanomaterials on their toxicity. J. Biomed. Mater. Res. Part A. 103(7): 2499–2507.

Liu, D., L. Shi, Q. Dai, X. Lin, R. Mehmood, Z. Gu and L. Dai. 2024. Functionalization of carbon nanotubes for multifunctional applications. Trends Chem.

Liu, H.H. and Y. Cohen. 2014. Multimedia environmental distribution of engineered nanomaterials. Environ. Sci. Technol. 48(6): 3281–3292.

Liu, Y., W. Li, F. Lao, Y. Liu, L. Wang, R. Bai etal. 2011. Intracellular dynamics of cationic and anionic polystyrene nanoparticles without direct interaction with mitotic spindle and chromosomes. Biomaterials 32(32): 8291–8303.

Loureiro, S., P.S. Tourinho, G. Cornelis, N.W. Van Den Brink, M. Díez-Ortiz, S. Vázquez-Campos et al. 2018. Nanomaterials as soil pollutants. pp. 161–190. *In*: Soil Pollution. Academic Press, London.

Ma, H., P.L. Williams and S.A. Diamond. 2013. Ecotoxicity of manufactured ZnO nanoparticles—a review. Environ. Pollut. 172: 76–85.

Ma, R., C. Levard, J.D. Judy, J.M. Unrine, M. Durenkamp, B. Martin et al. 2014. Fate of zinc oxide and silver nanoparticles in a pilot wastewater treatment plant and in processed biosolids. Environ. Sci. Technol. 48(1): 101–112.

Ma, Y., X. Wang, Y. Jia, X. Chen, H. Han and C. Li. 2014. Titanium dioxide-based nanomaterials for photocatalytic fuel generations. Chem. Rev. 114(19): 9987–10043.

Mazari, S.A., E. Ali, R. Abro, F.S.A. Khan, I. Ahmed, M. Ahmed et al. 2021. Nanomaterials: applications, waste-handling, environmental toxicities, and future challenges—a review. J. Environ. Chem. Eng. 9(2): 105028.

McKee, M.S. and J. Filser. 2016. Impacts of metal-based engineered nanomaterials on soil communities. Environ. Sci. Nano. 3(3): 506–533.

McNew, C.P. and E.J. LeBoeuf. 2015. The role of attached phase soil and sediment organic matter physicochemical properties on fullerene (nC60) attachment. Chemosphere. 139: 609–616.

Merrifield, R.C., K.P. Arkill, R.E. Palmer and J.R. Lead. 2017. A high resolution study of dynamic changes of Ce2O3 and CeO2 nanoparticles in complex environmental media. Environ. Sci. Technol. 51(14): 8010–8016.

Mueller, N.C. and B. Nowack. 2008. Exposure modeling of engineered nanoparticles in the environment. Environ. Sci. Technol. 42(12): 4447–4453.

Muraru, S. and M. Ionita. 2020. Towards performant design of carbon-based nanomotors for hydrogen separation through molecular dynamics simulations. Int. J. Mol. Sci. 21(24): 9588.

Nie, G., J. Zhao, R. He and Y. Tang. 2020. CuO nanoparticle exposure impairs the root tip cell walls of *Arabidopsis thaliana* seedlings. Water, Air, and Soil Pollut. 231: 1–11.

Odzak, N., D. Kistler and L. Sigg. 2017. Influence of daylight on the fate of silver and zinc oxide nanoparticles in natural aquatic environments. Environ. Pollut. 226: 1–11.

Pan, B. and B. Xing. 2012. Applications and implications of manufactured nanoparticles in soils: a review. Eur. J. Soil Sci. 63(4): 437–456.

Pathak, J., D.K. Singh, P.R. Singh, N. Kumari, J. Jaiswal, A. Gupta et al. 2024. Application of nanoparticles in agriculture: nano-based fertilizers, pesticides, herbicides, and nanobiosensors. pp. 305–331. *In*: Sinha, R. P. [ed.]. Molecular Impacts of Nanoparticles on Plants and Algae. Elsevier.

Peng, Z., X. Liu, W. Zhang, Z. Zeng, Z. Liu, C. Zhang et al. 2020. Advances in the application, toxicity and degradation of carbon nanomaterials in environment: A review. Environ. Int. 134: 105298.

Pradelus, D., A.P. Coutinho, L. Lassabatere, T. Winiarski and R. Angulo-Jaramillo. 2015. Combined effect of capillary barrier and layered slope on water, solute and nanoparticle transfer in an unsaturated soil at lysimeter scale. J. Contam. Hydrol. 181: 69–81.

Quik, J.T., M.C. Stuart, M. Wouterse, W. Peijnenburg, A.J. Hendriks and D. van de Meent. 2012. Natural colloids are the dominant factor in the sedimentation of nanoparticles. Environ. Toxicol. Chem. 31(5): 1019–1022.

Quik, J.T., I. Velzeboer, M. Wouterse, A.A. Koelmans and D. van de Meent. 2014. Heteroaggregation and sedimentation rates for nanomaterials in natural waters. Water Res. 48: 269–279.

Rawat, S., V.L. Pullagurala, I.O. Adisa, Y. Wang, J.R. Peralta-Videa and J.L. Gardea-Torresdey. 2018. Factors affecting fate and transport of engineered nanomaterials in terrestrial environments. Curr. Opin. Environ. Sci. Health. 6: 47–53.

Romero-Franco, M., H.A. Godwin, M. Bilal and Y. Cohen. 2017. Needs and challenges for assessing the environmental impacts of engineered nanomaterials (ENMs). Beilstein J. Nanotechnol. 8(1): 989–1014.

Rousk, J., K. Ackermann, S.F. Curling and D.L. Jones. 2012. Comparative toxicity of nanoparticulate CuO and ZnO to soil bacterial communities. PLoS One. 7(3): e34197.

Samaddar, P., Y.S. Ok, K.H. Kim, E.E. Kwon and D.C. Tsang. 2018. Synthesis of nanomaterials from various wastes and their new age applications. J. Clean. Prod. 197: 1190–1209.

Scheringer, M., A. Praetorius and E.S. Goldberg. 2014. Environmental fate and exposure modeling of nanomaterials. pp. 89–125. *In*: Frontiers of nanoscience. Elsevier.

Schlich, K. and K. Hund-Rinke. 2015. Influence of soil properties on the effect of silver nanomaterials on microbial activity in five soils. Environ. Pollut. 196: 321–330.

Sergaki, C., B. Lagunas, I. Lidbury, M.L. Gifford and P. Schafer. 2018. Challenges and approaches in microbiome research: from fundamental to applied. Front. Plant Sci. 9: 387738.

Shang, Y., M.K. Hasan, G.J. Ahammed, M. Li, H. Yin and J. Zhou. 2019. Applications of nanotechnology in plant growth and crop protection: a review. Molecules. 24(14): 2558.

Simon, J., E. Flahaut and M. Golzio. 2019. Overview of carbon nanotubes for biomedical applications. Materials. 12(4): 624.

Spurgeon, D.J., E. Lahive and C.L. Schultz. 2020. Nanomaterial transformations in the environment: effects of changing exposure forms on bioaccumulation and toxicity. Small. 16(36): 2000618.

Sun, Y., J. Liang, L. Tang, H. Li, Y. Zhu, D. Jiang et al. 2019. Nano-pesticides: A great challenge for biodiversity?. Nano Today. 28: 100757.

Suppan, S. 2013. Nanomaterials in soil. Institute for Agriculture and Trade Policy.

Suresh, A.K., D.A. Pelletier and M.J. Doktycz. 2013. Relating nanomaterial properties and microbial toxicity. Nanoscale 5(2): 463–474.

Tang, Z. and T. Cheng. 2018. Stability and aggregation of nanoscale titanium dioxide particle (nTiO2): effect of cation valence, humic acid, and clay colloids. Chemosphere. 192: 51–58.

Tiwari, A.J. and L.C. Marr. 2010. The role of atmospheric transformations in determining environmental impacts of carbonaceous nanoparticles. J. Environ. Qual. 39(6): 1883–1895.

Toncelli, C., K. Mylona, I. Kalantzi, A. Tsiola, P. Pitta, M. Tsapakis et al. 2017. Silver nanoparticles in seawater: A dynamic mass balance at part per trillion silver concentrations. Sci. Total Environ. 601: 15–21.

Tripathi, G., N. Kachhwaha, I. Dabi and J. Singh. 2007. Earthworms as bioengineers. pp. 187–270. *In*: Frontiers in Ecology Research. Nova Science Publisher.

siola, A., P. Pitta, A.J. Callol, M. Kagiorgi, I. Kalantzi, K. Mylona et al. 2017. The impact of silver nanoparticles on marine plankton dynamics: Dependence on coating, size and concentration. Sci. Total Environ. 601: 1838–1848.

Velzeboer, I., C. Kwadijk and A. Koelmans. 2014. Strong sorption of PCBs to nanoplastics, microplastics, carbon nanotubes, and fullerenes. Environ. Sci. Technol. 48: 4869–4876.

Wang, F., M.G. Bexiga, S. Anguissola, P. Boya, J.C. Simpson, A. Salvati et al. 2013. Time resolved study of cell death mechanisms induced by amine-modified polystyrene nanoparticles. Nanoscale. 5: 10868–10876.

Wang, S., H. Alenius, H. El-Nezami and P. Karisola. 2022. A new look at the effects of engineered ZnO and TiO2 nanoparticles: evidence from transcriptomics studies. Nanomaterials. 12: 1247.

Wang, Z., L. Zhang, J. Zhao and B. Xing. 2016. Environmental processes and toxicity of metallic nanoparticles in aquatic systems as affected by natural organic matter. Environ. Sci. Nano. 3: 240–255.

Wildgoose, G.G., C.E. Banks and R.G. Compton. 2006. Metal nanoparticles and related materials supported on carbon nanotubes: methods and applications. Small. 2: 182–193.

Xiao, R., Z. Luo, Z. Wei, S. Luo, R. Spinney, W. Yang et al. 2018. Activation of peroxymonosulfate/persulfate by nanomaterials for sulfate radical-based advanced oxidation technologies. Curr. Opin. Chem. Eng. 19: 51–58.

Yang, J., B. Hou, J. Wang, B. Tian, J. Bi, N. Wang et al. 2019. Nanomaterials for the removal of heavy metals from wastewater. Nanomaterials. 9: 424.

Yao, W., Y. Yan, L. Xue, C. Zhang, G. Li, Q. Zheng et al. 2013. Controlling the structures and photonic properties of organic nanomaterials by molecular design. Angew. Chem. Int. Ed. 52: 8713–8717.

Yecheskel, Y., I. Dror and B. Berkowitz. 2016. Transport of engineered nanoparticles in partially saturated sand columns. J. Hazard. Mater. 311: 254–262.

Zarzycki, P.K. 2022. Pure and functionalized carbon-based nanomaterials. Analytical, Biomedical, Civil and Environmental Engineering Applications.

Zhang, W., A.P. Schwab, J.C. White and X. Ma. 2018. Impact of nanoparticle surface properties on the attachment of cerium oxide nanoparticles to sand and kaolin. J. Environ. Qual. 47(1): 129–138.

Zhao, L., H. Zhang, J.C. White, X. Chen, H. Li, X. Qu et al. 2019. Metabolomics reveals that engineered nanomaterial exposure in soil alters both soil rhizosphere metabolite profiles and maize metabolic pathways. Environ. Sci. Nano 6(6): 1716–1727.

Zhu, Y. and Y. Huang. 2023. Source, fate and transport of ENMs in the environment, especially those that may eventually reach plant systems. pp. 25–49. *In*: Physicochemical Interactions of Engineered Nanoparticles and Plants.

CHAPTER-10

Nanomaterials for Contaminant Detection and Remediation in Soil

Akhtar Hameed, Rabia Rasheed, Rana Binyamin, Muhammad Ashfaq, Nadeem Ahmad* and *Asif Mehmood Arif*

1. Introduction

The greatest problem in the world today is pollution or contamination. Contamination of the soil is a severe problem that has an impact on the earth's environment and the health of people across the globe. Sources of contaminants can be derived from industrial processes, farming, and urbanization (Hanfi et al. 2020). They can also remain in the environment for several years and cause long-term adverse impacts on the ground and other forms of life (Alloway 2012, Fei et al. 2022). The bioaccumulation of hazardous trace elements is accelerated in soil settings by their continuous accumulation. Severe health issues in the skin, lungs, kidneys, and brain can result from human exposure to cadmium (Cd), lead (Pb), arsenic (As), and fluoride (F) from drinking water and eating contaminated food (Schaefer et al. 2020, Wang et al. 2019). Because of their recalcitrance and sensitivity to the physicochemical qualities of their surroundings for degradation, organic pollutants like industrial dyes are challenging to remove (Lellis et al. 2019). Similarly, the overuse of pesticides in agriculture has tainted soil and water supplies worldwide (de Souza et al. 2020). Inefficient hydrocarbon combustion, including anthropogenic and natural processes such as coal gasification, coal-tar pitching, and open burning, can create significant amounts of polycyclic aromatic hydrocarbons (PAHs) (Li et al. 2020). Personal care and pharmaceutical products (PPCPs) are emerging contaminants complicating wastewaters originating from homes, hospitals, and the pharmaceutical industrial sector (Meyer et al. 2019). Contaminants, including heavy metals, organic pollutants, and biological agents, pose significant risks to plants, animals, and humans (Nagajyoti et al. 2010). The process of their entry into soil is illustrated in Figure 1.

1.1 Importance of Detection and Remediation

Today, soil remediation and identification of specific contaminants remain essential to prevent harm to the health of people and the environment. These detection methods are important in determining the sources of contamination and the extent of contamination in order to minimize contamination risks (De Gisi et al. 2017). On the other hand, remediation techniques have to reduce the impact of

Institute of Plant Protection, MNS University of Agriculture, Multan 66000, Pakistan.
* Corresponding author: akhtar.hameed@mnsuam.edu.pk

DOI: 10.1201/9781003558354-10

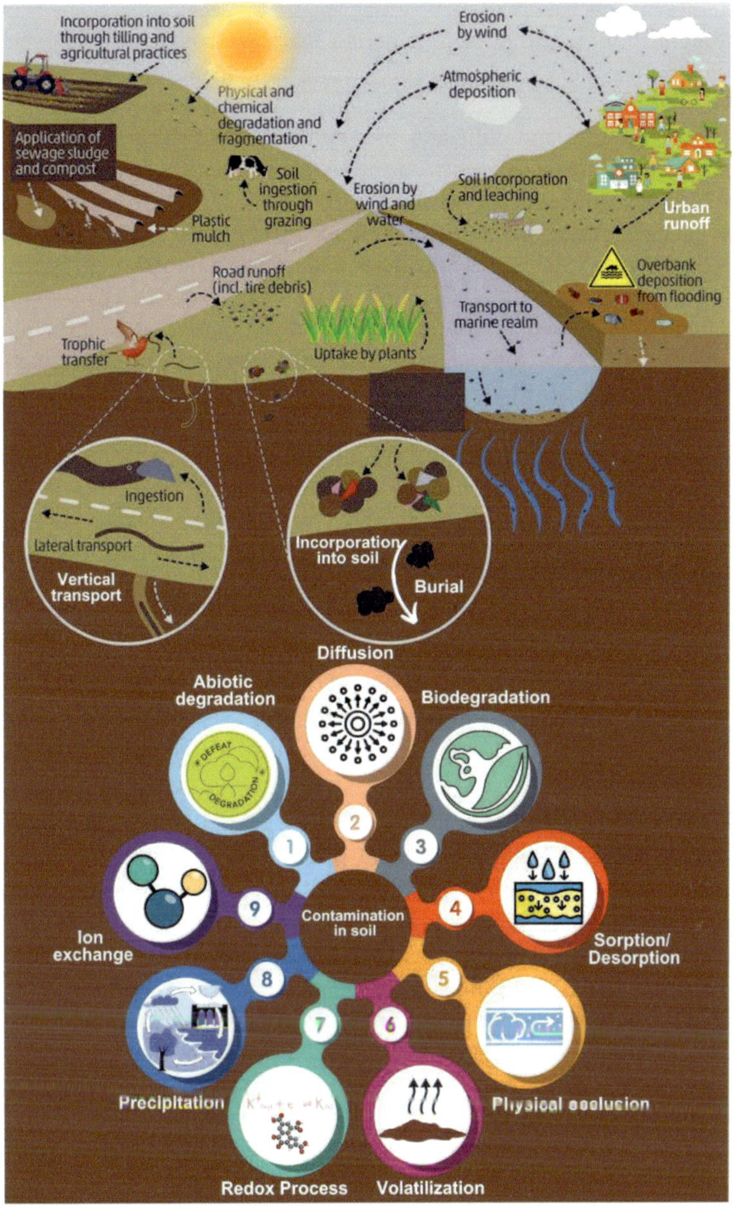

Figure 1. Processes of contaminants' entry and fate in soils

the pollutants, recover the properties and the role of the soil, and protect the environment against further harm (Grifoni et al. 2022). Other conventional methods used in detection and treatment of cancer have some drawbacks in terms of sensitivity, specificity, and treatment efficiency, hence the need for new approaches (Yadav et al. 2021).

1.2 Overview of Nanomaterials in Environmental Science

Currently, the physicochemical properties of NMs make them increasingly suitable for use in environmental applications. In the last few decades, nanotechnology has gained a lot of attention mainly because of the several unique properties of the material at the nanoscale level (New et al. 2023). Nanotechnology is the manipulation of matter at the atomic and molecular level, and the two principles of this concept are small particle size and large surface area. Such properties open up enormous opportunities for the use of these materials. Nanoscale materials are even more reactive and efficient than macroscopic ones because they contain more surface area per volume (New et al. 2023). These NMs may have specific surface chemistry that would enable functional groups to attach to specific molecules and pollutants in a more effective manner (Shah and Imae 2016). Besides size and shape, the chemical nature and surface area of NMs also influence the effectiveness of their pollutant removal. The versatility of surface modification chemistry and the ability to modify the physicochemical characteristics of the nanoparticles are major advantages over past mere sol-gel synthesis, which can mix many materials together in a single step to form hybrids or composites that are often better, more stable, and more selective (Ojea-Jiménez et al. 2012).

Nanotechnology is also used in the removal of contaminants in water, soil, and even air (Nwankwoala et al. 2018). As the major part of the contamination elimination processes, nanoparticles (NPs) consist of filtration, chemical transformations, photocatalytic properties, adsorption, and absorption. They include organic NPs, inorganic NPs, as well as polysaccharide-based NPs. Scientific research also shows that the integration of nanotechnology in the disinfection process is capable of enhancing the reduction of bacterial activity and thus the impact of such microorganisms on the environment (Xiu et al. 2012). Some studies by Alizadeh Fard et al. (2013), Bessa da Silva et al. (2016), and Gu et al. (2012) investigate the efficiency of employing TiO_2 NPs for PAHs, nutrient, and phenanthrene-contaminated soil to prevent algal bloom and photocatalytic degradation, respectively. The academic study conducted by Park et al. (2012) focused on the degradation of hetero-composites in wastewater by employing titanium oxide and titanium oxide with silver doping. Polymers with iron oxide nanoparticles coated with polyvinylpyrrolidone have been studied in the enhancement of the bioremediation of metals, while polymer-based nanoparticles have been applied for treatment of PAHs-contaminated soil (New et al. 2023).

2. Types of Contaminants in Soil

Contamination of soil presents a number of risks to the environment and to the health of those who reside in areas that are affected by contamination, originating from diverse sources, and there are many forms of contaminants. These contaminants are usually classified based on their nature as organic contaminants, inorganic contaminants, and biological contaminants. Organic pollutants are comprised of pesticides, hydrocarbons, and chemical industry byproducts, which arise from agricultural practices, oil leaks, and industrial discharges. These substances can sometimes remain in the soil and might have detrimental impacts on ecology and human health (Das et al. 2015). Heavy metals (for instance, lead, mercury, cadmium) and nitrates mainly result from the influence of mining, industrial effluents, and the use of fertilizers in agriculture. They can also build up in the soil and enter a state of bioavailability that may pose health risks to living organisms (Alloway 2012). Biological pollutants consist of microorganisms such as bacteria, viruses, fungi, and parasites, which find their way into soil through sewage, feces, and contaminated water. These pathogens pose latent health risks with possibilities of disease epidemics (Maier et al. 2009).

Types of contaminants are summarized in Table 1.

2.1 Organic Contaminants

Soil pollutants mainly comprise pesticide residues, oil, and numerous industrial chemicals. Agricultural chemicals, particularly the pesticides used in pest control, have been found to remain in the soil for long periods, hence causing long-term contamination. These chemicals can dissolve in water and contaminate it or be absorbed by crops and penetrate into the food cycle, where they are dangerous to humans and other animals (Zhang and Cheng 2017). Petroleum hydrocarbons from oil spills and industrial discharges are another group of organic pollutants classified within this group. These include polycyclic aromatic hydrocarbons (PAHs), which are some of the most dangerous substances since they cause cancer (Johnsen et al. 2005).

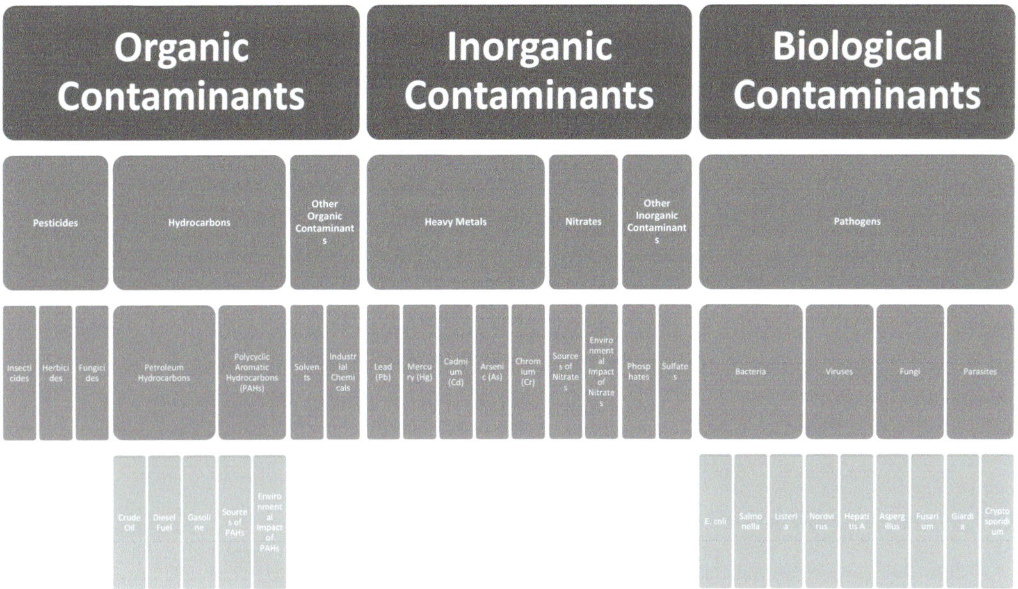

Figure 2. Overview of Different Types of Contaminants in Soil.

2.2 Sources of Organic Contaminants

- Agricultural activities (pesticides, herbicides)
- Industrial processes (chemical manufacturing, waste disposal)
- Petroleum spills and leaks
- Urban runoff

2.3 Impact on the Environment and Health

The impacts of organic contaminants on the soil microbial population and bioactivity can have adverse effects on soil fertility and crop yields in agriculture. Chemicals in plastics also influence health in dangerous ways by causing cancer, endocrine disruption, and neurological disorders, among others. Some of the most worrisome POPs are those that are relatively long-lasting, such as DDT and PCBs, due to their accumulation in organisms (Mishra et al. 2022).

2.4 Inorganic Contaminants

Inorganic pollutants represent a broad category of pollutants depending on the type, and these include heavy metals such as lead, cadmium, arsenic, and nitrates. Heavy metals enter the soil through various sources such as mining, industrial wastes, and the use of polluted fertilizers, as highlighted by New et al. (2023). These metals can remain in the soil for extended periods, thus reaching toxic levels in plants and animals (Adriano 2001). Nitrates, caused by either the use of fertilizers in farming or inadequate disposal of human and animal wastes, may find their way into groundwater sources, which may have adverse impacts on the quality of drinking water (New et al. 2023).

2.5 Sources of Inorganic Contaminants

- Mining and industrial activities
- Use of metal-containing pesticides and fertilizers
- Waste disposal and landfill leachates
- Agricultural runoff (nitrates)

2.6 Impact on the Environment and Health

Some heavy metals inhibit plant growth, decrease microbial activity within the soil, and can be absorbed into the food chain and accumulated in living organisms. Prolonged toxicity of heavy metals has been associated with numerous illnesses such as renal disorders, neurological ailments, and developmental complications among children (Tchounwou et al. 2012). Contamination of drinking water with nitrates is hazardous since it can lead to methemoglobinemia, commonly referred to as "blue baby syndrome," especially among infants (Ward et al. 2018).

2.7 Biological Contaminants

The biological contaminants in soil consist of microorganisms such as bacteria, viruses, and parasites. These microorganisms can be introduced through the use of untreated sewage sludge, manure, and irrigation water that may be contaminated. This is because pathogens that may be found in the soils are capable of infecting animals, plants, and human beings, hence presenting a health risk to human beings (Pepper et al. 2011).

2.8 Sources of Biological Contaminants

- Application of untreated sewage sludge and manure
- Contaminated irrigation water
- Runoff from agricultural fields
- Wildlife and domestic animal activities

2.9 Impact on the Environment and Health

The microbes present in the soil can be pathogens which may cause diseases affecting plant health and production. It also affects human health through direct impacts on health complications including gastro, respiratory health issues, and skin disorders. Hygiene and sanitation measures are important in order to reduce the hazards that human and animal waste poses to human health (Gerba and Smith 2005).

3. Overview of Nanomaterials

NMs are particles that possess at least one characteristic measure in the size range of 1–100 nm. Some of the observable characteristics arising from the small size of NMs include differences in physical, chemical, and biological properties from those of the bulk material. Some of these properties include enlarged surface area, higher reactivity, and quantum effects, and thus are well-suited for many applications, for instance, in environmental remediation (Auffan et al. 2009). NMs can be categorized based on whether they are one-, two-, three-, or multidimensional, as well as in terms of their chemical makeup, form, and source. The primary classifications include:

3.1 Dimensional Classification

3.1.1 Zero-Dimensional (0D)

These are materials in which all three dimensions are in the nanoscale category. Some examples include nanoparticles and quantum dots. These materials are employed in biomedical imaging and as vehicles for drug delivery owing to their characteristics in optics and electronics (Michalet et al. 2005).

Table 1. Summary Table of Types of Contaminants in Soil.

Contaminant	Sources	Impact	References
Organic Contaminants			
Persistent Organic Pollutants (POPs)	Pesticides like DDT, aldrin, dieldrin, endrin, chlordane, heptachlor, hexachlorobenzene, mirex, toxaphene Industrial chemicals like PCBs, PBDEs, PFOS By-products like dioxins, furans, PAHs	Persistence and bioaccumulation in the environment Toxic effects like endocrine disruption, cardiovascular diseases, cancers, diabetes, birth defects, immune and reproductive dysfunction	(Guo et al. 2019, Ritter et al. 1995)
Polycyclic Aromatic Hydrocarbons (PAHs)	Incomplete combustion of natural sources like coal and wood Man-made sources like vehicle exhaust, industrial processes	Lipophilicity and continuous release Classified as POPs under the Aarhus Protocol	(Guo et al. 2019, Ritter et al. 1995)
Per- and Polyfluoroalkyl Substances (PFASs)	Industrial processes	Persistence and bioaccumulation in the environment	(Guo et al. 2019)
Agrochemicals	Designed to be soluble in water and form water-soluble ions	Potential toxicological effects Persistence in the environment	(Bussian et al. 2021)
Pharmaceuticals	Antibiotics, anti-diabetics, anti-epileptics, anti-inflammatories, analgesics, antiulcer drugs, anticancer drugs, antidepressants, lipid regulators	Emerging contaminants of concern Limited understanding of fate and effects in soils	(Bussian et al. 2021)
Inorganic Contaminants			
Heavy Metal			
Arsenic (As)	Mining, smelting, pesticides, wood preservatives	Carcinogenic, damages nervous system, skin lesions	(Tyler and Allan 2014)
Cadmium (Cd)	Mining, smelting, fertilizers, batteries	Kidney damage, bone disease, cancer	(Genchi et al. 2020)
Chromium (Cr)	Leather tanning, wood preservation, electroplating	Allergic dermatitis, lung cancer, liver/kidney damage	(Genchi et al. 2020)
Copper (Cu)	Mining, smelting, pesticides, wood preservatives	Gastrointestinal distress, liver/ kidney damage	(Bussian et al. 2021)
Mercury (Hg)	Mining, industrial processes, pesticides	Neurological disorders, kidney damage	(Bussian et al. 2021)
Lead (Pb)	Mining, smelting, gasoline, paint	Developmental delays, neurological effects, kidney disease	(Cempel and Nikel 2006)
Nickel (Ni)	Mining, smelting, electroplating	Lung fibrosis, skin allergies, cardiovascular effects	(Wani et al. 2015)
Zinc (Zn)	Mining, smelting, galvanization	Nausea, stomach cramps, skin irritation	(Chasapis et al. 2020
Biological Contaminant			
Bacteria	Humans Hot water sources Infected animals/fleas Animal handlers, veterinarians	Tuberculosis Pneumonia Plague Anthrax	(EPA 2021, Kumar et al. 2021)
Fungi	Outdoor air Damp surfaces Bird droppings Mold-contaminated buildings	Asthma Rhinitis Histoplasmosis Cancer	(EPA 2024, Kumar et al. 2021)

Table 1 contd. ...

...Table 1 contd.

Contaminant	Sources	Impact	References
Biological Contaminant			
Viruses	Humans	Respiratory infections Chicken pox COVID-19	(EPA 2024, Kumar et al. 2021)
Algae	Outdoor air	Asthma Rhinitis	(Kumar et al. 2021)
Endotoxins	Bacteria	Toxic/inflammatory responses	(Kumar et al. 2021)
Mycotoxins	Fungi	Toxic/inflammatory responses	(Kumar et al. 2021)
VOCs	Fungi	Toxic/inflammatory responses	(Kumar et al. 2021)

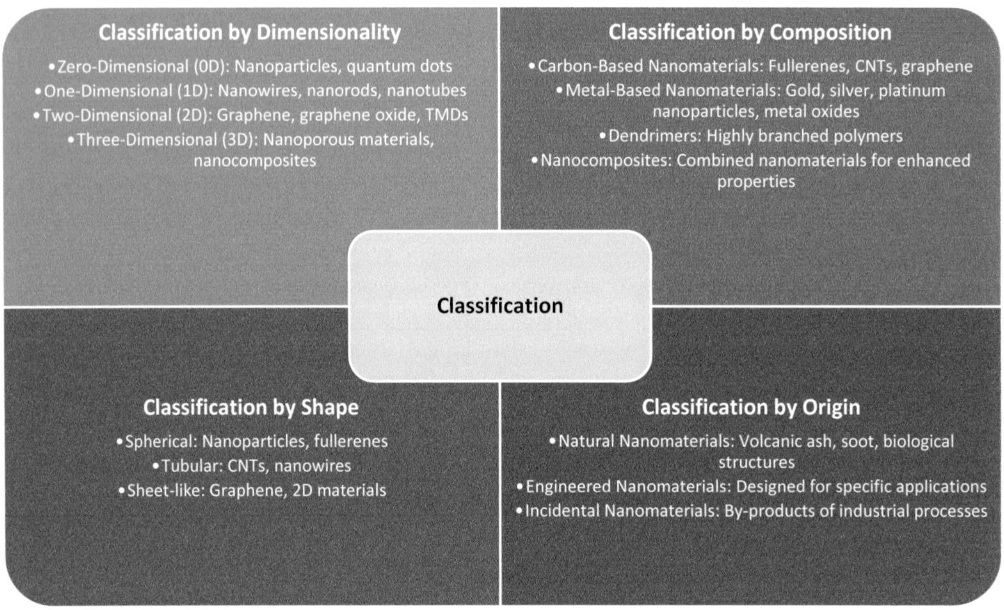

Figure 3. The classification of NMs based on their dimensionality, composition, shape, and origin.

3.1.2 One-Dimensional (1D)

These materials possess one dimensional characteristic which is out of the nanoscale spectrum. Examples include nanowires, nanorods, and nanotubes. Carbon nanotubes (CNTs) are one of the examples that are widely used in electronics and material science because of its strength and electric conductivity (Iijima 1991).

3.1.3 Two-Dimensional (2D)

These materials have two dimensions besides the nanoscale dimension. The first dimension is that the melting point of the CNTs varies with the diameter of the tubes. Some examples are graphene, graphene oxide, transition metal dichalcogenides (TMDs), etc. Structurally, graphene is electrically conductive and mechanically robust, which makes it ideal for use in sensors, batteries, and composite materials (Novoselov et al. 2004).

3.1.4 Three-Dimensional (3D)

These are large amounts of materials that contain one or many structures at the nano level, for instance, nanoporous materials and nanocomposites. These materials sometimes incorporate features of more than one nanomaterial to improve performance in uses like catalysis and energy storage (Tiwari et al. 2016).

4. Compositional Classification

4.1 Carbon-based NMs

They comprise fullerenes, carbon nanotubes, and graphene, among others. These materials are strong, electrically conductive, thermally stable, and, as a result, can find applications in many fields described by Geim and Novoselov (2007).

4.2 Metal-Based NMs

These are NPs derived from metals, for example, gold, silver, and platinum, and metal oxides, for example, TiO_2 and ZnO. These are used in catalysis, medical applications (particularly in imaging), and environmental applications because of their ability to absorb, emit, or reflect light and electricity (Jiang et al. 2008).

4.3 Dendrimers

These are highly branched tree-like polymers that can be designed to exhibit precisely the required properties. They are employed for drug delivery and as catalytic agents due to the structural characteristics and functionality of dendrimers (Tomalia et al. 2007).

4.4 Nanocomposites

These are structures with two or more NMs in which there are enhanced or improved characteristics. For example, a composite of graphene and metal nanoparticles can increase the strength of the material and also improve its catalytic properties.

5. Synthesis and Characterization Techniques

5.1 Synthesis Techniques

5.1.1 Top-Down Approaches

These methods involve breaking down bulk materials into nanoscale particles. Common top-down techniques include:

5.1.1.1 Mechanical Milling

It involves grinding bulk materials into nanoparticles using high-energy ball mills. This technique is extensively employed for synthesizing NMs of specific size and shape (Suryanarayana 2001).

5.1.1.2 Lithography

Pattern of nanostructures on substrate using light or electron beam. Methods including photolithography and electron beam lithography are important in making nanodevices and circuits (Chou et al. 1996).

5.1.2 Bottom-Up Approaches

These methods build NMs from atomic or molecular precursors. Key bottom-up techniques include:

5.1.2.1 Chemical Vapor Deposition (CVD)

A technique where gaseous reactants deposit a solid product onto a substrate. CVD is used for the synthesis of high-purity and high-performance NMs such as carbon nanotubes and graphene (Li et al. 2009).

5.1.2.2 Sol-Gel Process

This involves converting a solution into a gel phase, then drying the gel and subjecting it to heat treatment to produce NMs. This method is versatile and used for producing various metal oxides and composite NMs (Brinker and Scherer 2013).

5.1.2.3 Hydrothermal Synthesis

Uses high temperature and pressure in aqueous solutions to grow NMs. This technique is effective for producing well-crystallized and uniform nanoparticles (Byrappa and Yoshimura 2012, Klabunde and Richards 2009). According to their respective functions, different NMs may be divided into three categories: membranes (a pressure-driven method for treating wastewater and saltwater), catalysts (for the breakdown or transformation of pollutants), and adsorbents (for the adsorption of contaminants) (Anjum et al. 2019). The majority of investigations conducted at the bench scale, however, are focused on the remediation of aqueous systems, which is the main emphasis of the literature. The use of NMs for polluted soil cleanup has not received as much attention. A number of technological, sociological, and financial roadblocks have impeded the widespread use of nanotechnologies in environmental remediation, such as the high preparation and implementation costs of NMs and the inability of NMs to remove contaminants as effectively as needed in *in-situ* settings.

5.2 Characterization Techniques

5.2.1 Microscopy

5.2.1.1 Transmission Electron Microscopy (TEM)

Offers a clear microscopic picture of the NMs with internal structures and surface defects that are evident at the nanoscale. SSE-TEM is vital to examine the morphological features and the crystal structure of nanoparticles (Williams and Carter 2009).

5.2.1.2 Scanning Electron Microscopy (SEM)

Provides high magnification of the surface morphology. SEM is extensively applied in the analysis of geometrical characteristics of NMs and the morphology of their surface (Goldstein et al. 2017).

5.2.1.3 Force Microscopy (AFM)

Is able to determine the height and texture of the surface with high resolution, by moving a very fine point over the material. AFM is applied to the investigation of such phenomena as the surface roughness and mechanical properties (Binnig et al. 1986).

5.2.2 Spectroscopy

5.2.2.1 X-Ray Diffraction (XRD)

Analyzes the content of the nanomaterial based on the diffraction patterns of the X-ray radiation affecting the material and determines the crystalline structure and phase constitution of the nanomaterial. XRD is used in determination of quantitative crystalline phases present in the sample (Cullity 1956).

5.2.2.2 Fourier-Transform Infrared Spectroscopy (FTIR)

Describes the chemical and structural characteristics by detecting infrared light absorption by the NM samples. It is generally used in the identification of the functional groups of organic and inorganic compounds (Smith 2011).

5.2.2.3 Raman Spectroscopy

Examines the vibrational frequencies of molecules in NMs with the aim of providing details of the molecular mix and arrangement. Raman spectroscopy is proven to be very effective, specifically in analysis of carbon-based NMs such as graphene and carbon nanotubes (Ferrari and Basko 2013).

6. Surface Area and Porosity Analysis

6.1 Brunauer-Emmett-Teller (BET) Analysis

Measures the specific surface area and porosity of NMs by nitrogen gas adsorption. BET analysis is critical for understanding the adsorption properties and catalytic potential of NMs (Brunauer et al. 1938).

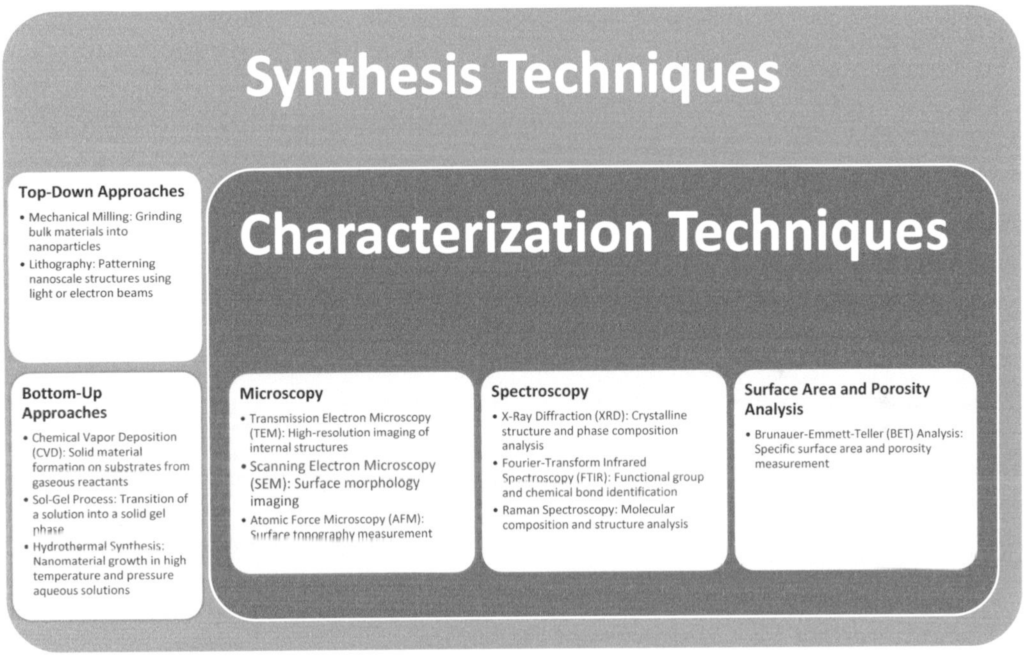

Figure 4. An overview of the synthesis and characterization techniques of NMs.

7. Properties and Advantages of NMs for Environmental Applications

7.1 Properties

NMs possess distinctive properties that make them highly effective for environmental applications:

1. **High Surface Area:** A significant aspect of NMs is that they have a higher surface area to volume ratio. This means that they are highly reactive and can easily get involved in the removal of contaminants. This property is applicable in areas of water purification as well as air purification (Roy et al. 2021).

2. **Enhanced Reactivity:** This can lead to increased chemical activity because NMs have more active sites on their surface area. This property makes them suitable for usage in catalytic activities such as the degradation of organic pollutants (Liu et al. 2017).
3. **Quantum Effects:** At the nanoscale, some quantum behavior can occur, which may alter certain properties of the material—for example, the optical, electronic, or magnetic properties. These effects may be used in the application of fluorescence and surface-enhanced Raman scattering (SERS) in the identification and analysis of pollutants within the environment (Alivisatos 1996).
4. **Versatility:** There is always the possibility of designing an NM for a particular type of functionality for the removal of certain contaminants. For example, changing the surface property of the adsorbent can enhance the selectivity and the reaction rate of the target pollutant (Khan 2018).

7.2 Advantages in Environmental Applications

- **Efficient Contaminant Removal:** The adsorption and degradation capability of pollutants—including heavy metals, organics, and microorganisms—is relatively high. For example, CNTs and GO demonstrated high performance in the removal of heavy metals from water (Peng et al. 2003).
- **Enhanced Sensitivity in Detection**: Environmental sensors can benefit from the use of NMs by having higher sensitivity and selectivity. For instance, quantum dots and gold nanoparticles are used as biosensors with high sensitivity in detecting water and soil pollutants (Medintz et al. 2005).
- **Versatility and Multifunctionality:** Graphene-based NMs can even be designed to achieve two or more functions, including sensing and cleaning. For instance, TiO_2 nanoparticles are capable of photodegrading organic pollutants while at the same time sensing their presence through techniques such as electrochemical detection (Chen and Mao 2007).
- **Reduced Chemical Usage:** The opportunity to apply NMs can help reduce the use of concentrated reagents commonly used in typical treatment procedures. This leads to the enhancement of environmentally sustainable management practices, as indicated by Liu et al. (2017).
- **Economic Benefits:** Even though, at first, nanomaterial expenses can be rather high, the outcomes—such as efficiency and effectiveness—can prove more economical in the long run. The decrease in the use of chemicals and energy, coupled with enhancement in remediation processes, can help reduce expenses in the long term (Roy et al. 2021).

8. Nanomaterials for Contaminant Detection

8.1 Mechanisms of Detection

The application of NMs has been found critical for the detection of environmental pollutants owing to their useful features like increased surface area, high reactivity, and the ability to distinguish contaminants based on their optical and electrical properties. The primary mechanisms through which NMs detect contaminants include:

1. **Adsorption:** Adsorption is one of the simplest processes through which NMs remove pollutants from the environment. Most NMs such as carbon nanotubes (CNTs), graphene, and metal-organic frameworks (MOFs) possess large surface areas as well as a large number of active sites, which make them excellent adsorbents. In these NMs, contaminants attach to their surface, enhancing their identification. For instance, graphene oxide has been utilized for adsorption of heavy metals such as lead and mercury from water, increasing the detection limit (Yang et al. 2007).

2. **Catalysis:** Some NMs are able to convert detectable signals in the process of catalyzing certain reactions online. For instance, gold and silver nanoparticles can reduce the dyes or other naked compounds that can change their colors and suggest contamination. These catalytic properties prove highly useful in the mainstream synthesis of colorimetric sensors for a range of pollutants within the context of the environment (Liu et al. 2017).
3. **Electrochemical Detection:** Because of their nanoscale, these materials facilitate electron transfer and are thus highly useful in electrochemical sensors. These detectors respond to variations such as current or potential when they interact with contaminants. For instance, CNTs or gold nanoparticles can be used to enhance the electrode surface, thereby increasing sensitivity and selectivity for the presence of contaminant agents such as heavy metals, pesticides, and organic pollutants (Chen and Mao 2007).
4. **Optical Detection:** Optical techniques can be applied for detecting contaminants, which include fluorescence, surface plasmon resonance, and Raman scattering, which are properties of NMs. For instance, quantum dots and fluorescent nanoparticles comprise systems that can be defeased with specific ligands that result in a change of fluorescence intensity in the presence of contaminants. Also, in SPR sensors, gold nanoparticles can be used to detect variations in refractive index upon contaminant binding and, therefore, are a good choice for a sensitive and specific sensor method (Gao et al. 2004).
5. **Surface-Enhanced Raman Scattering (SERS):** SERS is a very effective detection tool that amplifies the Raman scattering signals of molecules attached to metallic platforms like gold or silver nano-islands. It also allows for the quantification of very low concentrations of contaminants such as organic pollutants and biomolecules. SERS-based sensors can provide a fingerprint for the contaminants and, therefore, are highly selective and sensitive devices (Brown and Milton 2008).

9. Types of Nanomaterials Used

NMs have shown tremendous potential in environmental applications, particularly in the detection and remediation of contaminants. Various types of NMs are employed, each offering unique properties that make them suitable for specific applications.

9.1 Metal Nanoparticles

Common metal NPs include Au, Ag, and Pt due to their outstanding optical and electronic characteristics. For instance, gold nanoparticles are used in colorimetric sensors for the identification of both heavy metal and organic pollutants. These nanoparticles have size-induced optical features including LSPR and can be used for detection at high sensitivity (Jiang et al. 2008, Liu and Lu 2004).

9.2 Metal Oxide Nanoparticles

TiO_2 and ZnO nanoparticles are also recognized for their photocatalytic effects. TiO_2 nanoparticles find wide application in the photodegradation of various organic pollutants, turning them into less toxic forms in the presence of UV light (Li et al. 2003).

9.3 Nanotubes

9.3.1 Carbon Nanotubes (CNTs)

Both single-walled CNTs (SWCNTs) and multi-walled CNTs (MWCNTs) have attracted widespread attention due to their unique properties such as high surface area, high mechanical strength, and electrical conductivity. These characteristics allow CNTs to adsorb and detect various types of pollutants, including heavy metals and organic compounds (Cui et al. 2001, Peng et al. 2003).

9.3.2 Graphene and Graphene Oxide

Graphene and graphene oxide (GO) are lightweight materials with high surface area and wood-like electronic structure. Graphene- and GO-based sensors for heavy metal and organic pollutants are incorporated because of their high conductivity and their ability to be functionalized (Liu et al. 2011).

9.4 Quantum Dots

Quantum dots (QDs) are semiconductor nanocrystals that exhibit size-tunable fluorescence. These NMs are applied as specific fluorescent markers for determining different contaminants such as heavy metals and biological pathogens. QDs can be specifically ligated to achieve selective traceability of definite pollutants, which makes them valuable tools in environmental analysis (Medintz et al. 2005).

9.5 Nanocomposites

Nanocomposites are the mixing of two or more NMs in a bid to improve their properties. For example, nanocomposites based on graphene and metal nanoparticles or metal oxides can have high sensitivity and selectivity for contaminants and their removal. These composites make use of the combined properties of their constituents to enhance their performance (Njuguna et al. 2021).

10. Techniques and Methods

10.1 Sensors

10.1.1 Electrochemical Sensors

These sensors use NMs to improve electron transfer, offering an advantage in terms of sample detection. For instance, when electrodes are coated with gold nanoparticles, detection of mercury ions is enhanced because of the conductivity and surface characteristics of gold nanoparticles (Zhu et al. 2015).

10.1.2 Optical Sensors

Optical sensors rely on alterations in the optical characteristics of NMs. Gold nanoparticle-incorporated surface plasmon resonance (SPR) sensors cause changes in the resonance wavelength due to the binding of pesticides or other contaminants to the nanoparticle surface (Homola 2008).

10.2 Biosensors

10.2.1 Enzyme-based Biosensors

These biosensors use biological components with NMs. For example, CNTs with enzymes immobilized on their surface can sense pesticides and heavy metals based on the fact that they catalytically produce a signal that can be measured (Soldatkin et al. 2022).

10.2.2 DNA-based Biosensors

This method of analysis involves using DNAzymes that are conjugated with nanoparticles for the identification of metal ions. For instance, a DNAzyme-gold nanoparticle conjugate reagent can produce a color variation of the solution in response to lead ions for a quick and effective detection process (Liu and Lu 2004).

10.3 Fluorescent Probes

10.3.1 Quantum Dots

These are used as fluorescent markers due to their high and stable fluorescence intensity, which can be customized to interact with specific pollutants, including heavy metals, and indicate pollution levels (Medintz et al. 2005).

10.3.2 Up-conversion Nanoparticles

These NPs transform low-energy photons into high-energy emissions that allow the identification of contaminants in low light, especially in biological and environmental applications (Chen and Wang 2020).

10.4 Colorimetric Sensors

Gold nanoparticles can be easily used in colorimetric sensors since they change color upon aggregation with the analyte, enabling on-site and real-time analysis, such as detecting lead ions by turning red to blue (Liu and Lu 2004).

Some examples of the use of inorganic gold NPs include their use in the formation of colorimetric sensors for the determination of lead ions in water. When lead is present, the gold nanoparticles form clusters and the solution changes color from red to blue (Liu and Lu 2004). A biosensor for organophosphate pesticide detection has been developed using enzyme-coated CNTs. The enzyme contributes to the breakdown of the pesticide, and the shift in current is assessed (Soldatkin et al. 2022). Different approaches, such as using quantum dots, have been employed in the detection of polycyclic aromatic hydrocarbons (PAHs) in soil and water. The PAHs reduce the fluorescence intensity of the quantum dots, and since fluorescence intensity is proportional to the amount of contaminant, this offers a quantitative measure of contamination (Medintz et al. 2005). Silver nanoparticles have been applied in sensors to identify the presence of *E. coli* in water samples. In the presence of E. coli, the aggregation of silver nanoparticles takes place, which results in a color change (Cheng and Mao 2007).

11. Nanomaterials for Contaminant Remediation

11.1 Mechanisms of Remediation

NMs offer innovative solutions for the remediation of environmental contaminants due to their unique properties such as high surface area, enhanced reactivity, and specific interaction with pollutants. The primary mechanisms through which NMs remediate contaminants include.

11.2.1 Adsorption

Adsorption is regarded as the most efficient way of decontaminating the environment using NMs. Because of their high surface area and availability of many active sites, NMs such as activated carbon, graphene oxide, and CNTs have the capability to adsorb various pollutants including heavy metals, organics, and radionuclides. For instance, graphene oxide has been employed in the adsorption of various metal ions such as lead and mercury ions from effluents. Some of the metals that have been reported to adhere to graphene oxide include gold, platinum, palladium, and silver, among others. These metals adhere to GO due to the oxygen-containing functional groups (Perreault et al. 2015, Li et al. 2003).

11.2.2 Reduction and Oxidation (Redox) Reactions

NMs, due to their ability to change the redox status, can convert some of the dangerous contaminants into harmless byproducts. Among NWs, nZVI is extremely efficient in removing toxic metals and organic pollutants by electron transfer mechanisms. For instance, nZVI has been found to remediate in-depth chlorinated solvents like trichloroethylene (TCE) contaminated groundwater. The iron particles then transfer electrons to the chlorinated compounds and transform them into harmless ethane and chloride ions (Zhang 2003, Latif et al. 2020).

11.2.3 Catalytic Degradation

Some NMs aid in the degradation process by enhancing the rate of chemical changes to the contaminant molecules. For example, TiO_2 nanoparticles can break down organic pollutants through photocatalytic reactions under UV light. TiO_2 nanoparticles have been applied in the photodegradation of organic dyes in wastewater. The UV light activates the TiO_2 and releases ROS that oxidize the dye particles into harmless products including CO_2 and H_2O (Chen and Mao 2007).

11.2.4 Immobilization

Immobilization is a process that involves the immovable contaminants to NMs units to decrease their movement and biological action in the environment. This mechanism is especially effective for heavy metals and radioactive products. For instance, modified silica nanoparticles have been applied in the immobilization of heavy metals within soils. The functional groups available on the surface of the silica selectively anchor to the metal ions and restrict their release into the aqueous environment in bioactive form (Tang et al. 2014).

12. Types of Nanomaterials Used

The nanomaterials used in the detection and remediation of soil contaminants are summarized in Table 2.

12.1 Nanoscale Zero-Valent Iron (nZVI)

Some contaminants are removed by nZVI particles through redox reactions. Particularly in the case of chlorinated solvents, heavy metals, and nitrates, these nanofilters show superb performance. Due to strong interactions with their environment, nZVI has a large surface area and contains reactive sites (Zhang 2003). Therefore, successful applications have been reported for using nZVI in groundwater remediation and treatment of industrial effluents. Iron can remove chlorinated organic compounds, transform hexavalent chromium into trivalent chromium, and neutralize heavy metals (Tratnyek and Johnson 2006). nZVI was employed to remove TCE from a contaminated groundwater source. The investigation confirmed that nZVI could quickly decrease TCE levels, producing ethane and chloride ions, thus purifying the water in the aquifer (Gavaskar et al. 2005). However, nZVI particles are capable of aggregation, which in turn reduces their reactivity and mobility. Additives and stabilizers have been employed to disperse nZVI and improve their stability in environmental applications (He and Zhao 2007).

12.2 Nanocomposites

Nanocomposites are another class of NMs where several NMs are integrated to enhance their contaminant remediation capabilities. They are also reusable in certain ways, such as adsorbents, degrading pollutants through catalytic capabilities, and participating in redox reactions. For example, metal oxide-incorporated graphene-based nanocomposites can simultaneously adsorb and photodegrade polluted organic species (Xie et al. 2010). They are used in the removal of dyes, metals, and other organic compounds in water and soil. Due to their versatility, they can tackle multiple contamination scenarios (Njuguna et al. 2021).

Table 2. Summary Table of NMs for Contaminant Detection.

Nanomaterial	Application	Example	Advantages	Challenges	References
Gold Nanoparticles	Detection of heavy metals	Colorimetric sensors for lead detection	High sensitivity and selectivity	Potential toxicity and high cost	Liu and Lu 2014
Titanium Dioxide (TiO_2)	Photodegradation of pollutants	Photocatalytic degradation of organic pollutants	Dual function: detection and degradation	Stability issues and aggregation	Chen and Mao 2007
Carbon Nanotubes (CNTs)	Adsorption and detection of contaminants	Detection of heavy metals and organic compounds	High surface area, mechanical strength, and electrical conductivity	Toxicity concerns and environmental impact	Peng et al. 2005
Quantum Dots	Fluorescent detection	Fluorescent probes for heavy metals	High sensitivity and tunable fluorescence	Potential toxicity and environmental persistence	Medintz et al. 2005
Graphene and Graphene Oxide	Adsorption and detection of heavy metals	Detection of lead and mercury in water	High surface area, conductivity, and functionalization potential	Aggregation and cost	Wang and Chen 2015
Nanocomposites	Combined detection and remediation	Graphene-based nanocomposites with metal nanoparticles	Enhanced sensitivity and selectivity through synergistic effects	Complexity in design and synthesis	He et al. 2018
Silver Nanoparticles	Antimicrobial detection	Detection of biological contaminants	High sensitivity and antimicrobial properties	Potential environmental and health risks	Oberdörster et al. 2005

12.3 Carbon-Based NMs

12.3.1 Carbon Nanotubes (CNTs)

They have a large surface area and the results show that CNTs have high adsorption performance. It can filter out organic poisons, metals and radioactive elements from water through adsorption (Gupta et al. 2013). The work used multi-walled carbon nanotubes (MWCNTs) in the removal of phenol in water through the adsorption process. They argued that due to available large surface area for the pellets and non-polar nature of phenol molecules, MWCNTs have very high adsorption ability of phenol. The synthesis of CNTs could be costly and attention must also be given to its effects on the environment. On the other hand, functionalization of CNTs improves their dispersion and their adsorptive capacities, but it comes with added costs and drawbacks (Baughman et al. 2002).

12.3.2 Graphene and Graphene Oxide

These materials possess very high mechanical strength, a large surface area to volume ratio, and high conductivity. GO has the ability to adsorb heavy metals and organic pollutants in the presence of oxygen-containing functional groups (Liu et al. 2008). Graphene oxide–iron oxide nanocomposite was used as an adsorbent in this work with arsenic as the adsorbate in water. It proved that the adsorption process had a high capacity for adsorbing arsenic ions with a selectivity that lessened the concentration of the same ions in water (Das 2021). The large surface area of graphene and magnetism of iron oxide make the nanocomposite easy to separate from water after treatment, which minimizes issues of practical applicability (Marcelo et al. 2021).

12.3.3 Activated Carbon Nanofibers

The nanofibers are utilized for their high adsorption capacity and mechanical properties. They are proficient in the elimination of volatile organic compounds (VOCs) and in eradicating metallic elements in both air and water (Mobasser et al. 2022).

13. Methods and Technologies

13.1 Adsorption

Adsorption is one of the most widely used methods for contaminant remediation using NMs. NMs such as activated carbon nanofibers, carbon nanotubes (CNTs), and graphene oxide exhibit high surface areas and abundant active sites, making them effective adsorbents.

13.1.1 Activated Carbon Nanofibers

These nanofibers have the ability to adsorb high concentrations of VOCs and heavy metals. Activated carbon nanofibers are characterized by structures that afford a large surface area which can be used to adsorb contaminants from air and water (Mobasser et al. 2022).

13.1.2 Carbon Nanotubes (CNTs)

CNTs have the ability to adsorb heavy metals, organic pollutants, and radionuclides. High surface area and hydrophobic and π–π interactions promote the adsorption of phenol and lead from water solutions (Peng et al. 2005).

13.2 Catalysis

Catalysis involves NMs accelerating the breakdown of contaminants into less harmful substances through chemical reactions.

13.2.1 Nanoscale Zero-Valent Iron (nZVI)

nZVI particles exhibit great reactivity and are capable of degrading various pollutants including chlorinated organic compounds, nitroaromatic compounds, and heavy metals. The reduction reactions that involve nZVI work to deconstruct pollutants into non-hazardous or less hazardous forms (Zhang 2003).

13.2.2 Metal Oxide Nanoparticles

Of all oxide nanoparticles, TiO_2 nanoparticles are the most widely used photocatalysts in the field. The photocatalytic ability of TiO_2 is based on the generation of reactive oxygen species (ROS) upon exposure to UV light, which decomposes organic pollutants such as dyes and pesticides into non-harmful compounds (Chen and Mao 2007).

13.3 Photodegradation

Photodegradation utilizes the light energy absorbed by NMs to break down pollutants in water. For instance, the photocatalytic TiO_2 and ZnO NMs are among the most effective materials for the removal of organic pollutants.

13.3.1 Titanium Dioxide (TiO_2)

Nanocrystals of TiO_2 absorb UV radiation and create reactive oxygen species capable of breaking down organic compounds into carbon dioxide and water. It has been observed that this method is very efficient in degrading complex organic materials such as pharmaceutical wastes and pesticides (Chen and Mao 2007) (Table 3).

Table 3. Summary Table of NMs for Contaminant Remediation.

Nanomaterial	Mechanism	Example	Advantages	Challenges	References
Graphene Oxide	Adsorption	Adsorption of heavy metals like lead and mercury	High surface area, strong binding sites	Potential aggregation, cost	(Perreault et al. 2015)
Nanoscale Zero-Valent Iron (nZVI)	Redox Reactions	Reduction of trichloroethylene (TCE) in groundwater	Effective reduction of contaminants	Stability issues, potential toxicity	(Zhang 2003)
Titanium Dioxide (TiO_2)	Catalytic Degradation	Photodegradation of organic dyes	Dual function: detection and degradation	Aggregation, stability under environmental conditions	(Chen and Mao 2007)
Modified Silica	Immobilization	Immobilization of heavy metals in soil	Prevents leaching and bioavailability	Requires surface modification, cost	(Tang et al. 2014)

13.3.2 Zinc Oxide (ZnO)

Similarly, ZnO nanoparticles have photocatalytic activity under UV light irradiation for extended periods. They have been used for the degradation of many organic species like phenols and dyes in water treatment methods (Li et al. 2003).

One of the examples of nZVI usage was mentioned while discussing the enhancement of the removal rate of trichloroethylene (TCE) from groundwater. In this study, nZVI particles were delivered through an injection into the contaminated aquifer. This proved that nZVI is an effective reagent in the degradation of chlorinated solvents since TCE converted to ethane and chloride ions within a short period (Gavaskar et al. 2005, Semerád et al. 2021). Numerous studies have confirmed that graphene oxide (GO) is a very efficient adsorbent material in the removal of heavy metals from aqueous solutions. For instance, graphene oxide was used in water treatment where arsenic preferentially adsorbed with high capacity. Additionally, the functional groups available on the surface of graphene oxide enable adsorption of arsenic ions and minimize the concentration of As in treated water (Soldatkin et al. 2022). Titanium dioxide (TiO_2) nanoparticles have also been reported to be effective in the photodegradation of dyes in water. For instance, TiO_2 was employed in the photocatalytic degradation of methylene blue dye under UV light. The results, which indicated the complete disappearance of the dye within a few hours, clearly proved the ability of TiO_2 in treating water contaminated with dyes (Chen and Mao 2007, Bhattacharya and Selvaraj 2022).

14. Advantages

- Regarding the detectability of contaminants, NMs are analyzed and it reveals that these materials are very sensitive and selective in nature. This is because the large surface area and high reactivity of the materials can allow them to sense the presence of contaminants at low concentrations, which is very important when it comes to prevention or limiting contamination of the environment. For instance, there is utilization of gold nanoparticles and quantum dots in biosensors for the detection of low concentrations of heavy metals and organic pollutants (Yang et al. 2007).
- Due to the advanced optical and electronic characteristics of nanoscale substances, contamination can be detected immediately. Fluorescence and surface-enhanced Raman scattering (SERS), which are real-time monitoring techniques, can help in responding to cases of contamination (Thakur and Kumar 2022).
- In addition, many kinds of nanomaterial-based detection systems are small in size and can be hand-carried and easily operated, allowing them to be applied in the field (Liu and Lu 2004).

- These NMs can be further modified with different chemical groups or biomolecules to selectively interact with particular pollutants. This versatility enables the growth of adaptive sensors that are mindful of numerous pollutants that encompass heavy metals, pesticides, and pathogens (Darwish and Mohammadi 2018).
- Miniaturization plays important roles in environmental monitoring. Incorporation of NMs in portable and miniaturized detection devices enables easy operation. These devices are portable, which allows for determining the necessary or unwanted characteristics on the spot without setting up complex laboratory equipment and allows management decisions to be made quickly (Gao et al. 2004).
- NMs are used in both detection and removal of pollutants from the environment. For instance, titanium dioxide (TiO_2) nanoparticles, upon application, are capable of detecting the organic pollutant and degrading it through a photocatalytic reaction, which offers a dual-benefit solution (Chen and Mao 2007).

15. Challenges

- Toxicity and Environmental Impact: The major disadvantage of nanomaterials is that they may prove to be toxic to human beings and the environment. Problems associated with nanoparticles include their ability to easily pass through biological membranes and get assimilated into tissues, causing health hazards. Furthermore, the potential aggressiveness of NMs to the environment should also be seriously discussed, such as persistence and bioaccumulation levels of NMs (Oberdörster et al. 2005).
- NMs have a tendency to clump together, decreasing the quantity of material exposed on the surface, and thus, their reactivity may be diminished, thereby hampering their detection power. These findings require the use of stabilizing agents and either surface or bulk modifications to maintain dispersion and function in a given environment (He and Zhao 2007).
- The production of nanostructures and their functionalization process can at times be very costly, thus hampering the scalability of the materials. Cost issues need to be addressed in order to bring advanced production methods down to the level at which such NMs could become affordable to a wider section of society (Karn et al. 2009).
- Regarding the approval of NMs, there are difficulties tied to the absence of specific protocols for the examination and testing of NMs. It is critical to set clear objectives and assign legal acts to ensure the safe and efficient use of NMs in a given environment (EPA 2015).

16. Integrated Detection and Remediation Systems

A continuum of detection and remediation is an efficient model for the general prevention of contamination in the environment. These systems include features for identification of contaminants and subsequent purification in one complex system. The most significant benefits associated with such integrated systems include real-time monitoring and early detection of contamination, halving the time delay between contamination detection and removal, and enhancing the effectiveness of environmental management (Rasheed et al. 2019).

Integrated designs are derived from NMs that can sense and remediate threats, capture contaminants through adsorption or catalysis, and initiate processes such as degradation or immobilization. These functions incorporated into information systems form a useful and efficient approach toward consultation on contamination, thus making it a suitable means for handling environmental concerns.

16.1 Development and Design of Multifunctional NMs

- NMs like metal oxides, carbonaceous NMs, and nanocomposites are employed in applications as a result of their inherent qualities such as high reactivity, large surface area, and the ability to have surface chemistries altered (Saraswathi et al. 2019).
- Changing the nanoparticles chemically or biologically with certain receptors helps improve selectivity and sensitivity towards the target contaminant, through the use of antibodies or enzymes (Sapsford et al. 2013).
- The process of creating nanocomposites obtained by blending disparate forms of NMs can result in improvement of the properties of all the types of NMs used. For instance, the use of metal NPs in a carbon matrix together would enhance the detection sensitivity and catalytic activity for potential remediation (Zhu et al. 2019).
- The fact that NMs need to be produced at large volumes as well as be stable under the conditions to which they will be subjected is also very important. This also encompasses matters that relate to the accumulation and breakdown of NMs over time (He and Zhao 2007).

17. Examples of Integrated Systems in Practice

17.1 Graphene-based Sensors for Heavy Metal Detection and Removal

Graphene and other materials derived from graphene, such as graphene oxide, have been receiving much attention due to their multifunctional properties. One such application is the incorporation of graphene-based sensors to aid in the detection and removal of heavy metals within water supply sources. A report confirmed that graphene oxide with appropriate ligands has the ability to adsorptively recognize and analyze lead and mercury. Yet, these NMs can also be used for adsorption and reduction of the detected metals in water (Liu et al. 2008).

17.1.1 Titanium Dioxide (TiO_2) Photocatalysts for Organic Pollutants

The unique property of TiO_2 nanoparticles that has attracted the most popularity is photocatalytic activity. There are integrated systems that use TiO_2 and can separately recognize organic materials like dyes and some pharmaceuticals in water using UV fluorescence. The same TiO_2 nanoparticles can degrade these pollutants through photocatalytic oxidation upon detection and transform them into non-hazardous products such as CO_2 and H_2O. This dual usage has favored the employment of TiO_2-based systems in wastewater treatment processes (Chen and Mao 2007).

17.1.2 Nanoscale Zero-Valent Iron (nZVI) for Chlorinated Solvent Remediation

nZVI has applications in integrated remediation systems involving the detection and removal of chlorinated solvents in groundwater. The nZVI particles can identify these contaminants through reduction-oxidation reactions that alter the conductivity of the solution. Subsequently, nZVI plays a role in the dechlorination of solvents or the reduction of the solvents into non-hazardous components. It has been identified that this combination of artificial neural network and molecular dynamics simulation can work in field studies for groundwater remediation (Tratnyek and Johnson 2006)

17.1.3 Benefits

- Closed systems enable constant assessment of the contaminants in the environment and the beginning of the treatment process. This minimizes the time between detection and response, which helps in preventing the spread of the contaminants and minimizes their effects (Vaseashta et al. 2007).

- Integration of detection and remediation increases optimization of contamination control since it is done under one system. Dual-functional NMs mean that the same NMs can act as both, decreasing the need for additional detection and removal systems (de Dios and Díaz-García 2010).
- Integrated systems may be beneficial in eliminating high costs of monitoring the environment and cleaning it. Integration of operations in a single multifunctional material or device reduces the number of systems required, thus controlling costs of operation and maintenance (He and Zhao 2007).
- Targeted contaminants can be organic pollutants, heavy metals, and pathogens, and multifunctional NMs can be developed to eliminate a broad spectrum of contaminants. For this reason, integrated systems are appropriate for a wide range of environmental contexts (Bhateria and Singh 2022).

17.1.4 Limitations

- Fabrication of multifunctional NMs for application in detection and remediation requires design and molecular construction and engineering techniques. Balancing between the two is, however, hard to achieve in a way that both are optimally implemented without one being a tradeoff for the other, according to Yadav et al. (2021).
- One of the major issues to address about NMs is how stable they are under environmental influences. There are drawbacks that can stem from integration such as aggregation, leaching, or degradation, and therefore influence the performance of this composition. Improving the mechanical and chemical properties of these types of materials is important for their use in real-life situations (He and Zhao 2007).
- This is something which poses questions about the toxicity of NMs to the human beings and the natural environment. Due to the nanomaterials' unique characteristics, it is critical to effectively evaluate the risks that arise from the release and disposal of NMs to avoid the dangers they pose (Karn et al. 2009).
- Even in the utilization of NMs in the environment, regulation and standards on their use are not explicitly defined and hence may slow down the realization of integrated systems. There is a need to develop rules on the proper usage of the mentioned technologies to enable their adoption (Karn et al. 2009).

18. Environmental and Health Impacts Of NMs

18.1 Potential Risks and Toxicity of NMs

NMs, because of their tiny size and high activity level, pose various challenges with respect to their effects on the environment and human health. Some of the possible hazards and toxic effects of NMs result from properties such as chemical makeup, size, morphology, and the capability to produce ROSs (Oberdörster et al. 2005). The toxicity of NMs is usually influenced by the type of chemicals used and the nature of their surfaces. For example, metal oxide particles such as TiO_2 and ZnO can produce ROS, and their exposure to living organisms results in oxidative stress. Nanoparticle properties of a bare surface can either be reduced or enhanced through modifications like coating with organic molecules, as noted by Nel et al. (2006). Since nanoparticles that are smaller in size have a greater surface area-to-volume ratio, they are more reactive and may be more toxic. Also, the shape of nanoparticles—like rods, spheres, or tubes—can affect interactions with biological systems/media. For instance, carbon nanotubes (CNTs), having a needle-like structure, can pierce through cell membranes and create outright physical damage (Donaldson et al. 2006). Investigations have shown that a number of NMs can cause the generation of ROS that alter cellular components including DNA, proteins, and lipids, thus causing oxidative stress and inflammation.

This is particularly worrisome for NMs utilized in environmental applications since they may impact organisms not originally under consideration (Xia et al. 2008).

18.2 Environmental Fate and Transport of NMs

Information concerning the environmental behavior of NMs is crucial in risk assessment procedures. The different ways that NMs can be introduced into the environment include industrial processes, commercial products, and medical applications. Their behavior in the environment depends on processes such as aggregation, dissolution, and interactions with natural organic matter (Batley et al. 2013). Some limitations to the dispersion of nanoparticles in natural waters include van der Waals forces and the presence of natural organic matter. These can decrease transport and solubility, but may also promote sedimentation and concentration in aquatic sediments, thereby contributing to long-term hazards (Petosa et al. 2010).

A number of NMs, especially some metal-based nanoparticles, are known to dissolve and release ions that are toxic. For instance, silver nanoparticles may disintegrate to form free silver ions, which have been reported to be lethal to aquatic species (Fabrega et al. 2011).

Bioavailability and mobility: NMs can associate with natural organic matter in the environment, altering their stability. These interactions can either magnify or minimize their effects. For example, the interaction between nanoparticles and organic matter can lower the toxicity of nanoparticles by shielding them from biological membranes (Chen et al. 2012).

18.3 Strategies for Risk Assessment and Management

The potential hazards of NMs can thus only be avoided by the right approach towards identifying risks as well as implementing preventive measures to reduce their impacts on human beings as well as the environment. These strategies include the identification of hazards, the assessment of exposures, and risk characterization. Hazard identification focuses on determining the toxic effects of NMs by employing acute and chronic toxicity studies. Important endpoints include cytotoxicity, genotoxicity, and oxidative stress. Molecular biology techniques like omics data and high-throughput screening methods can aid in defining the cross-platform biological impacts of NMs (Warheit et al. 2007). Evaluating exposure to NMs entails considering the source, distribution and concentration of the NMs in the environment. This includes observation of emissions of NMs in industries, consumer products, and uses in medical sectors. Exposure assessment models that have been developed to predict the transport and behavior of NMs in the environment can also be useful in assessing probable exposure levels by utilizing environmental monitoring systems and predictive modeling tools (Gottschalk et al. 2009). The integration of hazard and exposure data determines the possible threat NMs can present to people. This is done by comparing the estimated exposure levels with the concentrations shown to elicit toxic effects in toxicity studies. Risk characterization can, therefore, help in decision-making on the formulation of safety standards and support the development of regulatory policies (Linkov et al. 2009).

18.4 Regulatory Frameworks and Guidelines

Regulation and guidelines are crucial for the protection and safety of NMs, considering their unique characteristics and hazards throughout their life cycle. Some test guidelines already exist, provided by organizations such as the Organisation for Economic Co-operation and Development (OECD) and the International Organization for Standardization (ISO). These guidelines establish procedural measures for assessing the risks associated with NMs concerning their physical and chemical characteristics, fate in the environment, and toxicity (Canton 2021).

The European Union has introduced certain regulations concerning NMs under the regulatory framework called Registration, Evaluation, Authorization and Restriction of Chemicals (REACH), along with the recent NICNAS which targets manufacturers. These regulations compel manufacturers/

importers of NMs to declare additional information about the safety of their products and any likely environmental impacts. Also, the European Food Safety Authority (EFSA) has guidelines on the risk assessment of NMs in food and feed (EC 2011). Similarly, in the United States, the EPA and FDA are responsible for the management of NMs under TSCA and FFDCA, respectively. The EPA has provided guidelines for the testing and evaluation of released NMs regarding their fate, transport, and toxicity (EPA 2015). Besides regulations on the use of NMs, proper measures for handling and disposal are also important. This includes adjustments in worksite design, personal protective equipment, and disposal techniques to decrease workers' exposure and emissions to the environment. Awareness and proper education of workers and researchers in NM industries are also vital components when it comes to the manipulation of these materials (Hodson et al. 2009).

19. Future Trends and Research Directions

19.1 Emerging NMs and Technologies

Nanotechnology is a dynamic field, and more NMs and nanotechnologies are being developed that hold potential for improving the sensing and treatment of pollutants in the environment. Graphene, molybdenum disulfide (MoS_2), and transition metal dichalcogenides (TMDs) have properties that enhance their applicability for environmental processes due to their mechanical, electronic, and optical characteristics. These 2D materials have exposed surface areas suitable for highly sensitive sensing and effective removal of even the lowest concentrations of pollutants (Chhowalla et al. 2013).

MOFs are crystalline materials containing metal cations and organic linkers in the form of a three-dimensional open framework. Due to their highly tunable porosity and large surface areas, these metal-organic frameworks are well suited for gas adsorption and storage, as well as for capturing and destroying environmental pollutants (Furukawa et al. 2013). Making hybrids or composites out of various NMs may help improve subsequent properties by acting cooperatively. For instance, dispersing metal nanoparticles or metal oxides in graphene composites can enhance both the detectability and catalytic efficiency in treatment (Yang et al. 2021).

19.2 Innovations in Detection and Remediation Techniques

Sensors that combine NMs with electronics and optics are being designed for the continuous measurement of pollutants in the environment. Such smart sensors can offer real-time responses and are also amenable to remote surveillance, which makes them very useful in environmental management (Bibri 2018).

New techniques that incorporate both photothermal and photoelectrochemical properties of nanostructures are emerging. These techniques employ light to increase the reactivity of NMs, which in turn increases the degradation of pollutants through mechanisms like photo-Fenton and photoelectrochemical oxidation (García-Segura and Brillas 2017).

Responding to critics, nanotechnologists are trying to create biodegradable NMs that disintegrate into harmless products once they have been used. These materials reduce adverse long-term environmental impacts and are best suited to areas of ecological concern (Bhushan 2017).

19.3 Interdisciplinary Approaches and Collaborations

Chemists, biologists, environmental scientists, and engineers should work together to create multifunctional NMs and systems. Often, interdisciplinary research can lead to innovations where the strengths of several fields are combined to produce better solutions (Roco and Bainbridge 2007).

Collaborations between universities, state agencies, and industry can enhance the process of using nanotechnologies for environmental purposes. Such partnerships can offer the financial, technical, and logistical support to translate innovations from bench to bedside (Karn et al. 2009).

The increase in understanding, use, and acceptance of nanotechnologies can be improved by involving local communities in environmental monitoring and cleanup activities. Community involvement ensures that the technologies meet the needs of a given community and that there is transparency in the implementation of these technologies (Gavelin 2007).

19.4 Long-Term Sustainability and Environmental Impact

The application of NMs should not only address present needs but also align with a sustainable future. Therefore, it is important to reduce the impacts of nanomaterials on society. LCAs offer the opportunity to assess the impacts of NMs throughout creation, usage, and disposal. Life cycle assessments are valuable in understanding all environmental and health risk implications of NMs at every stage of the product's life and help enhance more sustainable practices (Nizam et al. 2021). For instance, synthesizing NMs with safer chemicals rather than toxic ones, or using recyclable and non-hazardous chemicals, and adopting energy- and cost-effective green synthesis techniques can help minimize the potential hazards posed by nanomaterial production. For example, using green synthesis processes like plant-mediated synthesis or biosynthesis of nanoparticles by microorganisms minimizes the use of hazardous chemicals (Iravani 2011).

Overcoming the limitations of effectively regulating NMs is possible by developing proper regulatory and policy requirements that define usage conditions. It is suggested that regulations should focus on precautionary measures that could cover possible threats, set up safety requirements, as well as contribute to the creation of sustainable nanotechnologies (Karn et al. 2009). However, once the risks of NMs have been established, it is equally important to monitor their usage in the same manner as the impact they pose on the environment. The right action for addressing any new leaks or contamination that may present some risks is manageable using proper intervention measures. This needs to be changed for research to thrive and to develop NMs that can prevent the continued pollution of the environment in the future (Gottschalk et al. 2009).

20. Conclusion

NMs are quite useful in the detection and remediation of contaminated soils, as described in this review. NMs possess characteristics such as large surface area, high activity, and well-defined functions for solving environmental issues. Through adsorption, catalysis, electrochemical, and optical methods, contaminants can be identified; and through adsorption, catalysis, and photochemical methods, contaminants can be eliminated. Integrated systems are real-time based and can track contamination and respond to it as soon as it is observed. The use of NMs in soil management or treatment and protection of the environment is very promising, the reason being that NMs have the ability to identify contaminants at very low levels and remove the identified contaminants effectively, thereby helping to obtain clean soils and consequently a healthy environment. These technologies also contribute to the Sustainable Development Goals, as they encourage clean production and discourage the use of deleterious chemicals for soil analysis. Current and potential uses of NMs in the environmental field should be explored further to identify both the efficiency and drawbacks of the technology. Such deployment would require interdisciplinary effort from the scientific community, engineers, policymakers, and various stakeholders. This suggests that proper guidelines and safety measures should be put in place to prevent irresponsible use of the technology. Greener synthesis pathways and green chemistry principles can minimize adverse effects on the environment and promote sustainability. Public acceptance is another challenge because it requires educating the populace about the NM's advantages and liabilities. Experiments and pilot projects can show that nanomaterial based technologies are useful and feasible. It requires constant assessment and observation of the adverse effects that they have on the environment and health of the people.

21. Acknowledgement

The authors gratefully acknowledge the support of the Institute of Plant Protection, Muhammad Nawaz Sharif University of Agriculture, Multan for the support in this publication.

References

Adriano, D.C. 2001. Trace elements in terrestrial environments: biogeochemistry, bioavailability, and risks of metals. Springer, New York.

Alivisatos, A.P. 1996. Semiconductor clusters, nanocrystals, and quantum dots. Science. 271: 933–937.

Alizadeh Fard, M., B. Aminzadeh and H. Vahidi. 2013. Degradation of petroleum aromatic hydrocarbons using TiO2 nanopowder film. Environ. Technol. 34: 1183–1190.

Alloway, B.J. [ed.]. 2012. Heavy metals in soils: trace metals and metalloids in soils and their bioavailability. Springer Science and Business Media.

Auffan, M., J. Rose, J.Y. Bottero, G.V. Lowry, J.P. Jolivet and M.R. Wiesner. 2009. Towards a definition of inorganic nanoparticles from an environmental, health and safety perspective. Nat. Nanotechnol. 4: 634–641.

Batley, G.E., J.K. Kirby and M.J. McLaughlin. 2013. Fate and risks of nanomaterials in aquatic and terrestrial environments. Acc. Chem. Res. 46: 854–862.

Baughman, R.H., A.A. Zakhidov and W.A. De Heer. 2002. Carbon nanotubes—the route toward applications. Science. 297: 787–792.

Bhateria, R. and R. Singh. 2022. Nanotechnology enabled multifunctional materials for removal of toxicants from wastewater. pp. 233–254. *In*: Cost-efficient Wastewater Treatment Technologies: Engineered Systems. Springer International Publishing, Cham.

Bhattacharya, A. and A. Selvaraj. 2022. Carbonaceous-TiO2 photocatalyst for treatment of textile dye-contaminated wastewater. pp. 319–340. *In*: Advanced Oxidation Processes in Dye-Containing Wastewater: Volume 1. Springer Nature Singapore, Singapore.

Bhushan, B. [ed.]. 2017. Springer handbook of nanotechnology. Springer.

Bibri, S.E. 2018. The IoT for smart sustainable cities of the future: an analytical framework for sensor-based big data applications for environmental sustainability. Sustain. Cities Soc. 38: 230–253.

Binnig, G., C.F. Quate and C. Gerber. 1986. Atomic force microscope. Phys. Rev. Lett. 56: 930.

Brinker, C.J. and G.W. Scherer. 2013. Sol-gel science: the physics and chemistry of sol-gel processing. Academic Press.

Brown, R.J. and M.J. Milton. 2008. Nanostructures and nanostructured substrates for surface-enhanced Raman scattering (SERS). J. Raman Spectrosc. 39: 1313–1326.

Brunauer, S., P.H. Emmett and E. Teller. 1938. Adsorption of gases in multimolecular layers. J. Am. Chem. Soc. 60: 309–319.

Bussian, B.M., N.R. Eugenio, S.C. Wilson, A. Ceci, C. Parelho, D. Semenov et al. 2021. Main soil contaminants and their fate in the soil environment. *In*: Global assessment of soil pollution: Report. Food and Agriculture Organization of the United Nations (FAO).

Byrappa, K. and M. Yoshimura. 2012. Handbook of hydrothermal technology. William Andrew.

Canton, H. 2021. Organisation for economic co-operation and development—OECD. pp. 677–687. *In*: The Europa Directory of International Organizations 2021. Routledge.

Carter, D.B.W.C.B. 2009. Transmission electron microscopy: a textbook for materials science.

Cempel, M. and G.J.P.J.S. Nikel. 2006. Nickel: a review of its sources and environmental toxicology. Pol. J. Environ. Stud. 15: 3.

Chasapis, C.T., P.S.A. Ntoupa, C.A. Spiliopoulou and M.E. Stefanidou. 2020. Recent aspects of the effects of zinc on human health. Arch. Toxicol. 94: 1443–1460.

Chen, B. and F. Wang. 2020. Emerging frontiers of upconversion nanoparticles. Trends Chem. 2: 427–439.

Chen, K.L., S.E. Mylon and M. Elimelech. 2006. Aggregation kinetics of alginate-coated hematite nanoparticles in monovalent and divalent electrolytes. Environ. Sci. Technol. 40: 1516–1523.

Chen, X. and S.S. Mao. 2007. Titanium dioxide nanomaterials: synthesis, properties, modifications, and applications. Chem. Rev. 107: 2891–2959.

Chhowalla, M., H.S. Shin, G. Eda, L.J. Li, K.P. Loh and H. Zhang. 2013. The chemistry of two-dimensional layered transition metal dichalcogenide nanosheets. Nat. Chem. 5: 263–275.

Chou, S.Y., P.R. Krauss and P.J. Renstrom. 1996. Imprint lithography with 25-nanometer resolution. Science. 272: 85–87.

Cui, Y., Q. Wei, H. Park and C.M. Lieber. 2001. Nanowire nanosensors for highly sensitive and selective detection of biological and chemical species. Science. 293: 1289–1292.

Cullity, B.D. 1956. Elements of X-ray diffraction. Addison-Wesley Publishing.

da Silva, M.B., N. Abrantes, V. Nogueira, F. Goncalves and R. Pereira. 2016. TiO2 nanoparticles for the remediation of eutrophic shallow freshwater systems: efficiency and impacts on aquatic biota under a microcosm experiment. Aquat. Toxicol. 178: 58–71.

Darwish, M. and A. Mohammadi. 2018. Functionalized nanomaterial for environmental techniques. pp. 315–350. In: Nanotechnology in Environmental Science.

Das, T.K. 2021. Graphene oxide supported metal oxide nanohybrids for aqueous arsenic removal. North Dakota State University.

de Dios, A.S. and M.E. Díaz-García. 2010. Multifunctional nanoparticles: analytical prospects. Anal. Chim. Acta. 666: 1–22.

De Gisi, S., D. Minetto, G. Lofrano, G. Libralato, B. Conte, F. Todaro et al. 2017. Nano-scale zero valent iron (nZVI) treatment of marine sediments slightly polluted by heavy metals. Chem. Eng. Trans. 60: 139–144.

de Souza, R.M., D. Seibert, H.B. Quesada, F. de Jesus Bassetti, M.R. Fagundes-Klen and R. Bergamasco. 2020. Occurrence, impacts and general aspects of pesticides in surface water: a review. Process Saf. Environ. Prot. 135: 22–37.

Donaldson, K., V. Stone, C.L. Tran, W. Kreyling and P.J. Borm. 2004. Nanotoxicology. Occup. Environ. Med. 61: 727–728.

EC. 2011. European Commission recommendation on the definition of nanomaterial. Available at: European Commission.

EPA. 2015. Nanomaterial case study: nanoscale silver in disinfectant spray. Available at: EPA.

EPA. 2024. Biological pollutants' impact on indoor air quality. Environ. Prot. Agency. Available at: https://www.epa.gov/indoor-air-quality-iaq/biological-pollutants-impact-indoor-air-quality.

Fabrega, J., S.N. Luoma, C.R. Tyler, T.S. Galloway and J.R. Lead. 2011. Silver nanoparticles: behaviour and effects in the aquatic environment. Environ. Int. 37: 517–531.

Fei, X., Z. Lou, R. Xiao, Z. Ren and X. Lv. 2022. Source analysis and source-oriented risk assessment of heavy metal pollution in agricultural soils of different cultivated land qualities. J. Clean. Prod. 341: 130942.

Ferrari, A.C. and D.M. Basko. 2013. Raman spectroscopy as a versatile tool for studying the properties of graphene. Nat. Nanotechnol. 8: 235–246.

Furukawa, H., K.E. Cordova, M. O'Keeffe and O.M. Yaghi. 2013. The chemistry and applications of metal-organic frameworks. Science. 341: 1230444.

Gao, X., Y. Cui, R.M. Levenson, L.W. Chung and S. Nie. 2004. In vivo cancer targeting and imaging with semiconductor quantum dots. Nat. Biotechnol. 22: 969–976.

Garcia-Segura, S. and E. Brillas. 2017. Applied photoelectrocatalysis on the degradation of organic pollutants in wastewaters. J. Photochem. Photobiol. C. 31: 1–35.

Gavaskar, A.R., L. Tatar and W. Condit. 2005. Cost and performance report nanoscale zero-valent iron technologies for source remediation. Port Hueneme, CA: Naval Facilities Engineering Command.

Gavelin, K. 2007. Democratic technologies?: The final report of the Nanotechnology Engagement Group (NEG). Involve Foundation.

Geim, A.K. and K.S. Novoselov. 2007. The rise of graphene. Nat. Mater. 6: 183–191.

Genchi, G., M.S. Sinicropi, G. Lauria, A. Carocci and A. Catalano. 2020. The effects of cadmium toxicity. Int. J. Environ. Res. Public Health. 17: 3782.

Geng, Z., Y. Lin, X. Yu, Q. Shen, L. Ma, Z. Li et al. 2012. Highly efficient dye adsorption and removal: a functional hybrid of reduced graphene oxide–Fe_3O_4 nanoparticles as an easily regenerative adsorbent. J. Mater. Chem. 22: 3527–3535.

Gerba, C.P. and J.E. Smith. 2005. Sources of pathogenic microorganisms and their fate during land application of wastes. J. Environ. Qual. 34: 42–48.

Goldstein, J.I., D.E. Newbury, J.R. Michael, N.W. Ritchie, J.H.J. Scott and D.C. Joy. 2017. Scanning electron microscopy and X-ray microanalysis. Springer.

Gottschalk, F., T. Sonderer, R.W. Scholz and B. Nowack. 2009. Modeled environmental concentrations of engineered nanomaterials (TiO_2, ZnO, Ag, CNT, fullerenes) for different regions. Environ. Sci. Technol. 43: 9216–9222.

Grifoni, M., E. Franchi, D. Fusini, M. Vocciante, M. Barbafieri, F. Pedron et al. 2022. Soil remediation: Towards a resilient and adaptive approach to deal with the ever-changing environmental challenges. Environments. 9: 18.

Gu, J., D. Dong, L. Kong, Y. Zheng and X. Li. 2012. Photocatalytic degradation of phenanthrene on soil surfaces in the presence of nanometer anatase TiO_2 under UV-light. J. Environ. Sci. 24: 2122–2126.
Guerra, F.D., M.F. Attia, D.C. Whitehead and F. Alexis. 2018. Nanotechnology for environmental remediation: materials and applications. Molecules. 23: 1760.
Guo, W., B. Pan, S. Sakkiah, G. Yavas, W. Ge, W. Zou et al. 2019. Persistent organic pollutants in food: contamination sources, health effects and detection methods. Int. J. Environ. Res. Public Health. 16: 4361.
Gupta, V.K., R. Kumar, A. Nayak, T.A. Saleh and M.A. Barakat. 2013. Adsorptive removal of dyes from aqueous solution onto carbon nanotubes: a review. Adv. Colloid Interface Sci. 193: 24–34.
Hanfi, M.Y., M.Y. Mostafa and M.V. Zhukovsky. 2020. Heavy metal contamination in urban surface sediments: sources, distribution, contamination control, and remediation. Environ. Monit. Assess. 192: 1–21.
He, F. and D. Zhao. 2007. Manipulating the size and dispersibility of zerovalent iron nanoparticles by use of carboxymethyl cellulose stabilizers. Environ. Sci. Technol. 41: 6216–6221.
He, K., Z. Zeng, A. Chen, G. Zeng, R. Xiao, P. Xu et al. 2018. Advancement of Ag–graphene based nanocomposites: an overview of synthesis and its applications. Small. 14: 1800871.
Hodson, L., M. Methner and R.D. Zumwalde. 2009. Approaches to safe nanotechnology: managing the health and safety concerns associated with engineered nanomaterials.
Homola, J. 2008. Surface plasmon resonance sensors for detection of chemical and biological species. Chem. Rev. 108: 462–493.
Iijima, S. 1991. Helical microtubules of graphitic carbon. Nature. 354: 56–58.
Iravani, S. 2011. Green synthesis of metal nanoparticles using plants. Green Chem. 13: 2638–2650.
Jiang, W., B.Y. Kim, J.T. Rutka and W.C. Chan. 2008. Nanoparticle-mediated cellular response is size-dependent. Nature Nanotech. 3: 145–150.
Johnsen, A.R., L.Y. Wick and H. Harms. 2005. Principles of microbial PAH-degradation in soil. Environ. Pollut. 133: 71–84.
Karn, B., T. Kuiken and M. Otto. 2009. Nanotechnology and in situ remediation: a review of the benefits and potential risks. Environ. Health Perspect. 117: 1813–1831.
Khalid, S., M. Shahid, N.K. Niazi, B. Murtaza, I. Bibi and C. Dumat. 2017. A comparison of technologies for remediation of heavy metal contaminated soils. J. Geochem. Explor. 182: 247–268.
Khan, S.B. [ed.]. 2018. Nanomaterials for Environmental Applications and their Fascinating Attributes. Vol. 2. Bentham Science Publishers, Sharjah.
Klabunde, K.J. and R.M. Richards [eds.]. 2009. Nanoscale Materials in Chemistry. John Wiley and Sons, New York.
Kumar, P., M.A. Kausar, A.B. Singh and R. Singh. 2021. Biological contaminants in the indoor air environment and their impacts on human health. Air Quality, Atmosphere and Health. 14: 1723–1736.
Latif, A., D. Sheng, K. Sun, Y. Si, M. Azeem, A. Abbas et al. 2020. Remediation of heavy metals polluted environment using Fe-based nanoparticles: Mechanisms, influencing factors, and environmental implications. Environ. Pollut. 264: 114728.
Lellis, B., C.Z. Fávaro-Polonio, J.A. Pamphile and J.C. Polonio. 2019. Effects of textile dyes on health and the environment and bioremediation potential of living organisms. Biotech. Res. Innov. 3: 275–290.
Li, F., J. Chen, X. Hu, F. He, E. Bean, D.C. Tsang et al. 2020. Applications of carbonaceous adsorbents in the remediation of polycyclic aromatic hydrocarbon-contaminated sediments: A review. J. Clean. Prod. 255: 120263.
Li, X., W. Cai, J. An, S. Kim, J. Nah, D. Yang et al. 2009. Large-area synthesis of high-quality and uniform graphene films on copper foils. Science 324: 1312–1314.
Linkov, I., J. Steevens, G. Adlakha-Hutcheon, E. Bennett, M. Chappell, V. Colvin et al. 2009. Emerging methods and tools for environmental risk assessment, decision-making, and policy for nanomaterials: summary of NATO Advanced Research Workshop. J. Nanopart. Res. 11: 513–527.
Liu, B., F. Zeng, G. Wu and S. Wu. 2012. Nanoparticles as scaffolds for FRET-based ratiometric detection of mercury ions in water with QDs as donors. Analyst 137: 3717–3724.
Liu, J. and Y. Lu. 2004. Accelerated color change of gold nanoparticles assembled by DNAzymes for simple and fast colorimetric Pb2+ detection. J. Am. Chem. Soc. 126: 12298–12305.
Liu, J. and Y. Lu. 2004. Colorimetric biosensors based on DNAzyme-assembled gold nanoparticles. J. Fluoresc. 14: 343–354.
Liu, P., R. Qin, G. Fu and N. Zheng. 2017. Surface coordination chemistry of metal nanomaterials. J. Am. Chem. Soc. 139: 2122–2131.
Liu, Z., J.T. Robinson, X. Sun and H. Dai. 2008. PEGylated nanographene oxide for delivery of water-insoluble cancer drugs. J. Am. Chem. Soc. 130: 10876–10877.
Mahlambi, M.M., C.J. Ngila and B.B. Mamba. 2015. Recent developments in environmental photocatalytic degradation of organic pollutants: the case of titanium dioxide nanoparticles—a review. J. Nanomater. 2015: 790173.

Maier, R.M., I.L. Pepper and C.P. Gerba. 2009. Environmental Microbiology. Vol. 397. Academic Press, San Diego.

Marcelo, L.R., J.S. de Gois, A.A. da Silva and D.V. Cesar. 2021. Synthesis of iron-based magnetic nanocomposites and applications in adsorption processes for water treatment: a review. Environ. Chem. Lett. 19: 1229–1274.

Medintz, I.L., H.T. Uyeda, E.R. Goldman and H. Mattoussi. 2005. Quantum dot bioconjugates for imaging, labelling, and sensing. Nat. Mater. 4: 435–446.

Meyer, M.F., S.M. Powers and S.E. Hampton. 2019. An evidence synthesis of pharmaceuticals and personal care products (PPCPs) in the environment: imbalances among compounds, sewage treatment techniques, and ecosystem types. Environ. Sci. Technol. 53: 12961–12973.

Michalet, X., F.F. Pinaud, L.A. Bentolila, J.M. Tsay, S.J.J.L. Doose, J.J. Li et al. 2005. Quantum dots for live cells, in vivo imaging, and diagnostics. Science. 307: 538–544.

Mishra, A., M. Kumari, R. Kumar, K. Iqbal and I.S. Thakur. 2022. Persistent organic pollutants in the environment: risk assessment, hazards, and mitigation strategies. Bioresour. Technol. Rep. 19: 101143.

Mobasser, S., Y. Wager and T.M. Dittrich. 2022. Indoor air purification of volatile organic compounds (VOCs) using activated carbon, zeolite, and organosilica sorbents. Ind. Eng. Chem. Res. 61: 6791–6801.

Mueller, N.C., J. Braun, J. Bruns, M. Černík, P. Rissing, D. Rickerby et al. 2012. Application of nanoscale zero valent iron (NZVI) for groundwater remediation in Europe. Environ. Sci. Pollut. Res. 19: 550–558.

Nagajyoti, P.C., K.D. Lee and T.V.M. Sreekanth. 2010. Heavy metals, occurrence and toxicity for plants: a review. Environ. Chem. Lett. 8: 199–216.

Nel, A., T. Xia, L. Madler and N. Li. 2006. Toxic potential of materials at the nanolevel. Science. 311: 622–627.

New, W.X., J.E. Ogbezode and P. Gani. 2023. Nanoparticles in soil remediation: challenges and opportunities. Ind. and Dom. Waste Manag. 3: 127–140.

Nizam, N.U.M., M.M. Hanafiah and K.S. Woon. 2021. A content review of life cycle assessment of nanomaterials: current practices, challenges, and future prospects. Nanomaterials. 11: 3324.

Njuguna, J., F. Ansari, S. Sachse, V.M. Rodriguez, S. Siqqique and H. Zhu. 2021. Nanomaterials, nanofillers, and nanocomposites: types and properties. *In*: Health and Environmental Safety of Nanomaterials. pp. 3–37. Woodhead Publishing.

Novoselov, K.S., A.K. Geim, S.V. Morozov, D.E. Jiang, Y. Zhang, S.V. Dubonos et al. 2004. Electric field effect in atomically thin carbon films. Science. 306: 666–669.

Oberdörster, G., E. Oberdörster and J. Oberdörster. 2005. Nanotoxicology: an emerging discipline evolving from studies of ultrafine particles. Environ. Health Perspect. 113: 823–839.

Nagajyoti, P.C., K.D. Lee and T.V.M. Sreekanth. 2010. Heavy metals, occurrence and toxicity for plants: a review. Environ. Chem. Lett. 8: 199–216.

Nel, A., T. Xia, L. Madler and N. Li. 2006. Toxic potential of materials at the nanolevel. Science. 311: 622–627.

New, W.X., J.E. Ogbezode and P. Gani. 2023. Nanoparticles in soil remediation: challenges and opportunities. Ind. and Dom. Waste Manag. 3: 127–140.

Nizam, N.U.M., M.M. Hanafiah and K.S. Woon. 2021. A content review of life cycle assessment of nanomaterials: current practices, challenges, and future prospects. Nanomaterials. 11: 3324.

Njuguna, J., F. Ansari, S. Sachse, V.M. Rodriguez, S. Siqqique and H. Zhu. 2021. Nanomaterials, nanofillers, and nanocomposites: types and properties. *In*: Health and Environmental Safety of Nanomaterials. pp. 3–37. Woodhead Publishing.

Novoselov, K.S., A.K. Geim, S.V. Morozov, D.E. Jiang, Y. Zhang, S.V. Dubonos et al. 2004. Electric field effect in atomically thin carbon films. Science. 306: 666–669.

Oberdörster, G., E. Oberdörster and J. Oberdörster. 2005. Nanotoxicology: an emerging discipline evolving from studies of ultrafine particles. Environ. Health Perspect. 113: 823–839.

Ojea-Jiménez, I., X. López, J. Arbiol and V. Puntes. 2012. Citrate-coated gold nanoparticles as smart scavengers for mercury (II) removal from polluted waters. ACS Nano. 6: 2253–2260.

Park, J.Y. and I.H. Lee. 2014. Photocatalytic degradation of 2-chlorophenol using Ag-doped TiO2 nanofibers and a near-UV light-emitting diode system. J. Nanomaterials. 2014. 250803.

Peng, X., Z. Luan, Z. Di, Z. Zhang and C. Zhu. 2005. Carbon nanotubes-iron oxides magnetic composites as adsorbent for removal of Pb (II) and Cu (II) from water. Carbon. 43: 880–883.

Pepper, I., C.P. Gerba and M.L. Brusseau. 2011. Environmental and Pollution Science. Elsevier.

Perreault, F., A.F. De Faria and M. Elimelech. 2015. Environmental applications of graphene-based nanomaterials. Chem. Soc. Rev. 44: 5861–5896.

Petosa, A.R., D.P. Jaisi, I.R. Quevedo, M. Elimelech and N. Tufenkji. 2010. Aggregation and deposition of engineered nanomaterials in aquatic environments: role of physicochemical interactions. Environ. Sci. Technol. 44: 6532–6549.

Qu, X., P.J. Alvarez and Q. Li. 2013. Applications of nanotechnology in water and wastewater treatment. Water Res. 47: 3931–3946.

Rao, C.N.R. and K. Biswas. 2009. Characterization of nanomaterials by physical methods. Annu. Rev. Anal. Chem. 2: 435–462.

Rasheed, T., M. Bilal, F. Nabeel, M. Adeel and H.M. Iqbal. 2019. Environmentally-related contaminants of high concern: potential sources and analytical modalities for detection, quantification, and treatment. Environ. Int. 122: 52–66.

Ritter, L., K.R. Solomon, J. Forget, M. Stemeroff and C. O'Leary. 1995. Persistent organic pollutants: an assessment report on: DDT, aldrin, dieldrin, endrin, chlordane, heptachlor, hexachlorobenzene, mirex, toxaphene, polychlorinated biphenyls, dioxins and furans. International Programme on Chemical Safety (IPCS).

Roco, M.C. and W.S. Bainbridge [eds.]. 2007. Nanotechnology: Societal Implications: Maximizing Benefits for Humanity. Dordrecht, The Netherlands: Springer.

Roy, A., A. Sharma, S. Yadav, L.T. Jule and R. Krishnaraj. 2021. Nanomaterials for remediation of environmental pollutants. Bioinorg. Chem. Appl. 2021: 1764647.

Sapsford, K.E., W.R. Algar, L. Berti, K.B. Gemmill, B.J. Casey, E. Oh et al. 2013. Functionalizing nanoparticles with biological molecules: developing chemistries that facilitate nanotechnology. Chem. Rev. 113: 1904–2074.

Saraswathi, M.S.S.A., A. Nagendran and D. Rana. 2019. Tailored polymer nanocomposite membranes based on carbon, metal oxide and silicon nanomaterials: a review. J. Mater. Chem. A 7: 8723–8745.

Schaefer, H.R., S. Dennis and S. Fitzpatrick. 2020. Cadmium: Mitigation strategies to reduce dietary exposure. J. Food Sci. 85(2): 260–267.

Semerád, J., A. Ševců, N.H. Nguyen, P. Hrabák, R. Špánek, K. Bobčíková et al. 2021. Discovering the potential of an nZVI-biochar composite as a material for the nanobioremediation of chlorinated solvents in groundwater: degradation efficiency and effect on resident microorganisms. Chemosphere. 281: 130915.

Shah, K.J. and T. Imae. 2016. Selective gas capture ability of gas-adsorbent-incorporated cellulose nanofiber films. Biomacromolecules. 17(5): 1653–1661.

Smith, B.C. 2011. Fundamentals of Fourier transform infrared spectroscopy. CRC Press.

Soldatkin, O.O., O.V. Soldatkina, I.I. Piliponskiy, L.S. Rieznichenko, T.G. Gruzina, S.M. Dybkova et al. 2022. Application of gold nanoparticles for improvement of analytical characteristics of conductometric enzyme biosensors. Appl. Nanosci. 12(4): 995–1003.

Suryanarayana, C. 2001. Mechanical alloying and milling. Prog. Mater. Sci. 46(1-2): 1–184.

Tchounwou, P.B., C.G. Yedjou, A.K. Patlolla and D.J. Sutton. 2012. Heavy metal toxicity and the environment. In: Molecular, clinical and environmental toxicology: Volume 3: Environmental Toxicology, pp. 133–164.

Thakur, A. and A. Kumar. 2022. Recent advances on rapid detection and remediation of environmental pollutants utilizing nanomaterials-based (bio) sensors. Sci. Total Environ. 834: 155219.

Tiwari, A., Y.K. Mishra, H. Kobayashi and A.P. Turner [eds.]. 2016. Intelligent nanomaterials. John Wiley and Sons.

Tomalia, D.A., L.A. Reyna and S. Svenson. 2007. Dendrimers as multi-purpose nanodevices for oncology drug delivery and diagnostic imaging.

Tratnyek, P.G. and R.L. Johnson. 2006. Nanotechnologies for environmental cleanup. Nano Today. 1(2): 44–48.

Tyler, C.R. and A.M. Allan. 2014. The effects of arsenic exposure on neurological and cognitive dysfunction in human and rodent studies: a review. Curr. Environ. Health Rep. 1: 132–147.

Vaseashta, A., M. Vaclavikova, S. Vaseashta, G. Gallios, P. Roy and O. Pummakarnchana. 2007. Nanostructures in environmental pollution detection, monitoring, and remediation. Sci. Technol. Adv. Mater. 8(1-2): 47.

Wang, J. and B. Chen. 2015. Adsorption and coadsorption of organic pollutants and a heavy metal by graphene oxide and reduced graphene materials. Chem. Eng. J. 281: 379–388.

Wang, L., W.M. Wu, N.S. Bolan, D.C. Tsang, Y. Li, M. Qin et al. 2021. Environmental fate, toxicity and risk management strategies of nanoplastics in the environment: Current status and future perspectives. J. Hazard. Mater. 401: 123415.

Wang, M., F. Hossain, R. Sulaiman and X. Ren. 2019. Exposure to inorganic arsenic and lead and autism spectrum disorder in children: a systematic review and meta-analysis. Chem. Res. Toxicol. 32(10): 1904–1919.

Wani, A.L., A. Ara and J.A. Usmani. 2015. Lead toxicity: a review. Interdiscip. Toxicol. 8(2): 55–64.

Ward, M.H., R.R. Jones, J.D. Brender, T.M. De Kok, P.J. Weyer, B.T. Nolan et al. 2018. Drinking water nitrate and human health: an updated review. Int. J. Environ. Res. Public Health. 15(7): 1557.

Warheit, D.B., R.A. Hoke, C. Finlay, E.M. Donner, K.L. Reed and C.M. Sayes. 2007. Development of a base set of toxicity tests using ultrafine TiO2 particles as a component of nanoparticle risk management. Toxicol. Lett. 171(3): 99–110.

Xia, T., M. Kovochich, M. Liong, L. Madler, B. Gilbert, H. Shi et al. 2008. Comparison of the mechanism of toxicity of zinc oxide and cerium oxide nanoparticles based on dissolution and oxidative stress properties. ACS Nano. 2(10): 2121–2134.

Xie, J., S. Lee and X. Chen. 2010. Nanoparticle-based theranostic agents. Adv. Drug Deliv. Rev. 62(11): 1064–1079.

Xiu, Z.M., Q.B. Zhang, H.L. Puppala, V.L. Colvin and P.J. Alvarez. 2012. Negligible particle-specific antibacterial activity of silver nanoparticles. Nano Lett. 12(8): 4271–4275.

Yadav, N., V.K. Garg, A.K. Chhillar and J.S. Rana. 2021. Detection and remediation of pollutants to maintain ecosustainability employing nanotechnology: A review. Chemosphere. 280: 130792.

Yang, W., M. Pan, C. Huang, Z. Zhao, J. Wang and H. Zeng. 2021. Graphene oxide-based noble-metal nanoparticles composites for environmental application. Compos. Commun. 24: 100645.

Yang, W., P. Thordarson, J.J. Gooding, S.P. Ringer and F. Braet. 2007. Carbon nanotubes for biological and biomedical applications. Nanotechnology. 18(41): 412001.

Zhang, P. and Y. Chen. 2017. Polycyclic aromatic hydrocarbons contamination in surface soil of China: a review. Sci. Total Environ. 605: 1011–1020.

Zhang, W.X. 2003. Nanoscale iron particles for environmental remediation: an overview. J. Nanopart. Res. 5: 323–332.

Zhu, C., G. Yang, H. Li, D. Du and Y. Lin. 2015. Electrochemical sensors and biosensors based on nanomaterials and nanostructures. Anal. Chem. 87(1): 230–249.

Zhu, L., L. Meng, J. Shi, J. Li, X. Zhang and M. Feng. 2019. Metal-organic frameworks/carbon-based materials for environmental remediation: A state-of-the-art mini-review. J. Environ. Manag. 232: 964–977.

CHAPTER-11

Geographic Information System, Remote Sensing and Nanotechnology for Site-Specific and Efficient Nutrient Management

Muhammad Imran,[1] *Humair Ahmad Malik,*[2] *Hasan Riaz,*[3] *Akhtar Hameed,*[3] *Nadeem Ahmad*[3] and *Waseem Hassan*[1,*]

1. Introduction

Agriculture is the mainstay for human civilization; almost all necessary resources for sustaining life on this planet are provided by the agriculture sector. The world's population is rising rapidly and pressure on the agricultural sector to provide food, fiber and bioenergy is also increasing. Higher average yield utilizing economic resources while having a minimum environmental footprint requires adoption of innovative and sophisticated approaches. The yield of major crops grown in the countries have become stagnant through traditional practices (Godfray et al. 2010). Employing traditional practices have been recognized as inefficient and unsuitable for the environment. Traditional practices in the agriculture have led to soil degradation, nutrient losses and water pollution. Overuse or imbalanced use of nitrogenous fertilizers has contributed to water pollution (Rafique et al. 2023).

Precision farming has been considered one of the main solutions to the above challenges. It takes advantage of advanced technologies like digital mapping and nanotechnology, which lead to site-specific nutrient management. Crops are provided with inputs in precise amounts. Application of nutrients where and when needed leads to a reduction in fertilizer loss. Higher yields with decreased input costs are the results of site-specific nutrient management. Precision farming also helps mitigate the environmental deterioration caused by traditional farming practices carried out in different countries around the world (Mulla n.d.). Components of digital mapping technologies—comprising Geographic Information Systems (GIS), Remote Sensing (RS), and nanotechnology—have revolutionized traditional farming. In this way, the farming community is able to manage the

[1] Scientific officer, NARC, Islamabad.
[2] Department of Soil and Environmental Sciences, MNS-University of Agriculture, Multan, Pakistan.
[3] Institute of Plant Protection, MNS University of Agriculture, Multan 61000, Pakistan.
* Corresponding author: waseem.hassan@mnsuam.edu.pk

DOI: 10.1201/9781003558354-11

spatial variability of soils' physico-chemical properties and nutrient status. Spatial variability is an inherent property of soils. Tools of GIS and RS help to generate detailed maps that manifest the spatial variability of soils.

Digital maps generated through GIS and RS, combined with nanotechnology, lead to the development of innovative materials and delivery systems for agricultural inputs. Application of agricultural inputs using delivery systems developed with the help of nanotechnology at the nanoscale enables efficient use of resources, making their effect more significant. This chapter explores the use of innovative approaches such as GIS, Remote Sensing, and nanotechnology in agricultural production. It also highlights the relationships among these three technologies in the context of site-specific nutrient management. The latest advancements in these fields are also discussed in this chapter. Prospects and directions for research and adoption of these technologies for sustainable agricultural production are also considered.

2. Digital Mapping in Agriculture

Digital mapping is one of the basic tools in modern agriculture, enabling farmers, the scientific community, and policymakers to observe and manage the spatial variability present within fields or agricultural areas. Geographic Information Systems (GIS), RS, and soil sensors provide detailed, precise, and accurate information related to physico-chemical properties and other soil health parameters like nutrient status and environmental conditions. Farmers can make informed decisions with the help of prepared digital maps (Liu et al. 2021). Nutrient status of the area, along with other physico-chemical properties, helps farmers and scientists make site-specific decisions. Policymakers and marketing agencies can also make decisions regarding product sales and the import/export of inputs (Bongiovanni and Lowenberg-DeBoer 2004).

2.1 Geographic Information Systems (GIS)

A Geographic Information System (GIS) is a science or system that helps in the collection, analysis, storage, and visualization of spatial data. Attribute data is indicated with reference to spatial data. Manifestation of attribute data with spatial data helps to make site-specific decisions. GIS-based maps highlight variability within a field, such as differential nutrient status, soil texture, and moisture content (McBratney et al. 2005). Digital maps generated through GIS are vital for precision agriculture and serve as the basis for making site-specific, informed decisions. These maps lead to the application of inputs like fertilizers—where, when, and how much is needed ((PDF) Site Specific Nutrient Management for Precision Agriculture, n.d.). It has been estimated that the application of innovative approaches using GIS and RS has led to increases in crop yield by up to 30% and reductions in input costs by up to 20% (Denton et al. 2017).

Geo-statistics coupled with GIS help to encounter the variation due to spatial variability, which was not possible through classical statistical approaches, which take into account the local variability. It had previously led to losses in terms of inputs and yield. For example, the blanket fertilizer recommendations were usually made without taking care of spatial variability, which has led to fertilizer wastage. The appropriate term used for nutrient management using GIS is Site-Specific Nutrient Management. Site-specific nutrient management (SSNM) is becoming increasingly essential in modern agriculture due to the growing need for sustainable farming practices that optimize resource use while maintaining crop productivity. Traditional nutrient management methods often apply fertilizers uniformly across large fields, leading to inefficiencies, over-application, or under-application in certain areas (Karmakar et al. 2024). Digital mapping technology, coupled with precision agriculture techniques, provides a powerful solution by enabling farmers to tailor nutrient applications based on the specific needs of different parts of their fields. This chapter explores how digital mapping supports SSNM, the technologies involved, and the benefits for modern agriculture. Various steps involved in the site-specific nutrient management are as under:

2.1.1 Data Collection and Analysis

Data collection from multiple sources is the first step for digital mapping. Data is collected through multispectral and thermal sensors, which capture detailed images of fields showing variation in soil moisture, crop health, and nutrient status. These sensors are mounted on a drone (Shao et al. 2022). Satellite remote sensing provides imagery on a large scale and high resolution, which is analyzed and further processed using GIS algorithms. Analysis of imagery obtained from various sources using GIS leads to detection patterns of nutrient deficiencies. Soil sensors employed in various corners of the field monitor soil pH and moisture content and feed real-time data to generate digital maps, making data-driven agriculture possible.

The hyperspectral imaging at high resolution is still a very young instrument, and it just appeared rather recently, but it seems to be rather powerful, especially for investigations on the deeper level of nutrition. Based on the crop reflectance spectra, it is able to differentiate between several kinds of nutrient deficiencies, which results in a much higher accuracy in the issues identified at the μ level. Hence, ground-based sensors are used as weather monitoring stations and EC meters to produce local and real-time analysis of salinity, temperature, and rainfall of the soil. This proposition of McCall and Molyneux is an integrated approach that takes into account the macro and micro differences of the soils and crops across the whole production line. Once the data is collected, it is analyzed and superimposed using advanced GIS software, enabling the creation of complicated digital maps. Following the geographical examination of these maps, statistics are used to determine specific segments within the field that require certain actions. As more and more machine learning is integrated into GIS, it is now possible to predict trends and come up with recommendations for the application of fertilizers using historical and current data. Such comprehensive data collection includes not only the physical characteristics of the farm, such as size, types of crops produced, and availability of capital, but also the socioeconomic characteristics. This makes it possible to design Specific Nutrient Management Plans down to the individual field and farmer, taking into account the peculiar situation that he or she is in. Cooperating with GIS, remote sensing, and ground sensors, the process of data collection and analysis is accurate, comprehensive, and adaptive. As such, the farmers have knowledge that can be applied to real life, hence effective and sustainable nutrient management.

2.1.2 Creation of Nutrient Maps

Maps indicating differential nutrient status are the key component of SSNM. These maps divide the field into zones indicating the nutrient status of the field or agricultural produce. Maps can be generated on a large scale, like continents, countries, or districts. Small-scale maps indicating the nutrient status and other physicochemical properties of the field may also be generated. Observing these maps, farmers can understand the variability of different parameters in the field (Leena et al. 2020).

The process of generating nutrient maps involves several steps: the use of remote sensing data, soil sensors, and laboratory analysis. Satellite imagery offers broad estimates of nutrient status at a regional scale, while drones equipped with multispectral cameras offer detailed information at the field scale. The integration of multiple sources is possible, and it leads to the creation of highly accurate layered maps. Interpolation methods such as kriging and inverse distance weighting (IDW) are applied to estimate nutrient levels in areas not sampled. This also makes it possible to capture areas that have limited statistical data points to work with. They also reveal the areas of failure or abundance, and therefore one can organize ways to close the gap (Tunçay et al. 2016). Today, there are also dynamic map applications that allow for updates of the map on a real-time basis. For example, soil sensors based on IoT send nitrogen, phosphorus, and potassium levels to map the area continually and input the data into the mapping program automatically. These produce "live" nutrient maps that capture the most current situation at the field level. It is quite impressive to observe such numbers that represent the level of precision and responsiveness at the same time. The maps can also be of much importance to precision agriculture; therefore, farmers are allowed to apply the inputs in a variety of ways in zones. They operate in conjunction with VRT to ensure that the applied fertilizers, herbicides, and water will be used where they are needed and not a drop more, and thus save wastage. Also, maps are often composed of layers including topographic, soil, and agricultural production yields, etc. These multi-dimensional maps provide overall details of field variability and help and assist farmers not only in decisions on nutrients but also on irrigation, crop density, and insect control. The GIS platforms that are easy to use are now available, and so nutrient mapping can be conducted even by small-scale farmers. Viewing and decision-making on site have become possible with portable devices and smartphone applications. This way, a farmer will be in a position to act on-site, considering real-time information. In other words, nutrient maps are exciting innovations that act as a link between data accumulation and actual farming activities. Due to their ability to indicate regional differences, they enable efficient and appropriate application of resources in sustainable and productive agricultural systems.

2.1.3 Variable Rate Technology (VRT)

Practical application of inputs—where, how much, and when needed—is possible through VRT. Generated maps are uploaded to farm machinery like tractors, drills, or other spreaders that contain VRT. This VRT-enabled machinery makes fertilizer applications according to the demand or need of zones delineated on the digital maps. The judicious use of the Global Positioning System (GPS) by VRT technology and digital maps helps control fertilizer application to fields. Movement of machinery through the field using these technologies ensures the application of fertilizer to places where needed. The main benefit of VRT is that it enhances fertilizer use efficiency. Previously, over- or under-fertilization of crops due to uniform recommendations was common. Variable Rate Technology reduces these losses and thus ensures crop health through the application of site-specific fertilizer and other inputs (Pawase et al. 2023).

VRT operates in two main modes: the map-based application and the sensor-based application. The map-based system utilizes digital maps produced prior to application to provide the machines with input rate directions for each zone, ensuring that delivery is accurate. The sensor-based strategy, on the other hand, applies real-time detection of field factors such as soil nutrient contents and crop canopy reflectance to alter the suggested input rate during the course of application. VRT includes irrigation, herbicide, and insecticide application among its uses. For example, variable rate irrigation has a mobile application that uses soil moisture data to deliver water efficiently and minimize excess usage. In the same manner, herbicide application can be done selectively so that only areas most affected by weeds are treated, thereby reducing environmental chemical load. VRT benefits are quite high really when it comes to the economic thought process because it reduces input cost by minimizing wastes while improving yield from available resources. Several surveys have indicated that VRT can reduce fertilizer costs by as much as 20–30%, the results having been improved through proper accuracy in nutritional imbalance. VRT supports the culture of sustainable farming since the incidents of nutrient-polluting water sources are minimized. The two factors that contribute most to eutrophication and water pollution, namely nitrogen and phosphorus applications, are significantly reduced by VRT. The opportunity to incorporate VRT with self-propelled equipment and, more importantly, with artificial intelligence is vast. Autonomous tractors will perform operations following pre-planned routes and apply inputs on the field without the need for a human operator. The analysis of field data by AI will generate recommendations to make VRT systems smarter and more optimal. That is why VRT is widely appreciated—it can scale from small tractors to larger machines, and its modular design allows integration with existing equipment. Any small-scale farmer can afford this technology by incorporating it at a basic level using simple GPS, while large-scale commercial farmers can fully automate and integrate it into their systems. In conclusion, it is evident that Variable Rate Technology is one of the major leaps in modern agriculture that enables farmers to apply resources in ways that are precise, efficient, and environmentally sustainable. The increasing use of its capabilities underscores its role in advancing sustainable and productive agricultural systems.

2.1.4 Continuous Monitoring and Adjustments

It is worth mentioning here that digital mapping is not a one-time process; it requires continuous monitoring and adjustments throughout the growing season, as seasonal variation of nutrient content in crops is common. Nutrient demands of various crops differ at different growth stages. Again, soil sensors, drones, and satellite imagery help farmers and the scientific community adjust and monitor management strategies as needed. Farmers are able to address temporal nutrient deficiencies through the timely application of inputs (Yang 2020).

Ongoing assessment will help track the regularly changing crop and soil nutrient status. For example, a crop may need more nitrogen during the vegetative stage than during flowering and fruit-forming phases, which require more phosphorus and potassium. Growth and yield can be optimized by regularly updating digital maps based on real-time information. Hyperspectral and thermal imaging remote sensing provide sophisticated and processed data about crop health and nutrient content. They enable early detection of nutrient stress, allowing correction before deficiencies affect yield. Additionally, drones equipped with real-time sensors help evaluate large areas, as they can carry heavy equipment and deliver high-resolution information suitable for fast decision-making. Soil sensors are essential for monitoring important parameters that remain relatively constant over time, including soil pH, electrical conductivity, and nutrient concentrations. These sensors are often integrated with cloud-based interfaces, allowing farmers to monitor real-time data via applications on portable devices or computers. Alerts and recommendations are sent to farmers, enabling rapid action. Irrigation practices are also influenced by temporal changes in management. A weather station, along with moisture sensors, provides data on evapotranspiration rates and soil water content. This facilitates the simultaneous application of water and fertilizers—referred to as fertigation—to maximize the efficiency of both inputs. Fertigation minimizes nutrient leaching and improves total resource useIncreasingly, continuous monitoring systems are based on artificial intelligence (AI) and machine learning. Using historical and current data, these systems enable predictive rather than reactive management of fertilizer availability and crop requirements. They can suggest nutrient management adjustments based on weather forecasts, soil conditions, and crop growth stages, ensuring resources are applied at the most effective times and locations. Moreover, these innovative systems support environmental sustainability by enhancing fertilizer use efficiency, reducing nitrogen losses, and limiting nutrient runoff. When crop nutrient requirements are optimally met and soil supply is enhanced, farmers can increase yields while reducing environmental impacts. Therefore, ongoing review and adjustments are vital for accomplishing an effective system of site-specific nutrient management. Applying state-of-the-art technology, fresh data, and adaptive management approaches ensures that crops are properly managed throughout the growing season, improving yield and sustainability.

2.2 Benefits of Digital Mapping for SSNM

2.2.1 Enhanced Nutrient Efficiency

Precise and accurate application of nutrients to crops based on site-specific needs leads to enhanced nutrient use efficiency. Fertilizer losses are reduced. Moreover, better crop health and improved crop yield due to enhanced nutrient efficiency are expected. This can be achieved by soil sampling and digital mapping of the field to identify varying soil fertility levels and the corresponding nutrient needs across the field. Soil-targeted fertilizer application ensures that nutrients are applied where needed and in the right quantity, reducing waste and minimizing the transport of excess nutrients to water bodies. It also helps prevent the release of greenhouse gases associated with unnecessary fertilizer use. Enhanced utilization reduces the amount of nutrients that need to be borrowed or purchased by farmers, thereby lowering production costs (Abiri et al. 2023). Currently, nutrient application can be optimized so that the average amount of fertilizer used is minimized while productivity is maintained or increased. Furthermore, high fertilizer efficiency contributes to the long-term improvement of soil nutrient reserves. Effective nutrient management avoids the accumulation of specific nutrients that could otherwise degrade soil quality and impair future productivity. Integrating real-time sensor data with GIS concepts in nutrient management decisions improves nutrient management efficiency. With this, farmers can optimally use available resources and contribute to sustainable farming through constant monitoring of soil nutrient levels.

2.2.2 Environmental Sustainability

Previously, over-fertilization of crops remained an environmental challenge, leading to problems like water pollution (i.e., eutrophication). Site-specific nutrient management (SSNM) following digital mapping reduces the chances of nutrient runoff and leaching and thus minimizes the risk of pollution in surface water bodies and the groundwater table (Rafique et al. 2023). A reduced environmental footprint, sustainable agricultural operations, and improved soil health are the main advantages of SSNM. It also helps reduce soil degradation. Some known effects of excessive fertilization include soil acidification, salinization, and changes in nutrient content. In digital mapping, nutrient application is optimized, soil health and fertility is enhanced, and thus yields that do not deplete the soils fertility are encouraged. Additionally, SSNM assists in lowering greenhouse gas emission connected to agricultural business. Nitrogen not used by the crops can evaporate as nitrous oxide (N_2O), a greenhouse gas four hundred times more effective than CO_2. Since SSNM synchronizes the use of fertilizers with crop needs, this means that the total emissions will be lower, and hence the global climate change goals will be met. This is achieved through the incorporation of GIS and remote sensing that encourages consistency in the evaluation of resource use and hence sustainability. These technologies give practical recommendations that enable farmers to use techniques that encourage the use of resources and reduction of wastage, and ensure the enhancement of the balance in the environment to enhance sustainable farming in the future.

2.2.3 Cost Savings

Input costs are reduced through fertilizer application guided by digital mapping in SSNM. Unnecessary fertilizer use is minimized. Farmers save money by applying fertilizers where and when needed, which also reduces labor, fuel, and other costs associated with multiple applications typical of traditional practices used in many agricultural countries. The time required to apply fertilizers is also reduced with farm operations using VRT. SSNM, other than decreasing the price of fertilizer, also cuts down the labor, fuel and operating expenses related with other procedures of fertilizer application. VRT make it easier since there are few applications and no intervention, which saves time and labor. For example, a single well thought out VRT application can do the job of numerous runs across the field while consuming less fuel and causing less wear on the machinery. The time can therefore be utilized by the farmers to engage in other important farming activities that enhance overall productivity in the country.

2.2.4 Increased Crop Productivity

Optimal nutrient availability through SSNM and digital mapping leads to better crop growth and improved productivity and produce quality. Site-specific nutrient management promotes sustainable soil health, which in turn supports higher and more sustainable crop production. Digital nutrient mapping ensures nutrients are applied in accurate amounts, preventing both excess and deficiency, which can hinder plant development. Crops grown under SSNM are more uniformly nourished, less susceptible to pests and diseases, and of higher quality in terms of dietary value and marketability. In addition to higher yields, SSNM contributes to improving the medium- and long-term health of the soil. Nutrient imbalances—often caused by over-fertilization—are significantly reduced through SSNM. The resulting improvements in soil structure and nutrient cycling mean crops will continue to benefit from the soil's fertility and health in future growing seasons. Targeted nutrient supplementation under SSNM not only boosts yields in the short term but also supports resilient agricultural systems over the long run, benefiting both farmers and the broader agricultural sector (Chivenge et al. 2022).

2.3 Challenges in Implementing Digital Mapping for SSNM

Although significant advantages are offered by the SSNM technology, some challenges also exist for the adoption of this technology by the farming community. One of the main limitations for the adoption of SSNM is high cost. Capital equipment like drones, sensors, and VRT-mounted machinery is very difficult for small-scale farmers to afford. However, initial costs are decreasing as technology advances. Technical expertise and farmer training to adopt this innovative technology—i.e., to interpret maps and adjust nutrient management strategies accordingly—is another barrier (Ahmed et al. 2021). Moreover, the accuracy of datasets obtained from RS technologies may produce inaccurate information. Noise in the satellite imagery due to cloud cover, sensor malfunctioning, or adverse weather conditions is a known issue. Continuous developments in sensor technology and machine learning are helping to tackle the limitation. However, more research is required in this direction yet.

2.3.1 Remote Sensing

Remote sensing is a science that helps in getting information about objects without being in contact. Sensors mounted on drones or satellite imagery are used to get images and data of the area under observation from a distance. Crop health indicators are detected, like vegetation vigor, chlorophyll content, plant height, and canopy temperature, which are often not visible to the naked eye (Thenkabail et al. 2014). Data obtained from RS can be processed by making use of GIS to get meaningful information in terms of generating dynamic digital maps, which in turn provide real-time understanding of crop growth and development. Soil conditions are also observed. The use of GIS and RS together assists the farmers in detecting problems such as nutrient deficiencies or insect-pest attacks at an early stage through timely interventions.

2.3.2 Soil Sensors

Devices that are placed in the soil directly to estimate pH, moisture content, and nutrient status are called soil sensors. These sensors ensure regular real-time data, which is helpful for adjusting agricultural activities like irrigation and fertilization. Soil sensors can help detect the areas within the field that are suffering from drought stress. Nutrient deficiencies can also be detected on a site-specific basis. Farmers can apply fertilizer or irrigation to areas as suggested by the digital maps, which in turn get data from soil sensors. Irrigating areas suffering from drought stress will lead to enhanced water use efficiency. Fertilizing areas suffering from nutrient deficiencies will result in enhanced nutrient use efficiency (Mulla 2013). It has been observed that irrigation of crops in precision agriculture employing soil sensors leads to reduced water usage by up to 40% and enhanced nutrient use efficiency by up to 25% (Faqir et al. 2024).

2.4 Hyperspectral Remote Sensing and Vegetation Indices for Efficient Nutrient Management

The limitation of satellite remote sensing imagery is noise in images and the greater swath of the satellite, which leads to less detail of the area under observation. This limitation is solved by adopting hyperspectral remote sensing. Hyperspectral remote sensing captures images in many—i.e., hundreds of—narrow bands, which leads to more detailed and accurate information in different wavelengths, i.e., visible to near-infrared and shortwave infrared. Conventional imagery obtained from multispectral sensors measures broad wavelength bands, while hyperspectral sensors give a more detailed spectral signature for every pixel, permitting more accurate and precise observation of changes in crops (Wang et al. 2024). Nutrient deficiencies, moisture stress, plant diseases, and insect-pest infestations that affect crop growth negatively are more accurately detected by using

Table 1. Determination of vegetation indexes and nutrients by using remote sensing.

Vegetation Index	Bands Used	Primary Application	Associated Nutrient	References
Normalized Difference Vegetation Index (NDVI)	Red, Near-infrared	Measures plant health and vigor	Nitrogen	(Tucker 1979)
Chlorophyll Index (CI)	Red-edge, Near-Infrared	Estimates chlorophyll content	Nitrogen	(Gitelson and Merzlyak 1994)
Soil-Adjusted Vegetation Index (SAVI)	Red, Near-Infrared	Minimizes soil reflectance effects	General	(Huete 1988)
Photochemical Reflectance Index (PRI)	Green, Blue	Estimates photosynthetic activity	Nitrogen, light use efficiency.	(Gamon et al. 1992)

hyperspectral imagery compared to the broad wavelength imagery obtained from satellite remote sensing (Senbagavalli et al. 2023, Tucker 1979). Chlorophyll content can be successfully detected, which in turn can be closely related to nitrogen content in plants. Nitrogen is one of the major nutrients that plays an important role in the yield and quality of crops. Deficiency of nitrogen leads to decreased yield. Over-application of fertilizers also leads to crop damage and environmental issues. Processing of images obtained from hyperspectral remote sensing can help in the early detection of nitrogen deficiency. Here is the determination of vegetation indices and nutrients by using remote sensing summarized in Table 1.

2.4.1 Vegetation Indices (VIs) for Nutrient Management

Mathematical combinations of spectral reflectance that manifest the specific characteristic features of vegetation, like chlorophyll content, leaf area index, and biomass, are called vegetation indices (VIs) (Sishodia et al. 2020). Assessment of plant health, physicochemical properties, and other soil health indicators can be done by the judicious use of vegetation indices. A total of more than 50 vegetation indices are summarized in the Table, but some commonly used vegetation indices for nutrient management include:

2.4.2 Normalized Difference Vegetation Index (NDVI)

It is important in determining the extent of health of vegetation canopy through vigor of plants and is among the most often used vegetation indices. The Normalized Difference Vegetation Index is obtained from reflectance values in the red and the near-infrared bands. This method is commonly used for estimation of nitrogen deficiencies because the amount of chlorophyll in the plants depends on nitrogen content. The NDVI measures the changes in concentration of chlorophyll, which gives the color green in plants. Consequently, nitrogen deficiency can be quantified before its manifestations appear through the use of the NDVI. This will enable the farmer to apply fertilizer with much accuracy. The same applies to NDVI in that it allows checking crop status and stress caused by other factors, including water shortage or pest infestation (Zhang et al. 2019).

2.4.3 Chlorophyll Index (CI)

Chlorophyll Index (CI) has been designed to measure or estimate the chlorophyll concentration in plants. The NIRS measures the amount of chlorophyll in the plant because nitrogen plays a major role in chlorophyll synthesis. Chlorophyll is very sensitive to the level and is extracted from the red-edge and the near-infrared bands. CI can establish that a community is experiencing a scarce supply of nitrogen well in advance of the signs being observed. CI enables farmers to set the concentration of chlorophyll by regulating the nutrient formulation to the crop plants. Therefore, it is an earlier diagnosis to reduce losses from the yield and thus increase the use of fertilizers for an environmentally friendly agricultural practice and cultivation (Samreen et al. 2023).

2.4.4 Soil-Adjusted Vegetation Index (SAVI)

Soil Adjusted Vegetation Index (SAVI) is used to provide corrections to vegetation indices for dense vegetation and soil background. Working with different kinds of vegetation and soil gathers a lot of information about reflectance due to the soil background, making the SAVI designed in a way to reduce this influence. It is more effective in diagnosing nutrient deficiencies in crops that are growing, particularly where vegetation cover is poor. SAVI then subtracts the signal from the vegetation depending on the influence that the soil nonetheless has on the reflectance and therefore the health of the plants. This type of indicator is particularly informative where an element of the signal stems from reflected soil luminosity, for instance in arid and semiarid terrain. The SAVI technology is capable of identifying nutrient-lacking nitrogen, phosphorus, and other nutrients, and thus farmers are able to apply fertilizer and concentrate on areas that need attention in large fields.

2.4.5 Photochemical Reflectance Index (PRI)

The PRI is particularly utilized to assess photosynthesis activity. Yet it reacts to changes in light use efficiency, potentially indicating an objective sign of nutrient stress, such as nitrogen deficiency. This index is obtained from the green and blue reflectance bands of remote sensing images. It offers important features of how optimally the plants under investigation can metabolize sunlight for photosynthesis. Farmers can use PRI in order to identify a shift in plants even though the nutrient deficiencies have not yet had time to manifest. This is because early implementation and diagnosis result in advanced interventions, thus enhancing utilization of the fertilizer and production of crops. PRI can be used particularly in surveying crop health and stress in various climatic conditions.

2.5 Applications of Hyperspectral Remote Sensing and VIs in Nutrient Management

Hyperspectral remote sensing is more efficient compared to satellite RS for observing changes in the physico-chemical characteristics and nutrient deficiencies in the soil. The combination of hyperspectral remote sensing and VIs permits farmers, the scientific community, and other stakeholders to efficiently manage nutrients. The data and information provided by this integration of hyperspectral remote sensing are real-time and spatially detailed on crop health. More accurate site-specific nutrient applications are possible, which in turn reduces the wastage of resources and improves crop yield (Haboudane et al. 2008).

2.5.1 Early Detection of Nutrient Deficiencies

Subtle changes in the plant reflectance, which are governed by nutrient deficiencies before the symptoms are observed by the naked eye, can be detected by hyperspectral remote sensing. Vegetation indices like NDVI and CI derived from the imagery generated by hyperspectral remote sensing are more efficient in detecting temporal nutrient deficiencies. This is a proactive approach in which blanket fertilizer application is reduced and nutrient use efficiency is enhanced.

2.5.2 Optimizing Fertilizer Application

Spatially variable maps generated for physico-chemical properties and nutrient status using hyperspectral remote sensing permit variable rate technology (VRT) to be utilized for nutrient application to crop-growing areas according to need and demand. By mapping nutrient variability within fields, hyperspectral data allows for variable rate technology (VRT) to be employed in fertilizer application. VRT adjusts the rate of fertilizer application based on the specific needs of each field section, preventing over-application in areas with sufficient nutrients and under-application in deficient zones. This precision reduces environmental pollution caused by nutrient runoff and increases overall crop productivity.

2.5.3 Monitoring Crop Stress

Such stresses can include, for example, water stress, diseases, insect pests, and nutrient stress. For instance, clear symptoms of stress can be identified from vegetation indices, such as the Photochemical Reflectance Index, which is associated with photosystem activity. When photosynthetic efficiency is low, as detected through PRI, there is likely to be either water stress/deficit or nutrient stress/deficit, and corrective actions can then be taken. Such stressors can be identified before they cause significant damage and thus prevent yield and economic loss through interventions such as altering the frequency of irrigation or using pesticides. It is also useful for acquiring multi-scale plant health information at different spatial resolution levels, which is valuable when identifying stress-affected regions within the field. This can enable farmers to target only the affected parts of the field, thus saving input costs and avoiding negative environmental effects that would result from uniform application across the entire field. In conjunction with other forms of hyperspectral remote sensing, Phytospecs has developed thermal imaging that would augment the extraction of accurate stress detection and consequently refine nutrient management.

3. Nanotechnology for Efficient Nutrient Management

Manipulation of materials to the nanoscale is possible through the science of nanotechnology. When the size of a material is reduced to the nanoscale, unique physico-chemical characteristics appear. It is observed that the application of nanotechnology in the agricultural sector can revolutionize nutrient use efficiency and effectiveness through the development of nano-fertilizers, smart delivery systems, and nanosensors.

3.1 Nano-fertilizers

It has been observed that reduction in the size of fertilizer before application to soil enhances its activity and effectiveness. Nanotechnology-based fertilizers are known to be more efficient. Higher nutrient uptake by plants is well documented. Nano-fertilizers are known to release nutrients in a well-controlled manner compared to traditional fertilizers, which are readily soluble and more prone to leaching and runoff losses (Nazeer et al. 2023). Slow-release fertilizers ensure nutrient availability to crops for the whole growing season, leading to enhanced nutrient use efficiency and reduced nutrient losses. It is estimated that the use of nano-fertilizers in wheat production leads to an increase in yield of up to 15% while decreasing nitrogen losses by 40% (DeRosa et al. 2010).

3.2 Smart Delivery Systems

Delivering nutrients and other inputs according to need requires a sophisticated delivery system. Smart delivery systems use more advanced and efficient technologies to provide nutrients and other inputs to crops, according to particular environmental triggers like soil moisture and pH. These systems utilize the science of nanotechnology to increase precision and accuracy (Singh et al. 2024). Examples of smart delivery systems could be systems containing nano-based nutrients that are released only when the moisture level in the soil falls to a particular critical value. Smart delivery systems have the potential to revolutionize nutrient management by minimizing waste and maximizing efficiency, ultimately leading to more sustainable agricultural practices (Subramanian et al. 2015).

3.3 Nanosensors

Tools to measure or detect the changes in surroundings are called nanosensors. Nanosensors are employed in the agriculture sector to observe or detect changes in soil health indicators, nutrient status, and stresses due to drought or insect pests. Nanosensors provide real-time information to growers about the need for nutrients before the appearance of visual symptoms. Precise and timely information is available to farmers, and they can make well-informed decisions in time for the application of agricultural inputs. According to estimates by scientists, the use of nanosensors in agriculture could enhance crop yield by up to 20% (Solanki et al. 2015).

Nutrient management through nanotechnological interventions leads to enhanced nutrient use efficiency, reduced environmental footprint, and improved crop health.

4. Integration of Digital Mapping and Nanotechnology

Precision agriculture is becoming more advanced by using the combination of digital mapping and nanotechnology. Detailed spatial data and information provided by hyperspectral remote sensing and digital maps generated through GIS, and the use of nano-fertilizers through nano delivery systems for targeted delivery of agricultural inputs, allow farmers to implement site-specific nutrient management according to the status of the field (Shaw and Honeychurch 2022a, Gogos et al. 2012).

For instance, digital mapping can lead to maps indicating differential nutrient status in the soils or areas where nutrient deficiencies are appearing. Nanotechnology-based fertilizers could be applied through nano delivery systems more precisely and accurately. Enhanced nutrient use efficiency, better crop health, and reduced environmental pollution are guaranteed through integrated use of digital mapping and nanotechnology. According to a study carried out to examine the effect of integration of digital mapping and nanotechnology in rice production, an increase in rice yield of up to 25% and a reduction in fertilizer use were observed as compared to conventional approaches utilized for fertilizer application. Similarly, the combination of digital mapping and nanotechnology may help to control insect-pest attacks, diseases, and enhance water use efficiency (Shaw and Honeychurch 2022b).

5. Challenges and Future Directions

Besides the potential of digital mapping and nanotechnology for site-specific, efficient nutrient management (Lowenberg-DeBoer and Erickson 2019), there are several challenges to be addressed before its widespread adoption which are discussed below.

5.1 Cost and Accessibility

In addition to the potential that digital mapping and nanotechnology hold for efficient nutrient placement given a location, several challenges need to be addressed before these two are adopted. Technical, economic, and regulatory issues are evident. Probably the most significant technical issue is the separation of information into multiple categories including satellite imagery, hyperspectral sensors, and soil sensors, and the development of a single, usable platform to which farmers have access. These technologies generate huge volumes of data, which require processing capacity that may not be easily accessible to many farmers, especially those in the developing world (El-Ramady et al. 2023).

The implementation of such advances also comes with financial limitations because the initial cost for equipment like drones, sensors, and technical software is capital-intensive for most smallholder farms. As much as it would be cheaper in the long run, the expenses coupled with development and ongoing maintenance would prevent it from being accessible to the populace. Secondly, the skills required to take advantage of high-tech equipment are not a preserve of every farmer; this requires investment in training and other forms of support. In Saskatchewan, another area of interest is represented by the regulatory and policy matters common to all the insurance companies. The concerns arising from the use of nanotechnology in agriculture include safety, environment, and acceptance by all users. It is high time that regulations were set to ensure that those used in the agricultural sector do not pose any harm to human beings and the environment.

Future opportunities for digital mapping for nutrient management and nanotechnology may include the development of affordable and effective solutions that many regional farmers may find easy and practical to adopt. Moreover, the use of nanomaterial technologies will require follow-up research on their safety and efficiency for consistent use. Multi-stakeholder partnerships among academic institutions, government agencies, and private firms will be decisive in providing solutions to these challenges and enhancing the utilization of precision nutrient management strategies.

5.2 Technical Expertise

Successful and widespread adoption of these technologies—digital mapping and nanotechnology—requires a higher level of technical expertise and subject specialists. After proper training and capacity building of scientists, farmers should also be able to interpret complex data from GIS, RS, and other soil sensors incorporated into their fields. They also need to be trained in the basic principles of nanotechnology. Training and educating farmers and their advisors/facilitators is crucial to ensure the adoption of these technologies for sustainable agricultural production.

5.3 Environmental and Health Concerns

Although nanomaterials and smart delivery systems can play a significant role in enhancing productivity and the economical use of resources, concerns regarding human health and the environment are not well studied and need to be understood. Research work to study these concerns must be carried out to assess the risks (Medici et al. 2019).

Recently, people started to realize that the use of nanomaterials and smart delivery systems can raise productivity and resource utilization to unprecedented levels, but there is growing concern about their effects on people and ecology (Gebbers and Adamchuk 2010). Several issues have been raised over the permanence of nanoparticles in agricultural systems through the use of fertilizers, pesticides, and other agricultural inputs; their bioaccumulation in soils, water, and food chains; and their impacts on the nature and extent of biodiversity. Still, sufficient study into these areas has not been carried out, and extensive research is needed to determine all risks that may arise from the application of nanotechnology in agricultural practice (Medici et al. 2019). There is a specific need to analyze how nanomaterials interact with soil macro- and microorganisms, plants, and how they may reach water sources. The environmental behavior of nanoparticles can likely only be partially predicted because the toxicity of an NP depends on its size, shape, and chemistry. When taken up by crops, they may become part of the human food chain, which presents potential health risks.

More research is needed to assess these potential dangers in actual farm settings and to derive suitable precautionary guidelines to build a realistic safety regulatory mechanism. With this approach, the application of nanotechnology in agriculture will not exert deleterious effects on ecosystems or human health. These problems should be addressed by rigorous environmental impact assessments coupled with the proper usage conditions of nanomaterials to ensure that these technologies are effectively used in a sustainable manner. The illustration of the integration of GIS, Remote Sensing, and Nanotechnology for Precision Nutrient Management is mentioned in Figure 1.

6 Future Directions

Digital mapping, nanotechnology, GIS, and Remote Sensing (RS) seem promising for enhancing agricultural output in the long run. Future work will focus on achieving the standard practice of these aforementioned technologies and making affordable machinery and tools available to farmers, especially smallholder types existing in developing countries. This element of technology cost makes the application of precision agriculture more prevalent across the entire farming society. Researchers should look into how to develop sensors, drones, and easy-to-use software applications that are effective and durable so that farmers will be in a position to adopt these new technologies without incurring high expenses (El-Ramady et al. 2023).

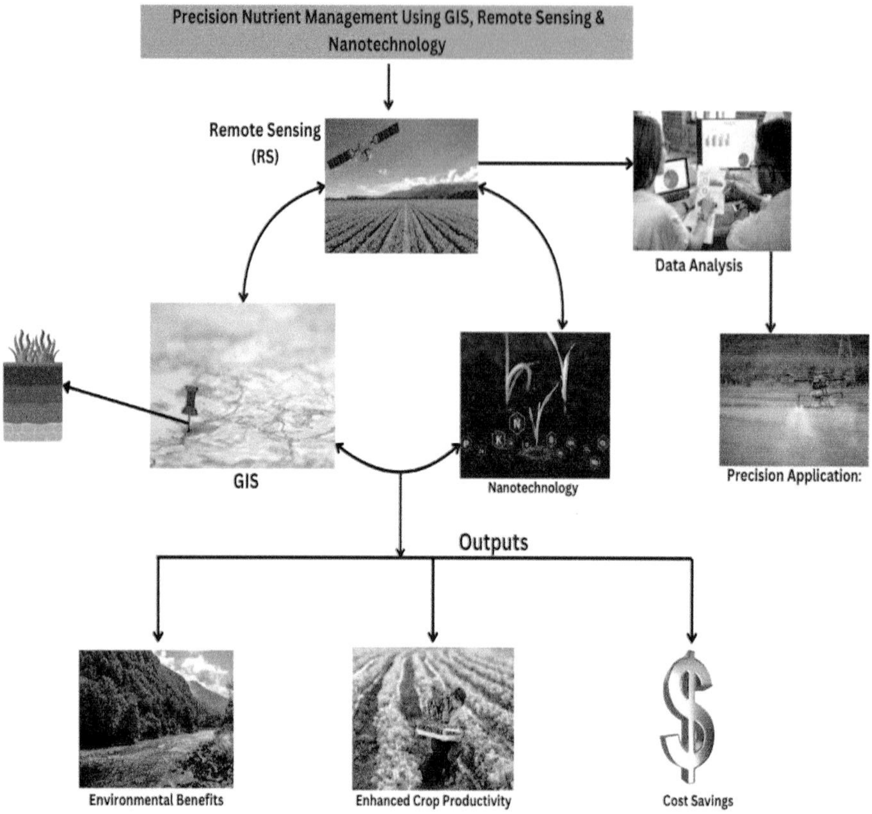

Figure 1. Integrating GIS, Remote Sensing, and Nanotechnology for Precision Nutrient Management.

However, with new technologies, more areas where current applications can be beneficial should be identified. One outstanding example is climate-smart agriculture, where applications of digital mapping technology or compact systems may be used to increase input efficiency and output in terms of climate change adaptation and mitigation (McBratney et al. 2005). They can also be used in carbon sequestration to determine where best to store soil carbon and how best to use fertilizer to promote sustainable farming.

Conservation of biological diversity might also be another applicable area for these technologies (Bongiovanni and Lowenberg-DeBoer 2004). Through the application of GIS and RS, the impacts of agricultural practices on ecosystems can be identified, and key suggestions to minimize these impacts can be provided. Such an integrated course of action might yield greater agro-productivity and protection of the natural resource base, and hence greater sustainability in the long run. In the big picture, subsequent studies are expected to increase the dissemination of such technologies in order to bring a positive impact on agricultural yields and environmental quality.

7. Conclusion

Digital mapping, GIS, RS, nanotechnology, and their integration can serve as game changers in site-specific nutrient management and precision agriculture. Use of resources like nutrients can be optimized by the farming community by adopting these cutting-edge technologies. Environmental safety and enhanced crop productivity are two great promises held by these technologies. The main challenges of high costs and data accuracy are concerns that should be resolved. Continuous evolution of precision agriculture will play a critical role in sustainable farming. Digital mapping, RS, nanotechnology, and their integration serve at the forefront of this revolutionary technology. These serve as powerful tools for site-specific nutrient management. Application of nutrients—how much, where, and when needed—can increase the yield and quality of crops on a sustainable basis. Widespread adoption of these technologies requires addressing the challenges of cost-effectiveness, technical expertise, and environmental concerns. Continuous research on the application of cutting-edge technologies in agriculture, GIS, RS, and nanotechnology will play a central role in achieving global food security (Foley et al. 2011). By explaining the power of these technologies for sustainable agricultural production, farmers will be able to meet the increasing demand for food for the rapidly expanding population while protecting the health of the planet.

8. Acknowledgement

We (Authors) are highly thankful to the University of Agriculture Faisalabad and Yangzhou University for their support in this publication. Authors also want to thank Professor Dr Abdul Wakeel from Soil Fertility and Plant Nutrition Laboratory, University of Agriculture Faisalabad, Pakistan, and Dr. Akhtar Hameed, Assistant Professor (Plant Pathology), Institute of Plant Protection for providing us this opportunity. We are highly thankful to our families who sacrificed their good times for the accomplishment of this milestone.

References

Abiri, R., N. Rizan, S.K. Balasundram, A.B. Shahbazi and H. Abdul-Hamid. 2023. Application of digital technologies for ensuring agricultural productivity. Heliyon. 9: e22601.

Adams, K., J. Carter and B. Miller. 2005. Soil fertility enhancement using organic amendments. Soil Fertil. Manag. 12: 30–40.

Ahmed, U., J.C.W. Lin, G. Srivastava and Y. Djenouri. 2021. A nutrient recommendation system for soil fertilization based on evolutionary computation. Comput. Electron. Agric. 189: 106407.

Chivenge, P., S. Zingore, K.S. Ezui, S. Njoroge, M.A. Bunquin, A. Dobermann and K. Saito. 2022. Progress in research on site-specific nutrient management for smallholder farmers in sub-Saharan Africa. Field Crops Res. 281: 108503.

Clark, D. and E. Lewis. 2018. Post-harvest losses in horticultural crops. HortTechnology 28: 250–258.

Doe, J.D. and A.B. Johnson. 2010. Impact of climate change on crop production. Agric. Sci. 72: 45–53.

Garcia, F., J. Lopez and M. Gonzalez. 2019. Integrated pest management in citrus production. J. Citrus Res. 33: 100–108.

Hallegraeff, G.M., J.A. Marshall, J. Valentine and S. Hardimen. 1998. Short cyst-dormancy period of an Australian isolate of the toxic dinoflagellate Alexandrium catenella. Mar. Freshw. Res. 49: 415–420.

Jones, M., K. Taylor and C. Thomas. 2020. Precision agriculture and its future prospects. Comput. Electron. Agric. 150: 89–98.

Lee, R., H. Kim and S. Park. 2022. Microbial interactions in the rhizosphere. Soil Biol. Biochem. 165: 112–120.

Nelson, B.C. 1995. Biotechnology applications in crop improvement. Plant Biotechnol. J. 10: 75–82.

Smith, J., P. Brown and L. White. 2001. Advances in plant disease management strategies. J. Plant Pathol. 55: 123–130.

Williams, P., T. Green and R. Black. 2015. The role of nanotechnology in sustainable agriculture. Environ. Nanotechnol. Monit. Manag. 10: 67–75.

Yang, C. 2020. Remote sensing and precision agriculture technologies for crop disease detection and management with a practical application example. Eng. 6: 528–532.

Zhang, Y., L. Guo, Y. Chen, T. Shi, M. Luo, Q.L. Ju, H. Zhang and S. Wang. 2019. Prediction of soil organic carbon based on Landsat 8 monthly NDVI data for the Jianghan Plain in Hubei Province, China. Remote Sens. 11: 1–16.

CHAPTER-12

Nano-Biochar

Applications and Implications for Soil Health

Muhammad Imran,[1] *Wasim Javed,*[2,*] *Abdul Ghaffar Khan,*[3] *Muhammad Kashif Irshad,*[4] *Muhammad Moaz Khursheed*[5] and *Sarmad Frogh Arshed*[6]

1. Introduction

Food security concerns due to the growing global population and urbanization have led to a need for productivity enhancements on agricultural land. By 2050, the world will need to produce 70% more food to feed 2.3 billion people (FAO 2009). This has led to excessive pressure on agricultural lands as intensive farming threatens the sustainable productivity of the soils due to rapid declines in soil organic matter, soil fertility, and soil health. Therefore, balancing these demands and environmental concerns remains a challenge that demands immediate attention to the efficient use of crop inputs, particularly new innovative and cost-effective organic fertilizers derived from wastes such as compost and biochar.

Biochar (BC), a carbon-rich material produced through the pyrolysis of organic matter or biomass, has garnered significant attention due to its potential for sustainable agriculture, waste management, soil enhancement, and mitigating climate change (Gyanwali et al. 2024). The agricultural significance of BC in enhancing soil fertility and crop productivity dates back to ancient civilizations (Mandal et al. 2021). It has been widely reported in the literature that BC is a promising soil organic amendment that offers benefits to improve soil fertility and health, crop productivity and mitigate environmental issues due to its unique properties (Pandian et al. 2024). In addition to its application in agriculture, the porous structure of biochar makes it useful for water treatment, enzyme immobilization, fuel additives, and photocatalysis (Gyanwali et al. 2024). However, due to the limited surface properties and multifunctionality of raw BC, recent studies on engineered nano-biochar (nBC) are being conducted to enhance these properties (Chaubey et al. 2024).

[1] Department of Soil and Environmental Sciences, MNS University of Agriculture, Multan, Pakistan.
[2] Water Management Research Center, University of Agriculture Faisalabad, Pakistan.
[3] Senior Scientific officer, Agriculture Department, Punjab.
[4] Department of Environmental Sciences, Government College University Faisalabad, Pakistan.
[5] Institute of Soil and Environmental Sciences, University of Agriculture Faisalabad, Pakistan.
[6] Department of Biochemistry and Biotechnology, MNS University of Agriculture, Multan, Pakistan.
* Corresponding author: wasim.javed@uaf.edu.pk

DOI: 10.1201/9781003558354-12

One of the emerging applications of BC is the production of nano-sized particles of nBC (particle size < 100 nm) from waste-derived raw BC as a novel engineered material. The smaller particle size of nBC significantly increases its surface area, resulting in enhanced physicochemical properties, greater efficacy, and benefits compared to conventional raw BC. During the last decade, extensive research has concentrated on converting raw BC to nBC, aiming to increase surface properties and expand its applications (Chaubey et al. 2024, Chausali et al. 2021). Therefore, applications of nBC are becoming very popular these days, as it combines the benefits of BC technology with nanotechnology.

Among the various applications of nBC materials, its soil application to improve soil health and quality is the most extensively investigated one. Recently, nBC has been explored for its great potential to enhance soil health and remediate contaminated soils. The smaller size and higher surface area of nBC enhance its adsorptive capacity, allowing it to effectively adsorb nutrients and remove contaminants from soil and water (Sani et al. 2023, Sarma et al. 2024). Additionally, the small size of nano-biochar enables it to penetrate soil pores and interact with the local microbial community, potentially stimulating microbial activity and diversity. Therefore, the advantages of a particular application of nBC depend on the unique properties of nBC.

2. Unique Properties of Nano-Biochar (NBC)

Nano-biochar inherits characteristics from its precursor raw BC, but it exhibits distinct enhanced properties compared to pristine BC. A comprehensive comparison between characteristic properties of precursor bulk BC and nBC has been presented by Chaubey et al. (2024). It was observed that nBC has superior properties as compared to its precursor BC. However, feedstock variations and pyrolysis conditions lead to significant differences in the main properties and compositions of nBC. The variations in feedstock material type can significantly affect the resulting BC particle sizes and nBC composition, including carbon, mineral, and ash contents, surface functional groups, and stability (Oleszczuk et al. 2016). For instance, nBC derived from cellulosic materials exhibits larger particle size and surface area than mineral-rich sources. Similarly, pyrolysis temperature affects the surface area, carbon content, and particle size of nBC (Anupama and Khare 2021, Xu et al. 2020). However, understanding these variations in key properties is essential for optimizing nBCs for their specific applications. As compared to precursor BC, nBC has several distinct properties as illustrated in Figure 1 and discussed below:

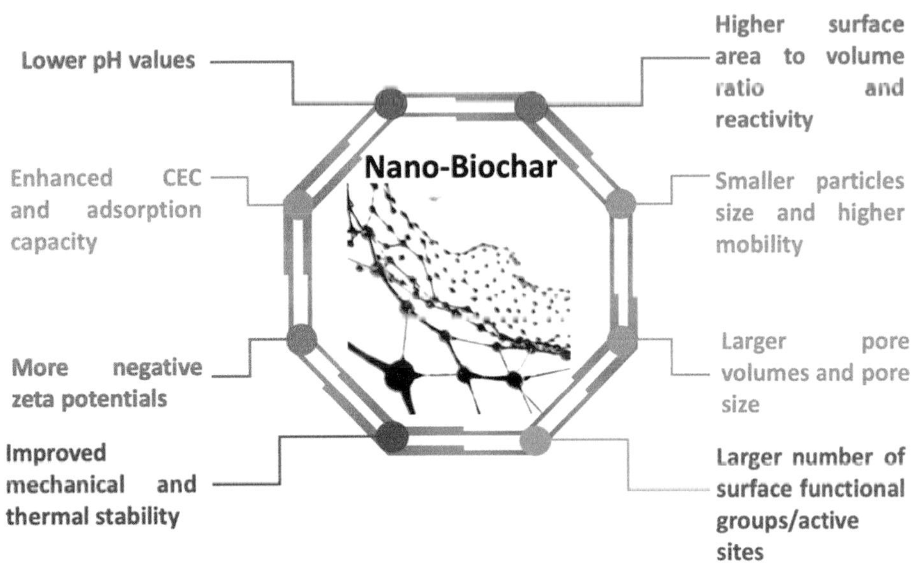

Figure 1. Unique characteristic properties of nano-biochar (nBC) compared to bulk BC.

2.1 Increased surface Area and Reactivity

The pore properties (e.g., surface area, pore volume, and pore size) of bulk-BCs and nBCs affect their various types of applications. Nano-BC has a much higher surface area-to-volume ratio than larger BC particles, enabling more efficient adsorption of nutrients and contaminants from soil and water (Xiao et al. 2020).

In literature, significant increases (up to 97-fold) in the surface area of nBC compared to bulk-BC were reported (Chaubey et al. 2024). For example, Xiang et al. (2020) found an increase in the specific surface area of nBC by 1.4–29.1 times and its adsorption capacity for pollutants by 1.3–13.0-fold. Similarly, the surface area of cow-bone BC also increased from 5.9 to 12.9 times after ball milling to obtain nano-scale BC particles, achieving 64–94% higher adsorption capacities for heavy metal contaminants (Xiao et al. 2020). The nBC particles are more reactive and have a greater tendency to adsorb protons (H^+) or react with water and other substances, due to their higher surface area and charge properties (Anupama and Khare 2021). The smaller size of nBC allows it to penetrate soil pores more readily and interact more intimately with soil particles, plant roots, and microbial communities, which can have important implications for soil health and fertility.

2.2 Smaller Particle Size and Increased Mobility

Bulk biochar, mostly used for agricultural and environmental applications, has a typical particle size of 0.04–20 mm (Gao and Wu 2014). The smaller particle size of BC in the µm range improves its ability to absorb substances, while further reducing BC particle sizes to the nanoscale (up to 100 nm) significantly enhances its characteristics, mobility, and effectiveness (Chausali et al. 2021, Ma et al. 2019).

A great variation in particle size of nBC (2–600 nm) was observed in the literature depending on the production process and raw material type used for pristine BC preparation (Anupama and Khare 2021). For example, Ma et al. (2019) reported a particle size of about 100–600 nm for nBC derived from corn and rice straw BC using a ball milling grinding technique, while Naghdi et al. (2017) obtained nBC with particle sizes of 212 nm at optimum milling conditions and as low as 60 nm with pretreatment of pinewood BC before ball milling. Moreover, nBC prepared by the sonication technique from biochar derived from various biomass feedstocks had an average particle size of about 100 nm (Lian et al. 2019, Song et al. 2019), whereas the average particle size of nBCs prepared from bulk-BC through gasification was about 70–80 nm (Goswami et al. 2020).

The smaller size of nBC particles can make them more mobile in soil, allowing for better transport, distribution, and interaction with soil components (Ma et al. 2019). However, this can also lead to potential concerns about the transport and fate of nBC in the environment.

2.3 Larger Pore Volume and Pore Size

The total pore volumes of ball-milled nBCs were significantly higher than those of their pristine BCs. It was also demonstrated that a higher pyrolysis temperature during the production of bulk BC increases pore volumes for both pristine BC and, thereby, nBC. For instance, the total pore volumes were increased from 0.017, 0.074, and 0.097 $cm^3\ g^{-1}$ of bulk-BC pyrolyzed at 300, 450, and 600°C, respectively, to 0.163, 0.367, and 0.453 $cm^3\ g^{-1}$ for ball-milled nBCs (Xiao et al. 2020). Similarly, Lian et al. (2019) also demonstrated that total pore volumes of bulk-BC derived from rice straw at 400 and 700°C increased from 0.089 and 0.101 $cm^3\ g^{-1}$ to 0.356 and 0.575 $cm^3\ g^{-1}$ for its nBC, respectively. Moreover, nBC prepared from rice hull biochar pyrolyzed at 600°C had a higher total pore volume of 0.062 $cm^3\ g^{-1}$ compared to its initial value of 0.036 $cm^3\ g^{-1}$ (Yue et al. 2019). Thus, larger pore volumes of nBC than regular BC show significant growth in surface area and pore networks by reducing the particle size.

2.4 Larger Number of Surface Functional Groups/Active Sites

The literature showed that nBC has significantly more surface functional groups including –COOH, –OH, and –CH groups compared to raw BC. For example, Song et al. (2019) determined an increase in oxygen-containing groups such as –COOH, OH, and –CH groups and aliphatic chains on the surfaces of nBC particles compared to large raw-BC particles. Similarly, Liu et al. (2018) also found higher O/C ratios of nBC compared to bulk BC, indicating the abundance of more oxygen-containing functional groups on nBC surfaces and higher surface functionality. Higher O/C ratios of nBC indicate its higher surface hydrophilicity due to the presence of more functional groups on the greater area of nBC surfaces (Lian et al. 2020). Yue et al. (2019) demonstrated that high-temperature nBC exhibits higher surface reactivity due to greater pore structure and Si content than the low-temperature.

2.6 Lower pH Values

It has been well reported that nBCs have lower pH values than their precursor BCs, which is due to the addition of oxygen-containing acidic groups (e.g. carboxyl, phenol, and hydroxyl) to nBCs during their size transformation (Chaubey et al. 2024). The larger surface area of nBC allows for greater oxidation and forms more acidic oxygen-containing functional groups on its surfaces. These acidic groups can release hydrogen ions (H^+), contributing to the lowering of the pH of nBC (Xiang et al. 2020). Xiang et al. (2020) reported that the pH values of nBCs produced from various hickory chip bulk BCs decreased by 0.9, 0.7, and 0.6 when pyrolyzed at 300, 450, and 600°C, respectively. Similarly, reductions in pH values ranging from 0.2 to 2.5 occurred depending on the pyrolysis temperature when nBC was produced from peanut shell bulk BC (Liu et al. 2018). Moreover, nBC made from wheat straw BC showed a reduced pH of 9.1 as compared to 9.9 pH for the bulk BC (Oleszczuk et al. 2016). Thus, the nanoscale sizing process significantly modifies the physicochemical properties of BC, rendering nBC more acidic than its bulk precursor.

2.7 Enhanced CEC and Adsorption Capacity

The higher cation exchange capacity (CEC) of nBC depends on the surface structure of the nBC particles with various functional groups that provide surface charges. Similarly, due to its large surface area, porosity, functional groups, and surface hydrophobicity, nBC also has increased adsorption potential for various nutrients and pollutants in soil and water (Rajput et al. 2022).

2.8 More Negative Zeta Potentials

Zeta potential measurements are used to estimate particle surface charges that show the stability behavior of the particles. High zeta potentials result in lower agglomeration and higher dispersion of nBC particles (Anupama and Khare 2021). The size reduction reduces the zeta potential of resulting nBC and makes it more negative due to an increase in oxygenated surface functional groups, which helps in removing positively charged contaminants. The zeta potential of bulk BC mostly ranges from –8 to –56 mV, which decreases to –17 to –65 mV after milling to nano sizes (Chaubey et al. 2024). Oleszczuk et al. (2016) found a significant reduction in zeta potential values from around –56 mV to a more negative –88 mV for nBCs derived from wheat straw, wicker, and miscanthus. Similarly, zeta potentials of wheat straw BC became more negative from –20.6 to –32.9 mV in nBC derived from pyrolysis at 550 °C, as reported by Wang et al. (2013). The zeta potentials of nBCs vary with particle size and type of feedstock, with nBCs being less stable than micro-BCs. For example, nBCs derived from feedstocks like wood and agricultural waste had higher zeta potential values due to more surface groups, while those from sewage sludge had the highest zeta potentials (Song et al. 2019), indicating differences in stability and properties based on the source material.

2.9 Improved Mechanical and Thermal Stability

The nanoscale size and porous structure with large surface area of nBC ensure improved mechanical and thermal stability. From a mechanical point of view, nBC has greater strength, flexibility, and resistance to mechanical stress, which provides the mechanical stability of soils and enhances soil resistance to erosion and compaction while in thermal terms, it ensures heat resistance to degradation and enhanced insulation (Liu et al. 2018). These stability traits make nBC a valuable material for agricultural and environmental applications.

3. Nano-Biochar Applications for Soil Health

The soil application of nBC is regarded as an innovative approach to enhance soil health and quality, soil productivity, and soil ecological functioning, as well as a significant tool to combat climate change issues. Nano-biochar represents a promising advancement in soil health management, supporting not only plant growth but also sustaining the ecological balance for long-term agricultural productivity and environmental health (Rajput et al. 2022).

Since nBC is derived from bulk BC, they have many common advantages for soil applications. Due to the unique properties of nBC, its soil application can more significantly provide multifaceted benefits for sustainable soil health (Chaubey et al. 2024, Pratap et al. 2021, Sarma et al. 2024), as discussed in the sections below.

3.1 nBC as Soil Amendment

The significance of nBC as a soil amendment can be discussed as its effect in amending soil physical, chemical, and biological properties, thereby the cumulative impact in promoting soil health, fertility, and crop productivity, as illustrated in Figure 2.

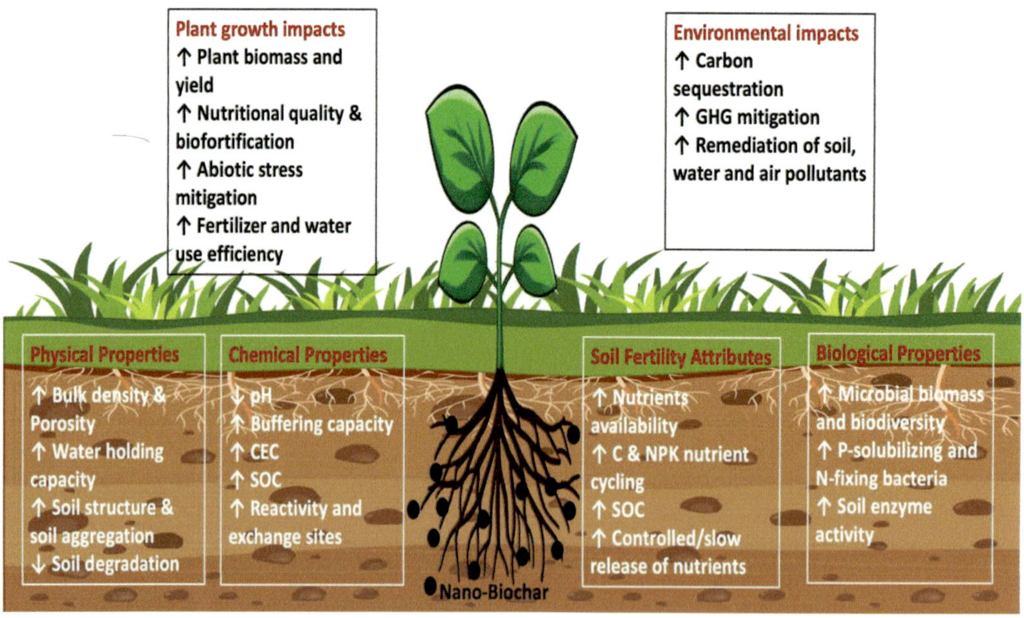

Figure 2. Effects of nBC as soil amendment on various soil attributes, environment and plant growth.

3.1.1 Impact on Soil Physical Properties

The BC advantages in soil are well established, as it improves the various physical properties of soil due to its carbon nature and porous structure (Pandian et al. 2024). Recently, nBC has been found more effective as a soil enhancer due to its improved physicochemical properties compared to bulk BC (Chaubey et al. 2024). Significant differences in particle size, larger surface area and functional groups, and more negative zeta potential of nBC are expected to impact soils significantly differently from bulk BC (Lian et al. 2019).

The most important impact of soil application of nBC amendment is the improvement of soil structure and aggregate stability (Rajput et al. 2022). These nBC particles act as binding agents to form soil aggregates, helping to stabilize soil structures and making soil less vulnerable to erosion (Chen et al. 2020). Application of nBC soil amendment can also boost soil bulk density and reduce soil compaction (Verheijen et al. 2019, Yang et al. 2020). Further, applying nBC increases soil porosity through modification of the soil pore structures by increasing micropores and decreasing macropores (Yang et al. 2020). The increase in soil porosity further improves the water retention capacity of soil, thereby reducing water runoff by up to 43% and soil erosion by 10% (Zhou et al. 2020).

Another benefit of nBC amendment can be the increase in the water infiltration rate of soil by improving pore sizes and stability of the soil aggregates (Chen et al. 2020). Verheijen et al. (2019) also reported that the application of small-sized BC improves soil water retention through increased pore volume, micropores, and uniform aggregate structure. The application of nBC is also observed to reduce the crack intensity in soil and improve soil due to the increase in bulk density and water holding capacity (Kumar et al. 2020). These studies have shown that nBC improves soil structure and aggregation, which can enhance water infiltration, soil moisture, aeration, and porosity.

The use of nano-sized nBC particles can improve the typical soil physical environment for the soil-plant system, but further long-term studies are needed to fully understand the benefits of nBC for various soil and crop management practices.

3.1.2 Impact on Soil Chemical Properties

The incorporation of nBC in the soil strongly alters soil chemistry, leading to dissolution, degradation, and formation of reactive surfaces and exchange sites. Due to these chemical properties, nBC serves as an excellent adsorbent for nutrients and pollutants in the soil profile (Sarma et al. 2024). It is also well reported that the addition of nBC significantly decreases the pH and electrical conductivity of soil. Also, the small particles of nBC considerably improve the organic matter content and saturation of soil (Rajput et al. 2022). The enhancement in exchange sites and surface functional groups of soils with the incorporation of nano-sized nBC particles greatly contributes to the sorption capacity of soil for nutrients and contaminants. Due to the nanoscale size and surface structure of nBC particles, nBC soil amendment greatly increases CEC and surface charges of amended soil (Pratap et al. 2021, Xu et al. 2020).

3.1.3 Impact on Soil Biological Properties

Soil microbial growth increases with nBC or BC supplementation, as it provides a favorable medium for the beneficial soil microbes to thrive and perform. Nano-BC can influence the soil microbiome, potentially stimulating the growth of beneficial microorganisms and altering the composition of the microbial community (Jiang et al. 2023). It greatly enhances soil biological properties, including microbial activity and diversity, which are crucial for nutrient cycling, soil fertility, and overall soil health (Udomkun et al. 2024). The porous structure, moisture content, and organic carbon of nBC can provide habitats and resources for a wide range of soil microbes, potentially stimulating their activity and diversity.

Increased microbial biomass and activity have been reported with nBC amendment, resulting in improved soil structure and resilience against degradation. Research indicates that the application of nano-biochar leads to a notable increase in microbial activity and diversity, particularly among actinomycetes and fungi, with populations rising by 19% to 38% and 3.7 to 9.3 times, respectively. This enhancement in microbial populations is accompanied by a significant increase in soil enzyme activities, including cellulase and neutral phosphatase, which can rise by as much as 328% (Sarma et al. 2024).

Similarly, Udomkun et al. (2024) also reported a significant increase in the abundance of beneficial microbial populations, including phosphate-solubilizing and nitrogen-fixing bacteria, by 1.8 and 1.9 times with nBC applications at 1% and 3% doses. This is attributed to nBC-induced impact on soil pH, carbon content, and moisture conditions, which are conducive to microbial activity. It is further found that the highly porous structure of nBC provides a habitat for microbes, promoting their growth and activity, which is crucial for nutrient cycling and soil fertility (Sarma et al. 2024).

Moreover, Liu et al. (2020) demonstrated that nBC soil application promotes the increase in microbial biomass, abundance, and diversity of the Actinobacteria and Bacteroidetes communities while reducing the Proteobacteria species. These changes in microbial community abundance are mainly attributed to nBC-induced enrichment of soil properties such as porosity, water-holding capacity, carbon content, and soil stability.

However, the effects of nBC as a soil amendment are more focused on the physicochemical properties of the soil, especially on improving soil fertility and quality. The biological effects on soil, especially on soil microorganisms and fauna, are an important aspect that needs to be further studied in the future.

3.2 nBC as Soil Fertilizer

The above discussion on the impacts of nBC on soil physical, chemical, and biological attributes has shown its suitability for soil and agricultural applications. Nano-BC, like BC derived from organic biomass, is rich in carbon and contains high concentrations of other macro- and micronutrients, making it a highly efficient organic fertilizer. nBC provides these essential nutrients to the soil solution as it decomposes over time. Moreover, nBC can effectively regulate the release, absorption, and mobilization of essential nutrients in soil (Khan et al. 2021).

Soil application of nBC has attracted great interest due to its higher nutrient-holding capacity and can be simply used as a nanofertilizer. Application of nBC amendment not only provides nutrients to soil but also significantly reduces nutrient loss from soil, thereby increasing nutrient use efficiency due to its high CEC, surface area, and water retention and adsorption capacities (Chen et al. 2020). For example, Zhou et al. (2020) reported that application of nBC (1%) reduced up to 43% nitrate (NO_3) loss from a sandy soil. A similar reduction in P and K loss through runoff was also reported with the application of nBC (Yang et al. 2020).

The nBC interactions with soil minerals have been described to enhance the mobilization of nutrients such as N and P (Mahmoud et al. 2020). Similarly, application of nBC also improves soil K concentration, as K in feedstock biomass is preserved in the resulting BC. nBC particles also adsorb K and P ions from the mobile soil solution pool onto their surfaces, improving their availability by making them accessible to plants for longer periods. The unique properties of nBC can also enhance nutrient cycling in the soil. By increasing CEC and adsorbing nutrients, nBC can improve the availability of essential plant nutrients (Jiang et al. 2023).

Several studies exhibited slow-release patterns of nutrients over a prolonged period, making nBC suitable as a slow-release fertilizer for sustainable soil fertility (Das and Ghosh 2023, Khan et al. 2021). Lateef et al. (2019) found that BC nanoparticles are efficient at slowly releasing cation and anion nutrients, supplying nutrients according to the growth stages of growing plants.

It is found that the highly porous structure of nBC provides a habitat for microbes, promoting their growth and activity, which is crucial for nutrient cycling and soil fertility. Sarma et al. (2024) demonstrated that an increased surface area and porosity of nBC-amended soil enhance its nutrient retention and slow-release capabilities, reducing nutrient leaching. The application of nBC has been shown to maintain acceptable levels of exchangeable nutrients while increasing pH and CEC, thereby promoting plant growth by up to 30%. Therefore, the use of nBC can be a more efficient and cost-effective approach for nutrient application in the soil-plant system compared to conventional BC, requiring less material for better results from its sustainable soil application.

3.3 nBC in Mitigating Soil Degradation

The application of nBC has the potential to mitigate soil degradation, which is a growing concern in many regions due to factors such as erosion, runoff, salinization, and desertification. For example, Chen et al. (2020) reported that coconut-derived nBC significantly decreased soil erosion (10–11%), runoff (40–74%), sediment yield (9–42%), and nitrate loss (14–60%). Similarly, Kumar et al. (2020) also found that the use of chalk and wheat-derived nBC in soil helps to retain more water, reduce evaporation, and prevent cracking in soil better than traditional methods, attributed to the higher porous structure and water retention in nBC materials.

3.4 nBC in Remediation of Soil Pollutants

Nano-BC, an emerging material in environmental remediation, offers a promising solution for reducing soil pollutants. With its nanoscale size and unique physicochemical properties, nBC enhances the adsorption and immobilization of various soil pollutants, including heavy metals, organic compounds, and pesticides, thereby reducing their availability and toxicity to plants.

3.4.1 Remediation of Heavy Metals

Soil metal contamination is a major concern because these metals can be absorbed by edible plants, accumulate in plant tissues, undergo biomagnification, and ultimately enter the human food chain, which can pose risks to human health. Among the various remediation strategies, carbon-based organic nanomaterials offer an emerging and innovative solution for environmental remediation of pollutants. Among them, BC or especially nBC is gaining attention due to its eco-friendly properties and benefits associated with nanotechnology. Toxic elements such as cadmium (Cd), lead (Pb), chromium (Cr), arsenic (As), and mercury (Hg) are particularly harmful, affecting a wide range of living things, including bacteria, plants, animals, and humans. Although BC has commonly been used to reduce metal toxicity, its effectiveness is limited due to the lack of nano-specific properties. Therefore, conversion of BC to nBC has proved more effective in increasing the remediation potential of soil pollutants.

Research shows that nBC reduces toxic heavy metals in soil by selectively binding metal ions through mechanisms such as surface ligands, electrostatic interactions, and ion exchange (Liu et al. 2020). Nano-BC with surface functional groups that resist protonation, or with negatively charged surfaces, has a high adsorption capacity due to the positive charge of heavy metals. Different studies revealed that the use of nBC is a very effective method for the adsorption of heavy metals. For example, Liu et al. (2020) reported that the soil application of nBC significantly reduced the Cd concentration by 86–95% in the plant tissues by converting the exchangeable and bioavailable Cd into an immobilized form. Similarly, Yue et al. (2019) also found reduced Cd uptake and phytotoxicity in the rice plant due to nBC's high affinity and adsorption for soil Cd and increased antioxidative enzymes in amended soil.

3.4.2 Remediation of Organic Contaminants

The unique properties of nBC make it extremely effective in adsorbing a variety of organic pollutants, including pesticides, polycyclic aromatic hydrocarbons (PAHs), polychlorinated biphenyls (PCBs), and dyes (Anupama and Khare 2021). When nBC is combined with the photocatalyst zinc oxide (ZnO), it can also promote the indirect degradation of phenol. This effect is attributed to the limited number of active surface sites and the low energy required to form adsorption bonds within the nBC-ZnO composite system (Chen et al. 2016). Xu et al. (2019) observed that N-doped nBC, a new form of nBC, exhibited an adsorption capacity for reactive red that was 18.8 times higher than that of bulk BC and twice that of regular nBC. This improvement is attributed to the N-doping process, which increases the zeta potential and point of zero charge of the nBC, thereby enhancing the adsorption of organic pollutants (Xu et al. 2019).

3.4.3 Remediation of Emerging Pollutants

Emerging pollutants include a variety of organic compounds that are highly polar, water-soluble, and resistant to biodegradation. These include pharmaceuticals, antibiotics, personal care products, fragrances, microplastics, industrial nanoparticles, dyes, detergents, and other micropollutants (Jiang et al. 2023). Sun et al. (2022) prepared ball-milled BC by using wheat and rice straw husk at different pyrolysis temperatures (300, 500, and 700°C) to assess the degradation of tetracycline and greenhouse gas (CO_2 and nitrous oxide) emissions to the atmosphere. The results showed that after 45 days of the incubation study, there was a slight decrease recorded in the pH of the soil where wheat straw BC was used as an amendment. Ball-milled BC that was pyrolyzed at 500 and 700°C significantly increased the degradation of tetracycline. The maximum tetracycline degradation (96%) was observed at a temperature of 500°C for wheat straw BC. Overall, the use of ball-milled BC from both straws improved microbial activity in soil and also reduced greenhouse gas emissions, especially CO_2 and nitrous oxide, to the atmosphere from different nitrogenous fertilizers.

3.4.4 Mechanisms of nBC Involved in the Remediation of Soil Contaminants

Figure 3 shows the various physicochemical interactions of the nBC and its mechanisms for pollutant immobilization and removal. The adsorption mechanism of nBC involves various processes, including physical adsorption, electrostatic interactions, complexation, precipitation, and ion exchange, all of which contribute to effective pollutant removal (Zhang et al. 2022). However, the specific adsorption mechanism is influenced by the specific properties of nBC, environmental conditions, and the type of pollutant (Antonangelo et al. 2019). The removal of organic pollutants and heavy metals by adsorbents like nBC or BC occurs through both physisorption and chemisorption. Chemisorption is most effective in the removal of heavy metals, while physisorption mainly helps

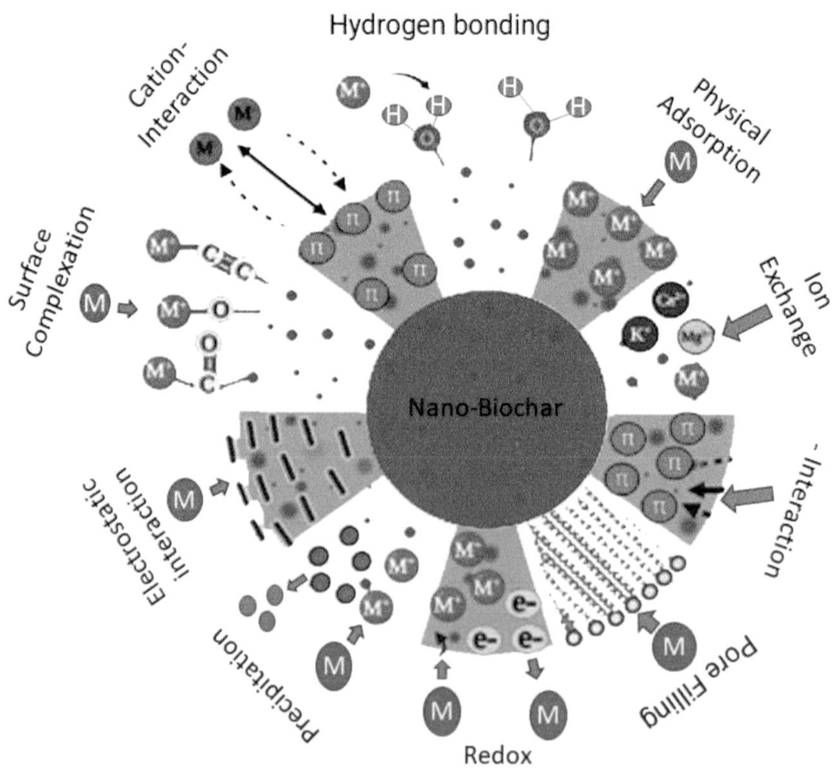

Figure 3. Various mechanisms involved in immobilization and remediation of soil pollutants by Nano-Biochar (nBC).

in the removal of organic pollutants (Pathak et al. 2024). Moreover, the removal mechanisms of heavy metals include complexation, ion exchange, precipitation, π-interactions, and electrostatic interactions, while the removal of organic pollutants involves electrostatic attraction, hydrogen bonding, π-interactions, complexation, and pore filling (Alsawy et al. 2022). In nBC particles, the introduction of acidic functional groups (such as hydroxyl and carboxyl groups) lowers the surface potential and point of zero charge (PZC), thereby increasing electrostatic adsorption. Additionally, pyrolysis temperature significantly affects the primary adsorption mechanism of nBC. For example, higher pyrolysis temperatures result in larger specific surface area and more porous structures, although they may reduce the presence of polarized surface chemical groups. These changes promote favorable conditions for physical adsorption. In contrast, low-temperature pyrolysis produces nBC with small surface areas but a large number of polarized surface chemical groups that enhance chemical adsorption (Zhuang et al. 2021). Understanding the prevailing physicochemical interactions is crucial to identify the properties of nBC that enable effective removal of heavy metals and various organic pollutants, including emerging pollutants (Khan et al. 2022).

3.5 nBC Environmental Applications

The incorporation of nBC into soil can also have a positive impact on the soil ecosystem and environment by promoting carbon sequestration, as it can effectively store carbon in the soil for long periods (Jiang et al. 2023). The process of capturing and storing atmospheric carbon dioxide in soil can be enhanced by nBC applications. Incorporating nBC into agricultural systems could contribute to climate change mitigation by reducing greenhouse gas emissions and enhancing carbon storage in soils. The great potential of nBC to remove various contaminants from soil and aqueous environments has been widely reported (Chaubey et al. 2024).

4. Implications of nBc for Soil Use

Despite its diverse applications, the use of nBC in agriculture also has some implications for environmental sustainability. The small nBC particles disperse more easily into the air than bulk biochar, which can result in dust clouds when used on farms with machinery (Gao and Wu 2014). While it can cause some lung issues in large quantities, it is generally safe unless exposure happens often (Dong et al. 2019). The nanoscale BC particles can move easily through soil, carrying harmful sorbed substances deeper into the soil profile. The high activity of nBC particles could be risky for the soil and water environments due to their ability to spread contaminants (Chaubey et al. 2024). Because of the higher sorption capacity and transport potential of nutrients and pollutants in nBC, there is a greater possibility of soil and water contamination, which is a burning issue similar to other engineered nanomaterials (Jiang et al. 2023, Wang et al. 2013).

The higher tendency of nBC to retain pollutants due to its larger surface area will have harmful impacts on a wide range of organisms. It is reported that microbes like algae may be at risk from the free radical content of nBC, oxidative stress, and nutrient depletion due to their adsorption on nBCs (Huang et al. 2021). Therefore, while the benefits of nBC are significant, it is important to consider possible implications and risks with its use, such as the risk of contaminant mobilization when inappropriately applied or overused, highlighting the need for careful management strategies tailored to specific soil types and crops (Sarma et al. 2024). Although nBC biochar offers promising applications for improving soil health and environmental sustainability, there are potential concerns that must be addressed. These include potential toxicity to soil organisms, the potential for nanoparticle accumulation and persistence in the environment, and uncertainty regarding long-term impacts on ecosystem function. So, careful assessment and regulation are needed to ensure the safe and responsible use of this emerging nBC technology.

5. Future Research Directions and Adaptation

As the field of nBC biochar research continues to evolve, there are several key areas that require further investigation for its successful application in agriculture. These include:

- To evaluate the optimal production methods and conditions to develop nBC with specific properties tailored to targeted applications in soil, water, and environmental remediation.
- To assess the long-term impacts of nBC on soil health, ecosystem function, and agricultural productivity through comprehensive field trials and monitoring studies.
- To explore the potential synergistic effects of combining nBC with other soil amendments or agricultural practices to optimize its performance and benefits.
- To fully understand the long-term implications of nBC use in sustainable agriculture.

The successful large-scale adoption of nBC technology will depend on overcoming various barriers, including:

- One key challenge is the need to develop scalable and cost-effective production methods for nBC material.
- Developing cost-effective and scalable production methods for nBC, as well as optimizing its integration into agricultural systems to maximize its benefits and minimize potential drawbacks.
- Addressing regulatory and policy gaps to ensure the responsible and sustainable use of nBC, including proper risk assessment, labeling, and consumer awareness.
- Furthermore, leveraging advance technologies such as precision agriculture and data analytics can optimize resource use, reduce waste, and improve crop yields through nBC adaptation, making it easier for farmers to transition to more sustainable methods.
- Ultimately, the widespread adoption of nBC will depend on effective education and outreach strategies to inform stakeholders, including farmers, policymakers, and the general public, about its potential benefits and potential risks.

6. Conclusions

Due to the distinct enhanced properties of nano-sized biochar (nBC), its soil application has been proven to be an innovative approach to improving soil health, quality and productivity, as well as combating climate change issues. Soil application of nBC has significant effects on soil health enhancement and soil contaminant remediation, contrasting it with traditional bulk biochar due to its superior physio-biochemical properties and paving the way for its application as a soil amendment for sustainable agriculture and the environment. Although nBC biochar offers promising applications for improving soil health and environmental sustainability, there are potential concerns and risks that must be addressed. These include potential toxicity to soil organisms, potential for nanoparticle accumulation and persistence in the environment, and uncertainty regarding long-term impacts on ecosystem function. So, careful assessment and regulation are needed to ensure the safe and responsible use of this emerging nBC technology. Future research studies are required to fully understand the long-term implications of nBC use in sustainable agriculture. However, nBC could be used as a beneficial soil amendment to boost soil health, soil fertility, and crop production in an economical and sustainable manner.

7. Acknowledgement

We (Authors) are highly thankful to the University of Agriculture Faisalabad and Yangzhou University for their support in this publication. Authors also want to thank Professor Dr Abdul Wakeel from Soil Fertility and Plant Nutrition Laboratory, University of Agriculture Faisalabad,

Pakistan, and Dr. Akhtar Hameed, Assistant Professor (Plant Pathology), Institute of Plant Protection for providing us this opportunity. We are highly thankful to our families who sacrificed their good times for the accomplishment of this milestone.

References

Anupama and P. Khare. 2021. A comprehensive evaluation of inherent properties and applications of nano-biochar prepared from different methods and feedstocks. J. Clean. Prod. 320: 128759.

Chaubey, A.K., T. Pratap, B. Preetiva, M. Patel, J.S. Singsit, C.U. Pittman et al. 2024. Definitive review of nanobiochar. ACS Omega. acsomega.3c07804.

Chausali, N., J. Saxena and R. Prasad. 2021. Nanobiochar and biochar based nanocomposites: advances and applications. J. Agric. Food Res. 5: 100191.

Chen, X., B. Zhou, Q. Wang, W. Tao and H. Lin. 2020. Nano-biochar reduced soil erosion and nitrate loss in sloping fields on the Loess Plateau of China. CATENA. 187: 104346.

FAO. 2009. How to feed the world 2050. High-Level Expert Forum. In: Global Agriculture Towards 2050. pp. 1–4. http://www.fao.org/fileadmin/templates/wsfs/docs/Issues_papers/HLEF2050_Global_Agriculture.pdf.

Gao, X. and H. Wu. 2014. Aerodynamic properties of biochar particles: effect of grinding and implications. Environ. Sci. Technol. Lett. 1: 60–64.

Goswami, L., K. Pakshirajan and G. Pugazhenthi. 2020. Biological treatment of biomass gasification wastewater using hydrocarbonoclastic bacterium Rhodococcus opacus in an up-flow packed bed bioreactor with a novel waste-derived nano-biochar based bio-support material. J. Clean. Prod. 256: 120253.

Gyanwali, P., R. Khanal, S. Pokhrel and K. Adhikari. 2024. Exploring the benefits of biochar: a review of production methods, characteristics, and applications in soil health and environment. Egypt. J. Soil Sci. 64: 10–20.

Kumar, H., W. Cai, J. Lai, P. Chen, S.P. Ganesan, S. Bordoloi et al. 2020. Influence of in-house produced biochars on cracks and retained water during drying-wetting cycles: comparison between conventional plant, animal, and nano-biochars. J. Soils Sediments. 20: 1983–1996.

Lian, F., W. Yu, Z. Wang and B. Xing. 2019. New insights into black carbon nanoparticle-induced dispersibility of goethite colloids and configuration-dependent sorption for phenanthrene. Environ. Sci. Technol. 53: 661–670.

Lian, F., W. Yu, Q. Zhou, S. Gu, Z. Wang and B. Xing. 2020. Size matters: nano-biochar triggers decomposition and transformation inhibition of antibiotic resistance genes in aqueous environments. Environ. Sci. Technol. 54: 8821–8829.

Liu, G., H. Zheng, Z. Jiang, J. Zhao, Z. Wang, B. Pan et al. 2018. Formation and physicochemical characteristics of nano biochar: insight into chemical and colloidal stability. Environ. Sci. Technol. 52: 10369–10379.

Liu, W., Y. Li, Y. Feng, J. Qiao, H. Zhao, J. Xie et al. 2020. The effectiveness of nanobiochar for reducing phytotoxicity and improving soil remediation in cadmium-contaminated soil. Sci. Rep. 10: 858.

Ma, S., F. Jing, S.P. Sohi and J. Chen. 2019. New insights into contrasting mechanisms for PAE adsorption on millimeter, micron- and nano-scale biochar. Environ. Sci. Pollut. Res. 26: 18636–18650.

Mahmoud, E., A. El Baroudy, N. Ali and M. Sleem. 2020. Spectroscopic studies on the phosphorus adsorption in salt-affected soils with or without nano-biochar additions. Environ. Res. 184: 109277.

Mandal, S., S. Pu, S. Adhikari, H. Ma, D.-H. Kim, Y. Bai et al. 2021. Progress and future prospects in biochar composites: application and reflection in the soil environment. Crit. Rev. Environ. Sci. Technol. 51: 219–271.

Naghdi, M., M. Taheran, S.K. Brar, T. Rouissi, M. Verma, R.Y. Surampalli et al. 2017. A green method for production of nanobiochar by ball milling-optimization and characterization. J. Clean. Prod. 164: 1394–1405.

Oleszczuk, P., W. Ćwikła-Bundyra, A. Bogusz, E. Skwarek and Y.S. Ok. 2016. Characterization of nanoparticles of biochars from different biomass. J. Anal. Appl. Pyrolysis. 121: 165–172.

Pandian, K., S. Vijayakumar, M.R.A.F. Mustaffa, P. Subramanian and S. Chitraputhirapillai. 2024. Biochar a sustainable soil conditioner for improving soil health, crop production and environment under changing climate: a review. Front. Soil Sci. 4: 1376159.

Pratap, T., M. Patel, C.U. Pittman, T.A. Nguyen and D. Mohan. 2021. Nanobiochar: a sustainable solution for agricultural and environmental applications. pp. 501–519. *In*: Nanomaterials for Soil Remediation. Elsevier.

Rajput, V.D., T. Minkina, B. Ahmed, V.K. Singh, S. Mandzhieva, S. Sushkova et al. 2022. Nano-biochar: a novel solution for sustainable agriculture and environmental remediation. Environ. Res. 210: 112891.

Sani, Md. N.H., M. Amin, A.B. Siddique, S.O. Nasif, B.B. Ghaley, L. Ge et al. 2023. Waste-derived nanobiochar: a new avenue towards sustainable agriculture, environment, and circular bioeconomy. Sci. Total Environ. 905: 166881.

Sarma, H.H., S.K. Borah, O. Das, S. Sonowal, N. Talukdar and B.C. Das. 2024. Nano biochar: properties, preparation, key features and its impact on soil and crop ecosystem. J. Adv. Biol. Biotechnol. 27: 66–78.

Song, B., M. Chen, L. Zhao, H. Qiu and X. Cao. 2019. Physicochemical property and colloidal stability of micron- and nano-particle biochar derived from a variety of feedstock sources. Sci. Total Environ. 661: 685–695.

Verheijen, F.G.A., A. Zhuravel, F.C. Silva, A. Amaro, M. Ben-Hur and J.J. Keizer. 2019. The influence of biochar particle size and concentration on bulk density and maximum water holding capacity of sandy vs sandy loam soil in a column experiment. Geoderma. 347: 194–202.

Wang, D., W. Zhang, X. Hao and D. Zhou. 2013. Transport of biochar particles in saturated granular media: effects of pyrolysis temperature and particle size. Environ. Sci. Technol. 47: 821–828.

Xiang, W., X. Zhang, K. Chen, J. Fang, F. He, X. Hu et al. 2020. Enhanced adsorption performance and governing mechanisms of ball-milled biochar for the removal of volatile organic compounds (VOCs). Chem. Eng. J. 385: 123842.

Xiao, J., R. Hu and G. Chen. 2020. Micro-nano-engineered nitrogenous bone biochar developed with a ball-milling technique for high-efficiency removal of aquatic Cd(II), Cu(II) and Pb(II). J. Hazard. Mater. 387: 121980.

Xu, C.-Y., Q.-R. Li, Z.C. Geng, F.N. Hu and S.W. Zhao. 2020. Surface properties and suspension stability of low-temperature pyrolyzed biochar nanoparticles: effects of solution chemistry and feedstock sources. Chemosphere. 259: 127510.

CHAPTER-13

Nanomaterial-Based Soil Amendments for Improved Crop Production

Amar Matloob,[1,*] *Farhena Aslam,*[2] *Muhammad Adeel Akram,*[3] *Hamna Iftikhar,*[3] *Diyan Rashid*[3] and *Ahmad Mehmood*[4]

1. Introduction

The basis of terrestrial life and agriculture is soil, a vibrant ecosystem teeming with life (Rinot et al. 2019). Along with air and water quality, soil health is also crucial for successful crop production and sustainable agriculture. However, soil health is increasingly threatened by both natural and human-induced factors, and the decline in soil health is causing grave concerns regarding food security, especially in the backdrop of climate change. Intensive farming practices, including excessive use of synthetic pesticides and fertilizers, coupled with mining, industrial activities, and climate change, are contributing to the decline in soil health (Bünemann et al. 2018). Soil salinization, erosion, degradation, and contamination are the associated challenges that impact food systems and ecosystems. The escalating demand for food and other ecosystem services, with a concurrent increase in population and unprecedented urbanization, is putting huge pressure on dwindling soil resources. A decline in soil health, which is often considered equivalent to dynamic soil quality (Moebius-Clune 2017), can limit the soil's ability to perform critical ecological functions resulting in low productivity (Dubey and Mailapalli 2016). van Dijk and Meijerink (2014) noted that 60–70% of used agrochemicals are ineffective because of processes including volatilization, leaching, and runoff. This underlines the idea of applying sustainable farming practices. Nanotechnology, with its implications across various dimensions of agriculture, can foster soil health in an environmentally benign manner. Latest research has indicated that nanoparticles may benefit the quality of the soil. A group of microorganisms - in other words, a microbiome - essentially governs soil, a storehouse of roughly 25 million species (Costa et al. 2018). Nanomaterials (NMs) are well known for their effect on synergy among soil physical, biological, and chemical properties, thereby modulating soil health and crop productivity (Sharma and Kumar 2024). The possibilities of overcoming the negative

[1] Department of Climate Change, MNS University of Agriculture, Multan-Punjab, Pakistan.
[2] Department of Agronomy, University of agriculture, Faisalabad.
[3] Department of Soil and Environmental Sciences, MNS University of Agriculture, Multan Punjab, Pakistan.
[4] Climate Change and Sustainable Development program, School of Interdisciplinary Engineering and Sciences (SINES), National University of Sciences and Technology (NUST), Islamabad, Pakistan.
* Corresponding author: amar.matloob@mnsuam.edu.pk

DOI: 10.1201/9781003558354-13

impacts of contemporary farming and restoring soil ecological functions are hinged on NMs. Studies show that nanoparticles have a positive effect on soil microbial communities, which in turn provides stability to the soil ecosystem for sustainable agriculture (Dinesh et al. 2012, Sharma and Kumar 2024). The accruing interest in sustainable agriculture, with emphasis on judicious management of natural resources, has paved the way for NMs to enhance soil health as a means of improving the quality of the earth's soils. The NPs synthesized through biological approaches with the help of nontoxic materials can be eco-efficient, possibly spurring an NMs-led green revolution. Although many studies have reported the beneficial uses of NMs in an agricultural context, long-term studies regarding various aspects—such as the ecological behavior of NMs, their effect on agricultural ecosystems with emphasis on NMs' transformation in soil, interaction with plant metabolism, influence on soil biota, and toxicity at higher doses—are lacking. Due deliberations are needed to make nanotechnology a safer tool while addressing its use for soil health and crop improvement. This chapter presents the critical role of NMs in improving soil health and crop productivity in the quest for sustainable agriculture. Current knowledge is exemplified, while existing knowledge gaps are addressed, and future research thrusts are also elaborated.

2. Types of Nanoparticles Used as Soil Amendments

A nanomaterial is a class of material that lies between atoms, molecules, and macroscopic materials. Its spatial size may range from 1–100 nm. It possesses a rare composition, an extensive surface area, and strong catalytic properties on the surface (Cao 2004). The unique structural and functional properties of NMs offer an array of opportunities and widespread implications across disciplines and also differentiate them from their bulk counterparts (Khan et al. 2022). It is believed that soil is the most widely available, reasonably priced natural building material having diverse uses in engineering. This organic substance is especially pertinent to nanotechnology. The US National Research Council noted that investigations into the initial steps in the field of nanotechnology were achieved with clay-sized particles (< 0.002 mm) in geotechnical engineering.

2.1 Metallic Nanoparticles

Because of their special physicochemical characteristics, metallic nanoparticles have become a focal point in soil research. Their high surface area, tiny size, and catalytic activity make them attractive options for enhancing soil and plant interaction. Nanoparticles of iron increase microbial activity and nutrient availability to boost soil productivity. They promote the uptake of contaminants by plants to aid in the phytoremedial treatment of polluted soil, besides boosting water retention and soil structure (Aragaw et al. 2021). Nanoparticles of silver have potent antibacterial qualities, inhibiting soilborne infections, increasing plant growth, and enhancing seed germination. Their possible toxicity to the environment is an issue, though (Zorraquín-Peña et al. 2020). Nanoparticles of zinc are known to improve the solubility and subsequent uptake of other nutrients and zinc by plants. They can boost plant development and productivity to demonstrate the efficiency of phytoremediation of heavy metal-contaminated soils (Zhou et al. 2020). Nanoparticles of copper can help mitigate soil-borne diseases due to their bactericidal capabilities. Despite their usefulness, the selectivity and effect of these nanomaterials on non-target organisms in the vicinity of the application area vindicate further research (Kahru and Dubourguier 2010).

2.2 Carbon-based Nanoparticles (Carbon Nanotubes)

Carbon atoms are arranged in a hexagonal lattice to generate carbon nanotubes (CNTs), which have remarkable mechanical properties such as high strength (Chong 2004). Studies have attempted to establish the feasibility of enhancing favorable qualities of soil using CNTs (Baig et al. 2022). Moreover, activities of soil enzymes like urease and sucrose were promoted in response to CNTs, accompanied by an increase in soil organic carbon and available potassium (Tiwari et al. 2014). As for the positive effects of the CNTs on the properties of the soil, studies have proved that increasing the quantity of cohesion and decreasing the value of the friction angle would significantly improve the compressive strength of the soil (Arabani et al. 2012). Exposure of tobacco cells to CNTs stimulated cell division at lower concentrations, as genes involved in cell division and water uptake were upregulated (Khodakovskaya et al. 2012). While CNTs have positive impacts regarding the improvement of soil quality, concerns have been raised about how they behave in the soil and whether they could be injurious to life forms. Although literature reveals that CNTs do not pose a significant effect on soil microorganisms (Tong et al. 2007, Johansen et al. 2008), more investigations are needed to better understand the successive impact of CNTs on soil microbial communities.

2.3 Colloidal Silica

Against the backdrop of improved soil health, colloidal silica—a stable suspension of silica nanoparticles in water—offers several benefits. These particles, which usually have a size between 7 and 22 nm, are extremely durable, nontoxic, and chemically inert (Chang et al. 1994). One important property of colloidal silica is its capacity to gel—that is, change from a liquid to a gel-like state resembling a solid by lowering the repulsive interactions between particles (Dietler et al. 1986). By creating a three-dimensional network, this gelation process improves the quality of the soil. Studies have indicated that colloidal silica is a useful supplement for enhancing soil structure. The unconfined compressive strength of sand increased linearly with increasing colloidal silica content (Ardi and Niroumand 2023). These results were supported by previous research (Persoff et al. 1999, Gallagher and Mitchell 2002), which showed that treating soil with colloidal silica significantly increased its strength. Furthermore, the application of colloidal silica improved the cyclic shear strength.

Colloidal silica has demonstrated potential in reducing liquefaction risk in addition to enhancing strength. Full-scale field, laboratory, and model testing confirmed that colloidal silica improves liquefaction resistance (Mollamahmutoglu and Yilmaz 2010), as revealed by the connection between the strength of the resultant nano-sand composite and gel formation. Influence of nano-additives like silica on soil physical properties such as shear strength, permeability, bearing capacity, and compressibility was reviewed by Shahin et al. (2015). Silica-based NPs can favor plant defense systems and confer salinity and drought tolerance in plants (AlSaeedi et al. 2019, Akhtar and Ilyas 2022).

2.4 Bentonite

Bentonite is a clay mineral made of weathered volcanic ash, mainly montmorillonite (Darvishi and Morsali 2011). According to Abdou et al. (2013), nano-bentonite has superior swelling, moisture absorption, and purity compared to natural bentonite. It is widely used in soil engineering to enhance soil characteristics. It works well as a thixotropy and high-elastic-property gel formant, which makes it a useful addition to grout and drilling fluids. For example, bentonite suspensions have been utilized to improve soil resilience to liquefaction and to minimize fluid loss during drilling operations (El Mohtar et al. 2008). Additives such as sodium pyrophosphate have been introduced to enhance the permeability of bentonite suspensions for grouting applications (Rugg et al. 2011). Studies conducted in lab and field settings have shown the benefits of bentonite in reducing the risk of liquefaction. Using numerical simulations, El Mohtar et al. (2013) effectively modeled the liquefaction mitigation effects of bentonite suspensions. All things considered, bentonite's special qualities make it an invaluable instrument for remediation and soil enhancement. To completely comprehend its long-term behavior in various soil conditions and to maximize its application for specific purposes, more research is necessary.

2.5 Laponite

Alternative NMs such as synthetic layered silicate nanoparticles, as in laponite, have certain characteristics that can enhance soil quality. More specifically, even if it is slightly smaller in size compared to bentonite, laponite has certain desirable characteristics. Similarly to montmorillonite, it has a structure that allows fast solubility in water, forming a transparent aqueous suspension. The rheological properties of laponite are noteworthy; therefore, depending on the ionic strength, it is either in a low-viscosity suspension form or in a gel form (Tanaka et al. 2004). Due to this tendency and its nanoparticle size, laponite holds potential to strengthen sands against the process of liquefaction. From literature, the study by El Mohtar et al. (2008) shows that laponite proves to have better results than bentonite on this criterion. Besides, Huang and Wang (2016) attributed the improvement of geotechnical properties of sand suspensions containing laponite to the differences in the cellular structures. Thus, outstanding characteristics of laponite as a small-sized particle with colloidal stability and the ability to form gels mean that it can contribute to the field of soil improvement. One major aspect that has been considered promising is its contribution to the diminution of liquefaction problems. That is why further investigation is required to understand its indefinite behavior in different types of soils and to use it to address specific engineering challenges.

3. Mechanisms of Action of Nanomaterials in Soil

3.1 Availability and Uptake of Nutrients

Due to their distinct physicochemical characteristics, NMs can greatly improve the nutrient relations (availability, uptake, translocation, and assimilation behavior, etc.) in crop plants. They can be a potential alternative to conventional fertilizers due to their capability of targeted delivery, gradual and slow release of nutrients with greater precision (Rahmati and Kousehlou 2023). Due to their greater capacity for ion exchange and high affinity for adsorption owing to their diverse pore sizes (micro, meso, and macro pores), zeolitic nano-fertilizers can provide nutrients over a prolonged period (up to 50 days) compared with conventional urea (Sharma et al. 2022). From a crop nutrition perspective, the role of NMs lies primarily as nano-fertilizers delivering essential micro- or macronutrients. Moreover, they can also serve as nano-carriers for other agro-chemicals (Babu et al. 2022). Nano-fertilizers can ideally match nutrient release patterns with crop requirements, thereby reducing negative implications arising from the use of synthetic fertilizers such as emission of greenhouse gases and leaching (Sharma and Kumar 2024), thus improving nutrient use efficiency (Raliya et al. 2018). They attract nutrients like potassium, phosphate, and nitrogen in the soil primarily by imposing a strong adsorption site due to their extremely large surface areas. Therefore, increased nutrient accessibility at the plant root system allows for faster nutrient intake, which enhances growth and development (Nowack and Bucheli 2007). Application of 1 mg kg^{-1} of chitosan nanoparticles to potato crops significantly improved nutrient availability (Kondal et al. 2021). Another study reported the positive effect of carbon-based NMs on maize growth due to increased nutrient availability mediated by enhanced activities of soil enzymes (Zhao et al. 2021). In addition, there are specific forms of NMs, for example, of iron, zinc, or phosphorus, which are able to deliver these sources of macro- and micronutrients directly for plant nutrition, thus increasing the efficiency of absorption from the soil. For instance, it has been illustrated that the solubility of phosphorus is enhanced by the use of iron nanoparticles, thereby favoring plant uptake of phosphorus (Khin et al. 2012). Additionally, one of the most essential factors regarding nutrient uptake is that NMs stimulate root production, since plants with deeper root systems explore the deeper soil layers to acquire more nutrients. It is also necessary to point out that interaction of root exudates with NMs can influence the mobilization and uptake of nutrients. For instance, a synergistic relationship of organic acids generated by the roots with metal oxide nanoparticles can release bound nutrients (Prakash and Chandran 2023).

3.2 Structure and Aeration of Soil

As an important characteristic of the soil, the structure of the soil is influenced by NMs. They help in the creation and consolidation of stable soil aggregates since they serve as a nucleus. Such aggregates are highly essential for the improvement of drainage, aeration, and infiltration through the creation of new pores in the soil. Germination, sprouting, root formation in plants, microbial growth, and presumably soil fertility in totality all entail the feature of enhancement of the soil structure (New et al. 2023). Besides, such entities as nanoparticles are capable of preserving organic matter stability. They assist in the creation of stable soil aggregates through chemical binding with the organic matter and thus hinder the erosion or leaching of any organic matter. Macro-aggregate stability and SOM have a direct relationship with sustainable production; therefore, these conclusions make sense and are significant in the field of agriculture (Tripathi et al. 2023). Application of silica nanoparticles increased soil porosity and improved soil structure (Rajput et al. 2021). Their role in enhancing soil aeration is also well documented since they create nano-sized channels or pores between soil aggregates (Sharma and Kumar 2024). Nevertheless, the effect of nanoparticles on soil structure and other physical properties also depends on the specific nature of the NMs. For example, a study documented the positive effect of nano-magnesium on soil porosity and soil structure as compared to nano-iron, which negatively affected the aforementioned soil properties (Bayat et al. 2018).

3.3 Management of Water Retention

Another significant advantage of NMs is their ability to significantly alter soil water retention (Rahmati and Kousehlou 2023). It is worth mentioning that some nanoparticles can absorb considerable volumes of water—this is true for nanoparticles that possess the highest hydrophilic characteristics. Therefore, this ability to absorb water may be useful in areas that experience water scarcity and practice strict water rationing, or in those areas characterized by frequent occurrences of drought spells. Utilization of NMs can possibly reduce water evaporation and, therefore, the effectiveness of water retention and use is considerably improved (Belal and El-Ramady 2016). Application of nano-clinoptilolite to a sandy loam soil increased water retention by modulating pore size distribution (Nakhli et al. 2017). Small nanoparticles were more effective in improving soil water storage and availability of water. The positive influence of nano-bentonite in improving water holding capacity and increasing the availability of water to crops is also documented elsewhere. Pot and field studies revealed that its use resulted in higher yields of wheat with a concurrent increase in NPK contents in both grain and straw (El-Nagar and Sary 2021). Nanoclays, with their enormous water retention potential (25% more water holding in coarse-textured sandy soils), have transformed the fate of desert agriculture in Egypt, where two-thirds of irrigation water was saved due to the use of nanoclays, resulting in a 416% increase in crop yields (Olesen 2010). Such positive effects are attributed to the formation of cohesive masses of soil, increased storage, and availability of nutrients and water. Likewise, some NMs affect the probabilities of water entry into the soil profile or the rates of infiltration. This can help refill the water reserves and reduce incidences of soil erosion. Indeed, climate change and heavy rainfall are related, thus the above notion that making a soil system ready for both will require an increase in infiltration and the ability to retain water.

3.4 Controlling Soil Acidity and pH

The pH of the soil has a significant impact on plant development, microbial activity, and nutrient availability (Fernández and Hoeft 2009). The NM-mediated buffering effects can help stabilize the pH of soil. Through their interactions with soil organic matter and ions, they can modulate soil pH. Significant pH changes can have a negative impact on soil processes and plant health; therefore, this buffering capacity helps to prevent such alterations (Nowack and Bucheli 2007).

Additionally, some NMs can mitigate soil acidity, which is a significant issue in many agricultural soils. For example, metal oxide-based NMs such as titanium oxide, cerium oxide, copper oxide, and iron oxide can increase EC and pH by releasing metal ions into the soil solution (Zahra et al. 2015, Majumdar et al. 2016). These substances lessen acidity, which increases microbial activity and nutrient availability, and thus improves the growing conditions for plants (Rajput et al. 2022). This is especially crucial in areas with acidic soils and for crops that are sensitive to acid (Khan et al. 2022). On the other hand, in alkaline soil, NMs were effective in reducing soil pH and nutrient mobilization by promoting root exudation (Zahra et al. 2015). Depending upon the type of NMs, soil pH could either increase (Alsharef et al. 2016) or decrease (Huang et al. 2022, Verma et al. 2022), supporting the notion of their buffering effect.

3.5 Impact on Microbial Communities in the Soil

Soil microbial communities and their functioning remain the most sensitive soil properties in response to NMs application. Research on the impact of NMs on soil microbial communities is a complex and evolving field. While certain NMs can have adverse effects on microbial activity, other studies have demonstrated that they can promote the growth of beneficial microbes involved in nutrient cycling and plant growth enhancement (Rahmati and Kousehlou 2023). The positive aspect covers their ability to deliver microbes or microbial biosorbents by acting as nano-carriers for the remediation of heavy metals in polluted soils (Ayangbenro et al. 2017). Silicon nanoparticles have been reported to improve the colonization of rhizospheric microbes and biomass, sustaining soil health (Gajić et al. 2018, Srivastava et al. 2021). Likewise, Bhattacharjya et al. (2021) documented enhanced activities of soil enzymes and microbial biomass in response to increased titanium oxide concentrations. Nevertheless, the response of soil bacteria to NMs is influenced by various factors, including soil characteristics, environmental conditions, and nanoparticles, i.e., size, type, surface area, charge, shape, concentration, and exposure time (Sun et al. 2022, Sharma and Kumar 2024). For example, zinc nanoparticles were more toxic to soil microbes as compared to titanium nanoparticles (Ge et al. 2011). The detrimental effects of NMs on soil microbes are microbial community specific. Following exposure to NMs, the population of methane oxidizers and nitrogen-fixing bacteria was severely declined; on the other hand, microbes involved in the decomposition of organic pollutants showed an increase in their population (Moll et al. 2017). Likewise, activities of β-proteobacteria and ammonia oxidizers were drastically reduced in response to silver nanoparticles, while populations of acido- and actinobacteria as well as bacteroidetes showed a significant surge (Grün et al. 2019). The CNTs reduced the microbial carbon and nitrogen biomass in soil (Chung et al. 2011). It is crucial to recognize that the structure and diversity of the soil microbiome are influenced by NMs. It is essential to evaluate the potential consequences on the functions and services provided by the soil ecosystem, even if certain changes may be beneficial. Use of nanotechnology-based fertilizers is advantageous over conventional, as summarized in Table 1.

4. Application Techniques

The NMs today are seen as opportunities, and their use for improving soil quality and crop yields warrants the need for efficient application techniques. In the recent past, nanoparticles' usage in agriculture has attracted a lot of interest due to their benefits such as enhanced and accelerated crop growth, better quality soil, and defense against adverse environmental conditions. Application techniques can affect the delivery, release, uptake, and subsequent fate of NMs in soil. Following are some nanoparticles application techniques for soil amendment.

4.1 Soil Incorporation Methods

This is the dominant strategy used to apply the NMs in the soil amendment, where the NM is directly mixed into the soil matrix. The technique is used with the objective of increasing desired soil properties, promoting nutrient availability, and triggering plant growth. There are several ways through which the NMs can be well incorporated into the soil matrix.

Table 1. Comparison: Applications of Conventional and Nanotechnology-based Fertilizers.

Sr. No.	Properties	Nanofertilizers-enabled Technologies	Conventional Technology
1	Dispersion and solubility of mineral micronutrients	Mineral micronutrient formulations at the nanoscale may decrease soil absorption and fixation, increase bioavailability, and improve the solubility and dispersion of insoluble minerals in soil.	Reduced solubility and big particle size resulting in decreased bioavailability for plants
2	Nutrient uptake efficiency	In agricultural production, nanostructure formulation may improve fertilizer efficiency and soil nutrient uptake ratio while conserving fertilizer resources.	For roots, bulk composite is unavailable and reduces efficiency.
3	Modes of controlled release	Water soluble fertilizers can have their nutrient release rate and pattern precisely regulated by encasing them in semi-permeable membrane envelopes covered with sulfur, waxes, and resin-polymer.	Overuse of fertilizers can lead to toxicity and upset the soil's natural equilibrium.
4	Duration of nutrient release that is effective	The creation of nanostructures can increase the amount of time that fertilizers effectively feed nutrients to the soil.	Utilized by the plants immediately after delivery, the remainder is transformed by the soil into insoluble salts.
5	Rate of nitrogen loss from fertilizers	The creation of nanostructures can slow down the rate at which nutrients from fertilizers seep or leach into the soil.	High rate of loss due to drift, rain, and leaching

(Singh and Singh 2021)

4.2 Direct Application

It is the simplest way for dispersing the NMs into the soil, and in the process, these materials get in contact with the plant roots and soil. The use of mechanical mixers or tillers leads to the enhancement of the homogeneity of the mixture. According to Bottero et al. (2017), direct application is a very important pathway to be considered in evaluating the efficacy of NMs such as copper, zinc, and silver oxides, etc. Common NMs used in this method are nano-silica, nano-clay, and nano-hydroxyapatite. These nanoparticles improve soil structure and other physical properties. Hence, direct application of NMs is an efficient and simple way of enhancing soil productivity.

4.2.1 Surface Application Followed by Tillage

Dry or liquid formulations of NMs can be mixed into soil by tillage. This method is applicable for large-scale use and ensures that nanoparticles are mixed into the soil surface, where they can influence root development and nutrient uptake.

4.2.2 Soil Injection

This method uses specialized types of equipment for injecting the NMs into the root zone or the intended area or specific depth. Many researchers have studied the injection method as very effective in improving soil properties. According to Barough et al. (2021), NMs have a positive effect on compressive strength, and the use of nano-silica is more effective than other nanoparticles.

4.2.3 Nano-composting

Mixing nanoparticles with compost provides a slow-release mechanism for nutrients and enhances the microbial activity of the amended soil. For example, Biochar-Nano-Metal oxide composite can significantly improve soil nutrient levels. According to the composting process, studies indicated that lower nanoparticle concentrations are effective (Gattupalli et al. 2023). Composting using NPs reduces levels of toxicity. Organic matter masks nanoparticles and thus shields the extracellular enzymes. The toxicity of nanoparticles during composting is a function of their size and concentration. Their use can mimic enzymes, thus resulting in the degradation of the material and improvement of the compost quality. The entry of NPs into compost and soil is illustrated in Figure 1.

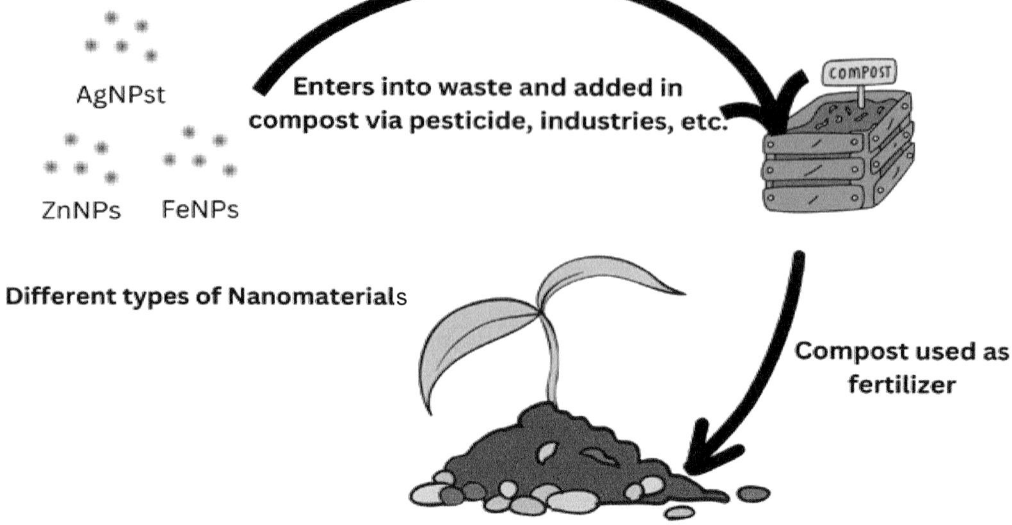

Figure 1. Entry of NPs into the composts and soil. Modified after Gattupalli et al. (2023).

4.2.4 Pelleting and Slurry Application

The NMs can be added to carrier materials such as clay or compost to produce pellets. These pellets can then be spread on the surface of soil and tilled in when necessary. In some cases, it is possible to place slurry that contains nanoparticles onto the surface of the soil. This method is widely used for remediation with the aim of stabilizing pollutants in the soil matrix. However, it calls for careful management regarding water drainage and the risk of environmental pollution.

4.2.5 Encapsulation and Slow-release System

Nanomaterials can be encapsulated in biodegradable materials that provide controlled release of nutrients and other substances, reducing the frequency of application and ensuring the sustainable release of applied substances over time. Their construction mechanism is illustrated in Figure 2.

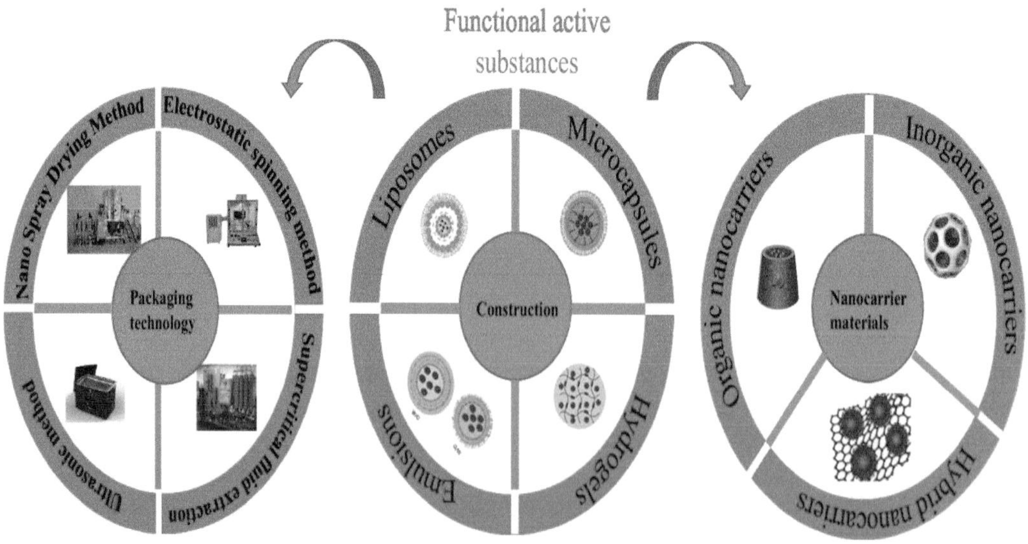

Figure 2. Construction of Nano-slow-release System. Source: Cao et al. (2023). © 2023 Cao, Gao, Wang, Liu, Zhang, Ping, Liu, Chen, Xu, Huang and Liu.

As per Cao et al. (2023), classification of antibacterial substances can be based on the type of nanocarrier materials—inorganic, organic, or hybrid. Examples of inorganic carriers are CNTs and silver nanoparticles. Organic carriers include proteins and polysaccharides. Composites can be made of inorganic and organic fractions that may make them unique from other types of nanomaterials. The selection of nanocarriers is based on bioavailability, methods of delivery, biodistribution, controllability, biodegradability, and biocompatibility so that the antibacterial substance will be well enclosed in the nanocarrier.

4.3 Foliar Application

In the foliar application, we spray NMs directly onto the leaves or aerial portions of the target plants. This method improves the performance of crop protection strategies as compared to the traditional soil-to-root application method because NM, if sprayed on the foliage of plants, enters into the leaves through stomata and is transported to the different parts of plants via symplastic and apoplastic pathways (Hong et al. 2021). Many factors affect the absorption, transportation, and accumulation of NMs. These include humidity, temperature, nanoparticle properties, and physiological characteristics.

4.3.1 Advantages of Foliar Application

Nanoparticles are directly absorbed by the plants, minimizing the chances of losses such as through soil adsorption, degradation, and leaching. This leads to improved nutrient use efficiency. In environment-sensitive areas, it reduces the risk of soil contamination by directly applying material to the targeted plant. This is also beneficial for nutrients that are poorly absorbed by the plants or when rapid uptake is required. Some of the nanoparticles can induce changes in the stress state, for example, during a dry spell or where nutrients are scarce, by manipulating plant physiology. The overall advantages of foliar application include its low cost, quick action and results, freedom from soil fixation, independence from root uptake, use of a small quantity of material, and increased quality and yield. It is illustrated in Figure 3.

Figure 3. Foliar application of NMs (Ahmad et al. 2021).

4.3.2 Method used for Foliar Application

Following are some methods that can be used in the foliar application of NMs.

4.3.2.1 Knapsack sprayer

This is a standard spray application tool that allows to carry approximately 4 gallons of sprayable material.

4.3.2.2 Boom sprayer

This type of sprayer uses a boom for the application of material and is usually used in large fields and offers more efficient coverage compared to a back sack sprayer.

4.3.2.3 Drone application

The use of Unmanned Aerial Vehicles (UAVs) for the application of NMS on the foliage of plants is an emerging technology that offers efficient foliar application of NMs.

4.3.2.4 Electrostatic spraying

This method involves charging particles of NMs with an electric charge that enhances their attraction to the surface of the leaves. It ensures a more uniform coverage and reduces the quantity of material needed.

4.4 Seed Coating with Nanomaterial

Seed coating is a technique used in agriculture and horticulture in which seeds are coated with active ingredients such as microbial inoculants. The purpose is to increase the rate of germination, growth, and resistance to biotic and abiotic stressors. It offers a novel approach to seedling establishment and early plant growth. Nanotechnology or nanoparticles are increasingly used in agriculture, especially NMs that improve crop yield with better nutrient and pest control mechanisms. Web references indicate that some of the NMs like FeO_2 and ZnO_2 serve as micronutrients and possess antimicrobial properties. Their special features, such as activity and biological compatibility, have increased their usage to a level that may often harm the environment (Zaim et al. 2023). According to Gupta et al. (2024), priming of seeds with NMs not only improves the germination rate but also offers the potential to mitigate biotic and abiotic stress. Furthermore, it also improves the water uptake of seeds through the formation of nanopores that enable the cells to maintain their homeostasis by activating reactive oxygen species-scavenging antioxidant systems in the seed. Nanopriming of tomato seeds with selenium improved germination and seed vigor (García-Locascio et al. 2024). Similarly, iron oxide NM coating on seeds significantly improved the germination rate, seed vigor, and germination index (Lin et al. 2024).

Coating seeds with NMs can result in improved seed germination, as the NMs are directly coated on the seeds, allowing targeted delivery of nutrients crucial for germination and early seedling development. Coating NMs on seeds protects them against several bacterial and fungal diseases. Certain NMs provide tolerance against different stresses, particularly valuable in areas with harsh climate conditions and unpredictable weather patterns.

4.5 Application of NMs in Irrigation System

Nanotechnology in conjunction with irrigation water is one of the possible concepts to raise yield and make cultivation more sustainable. It requires the suspension of NMs directly into the water. This could deliver nutrients, water, and protective substances to the plants. NMs can be introduced into water channels by some of the following mechanisms: NMs can be mixed directly into the irrigation channels and reservoirs to ensure the homogeneous distribution of material throughout the field. Alternatively, a specialized injection system can also be used for injecting the NMs into the irrigation line at specific positions that allow the controlled dosing and minimize wastage. Another method involves direct incorporation of NMs into pressurized irrigation systems such as drip and sprinkler that ensure the delivery to the root zone and enable foliar application alongside soil delivery. Benefits of incorporating nanomaterials into irrigation systems include enhanced nutrient delivery, controlled release of fertilizers, increased water use efficiency, reduction in water loss, enhanced pest and disease management, and environmental sustainability. However, there are several considerations and challenges associated with the incorporation of NMs in irrigation systems, such as long-term impacts on the environment, concerns about accumulation in water and soil, economic feasibility, regulatory compliance, and the need for education and training of farmers.

5. Enhancing Soil and Crop Productivity with NMs

5.1 Improving Soil Health with NMs

Soil health is essential for the robust growth of plants and for sustainable agriculture to succeed. However, the primary obstacle to global food security is soil degradation, which reduces the beneficial physical, chemical, and biological characteristics of the soil. For such problems, traditional soil management techniques frequently run into difficulties. While traditional methods of controlling soil fertility were once thought to be effective, in recent years, nanotechnology has emerged as a promising tool for improving soil fertility and quality (Sharma and Kumar 2024). Soil physico-chemical properties such as bulk density, structural stability, cation exchange capacity, electrical conductivity, nutrient cycling, and soil pH that influence crop growth and yield can be modified using NMs (Singh et al. 2018). These are a few of the soil properties that, through their interactions with plant roots, form an intricate system that is necessary to increase agricultural productivity. Rafiq et al. (2017) demonstrated that NMs can improve the characteristics of the soil in addition to maintaining crop yield and the sustainability of the land. Appraisal of physico-chemical and biological characteristics of soils in response to NMs application is crucial from a soil health perspective. In an attempt to develop efficient iron-based NMs that are least affected by soil pH and are not harmful to soil microbes, Das et al. (2016) synthesized a novel orthorhombic Fe–oxalate capped-Fe-oxide (Fe_3O_4) (OCIO) NM with prolific applications in agriculture. Regarding the release of Fe, the resulting NM was effective across a range of soil pH (4 to 9), depicting a uniform release pattern. Moreover, it was safe for soil bacteria, with no harm to seed germination as well. Its application at 20 g kg^{-1} soil enhanced soil N, P, and organic C contents, besides promoting soil enzymatic functions and water holding capacity. On the other hand, it reduced the bulk density of the amended soil. Such improvements in soil properties were conducive to enhanced growth and productivity of tomatoes. Another study reported that nano-submicron mineral-based soil conditioners have potential for increasing pH and porosity and reducing the aluminum concentration and associated harmful effects of aluminum in the soil, while it was effective in averting soil bulk density and cadmium concentration in the rice plants (Liu et al. 2017).

The usage of biochar and nanocomposites based on biochar as soil supplements has significantly increased. These materials are useful in terms of their pore structure, functional groups, and catalytic characteristics (Rasse et al. 2022). Among their many applications, reduction in greenhouse gases, improvement of water and nutrient availability, protection against soil erosion, and increased carbon sequestration are important ones. However, additional research must be done for improved synthesis to increase the nutritional profile, as current biochar products manifest low sorption capacity for the retention of nitrogen forms in the quest for biochar-based fertilizers with high nitrogen contents. Nanoscale porosity is crucial regarding sorption dynamics.

The threshold concentration at which a particular NM will induce environmental toxicology remains a germane issue. Thus, green synthesis involving the use of microbes or plant extracts has been acknowledged as a superior choice over traditional approaches (Saratale et al. 2018). Green synthesized silver nanoparticles (GSNPs) derived from *Thuja occidentalis* leaf extract, for example, have been shown to enhance the physicochemical properties of soil, such as pH, water holding capacity, cation exchange capacity, total organic carbon, and nutrient availability (Das et al. 2018). Further studies are necessary to evaluate the long-term impacts of GSNPs on agricultural productivity and soil ecosystem functioning. Besides directly modulating soil properties, NMs can also aid in monitoring and detection of soil (moisture, temperature, pH, nutrient status, etc.) and environmental parameters (soil and water pollutants such as organic contaminants, heavy metals, pesticides, pathogens, etc.) by acting as biosensors. For example, nitrate and phosphate levels in soil and water can be detected by using CNTs nanosensors, contributing to optimized use of applied fertilizers and minimizing environmental pollution (Wang et al. 2008).

5.2 Impacts of NMs on Soil Microbial Diversity

The soil constitutes a diverse ecosystem that is home to a plethora of microorganisms crucial for maintaining the fertility and health of the soil. Soil microbial diversity is recognized as a primary factor governing soil health. These microbes play a pivotal role in the growth of plants by influencing nutrient fixation, mobilization and sequestration, cycling, decomposition processes, and plant defense mechanisms. Through the release of specific root exudates, plants can assign specific tasks to microbial communities, and microbial diversity in soil is usually an outcome of the interplay between microbial quorum sensing and plant root exudates, resulting in adaptations and complex microbiomes. However, the current state of pollution, climate change, and intensive farming practices has resulted in a decline in the diversity and presence of microbial life in the soil. This has prompted the development of nanotechnology as a potential solution to address soil challenges. The NMs possess a large surface area, reactive nature, and small size, which enable them to interact with microorganisms and other substances in the soil in a multifaceted manner. Through their effect on soil biochemical processes, NMs can alter the soil environment and consequently the quantity and quality of plant root exudates, thus modulating soil microbial ecology and plant growth functions. Although NMs are beneficial to agriculture, research is still being conducted to determine their impact on soil microbes. Studies have revealed that certain NMs can increase the variety and subsequently the activity of microorganisms in the soil. For example, Rafiq et al. (2017) discovered that the pyrolysis of yak dung in combination with attapulgite (magnesium aluminum phyllosilicate) clay increased the population of beneficial microorganisms in the soil, which subsequently enhanced the nutrient cycling of crops. Similarly, another study by Ye et al. (2017) found that the application of redox-active clay-covered bamboo encouraged the growth of bacteria that fix carbon and release sulfur and iron. This is significant because these bacteria aid in the fixation of carbon and the availability of nutrients. However, the effects of metal-based NMs on soil microorganisms are not always straightforward. Some research has indicated that certain metal-based NMs can have harmful effects on microbial viability and richness (McKee and Filser 2016). The toxicity of NMs to microorganisms can be influenced by various factors, including concentration, size, form, and surface coating. For example, the toxicity of silver nanoparticles (AgNPs) to *Nitrosomonas europaea* was affected by both particle size and concentration (Wang et al. 2017). Therefore, it is essential to consider both the characteristics of NMs and the abiotic environment in order to understand how NMs and soil conditions will affect different microbial communities. Some examples of NMs that have less detrimental effects on microbial richness include orthorhombic Fe OCIO NMs and pH-responsive polysuccinimide NMs (Xin et al. 2018). Also, by combining NMs with other soil amendments, such as organic matter, the detrimental effects mentioned above can be mitigated. According to Bai et al. (2017), the results of flora and fauna in soils treated with carbon-based nanomaterials (graphene, graphene oxide, and carbon nanotubes) showed an increase in the number of microarthropods and a corresponding enhancement in their diversity. Based on these findings, it may be concluded that NMs may have a beneficial indirect influence on microarthropods by enhancing soil quality and increasing food sources. Under adverse soil conditions, NMs can induce plant growth through enhanced activities of root-associated microbes and increments in nutrient recycling and phytostimulation processes (Rajput et al. 2023). Consequently, it can be inferred that NMs have a variety of intricate and context-specific effects on soil microbial richness. However, more research is necessary to fully understand the mechanisms underlying these impacts and to determine the best ways to utilize NMs for the sustainability of agroecosystems. It is essential to accurately assess NM properties, the microbial population in the soil, and other factors for the ethical and effective application of nanotechnology for soil management.

5.3 Optimizing Soil Nutrients with NMs

To ensure the growth and development of crops, it is essential that the soil contains the necessary amount of required nutrients, as these are crucial for fertility. Certain soil microorganisms produce enzymes that facilitate the recycling and release of plant nutrients, which in turn supports plant growth. However, the use of chemical fertilizers can have negative effects on plant diversity and the soil. NMs can be used to improve soil fertility and increase nutrient availability. By enhancing the enzymatic activity of the soil, NMs can improve plants' uptake of nutrients and minerals. Studies have shown that exposure to orthorhombic Fe OCIO nanoparticles can increase soil N and P levels and elevate urease and phosphatase activity (Das et al. 2016). Similarly, investigations by Chaudhary et al. (2021) revealed that the activities of dehydrogenase, fluorescein diacetate hydrolase, β-glucosidase, dispersible amylase, acid phosphomonoesterase, and arylesterase in the soil were increased in response to NMs. The ability of MO-NPs to increase nutrient bioavailability has also been confirmed. Using ZnO NPs, the mineral contents of N, P, and K in the soil were increased (Aziz et al. 2019). According to Raliya et al. (2016), plant P uptake was improved by biosynthesized ZnO NPs because they increased phosphatase and phytase activity, which in turn enhanced phosphorus solubility.

Nonetheless, it has also been observed that a number of variables, including NM type, size, and concentration as well as soil characteristics, will affect how NMs interact with soil enzymes. To further emphasize the need for proper application of NMs, research also revealed that excessive concentrations of gold nanoparticles dramatically reduce soil enzyme activity (Asadishad et al. 2017). Nanofertilizers can revolutionize nutrient management as they permit nanoemulsions of macro- and micronutrients and improve the release and accessibility for plant growth. While the research's conclusions suggest that NMs can aid in boosting the soil's nutrient availability, additional research is needed to fully understand the impact of NMs, microbial genera, and nutrient cycling processes, as well as to compare and analyze their synergistic interactions. Improved soil fertility and sustainable agriculture are predicted to result from better application of NM properties and application methods.

5.4 Environmental Impacts of Using NMs in Soil

Similar to how NMs affect the environment, many people are concerned about what happens to them when they end up in the soil. However, given how well these materials work in agricultural techniques, there are certain drawbacks that might be linked to the aforementioned unique characteristics. These include the persistent problems of NM enrichment in soil and environmental pollution. Among these are metal oxides and carbon nanotubes, which are claimed to have a tendency to stay in the soil for extended periods and hence impact the soil's biota and ecosystem processes (Nowack and Bucheli 2007). Furthermore, these particles have the ability to contaminate fish and humans through leaching into groundwater (Khan et al. 2022). Another aspect is how specific NMs impact the efficiency of microorganisms in the context of the soil's nutrient cycling and health. Data indicate that certain nanoparticle types possess bactericidal activity, potentially upsetting the equilibrium of microorganisms in soil (Ameen et al. 2021). Furthermore, it is recognized that depending on the NMs, soil composition and characteristics may alter as a result of their interaction with soil organic matter (Nowack and Bucheli 2007). To mitigate the aforementioned issues, more thorough risk analysis must be conducted, and the application of NMs in agriculture must follow established guidelines while engaging in sustainable farming practices. More investigation is particularly needed into the ultimate behavior or fate of these NMs in the soil environment and any potential effects they may have on humans and/or biota.

6. Conclusion and Future Perspectives

Adding NMs to soil for agricultural purposes is one way to potentially address the global issues of food security and soil degradation. NMs can be produced by modifying their unique physicochemical characteristics to enhance soil structure, maximize water use efficiency, and supply essential nutrients. This potential to revolutionize agricultural practices has attracted significant investment and scientific interest. However, the application of NMs in agriculture is still relatively new, and there are still a number of important knowledge gaps. Research suggests that NMs may contribute to enhanced agricultural development and productivity, but concerns about potential negative effects on the environment and human health cannot be set aside. The behavior of NMs in the intricate soil matrix and their interactions with soil organic matter, clay minerals, and microorganisms need further insights for better understanding and devise safe application. Furthermore, the effects of prolonged exposure to NMs on soil ecosystems and food safety are unknown. Because of their potential toxicity, metal-based NMs need particular attention. According to studies, NMs based on copper and silver can negatively impact soil microbial communities, which can cause disturbances in crucial ecological processes. Furthermore, there is a chance that these NMs will contaminate waterways and subsequently people and animals will be exposed due to NMs' mobility and durability in the environment. Comprehensive and long-term research is necessary to achieve the full potential of NM-based agriculture while reducing potential negative impacts.

The following areas should be the focus of future research: Formulating NMs with desired qualities while limiting their potential side effects is the key for creating safe and effective NMs. Moreover, examining NMs behavior in the soil, particularly how they move through soil and water systems and how they interact with other soil elements are also critical aspects necessitating research. Evaluating the possible effects of NMs use in agriculture on the environment and human health by carrying out comprehensive risk assessments, devising regulatory frameworks to ensure the proper and safe application of NMs in agriculture is also needed. By abridging these knowledge gaps and taking required measures, the agriculture industry can benefit from nanotechnology.

7. Acknowledgement

The authors are thankful to the Department of Soil and Environmental Sciences, MNS University of Agriculture, Multan for the completion of this work.

References

Abdou, M.I., A.M. Al-Sabagh and M.M. Drdir. 2013. Evaluation of Egyptian bentonite and nano-bentonite as drilling mud. Egypt. J. Petrol. 22: 53–59.

Ameen, F., K. Alsamhary, J.A. Alabdullatif and S. ALNadhari. 2021. A review on metal-based nanoparticles and their toxicity to beneficial soil bacteria and fungi. Ecotoxicol. Environ. Saf. 213: 112027.

Arabani, M., A.K. Haghi and Y. Moradi. 2012. Evaluation of mechanical properties improvement of clayey sand by using carbon nanotubes. In: Proceedings of the 4th Int. Conf. Nanostruct. (ICNS4), Kish Island, Iran, 12–14 March 2012.

Aragaw, T.A., F.M. Bogale and B.A. Aragaw. 2021. Iron-based nanoparticles in wastewater treatment: A review on synthesis methods, applications, and removal mechanisms. J. Saudi Chem. Soc. 25: 101280.

Ayangbenro, A.S. and O.O. Babalola. 2017. A new strategy for heavy metal polluted environments: A review of microbial biosorbents. Int. J. Environ. Res. Public Health. 14: 94.

Asadishad, B., S. Chahal, V. Cianciarelli, K. Zhou and N. Tufenkji. 2017. Effect of gold nanoparticles on extracellular nutrient-cycling enzyme activity and bacterial community in soil slurries: Role of nanoparticle size and surface coating. Environ. Sci. Nano. 4: 907–918.

Alsharef, J., M.R. Taha, A.A. Firoozi and P. Govindasamy. 2016. Potential of using nanocarbons to stabilize weak soils. Appl. Environ. Soil Sci. 2016: 5060531.

Aziz, Y., G.A. Shah and M.I. Rashid. 2019. ZnO nanoparticles and zeolite influence soil nutrient availability but do not affect herbage nitrogen uptake from biogas slurry. Chemosphere. 216: 564–575.

Alsaeedi, A., H. El-Ramady, T. Alshaal, M. El-Garawany, N. Elhawat and A. Al-Otaibi. 2019. Silica nanoparticles boost growth and productivity of cucumber under water deficit and salinity stresses by balancing nutrients uptake. Plant Physiol. Biochem. 139: 1–10.

Akhtar, N. and N. Ilyas. 2022. Role of nanosilica to boost the activities of metabolites in Triticum aestivum facing drought stress. Plant Soil. 477: 99–115.

Baig, N., A.M. Alowaid, I. Abdulazeez, B. Salhi, M. Sajid and I. Kammakakam. 2022. Designing of nanotextured inorganic-organic hybrid PVDF membrane for efficient separation of the oil-in-water emulsions. Chemosphere. 308: 136531.

Bayat, H., Z. Kolahchi, S. Valaey, M. Rastgou and S. Mahdavi. 2018. Novel impacts of nanoparticles on soil properties: Tensile strength of aggregates and compression characteristics of soil. Arch. Agron. Soil Sci. 64: 776–789.

Bai, X., S. Zhao and L. Duo. 2017. Impacts of carbon nanomaterials on the diversity of microarthropods in turfgrass soil. Sci. Rep. 7: 1779.

Belal, E. and H. El-Ramady. 2016. Nanoparticles in water, soils and agriculture. pp. 311–358. *In*: Ranjan, S., N. Dasgupta and E. Lichtfouse [eds.]. Nanoscience in Food and Agriculture 2, Sustainable Agriculture Reviews 21. Springer Int. Publ., Switzerland.

Bhattacharjya, S., T. Adhikari, A. Sahu and A.K. Patra. 2021. Ecotoxicological effect of TiO2 nanoparticles on different soil enzymes and microbial community. Ecotoxicology. 30: 719–732.

Bottero, J.Y., M.R. Wiesner, J. Labille, M. Auffan, V. Vidal and C. Santaella. 2017. Life cycle of nanoparticles in the environment: Nanomaterials and agriculture application - Life cycle through their direct application and indirect application in biosludge-amended soils. pp. 333–346. *In*: Axelos, M.A. and M.V. de Voorde [eds.]. Nanotechnology in Agriculture and Food Science, First Edition. Wiley-VCH Verlag GmbH & Co. KGaA.

Bünemann, E.K., G. Bongiorno, Z. Bai, R.E. Creamer, G. De Deyn, R. de Goede et al. 2018. Soil quality – A critical review. Soil Biol. Biochem. 120: 105–125.

Cao, J., M. Gao, J. Wang, Y. Liu, X. Zhang, Y. Ping et al. 2023. Construction of nano slow-release systems for antibacterial active substances and its applications: A comprehensive review. Front. Nutr. 10: 1109204.

Chang, S.Y., L. Liu and S.A. Asher. 1994. Preparation and properties of tailored morphology, monodisperse colloidal silica-cadmium sulfide nanocomposites. J. Am. Chem. Soc. 116: 6739–6744.

Chung, H., Y. Son, T.K. Yoon, S. Kim and W. Kim. 2011. The effect of multi-walled carbon nanotubes on soil microbial activity. Ecotoxicol. Environ. Safety. 74: 569–575.

haudhary, P., P. Khati, A. Chaudhary, S. Gangola, R. Kumar and A. Sharma. 2021. Bioinoculation using indigenous Bacillus spp. improves growth and yield of Zea mays under the influence of nanozeolite. 3 Biotech. 11: 11.

Chong, K.P. 2004. Nano science and engineering in mechanics and materials. J. Phys. Chem. Solids. 65: 1501–1506.

Costa, J., R.S. Oliveira, I. Tiago, Y. Ma, C. Galhano, H. Freitas and P. Castro. 2018. Soil microorganisms. pp. 457–482. *In*: Sánchez-Moreiras, A.M. and M.J. Reigosa [eds.]. Advances in Plant Ecophysiology Techniques. Springer Int. Publ., Switzerland.

Darvishi, Z. and A. Morsali. 2011. Synthesis and characterization of nano-bentonite by sonochemical method. Ultrason. Sonochem. 18: 238–242.

Das, P., S. Baruam, S. Sarkar, N. Karak, P. Bhattacharyya, N. Raza et al. 2018. Plant extract–mediated green silver nanoparticles: Efficacy as soil conditioner and plant growth promoter. J. Hazard. Mater. 346: 62–72.

Das, P., K. Sarmah, N. Hussain, S. Pratihar, S. Das, P. Bhattacharyya et al. 2016. Novel synthesis of an iron oxalate capped iron oxide nanomaterial: A unique soil conditioner and slow release eco-friendly source of iron sustenance in plants. RSC Adv. 6: 103012–103025.

Dietler, G., C. Aubert, D.S. Cannell and P. Wiltzius. 1986. Gelation of colloidal silica. Phys. Rev. Lett. 57: 3117–3120.

Dinesh, R., M. Anandaraj, V. Srinivasan and S. Hamza. 2012. Engineered nanoparticles in the soil and their potential implications to microbial activity. Geoderma. 173: 19–27.

Dubey, A. and D. Mailapalli. 2016. Nanofertilisers, nanopesticides, nanosensors of pest and nanotoxicity in agriculture. pp. 307–330. *In*: Lichtfouse, E. [ed.]. Sustainable Agriculture Reviews 19. Springer Int. Publ., Switzerland.

El Mohtar, C., J. Clarke, A. Bobet, M. Santagata, V. Drnevich and C. Johnston. 2008. Cyclic response of a sand with thixotropic pore fluid. *In*: Zeng, D., M.T. Manzari and D.R. Hiltunen [eds.]. Geotechnical Earthquake Engineering and Soil Dynamics. Geotech. Special Publ. 181 (CD-ROM). Reston, VA, USA, ASCE.

El Mohtar, C., A. Bobet and M. Santagata. 2013. Liquefaction mitigation using bentonite suspensions. J. Geotech. Geoenviron. Eng. 139: 1369–1380.

El-Nagar, D.A. and D.H. Sary. 2021. Synthesis and characterization of nano bentonite and its effect on some properties of sandy soils. Soil Till. Res. 208: 104872.

Fakhim, B., A. Hassani, A. Rashidi and P. Ghodousi. 2013. Predicting the impact of multiwalled carbon nanotubes on the cement hydration products and durability of cementitious matrix using artificial neural network modeling technique. The Scientific World J. 2013: 103713.

Ahmad, F., A. Khaliq, B. Qiu, M. Sultan and J. Ma. 2021. Advancements of spraying technology in agriculture. *In*: Ahmad, F. and S. Muhammad [eds.]. Technology in Agriculture. IntechOpen, Rijeka.

Fernández, F.G. and R.G. Hoeft. 2009. Managing soil pH and crop nutrients. Ill. Agron. Handb. 24: 91–112.

Gallagher, P.M. and J.K. Mitchell. 2002. Influence of colloidal silica grout on liquefaction potential and cyclic undrained behavior of loose sand. Soil Dyn. Earthq. Eng. 22: 1017–1026.

García-Locascio, E., E.I. Valenzuela and P. Cervantes-Avilés. 2024. Impact of seed priming with selenium nanoparticles on germination and seedlings growth of tomato. Sci. Rep. 14: 6726.

Gajic´, G., L. Djurdjevic´, O. Kostic´, S. Jaric´, M. Mitrovic´ and P. Pavlovic´. 2018. Ecological potential of plants for phytoremediation and ecorestoration of fly ash deposits and mine wastes. Front. Environ. Sci. 6: 124.

Gattupalli, M., K. Dashora, M. Mishra, Z. Javed and G.D. Tripathi. 2023. Microbial bioprocess performance in nanoparticle-mediated composting. Crit. Rev. Biotechnol. 43: 1193–1210.

Gupta, N., S.K. Rai, R. Kumar, P.M. Singh, T. Chaubey, V. Singh et al. 2024. Seed priming with engineered nanomaterials for mitigating abiotic stress in plants. pp. 229–247. *In*: Pudake, R.N., R.M. Tripathi and S.S. Gill [eds.]. Nanotechnology for Abiotic Stress Tolerance and Management in Crop Plants. Academic Press, Elsevier, USA.

Grün, A-L., W. Manz, Y.L. Kohl, F. Meier, S. Straskraba, C. Jost et al. 2019. Impact of silver nanoparticles (AgNP) on soil microbial community depending on functionalization, concentration, exposure time, and soil texture. Environ. Sci. Eur. 31: 15.

Ge, Y., J.P. Schimel and P.A. Holden. 2011. Evidence for negative effects of TiO2 and ZnO nanoparticles on soil bacterial communities. Environ. Sci. Technol. 45: 1659–1664.

Hong, J., C. Wang, D.C. Wagner, J.L.G. Torresdey, F. He and C.M. Rico. 2021. Foliar application of nanoparticles: mechanisms of absorption, transfer, and multiple impacts. Environ. Sci. Nano 8: 1196–1210.

Huang, Y. and L. Wang. 2016. Experimental studies on nanomaterials for soil improvement: a review. Environ. Earth Sci. 75: 497.

Huang, H., J. Chen, S. Liu and P. Shengyan. 2022. Impact of ZnO nanoparticles on soil lead bioavailability and microbial properties. Sci. Total Environ. 806: 150299.

Barough, J.M., S. Çelik and M. Oltulu. 2021. Investigation into the effect of nanomaterial injection on improving the geotechnical properties of granular soils. Iran. J. Sci. Technol., Trans. Civ. Eng. 46: 3163–3179.

Johansen, A., A.L. Pedersen, K.A. Jensen, U. Karlson, B.M. Hansen, J.J. Scott-Fordsmand et al. 2008. Effects of C60 fullerene nanoparticles on soil bacteria and protozoans. Environ. Toxicol. Chem. 27: 1895–1903.

Khan, S.T., S.F. Adil, M.R. Shaik, H.Z. Alkhathlan, M. Khan and M. Khan. 2022. Engineered nanomaterials in soil: their impact on soil microbiome and plant health. Plants 11: 109.

Grün, A.L., W. Manz, Y.L. Kohl, F. Meier, S. Straskraba, C. Jost et al. 2019. Impact of silver nanoparticles (AgNP) on soil microbial community depending on functionalization, concentration, exposure time, and soil texture. Environ. Sci. Eur. 31: 15.

Ge, Y., J.P. Schimel and P.A. Holden. 2011. Evidence for negative effects of TiO2 and ZnO nanoparticles on soil bacterial communities. Environ. Sci. Technol. 45: 1659–1664.

Hong, J., C. Wang, D.C. Wagner, J.L.G. Torresdey, F. He and C.M. Rico. 2021. Foliar application of nanoparticles: mechanisms of absorption, transfer, and multiple impacts. Environ. Sci. Nano. 8: 1196–1210.

Huang, Y. and L. Wang. 2016. Experimental studies on nanomaterials for soil improvement: a review. Environ. Earth Sci. 75: 497.

Huang, H., J. Chen, S. Liu and P. Shengyan. 2022. Impact of ZnO nanoparticles on soil lead bioavailability and microbial properties. Sci. Total Environ. 806: 150299.

Barough, J.M., S. Çelik and M. Oltulu. 2021. Investigation into the effect of nanomaterial injection on improving the geotechnical properties of granular soils. Iran. J. Sci. Technol., Trans. Civ. Eng. 46: 3163–3179.

Johansen, A., A.L. Pedersen, K.A. Jensen, U. Karlson, B.M. Hansen, J.J. Scott-Fordsmand et al. 2008. Effects of C60 fullerene nanoparticles on soil bacteria and protozoans. Environ. Toxicol. Chem. 27: 1895–1903.

Khan, S.T., S.F. Adil, M.R. Shaik, H.Z. Alkhathlan, M. Khan and M. Khan. 2022. Engineered nanomaterials in soil: their impact on soil microbiome and plant health. Plants. 11: 109.

Khodakovskaya, M.V., K. de Silva, A.S. Biris, E. Dervishi and H. Villagarcia. 2012. Carbon nanotubes induce growth enhancement of tobacco cells. ACS Nano 6: 2128–2135.

Kahru, A. and H.C. Dubourguier. 2010. From ecotoxicology to nanoecotoxicology. Toxicology. 269: 105–119.

Khin, M., A.S. Nair, V.J. Babu, R. Murugan and S. Ramakrishna. 2012. A review on nanomaterials for environmental remediation. Energy Environ. Sci. 5: 8075.

Kondal, R., A. Kalia, O. Krejcar, K. Kuca, S.P. Sharma, K. Luthra et al. 2021. Chitosan-urea nanocomposite for improved fertilizer applications: The effect on the soil enzymatic activities and microflora dynamics in N cycle of potatoes (*Solanum tuberosum* L.). Polymers. 13: 2887.

Lin, T., X. Chen, Y. Ren, B. Qing, M. Zhang, Z. Mo et al. 2024. Effects of iron oxide nanocoatings on the seed germination, seedling growth, and antioxidant response of aromatic rice grown in the presence of different concentrations of rice straw extracts. J. Nanoparticle Res. 26: 78.

Liu, S., X. Qi, C. Han, J. Liu, X. Sheng, H. Li et al. 2017. Novel nano-submicron mineral-based soil conditioner for sustainable agricultural development. J. Cleaner Prod. 149: 896–903.

Moebius-Clune, B. 2017. Soil health initiatives of the USDA natural resources conservation service (NRCS). No. 1958-2017-2580.

Mrdi, M.S. and H. Niroumand. 2023. Nano soil-improvement: A novel approach for enhancing geotechnical properties of soil. Sci. Rep. 13: 37918.

McKee, M.S. and J. Filser. 2016. Impacts of metal-based engineered nanomaterials on soil communities. Environ. Sci. Nano. 3: 506–533.

Moll, J., F. Klingenfuss, F. Widmer, A. Gogos, T.D. Bucheli, M. Hartmann et al. 2017. Effects of titanium dioxide nanoparticles on soil microbial communities and wheat biomass. Soil Biol. Biochem. 111: 85–93.

Majumdar, S., J.R. Peralta-Videa, J. Trujillo-Reyes, Y. Sun, A.C. Barrios, G. Niu et al. 2016. Soil organic matter influences cerium translocation and physiological processes in kidney bean plants exposed to cerium oxide nanoparticles. Sci. Total Environ. 569: 201–211.

Mollamahmutoglu, M. and Y. Yilmaz. 2010. Pre- and post-cyclic loading strength of silica-grouted sand. Proc. ICE Geotech. Eng. 163: 343–348.

New, W., J.E. Ogbezode and P. Gani. 2023. Nanoparticles in soil remediation: Challenges and opportunities. Ind. Domest. Waste Manag. 3: 127–140.

Nakhli, S.A.A., M. Delkash, B.E. Bakhshayesh and H. Kazemian. 2017. Application of zeolites for sustainable agriculture: A review on water and nutrient retention. Water Air Soil Pollut. 228: 464.

Nowack, B. and T.D. Bucheli. 2007. Occurrence, behavior and effects of nanoparticles in the environment. Environ. Pollut. 150: 5–22.

Olesen, K.P. 2010. Turning sandy soil to farmland: 66% water saved in sandy soil treated with NanoClay. Desert Control Institute Inc. 1: 1–10.

Persoff, P., J. Apps, G. Moridis and J.M. Whang. 1999. Effect of dilution and contaminants on sand grouted with colloidal silica. J. Geotechn. Geoenviron. Eng. 125: 461–469.

Prakash, P. and S. Chandran. 2023. Nano-phytoremediation of heavy metals from soil: A critical review. Pollutants. 3: 360–380.

Rafiq, M.K., S.D. Joseph, F. Li, Y. Bai, Z. Shang, A. Rawal et al. 2017. Pyrolysis of attapulgite clay blended with yak dung enhances pasture growth and soil health: Characterization and initial field trials. Sci. Total Environ. 607-608: 184–194.

Rahmati, M. and M. Kousehlou. 2023. Nanoparticles as soil amendments. pp. 163–179. *In*: Fotopoulos, V. and Gohari, G. [eds.]. Engineered Nanoparticles in Agriculture: From Laboratory to Field. Walter de Gruyter GmbH, Berlin, Germany.

Raliya, R., J.C. Tarafdar and P. Biswas. 2016. Enhancing the mobilization of native phosphorus in mung bean rhizosphere using ZnO nanoparticles synthesized by soil fungi. J. Agric. Food Chem. 64: 3111–3118.

Raliya, R., V. Saharan, C. Dimkpa and P. Biswas. 2018. Nanofertilizer for precision and sustainable agriculture: Current state and future perspectives. J. Agric. Food Chem. 66: 6487–6503.

Rajput, V.D., T. Minkina, M. Feizi, A. Kumari, M. Khan, S. Mandzhieva et al. 2021. Effects of silicon and silicon-based nanoparticles on rhizosphere microbiome, plant stress and growth. Biology. 10: 791.

Rajput, V.D., T. Minkina, S.K. Upadhyay, A. Kumari, A. Ranjan, S. Mandzhieva et al. 2022. Nanotechnology in the restoration of polluted soil. Nanomaterials. 12: 769.

Srivastava, S., A. Shukla, V.D. Rajput, K. Kumar, T. Minkina, S. Mandzhieva et al. 2021. Arsenic remediation through sustainable phytoremediation approaches. Minerals. 11: 936.

Rasse, D.P., S. Weldon, E.J. Joner, S. Joseph, C.I. Kammann, X. Liu et al. 2022. Enhancing plant N uptake with biochar-based fertilizers: Limitation of sorption and prospects. Plant and Soil. 475: 213–236.

Rinot, O., G.J. Levy, Y. Steinberger, T. Svoray and G. Eshel. 2019. Soil health assessment: A critical review of current methodologies and a proposed new approach. Sci. Total Environ. 648: 1484–1491.

Rugg, D., J. Yoon, H. Hwang and C.S. El Mohta. 2011. Undrained shearing properties of sand permeated with a bentonite suspension for static liquefaction mitigation. Geo-Frontiers. 2011: 677–686.

Shahin, S.S., L.A.E. Fayed and H. Ahmad. 2015. Review of nano additives in stabilization of soil. In: Proceedings of the 7th International Conference on Nano-Technology in Construction.

Sharma, R. and V. Kumar. 2024. Nano enabled agriculture for sustainable soil. Waste Management Bulletin. 2: 152161.

Sharma, V., B. Javed, H. Byrne, J. Curtin and F. Tian. 2022. Zeolites as carriers of nano-fertilizers: From structures and principles to prospects and challenges. Appl. Nano. 3: 163–186.

Saratale, R.G., G.D. Saratale, H.S. Shin, J.M. Jacob, A. Pugazhendhi, M. Bhaisare et al. 2018. New insights on the green synthesis of metallic nanoparticles using plant and waste biomaterials: Current knowledge, their agricultural and environmental applications. Environ. Sci. Pollut. Res. 25: 10164–10183.

Sathurusinghe, P., W.M.N.A.P.B. Herath, H.A.R. Subhashini, A.M.C.S.S. Kumara and K.R.B. Herath. 2012. Stress concentrations in single walled carbon nanotube reinforced metal and polymer composites under uniaxial loading. Int. J. Adv. Struc. Geotechn. Engin. 1: 2319–5347.

Singh, A.K., P. Pal, V. Gupta, T.P. Yadav, V. Gupta and S.P. Singh. 2018. Green synthesis, characterization and antimicrobial activity of zinc oxide quantum dots using Eclipta alba. Mater Chem and Phys 203: 40–48.

Singh, P. and A.P. Singh. (2021). Nanomaterials in soil health management and crop production: potentials and limitations. pp. 1–25. *In*: Kharissova O.V., L.M. Torres-Martínez and B.I. Kharisov (eds.). Handbook of Nanomaterials and Nanocomposites for Energy and Environmental Applications. Springer Nature, Switzerland.

Sun, C., K. Hu, D. Mu, Z. Wang and X. Yu. 2022. The widespread use of nanomaterials: The effects on the function and diversity of environmental microbial communities. Microorganisms. 10: 2080.

Tiwari, D.K., N. Dasgupta-Schubert, L.M.V. Cendejas, J. Villegas, L.C. Montoya and S.E.B. García. 2014. Interfacing carbon nanotubes (CNT) with plants: Enhancement of growth, water and ionic nutrient uptake in maize (*Zea mays*) and implications for nanoagriculture. Appl. Nanosci. 4: 577–591.

Taha, M. and O. Taha. 2012. Influence of nano-material on the expansive and shrinkage soil behavior. J. Nanopart. Res. 14: 1190.

Tanaka, H., J. Meunier and D. Bonn. 2004. Nonergodic states of charged colloidal suspensions: Repulsive and attractive glasses and gels. Phys. Rev. E 69: 031404.

Tong, Z., M. Bischoff, L. Nies, B. Applegate and R.F. Turco. 2007. Impact of fullerene (C60) on a soil microbial community. Environ. Sci. Technol. 41: 2985–2991.

Tripathi, S., S. Mahra, J. Victoria, K. Tiwari, S. Rana, D.K. Tripathi et al. 2023. Recent advances and perspectives of nanomaterials in agricultural management and associated environmental risk: A review. Nanomaterials. 13: 1604.

van Dijk, M. and G. Meijerink. 2014. A review of food security scenario studies: Gaps and ways forward. pp. 30–32. *In*: Achterbosch, T.J., M. van Dorp, W.F. van Driel, J.J. Groot, J. van der Lee, A. Verhagen et al. [eds.]. The Food Puzzle: Pathways to Securing Food for All. Wageningen UR.

Verma, Y., S.K. Singh, H.S. Jatav, V.D. Rajput and T. Minkina. 2022. Interaction of zinc oxide nanoparticles with soil: Insights into the chemical and biological properties. Environ. Geochem. Health. 44: 221–234.

Wang, J., K. Shu, L. Zhang and Y. Si. (2017). Effects of Silver Nanoparticles on Soil Microbial Communities and Bacterial Nitrification in Suburban Vegetable Soils. Pedosphere 27(3): 482–490. https://doi.org/10.1016/S1002-0160(17)60344-8.

Xin, X., Z. He, M.R Hill, R.P. Niedz, X. Jiang and B.S. Sumerlin. 2018. Efficiency of Biodegradable and pH-Responsive Polysuccinimide Nanoparticles (PSI-NPs) as Smart Nanodelivery Systems in Grapefruit: *In Vitro* Cellular Investigation. Macromol. Biosci. 18: 1800159.

Ye, J., Joseph, S.D., Ji, M., Nielsen, S., Mitchell, D.R.G., Donne, S., Horvat, J., Wang, J., Munroe, P., and Thomas, T. 2017. Chemolithotrophic processes in the bacterial communities on the surface of mineral-enriched biochars. ISME J. 11(5): 1087–1101. https://doi.org/10.1038/ismej.2016.187.

Zaim, N.S.H.B.H., Tan, H.L., Rahman, S.M.A., Bakar, N.F.A., Osman, M.S., Thakur, V.K. et al. 2023. Recent advances in seed coating treatment using nanoparticles and nanofibers for enhanced seed germination and protection. J. Plant Growth Regulation. 42(12): 7374–7402. https://doi.org/10.1007/s00344-023-11038-4.

Zhou, P., Adeel, M., Shakoor, N., Guo, M., Hao, Y., Azeem, I. et al. 2020. Application of nanoparticles alleviates heavy metals stress and promotes plant growth: An overview. Nanomaterials. 11: 26. https://doi.org/10.3390/nano11010026.

Zhao, F., Xin, X., Cao, Y., Su, D., Ji, P., Zhu, Z. et al. 2021. Use of carbon nanoparticles to improve soil fertility, crop growth and nutrient uptake by corn (*Zea mays* L.). Nanomaterials. 11: 2717. https://doi.org/10.3390/nano11102717.

Zorraquín-Peña, I., Cueva, C., Bartolomé, B. and Moreno-Arribas, M.V. 2020. Silver nanoparticles against foodborne bacteria: Effects at intestinal level and health limitations. Microorganisms. 8(1): 132. https://doi.org/10.3390/microorganisms8010132.

Zahra, Z., Arshad, M., Rafique, R., Mahmood, A., Habib, A., Qazi, I.A. et al. 2015. Metallic nanoparticle (TiO_2 and Fe_3O_4) application modifies rhizosphere phosphorus availability and uptake by Lactuca sativa. J. Agric. Food Chem. 63: 6876–6882. https://doi.org/10.1021/acs.jafc.5b01611.

CHAPTER-14

Nanotechnology for Soil Erosion Control and Land Rehabilitation

Faizan Ali,[1] *Akhtar Hameed,*[2,*] *Abdul Rehman,*[3] *Nasir Ahmad Rajput,*[3] *Muhammad Atiq*[3] and *Nighat Raza*[4]

1. Introduction

Soil erosion and land degradation are pressing global concerns that pose severe threats to agricultural productivity, ecosystem stability, and overall food security. Soil degradation refers to the decline in soil quality, driven by a combination of physical, chemical, and biological factors. These include erosion, nutrient depletion, salinization, loss of organic matter, and contamination by pollutants. Among these, water and wind erosion are the primary mechanisms responsible for the loss of fertile topsoil, directly affecting land productivity and sustainability. It is estimated that soil erosion affects approximately 65% of the world's soil resources, leading to the annual loss of around 36 billion tons of fertile soil. This translates into significant economic burdens, with global costs reaching nearly $400 billion per year due to reduced agricultural yields, increased costs of food production, and the need for restoration efforts (Yin et al. 2022, Rhodes 2014). The underlying causes of soil degradation are multifaceted and interconnected. Deforestation plays a significant role by removing vegetation that stabilizes the soil and prevents erosion. Intensive agricultural practices, such as monocropping, overgrazing, and excessive tillage, contribute to soil depletion and structural breakdown. Urbanization further exacerbates land degradation by sealing soil surfaces with impervious materials, disrupting natural hydrological cycles, and increasing runoff. Climate change acts as a compounding factor, altering precipitation patterns and increasing the frequency of extreme weather events, such as heavy rainfall and droughts, which intensify soil erosion and nutrient loss. If left unaddressed, soil degradation can lead to desertification, loss of biodiversity, reduced water retention capacity, and diminished agricultural productivity. These consequences have severe implications for global food security and sustainable land management. Therefore, effective soil conservation and land rehabilitation strategies are critical for mitigating the negative impacts of soil degradation and ensuring long-term ecological balance.

[1] Department of Agricultural Science & Engineering, College of Agriculture, Tennessee State University, Nashville, TN 37209, USA.
[2] Institute of Plant Protection, MNS University of Agriculture Multan 61000, Pakistan.
[3] Department of Plant Pathology, University of Agriculture Faisalabad 38000, Pakistan.
[4] Department of Food Science and Technology, MNS University of Agriculture Multan 61000, Pakistan
* Corresponding author: akhtar.hameed@mnsuam.edu.pk

DOI: 10.1201/9781003558354-14

Land rehabilitation is a strategic approach aimed at restoring degraded lands to a productive and sustainable state. This process is essential for maintaining ecosystem health, supporting biodiversity, and ensuring the resilience of human livelihoods that depend on soil fertility. Land rehabilitation involves a combination of physical, biological, and chemical techniques to improve soil structure, enhance water retention, and restore lost nutrients. Key techniques for land rehabilitation include reforestation, soil conservation practices, and ecological restoration. Reforestation and afforestation initiatives help stabilize soil by increasing vegetation cover, reducing runoff, and improving organic matter content. Soil conservation practices such as terracing, contour farming, and cover cropping help mitigate erosion while improving soil fertility. Additionally, ecological restoration strategies, such as agroforestry and regenerative agriculture, integrate natural ecological processes to restore degraded lands sustainably.

The economic benefits of land rehabilitation are substantial. Research suggests that for every dollar invested in land restoration, up to $30 can be generated in economic returns through enhanced agricultural productivity, improved water quality, and ecosystem services (Jenkins et al. 2010). Furthermore, land rehabilitation is a key component in combating climate change, as restored lands contribute to carbon sequestration, reducing atmospheric greenhouse gas concentrations.

On a global scale, achieving Land Degradation Neutrality (LDN) by 2030 is a priority under international frameworks such as the United Nations Convention to Combat Desertification (UNCCD) and the UN Decade on Ecosystem Restoration (2021–2030) (Sharma et al. 2022). These initiatives emphasize the importance of sustainable land management practices, advocating for policies and investments that support land rehabilitation efforts worldwide. By integrating scientific advancements with policy frameworks, stakeholders can work towards reversing land degradation trends and promoting sustainable agricultural and environmental practices. Nanotechnology has emerged as a transformative tool in environmental protection, offering innovative solutions for addressing soil degradation, pollution control, and sustainable land management. Engineered nanomaterials (ENMs) have unique physicochemical properties that make them highly effective in soil remediation, water purification, and pollutant degradation (Huang et al. 2024). Their high surface area, reactivity, and tunable properties allow them to interact with environmental contaminants at a molecular level, leading to enhanced efficiency in remediation processes.

In the context of soil degradation, nanotechnology plays a crucial role in improving soil health and fertility. Nanomaterials such as nano-clays, nano-fertilizers, and biochar-based nanoparticles enhance soil structure, promote microbial activity, and increase nutrient availability (Rashid et al. 2023). For example, nano-fertilizers have been shown to improve nutrient use efficiency, reducing the overuse of chemical fertilizers that contribute to soil degradation and water pollution. Additionally, nano-based soil amendments help retain moisture, reducing the impacts of drought and desertification (Abd-Elsalam et al. 2024). Nanotechnology also offers effective solutions for pollution control, particularly in mitigating heavy metal contamination and pesticide residues in agricultural soils. Nanoparticles such as zero-valent iron (nZVI), carbon nanotubes, and titanium dioxide (TiO_2) are widely used in soil remediation to adsorb and neutralize toxic substances (Li et al. 2020). These materials can break down organic pollutants through photocatalytic and redox reactions, restoring soil quality and preventing harmful substances from entering food chains. Beyond soil health, nanotechnology is instrumental in water treatment and purification. Nano-filtration membranes and adsorption-based nanomaterials are highly effective in removing contaminants such as heavy metals, organic pollutants, and pathogens from water sources. This ensures a clean and sustainable water supply, which is essential for agricultural productivity and land rehabilitation efforts.

Despite its promising applications, concerns about the environmental and health risks of engineered nanomaterials persist (Hristozov and Malsch 2009). The potential toxicity and bioaccumulation of nanomaterials in ecosystems remain key areas of research. However, studies indicate that certain natural transformations, such as agglomeration, oxidation, and biodegradation, can reduce the impact of nanomaterials after their release into the environment.

Regulatory frameworks and risk assessment strategies are continuously evolving to ensure the safe and sustainable use of nanotechnology in environmental applications (Hull and Bowman 2014). The integration of nanotechnology with sustainable land management practices presents an opportunity to enhance environmental protection efforts while minimizing risks. As research advances, the development of eco-friendly nanomaterials and green nanotechnology approaches will further enhance their applicability in combating soil degradation, restoring land productivity, and mitigating climate change impacts. By harnessing the potential of nanotechnology, policymakers, scientists, and industry stakeholders can collaborate to achieve long-term sustainability in land rehabilitation and environmental conservation.

2. Causes and Impacts of Soil Erosion

Natural processes are significant contributors to soil erosion, often influenced by climatic and geological factors (Borrelli et al. 2017). Rainfall, surface runoff, and river flow are the primary drivers of water erosion. Raindrops dislodge soil particles (splash erosion), while runoff transports them, forming rills and gullies. This is most severe in areas with heavy rainfall, steep slopes, or poor vegetation cover. Prolonged water erosion can lead to severe degradation, reducing soil fertility and increasing the risk of landslides (Peng and Wang 2012). Additionally, coastal areas experience erosion due to wave action, which further accelerates land loss and habitat destruction. Flooding caused by excessive rainfall can also result in large-scale soil displacement, affecting both agricultural lands and natural ecosystems (Cui et al. 2017). Strong winds lift and transport loose soil particles, particularly in arid and semi-arid regions with sparse vegetation. Dust storms exacerbate this issue, leading to desertification in vulnerable areas. Wind erosion can remove significant amounts of topsoil, depleting essential nutrients necessary for plant growth. The phenomenon is particularly severe in dry agricultural regions where over-cultivation and lack of soil conservation measures intensify erosion rates. In some cases, fine dust particles are transported over long distances, impacting air quality and human health.

In colder climates, thawing ice releases sediment that is carried by water. Permafrost melting due to rising temperatures further destabilizes soil structures, increasing erosion risks (Vincent et al. 2017). Gravity also contributes by pulling soil down slopes, especially on steep terrains, causing landslides and mass soil movement, which can be devastating in mountainous and hilly regions (Liew 2023). These processes contribute to the degradation of natural landscapes and the loss of fertile land necessary for agriculture and vegetation growth. Abnormal weather patterns, such as intense rainfall or prolonged droughts, amplify erosion by destabilizing soil and reducing vegetation cover. Increased global temperatures lead to more frequent extreme weather events, which accelerate erosion rates and further threaten land sustainability. Climate change also affects soil moisture levels, leading to increased susceptibility to both wind and water erosion. Rising sea levels further contribute to coastal erosion, endangering infrastructure, communities, and agricultural lands in low-lying areas (Griggs and Reguero 2021). The removal of trees for agriculture or urban development exposes soil to wind and water erosion. Tree roots that stabilize the soil are lost, increasing vulnerability (Jafari et al. 2022). Deforestation also reduces the amount of organic matter in the soil, which plays a crucial role in maintaining soil structure and moisture retention. Large-scale deforestation for commercial logging and expansion of urban settlements leads to significant land degradation, increasing the risk of landslides and reduced land productivity (Amoakwah et al. 2022).

Intensive farming practices disturb the soil through plowing, monocropping, and leaving land fallow during off-seasons. Over-irrigation also contributes to salinization and erosion. Unsustainable farming techniques, such as excessive use of chemical fertilizers and pesticides, further degrade soil health, making it more susceptible to erosion. Practices like slash-and-burn agriculture exacerbate land degradation by destroying the protective vegetation cover. The expansion of farmland into marginal areas also increases the likelihood of soil degradation and desertification. Excessive grazing by livestock removes protective vegetation cover. Trampling by hooves compacts the soil, making it more prone to erosion (Mayel et al. 2021). Over time, overgrazed lands lose their capacity to support plant growth, leading to desertification and reduced soil productivity. In dry regions, overgrazing exacerbates wind erosion, leading to the loss of valuable topsoil and declining pasture quality. Land clearing for infrastructure development exposes large areas of soil to erosive forces. Mining and quarrying further disturb the topsoil. The expansion of urban areas leads to soil sealing, where natural surfaces are covered with concrete and asphalt, reducing the land's ability to absorb water and increasing surface runoff, which accelerates erosion in nearby regions (Gramaglia et al. 2024). Construction activities disrupt soil stability, increasing susceptibility to landslides and sedimentation in water bodies. Human-induced climate change exacerbates natural erosion processes by altering precipitation patterns and increasing extreme weather events. Rising temperatures and shifting rainfall patterns disrupt traditional land use practices, increasing soil vulnerability and reducing resilience against erosion. The increased occurrence of droughts and heavy storms significantly affects soil stability, leading to higher erosion rates and land degradation over time.

The consequences of soil erosion extend beyond land degradation, affecting ecosystems and human livelihoods (Table 1). Nutrient-rich topsoil is stripped away, reducing agricultural productivity and threatening food security. The decline in soil fertility necessitates increased use of fertilizers, which can have long-term negative effects on soil structure and sustainability. Loss of soil organic matter further reduces the soil's ability to retain water and nutrients, leading to declining agricultural yields over time. Sediments from eroded soils carry pollutants into waterways, degrading water quality and harming aquatic ecosystems. Excessive sedimentation can reduce the capacity of reservoirs and irrigation channels, increasing the risk of flooding and reducing water availability for agricultural use. Pollutants such as pesticides and heavy metals attached to eroded soil particles further contaminate water bodies, posing health risks to humans and wildlife. The loss of vegetation leads to barren landscapes prone to desertification, disrupting natural water cycles. Once soil is degraded beyond a certain point, it becomes nearly impossible to restore, leading to the expansion of unproductive lands. Desertification threatens food security in arid and semi-arid regions, reducing the availability of arable land and fresh water. Habitat destruction caused by erosion diminishes biodiversity, affecting ecosystem services like pollination and carbon sequestration. The decline in soil health also impacts belowground biodiversity, reducing microbial activity crucial for maintaining soil fertility. Wildlife populations dependent on stable ecosystems are forced to migrate or face extinction due to habitat degradation. Farmers face declining crop yields due to nutrient depletion, impacting food supply and rural economies. In developing regions, this can lead to increased food

Table 1. Key causes leading to soil erosion.

Cause	Description	Impact
Deforestation	Removal of trees and vegetation that protect soil from wind and water.	Increased vulnerability to erosion, loss of topsoil, and reduced soil fertility.
Overgrazing	Livestock grazing beyond the land's carrying capacity, leading to vegetation loss.	Soil compaction, reduced vegetation cover, and heightened risk of wind and water erosion.
Agricultural Practices	Poor farming practices such as monoculture, improper plowing, and tillage.	Loss of soil structure, reduced organic matter, increased runoff, and nutrient depletion.
Urbanization	Expansion of cities and infrastructure that alters natural land cover.	Soil sealing, loss of arable land, and increased surface runoff leading to erosion.
Climate Change	Changes in rainfall patterns, frequency of storms, and temperature extremes.	Intensified erosion from heavy rainfall events and reduced soil moisture retention.
Improper Irrigation	Over-irrigation or improper irrigation methods that lead to waterlogging or runoff.	Soil salinization, erosion due to water runoff, and reduced agricultural productivity.
Mining and Quarrying	Extraction of minerals and materials from the land.	Soil degradation, increased sedimentation in water bodies, and loss of biodiversity.
Natural Disasters	Events like floods, hurricanes, and landslides can rapidly alter the landscape.	Sudden, severe erosion leading to loss of topsoil, destruction of infrastructure, and habitat loss.

insecurity and malnutrition (Farooq et al. 2022). The reliance on external inputs such as fertilizers and irrigation increases production costs, further burdening farmers. In regions dependent on agriculture, reduced productivity exacerbates poverty and food insecurity. Communities reliant on subsistence farming are particularly vulnerable to the effects of soil erosion, as degraded land limits their ability to produce sufficient food and income. Displacement due to loss of arable land further contributes to economic instability and migration challenges. Soil erosion costs billions annually in lost productivity and restoration efforts. For instance, the U.S. alone incurs $44 billion in losses annually due to erosion-related issues. Globally, investments in soil conservation and sustainable land management are necessary to mitigate economic losses and protect livelihoods dependent on healthy soils. Governments and environmental organizations are increasingly recognizing the need for policies that promote sustainable land management practices to combat the long-term effects of soil erosion.

3. Fundamentals of Nanotechnology

Nanotechnology has emerged as a groundbreaking scientific field with vast potential across multiple disciplines, including agriculture and environmental management. By manipulating materials at the nanoscale, scientists can harness unique physical, chemical, and biological properties that differ significantly from bulk materials. These properties make nanomaterials particularly effective for addressing critical challenges in soil and water management, such as pollution remediation, nutrient delivery, and soil stabilization. This section explores the fundamental principles of nanotechnology, the different types of nanomaterials, and their specific applications in enhancing soil and water quality.

3.1 Definition and Principles of Nanotechnology

Nanotechnology is the manipulation, design, and application of materials at the nanoscale, typically ranging from 1 to 100 nanometers (nm) (Binns 2021). At this scale, materials exhibit distinctive physical, chemical, and biological properties that significantly differ from their macroscopic counterparts. These unique characteristics emerge primarily due to the increased surface-area-to-volume ratio and the dominance of quantum effects, which influence the behavior of nanoscale materials in ways not observed at larger scales.

Nanotechnology employs two fundamental approaches for material synthesis and modification (Abid et al. 2022):

Bottom-up approach: This method involves assembling structures atom by atom or molecule by molecule, using principles of molecular self-assembly and molecular recognition. It enables precise control over material properties, allowing for the development of highly specialized nanostructures.

Top-down approach: In contrast, this technique reduces larger structures to the nanoscale through physical or chemical means, such as lithography, etching, or milling. While this approach is effective for fabricating nanoscale devices, it often lacks atomic-level precision.

Nanotechnology bridges classical and quantum mechanics, enabling groundbreaking advancements in various fields, including medicine, agriculture, and environmental management. Its ability to manipulate matter at the atomic and molecular levels has led to revolutionary applications, particularly in soil and water management, where nanomaterials are increasingly being used to enhance soil stability, improve water filtration, and remediate environmental pollutants.

Figure 1 describes the main approaches for the synthesis of the NPs.

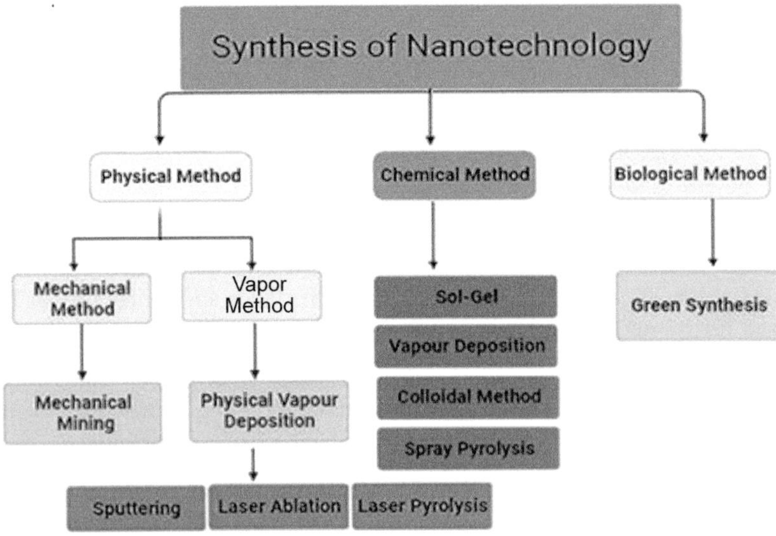

Figure 1. Key approaches for synthesis of nanotechnology/nanoparticles.

3.2 Types of Nanomaterials

Nanomaterials are broadly classified based on their dimensionality and composition (Tiwari et al. 2012):

3.2.1 Dimensional Classification

Zero-dimensional (0D): These materials have all three dimensions confined within the nanoscale. Examples include nanoparticles and nanoclusters, which exhibit high reactivity and unique optical properties.

One-dimensional (1D): These materials have one extended dimension beyond the nanoscale, such as nanorods, nanotubes, and nanowires. Their elongated structure enhances mechanical and electronic properties, making them useful for environmental sensing and filtration.

Two-dimensional (2D): These materials, including nanosheets and nanofilms, have two dimensions extending beyond the nanoscale. They possess exceptional surface area and electronic conductivity, beneficial for applications such as pollutant adsorption and soil stabilization.

Three-dimensional (3D): These nanostructures contain nanoscale features but are not entirely confined within the nanoscale. Examples include nanocomposites, which integrate nanoscale components into bulk materials to enhance their mechanical, thermal, or chemical properties.

3.2.2 Material-Based Classification

Carbon-based nanomaterials: These include fullerenes, graphene, and carbon nanotubes, known for their exceptional strength, electrical conductivity, and adsorption capabilities. They are widely used in pollutant removal and soil remediation (Díez-Pascual 2021).

Metal-based nanomaterials: These comprise nanoparticles made from metals such as gold, silver, and various metal oxides. They exhibit catalytic and antimicrobial properties, making them useful for environmental decontamination (Makvandi et al. 2020).

Dendrimers: These are branched polymers with highly functionalized surfaces, enabling precise drug delivery and targeted environmental applications, such as controlled-release fertilizers (Sherje et al. 2018).

Composites: These materials combine nanoparticles with bulk materials to enhance properties like durability, permeability, or reactivity. They are extensively used in improving soil structure and water purification technologies (Hanemann and Szabó 2010).

3.3 Properties of Nanomaterials Relevant to Soil and Water Management

Nanomaterials possess a range of unique properties that make them highly effective for soil and water management:

High Surface Area: Due to their nanoscale dimensions, these materials provide an immense reactive surface area, enhancing their ability to interact with soil particles and contaminants. This property is particularly useful for pollutant adsorption and nutrient retention (Mukhopadhyay et al. 2022).

Adsorption Capabilities: Carbon-based nanomaterials, such as carbon nanotubes and graphene, exhibit strong adsorption capacities for organic pollutants, including pesticides and heavy metals. Their high affinity for pollutants enables efficient remediation of contaminated soil and water (Rajput et al. 2022).

Catalytic Activity: Many metal-based nanoparticles, such as zero-valent iron (nZVI), serve as catalysts for chemical reactions that break down or immobilize pollutants. These nanoparticles play a crucial role in soil and groundwater decontamination efforts (Farooqi et al. 2022).

Soil Stabilization: Nanoparticles can penetrate soil pores and enhance binding strength between soil particles, improving soil structure and reducing susceptibility to erosion. This process strengthens the soil without requiring extensive mechanical intervention (Alkhaza'leh et al. 2025).

Controlled Release Mechanisms: Nanotechnology enables the development of smart delivery systems for fertilizers and soil amendments. Nanoparticles can encapsulate nutrients or remediation agents, releasing them gradually at targeted locations, thereby minimizing waste and maximizing efficiency (Vejan et al. 2021).

4. Applications of Nanotechnology in Soil Erosion Control

Soil erosion is a major environmental challenge that affects agricultural productivity, water quality, and ecosystem stability. Conventional methods for controlling soil erosion, such as mechanical barriers, chemical treatments, and vegetation cover, often have limitations in terms of efficiency, environmental sustainability, and long-term effectiveness. The advent of nanotechnology offers a promising approach to enhancing soil stability, improving water retention, and providing protective barriers against erosive forces. The unique properties of nanomaterials, including their high surface-area-to-volume ratio, reactivity, and tunable physical and chemical characteristics, make them well-suited for addressing soil erosion concerns. By integrating nanotechnology into soil management strategies, it is possible to develop innovative, cost-effective, and environmentally friendly solutions to mitigate the adverse impacts of erosion. This chapter explores various applications of nanotechnology in soil erosion control, focusing on nanomaterials for soil stabilization, water retention, and protective barriers.

4.1 Nanomaterials for Soil Stabilization

Nanotechnology offers innovative solutions for soil stabilization by enhancing soil strength, reducing erosion, and improving geotechnical properties. Nanomaterials such as nanoclays, nanosilica, and nano-cement have demonstrated significant potential in improving soil mechanical properties and long-term stability.

Nanoclays and Nanosilica: These nanomaterials fill voids in soil, creating a denser and more cohesive matrix. For example, nanoclay improves soil compaction, shear strength, and permeability characteristics, while nanosilica enhances compressive strength and reduces porosity. Studies have shown that incorporating nanosilica and nanoclay can increase unconfined compressive strength by up to 300% after curing. Additionally, nanosilica enhances the durability of soil by reducing water absorption, which in turn mitigates the impact of water-induced soil weakening (Ghadr et al. 2022, Niroumand et al. 2023).

Nano-cement: Nanocement suspensions improve soil binding and reduce breaking strain under stress (Harsh et al. 2023). The concentration of nanocement directly influences the mechanical stability of treated soils, offering a cost-effective alternative to traditional cementitious methods. The use of nano-cement not only increases the load-bearing capacity of soil but also enhances its resistance to external forces such as heavy rainfall, thereby preventing erosion and soil degradation (Vasconcelos 2024).

Eco-friendly Alternatives: Compared to conventional stabilization techniques, nanomaterials require less material and cause minimal environmental disturbance, making them a sustainable choice for embankments, slopes, and civil infrastructure. The reduced material requirement not only lowers costs but also minimizes carbon emissions associated with traditional soil stabilization methods. Furthermore, the integration of nanomaterials with bio-based stabilizers enhances their ecological compatibility, making them a viable alternative to synthetic soil stabilizers.

4.2 Nanotechnology in Water Retention and Drainage Management

Nanotechnology plays a critical role in improving soil water retention and managing drainage. Soil moisture availability is crucial for plant growth, and maintaining an optimal balance of water retention and drainage is essential for sustainable agricultural practices. Nanomaterials such as nanoclays, hydrogels, and nanoporous materials contribute to improved soil water management.

Nanoclays and Hydrogels: These materials enhance the moisture retention capacity of soils by over 10%, which is particularly beneficial in arid or drought-prone regions. They improve water infiltration and storage while reducing evaporation losses, leading to increased agricultural productivity by up to 20%. Hydrogels, in particular, act as superabsorbent materials capable of retaining water many times their weight, gradually releasing moisture to plant roots and maintaining soil hydration during dry periods (Krasnopeeva et al. 2022).

Nanoporous Materials: Nanoporous coatings applied to irrigation systems reduce water loss through evaporation and improve water distribution efficiency. These coatings also minimize energy consumption for water transport by reducing drag and scaling. Additionally, nanoporous materials can be engineered to selectively filter contaminants from water, ensuring the delivery of clean and nutrient-rich water to crops (Sharma et al. 2021).

Sustainability Benefits: By retaining more water in the soil profile, nanotechnology reduces the need for frequent irrigation, conserving water resources and supporting climate-resilient agriculture (Hatti et al. 2020). The application of nanomaterials in water management also improves soil aeration, preventing waterlogging and enhancing root growth. As a result, crops experience reduced stress from excessive water fluctuations, leading to higher yields and improved food security.

4.3 Nano-coatings and Barriers for Soil Protection

Nano-coatings provide a protective layer on soil surfaces or infrastructure components to mitigate erosion. These coatings create durable, resilient barriers that protect soil from external forces such as wind, water, and heavy rainfall. By leveraging nanotechnology, soil protection measures can be enhanced for long-term effectiveness and sustainability.

Surface Nano-coatings: Ultra-thin layers made from materials like silica, titanium dioxide, or graphene create durable barriers that protect soil from wind and water erosion (Behera et al. 2020).

These coatings are highly effective in harsh environmental conditions due to their strong molecular bonding with substrates. Graphene-based coatings, in particular, exhibit excellent mechanical strength, chemical stability, and resistance to environmental degradation, making them ideal for soil preservation.

Hydrophobic Barriers: Nano-coatings with hydrophobic properties prevent water infiltration into vulnerable areas, reducing erosion caused by heavy rainfall or surface runoff. These coatings also extend the lifespan of soil protection structures by preventing corrosion or biofouling (Garg et al. 2024). Hydrophobic nano-coatings are especially useful in sandy and loose soils, where excessive water infiltration can lead to instability and erosion.

Eco-friendly Application: Nano-coatings are resource-efficient; small quantities can cover large areas with minimal environmental impact. Their room-temperature curing process further reduces energy consumption during application. Additionally, advancements in biodegradable nano-coatings ensure minimal ecological footprint, promoting sustainable land management practices (Sah et al. 2024).

Conclusively, nanotechnology provides advanced tools for combating soil erosion through stabilization, water management, and protective barriers. These applications not only enhance soil resilience but also promote sustainable environmental practices. By harnessing the potential of nanomaterials, researchers and practitioners can develop innovative strategies for soil conservation, ensuring long-term agricultural productivity and environmental sustainability.

5. Nanotechnology for Land Rehabilitation

Land degradation and soil infertility are critical challenges threatening global food security, ecosystem sustainability, and environmental health. Unsustainable agricultural practices, industrial pollution, and climate change contribute to the deterioration of soil quality, necessitating innovative approaches for land rehabilitation. Nanotechnology has emerged as a promising solution for restoring soil health, improving fertility, and reclaiming contaminated lands through precision interventions. The application of nanomaterials in land rehabilitation offers unique advantages, such as controlled nutrient delivery, soil remediation, and enhanced plant growth, ensuring sustainable agricultural productivity and environmental restoration (Munir et al. 2024) (Table 2).

5.1 Nanoparticles for Nutrient Delivery and Soil Fertility Improvement

One of the most promising applications of nanotechnology in land rehabilitation is the development of nanofertilizers. These advanced fertilizers utilize nanoparticles to optimize nutrient delivery, ensuring efficient uptake by plants while minimizing environmental losses (Babu et al. 2022). Traditional fertilizers often suffer from inefficiencies due to leaching, volatilization, or poor absorption by plants. Nanofertilizers, on the other hand, enhance nutrient use efficiency by offering slow and controlled nutrient release, ensuring that essential elements remain available in the root zone for extended periods. For instance, nano-urea and zinc oxide (ZnO) nanoparticles have demonstrated remarkable improvements in plant growth and yield compared to conventional fertilizers (Adil et al. 2022). Studies indicate that nano-urea can significantly enhance nitrogen uptake efficiency, reducing the overall quantity of fertilizer needed while maintaining or even improving crop yields. Additionally, these nanoparticles influence soil microbial activity by stimulating beneficial microorganisms that contribute to nutrient cycling and organic matter decomposition (Chen et al. 2022). Titanium dioxide (TiO_2) nanoparticles, for example, have been found to enhance enzymatic activities such as dehydrogenase and phosphatase, which play a crucial role in soil fertility and plant nutrient absorption (Kaur et al. 2022). Furthermore, the adoption of nanofertilizers reduces the environmental impact of traditional fertilization methods. By minimizing over-application and

Table 2. Application of Nanotechnology for Soil Rehabilitation.

Nanoparticle	**Target**	**Result**	**References**
Iron sulfide	Mercury pollutants	Immobilizing *in-situ* mercury ions	(Xiong et al. 2009)
Nickle-iron bimetallic	Decabromodiphenyl ether (BDE209) pollutant	Degrade BDE209 in soil at ambient temperature	(Xie et al. 2014)
Nano iron	Trichloroethylene (TCE)	Removing TCE from the soil matrix	(Yang and Chang 2011)
Nano zerovalent iron (NZVI)	Organochlorine insecticides (DDT)	Efficient for DDT degradation in spiked soil	(El-Temsah and Joner 2013)
Nanocrystalline hydroxyapatite (nHA)	Cadmium and lead	Able to reduce water-soluble, bioaccessible, and phytoavailable Pb/Cd	(He et al. 2013)
Iron phosphate (vivianite)	Copper (Cu(II))	Availability of Cu in soil was decreased by the formation of copper phosphate minerals through precipitation and adsorption	(Liu and Zhao 2007)
Nanoscale zerovalent iron	Chromium (Cr(VI))	Immobilization of Cr(VI)	(Xu and Zhao 2007)
Titanium dioxide	Soil microbiome	Enhanced soil microbial activity	(Ihtisham et al. 2021)
Aluminum-based NPs	Soil texture improvement	Reduced leaching of agrochemicals, Increased biomass and nutrient uptake	(Landa 2021)
Magnetite nanoparticles	Nutrition profile	Increased soil fertility, Reduced heavy metal uptake by plants	(Khan et al. 2016)
Cerium dioxide nanoparticles	Soil texture and profile	Increased soil pH and nutrient availability, Reduced pesticide accumulation	(Majumdar et al. 2016)

runoff, nanotechnology helps mitigate water pollution, particularly in regions where excessive fertilizer use contributes to eutrophication of water bodies. This sustainable approach ensures that agricultural productivity is maintained without compromising ecosystem integrity.

5.2 Role of Nano-Biochar in Land Reclamation

Land reclamation is essential for restoring degraded or contaminated soils, and nano-biochar has emerged as an effective tool for this purpose. Nano-biochar, a modified form of biochar infused with nanoparticles, has demonstrated exceptional capabilities in improving soil properties and remediating polluted lands. One of its key advantages is the immobilization of heavy metals, which pose significant risks to soil health and plant growth. The high surface area and reactive functional groups of nano-biochar enable it to adsorb and immobilize toxic metals such as cadmium (Cd) and lead (Pb), reducing their bioavailability (Ma et al. 2023). This process significantly decreases phytotoxicity, making the soil more suitable for plant cultivation. Additionally, nano-biochar enhances soil microbial biomass and diversity by creating a more hospitable environment for beneficial microorganisms such as Actinobacteria and Bacteroidetes (Rajput and Minkina 2025). These microbes play a vital role in nutrient cycling, organic matter decomposition, and overall soil restoration. Beyond contaminant immobilization, nano-biochar contributes to soil structure improvement. Its porous nature enhances soil aeration, water retention, and cation exchange capacity, all of which are essential for revitalizing degraded soils. The integration of nano-biochar in land rehabilitation programs presents a sustainable and cost-effective strategy for restoring soil health and ensuring long-term agricultural productivity.

5.3 Enhancing Vegetative Growth with Nano-Fertilizers

The application of nanotechnology in plant nutrition extends beyond soil remediation to directly enhancing vegetative growth. Nano-fertilizers have been shown to optimize nutrient uptake efficiency, leading to improved plant health and productivity. By facilitating better root absorption and targeted nutrient delivery, these fertilizers help maximize photosynthesis, phytohormone biosynthesis, and antioxidant activity in plants. For example, ZnO nanoparticles have been reported to increase crop yields and improve nutritional quality in crops such as spinach and wheat (Zafar et al. 2022). These nanoparticles provide essential micronutrients in a bioavailable form, ensuring that plants receive the necessary elements for optimal growth. Additionally, plants treated with nano-fertilizers exhibit increased resistance to abiotic stresses such as drought and salinity (El-Saadony et al. 2022). By improving water-use efficiency and nutrient availability, nanotechnology enhances plant resilience in challenging environmental conditions. Moreover, the use of nano-fertilizers aligns with sustainable agricultural practices by reducing the dependency on bulk fertilizers. Conventional fertilizers often require large quantities to achieve desired results, leading to excessive resource consumption and environmental degradation. In contrast, nano-fertilizers allow for precise and efficient nutrient management, reducing input costs while maintaining high agricultural productivity. Nanotechnology provides transformative tools for land rehabilitation by improving soil fertility, reclaiming contaminated lands through nano-biochar, and enhancing vegetative growth with nano-fertilizers. These applications not only contribute to agricultural sustainability but also support broader environmental restoration efforts. By leveraging nanotechnology in soil management, it is possible to create resilient ecosystems, reduce environmental pollution, and ensure food security for future generations (Shoukat et al. 2024). As research and innovation in this field continue to advance, the integration of nanotechnology in land rehabilitation strategies holds immense potential for addressing global challenges related to soil degradation and land productivity.

6. Challenges and Limitations of Nanotechnology in Soil Erosion Control and Land Rehabilitation

Despite the remarkable potential of nanotechnology in mitigating soil erosion and rehabilitating degraded lands, several challenges and limitations hinder its widespread adoption. These challenges are multifaceted, ranging from environmental and health concerns to economic feasibility and regulatory hurdles. Addressing these issues is crucial for the responsible and sustainable integration of nanotechnology into environmental and agricultural applications.

6.1 Environmental and Health Concerns

The application of nanotechnology in soil stabilization, nutrient management, and erosion control introduces potential environmental and health risks that must be carefully evaluated. One of the primary concerns regarding the use of nanomaterials is their potential toxicity. Due to their nanoscale size and high surface reactivity, certain nanoparticles—such as titanium dioxide (TiO_2), carbon nanotubes, and some metal-based nanoparticles—can penetrate biological membranes and interact with cellular components (Medici et al. 2021). Studies have indicated that prolonged exposure to these nanoparticles may induce oxidative stress, inflammation, and even fibrosis in animal models. Additionally, their ability to bioaccumulate raises concerns about long-term health effects, particularly in agricultural workers handling these materials or populations consuming food grown in nano-enhanced soils. Another significant issue is the unintended release of nanoparticles into ecosystems. Once introduced into soil or water bodies, nanoparticles may persist due to their stability, affecting microbial communities and soil fauna. There is growing concern about their potential to disrupt soil biogeochemical cycles or alter the availability of essential nutrients. Furthermore, nanoparticles in water systems may be ingested by aquatic organisms, leading to unknown consequences for biodiversity and food chains. Given the limited data on their long-term ecological impact, researchers and policymakers must adopt a precautionary approach when implementing nanotechnology in environmental applications. Individuals involved in nanomaterial production, application, and disposal are at risk of exposure through inhalation, ingestion, or dermal contact. Fine nanoparticles can be airborne, posing inhalation hazards similar to ultrafine particulates found in industrial emissions. Proper handling procedures, personal protective equipment (PPE), and workplace ventilation systems are necessary to minimize risks. However, due to a lack of comprehensive safety guidelines, many industries and agricultural sectors are yet to establish best practices for worker protection.

6.2 Economic Feasibility and Scalability

While nanotechnology presents innovative solutions, its economic viability remains a significant barrier to widespread implementation (Pokrajac et al. 2021). Nanomaterials require sophisticated synthesis methods, specialized equipment, and highly skilled personnel, leading to high production costs. The need for precise control over nanoparticle size, purity, and functionality adds further expenses. Compared to conventional soil stabilizers and fertilizers, nanotechnology-based products may not always be cost-effective, particularly for large-scale agricultural and land rehabilitation projects (Marella et al. 2021). The high cost of raw materials and processing often discourages adoption by farmers and environmental agencies with budget constraints. Transitioning from laboratory research to commercial-scale applications remains a formidable challenge. Most nanotechnology innovations are developed at small scales under controlled conditions, and scaling up these processes for widespread deployment requires significant infrastructure investment. Continuous production techniques, such as roll-to-roll manufacturing and large-scale nanoparticle synthesis reactors, are still under development and require further refinement to ensure cost-effective and high-throughput manufacturing. Despite promising research, nanotechnology-based environmental solutions face slow market penetration due to various barriers. The agriculture and environmental sectors have well-established conventional methods for soil stabilization, water retention, and nutrient management, making it difficult for nanotechnology to replace traditional approaches. Furthermore, many small and medium enterprises lack the resources to invest in nanotechnology due to high initial costs and uncertain return on investment. Widespread adoption will require targeted financial incentives, subsidies, and awareness campaigns to encourage industries and stakeholders to transition to nano-based solutions.

6.3 Regulatory and Ethical Issues

The rapid advancements in nanotechnology have outpaced the development of regulatory frameworks, raising concerns about safety assessments, ethical considerations, and policy coordination (Amutha et al. 2024). Most current environmental and agricultural policies were not designed to address the unique properties of nanomaterials. As a result, there are significant gaps in regulatory oversight regarding the safe production, application, and disposal of nanoparticles. The absence of standardized testing protocols for nanoparticle toxicity and environmental impact assessments further complicates regulatory enforcement (Allan et al. 2021). Government agencies and international organizations must collaborate to establish clear guidelines and thresholds for nanomaterial use to ensure safety without stifling innovation. Public skepticism and ethical debates surrounding nanotechnology also pose hurdles to its adoption. Concerns regarding unforeseen consequences, potential environmental contamination, and human health risks fuel resistance among stakeholders. Ethical dilemmas arise when balancing technological progress with environmental stewardship and social responsibility. There is a need for increased transparency, public engagement, and ethical considerations in policymaking to foster responsible innovation and gain public trust. The global nature of nanotechnology research necessitates international coordination to harmonize policies and safety standards. However, differences in regulatory approaches between countries create inconsistencies in risk assessment and market entry requirements. Collaborative efforts among governments, research institutions, and industry leaders are essential to create a unified framework for safe and sustainable nanotechnology applications in soil conservation and land rehabilitation.

While nanotechnology holds immense promise in addressing soil erosion, enhancing soil fertility, and reclaiming degraded lands, its practical implementation is hindered by environmental risks, economic constraints, and regulatory uncertainties. Addressing these challenges requires a multidisciplinary approach involving scientific research, industry investment, and policy development. By improving safety assessments, reducing production costs, and establishing clear regulatory guidelines, nanotechnology can be integrated into environmental management strategies in a responsible and sustainable manner. Ensuring the safe and ethical application of nanotechnology will be crucial in realizing its full potential for long-term environmental sustainability and agricultural resilience.

7. Future Directions in Nanotechnology for Soil and Land Management

Nanotechnology has emerged as a promising field with transformative potential in soil and land management. By harnessing the unique properties of nanomaterials, such as their high surface area, reactivity, and ability to interact at the molecular level, nanotechnology offers innovative solutions for improving soil health, enhancing nutrient delivery, and addressing environmental challenges like contamination and erosion. The future of nanotechnology in soil and land management promises to revolutionize sustainable agricultural practices, mitigate the effects of climate change, and optimize resource use efficiency. As research progresses, the development of nanomaterials tailored for specific soil and land conditions will pave the way for more precise, effective, and eco-friendly land management strategies.

7.1 Innovations in Nano-engineering for Soil and Land Management

Advancements in nano-engineering are shaping the future of soil and land management by introducing more effective and sustainable solutions. Researchers are actively working on designing nanomaterials with specific properties tailored for various applications such as nutrient delivery, pollutant immobilization, and soil stabilization. These innovations hold immense potential for improving agricultural productivity, rehabilitating degraded lands, and mitigating environmental pollution. One of the most promising innovations is the development of customized nanomaterials that enable precision soil management. Intelligent nanosystems, for instance, are being designed to control nutrient release, reducing losses due to leaching while ensuring optimal nutrient availability to plants. For example, smart nano-fertilizers can release nutrients in response to specific environmental cues, such as soil moisture or pH levels, thereby enhancing nutrient use efficiency and minimizing waste. Additionally, nanophytoremediation, an emerging field combining nanotechnology and plant-based remediation, is showing great potential for decontaminating polluted soils. By integrating nanoparticles with phytoremediation techniques, researchers can enhance plant uptake of contaminants, accelerate detoxification processes, and promote plant growth in otherwise inhospitable environments. This approach has demonstrated success in removing heavy metals, organic pollutants, and even microplastics from soil ecosystems. Another significant breakthrough in nano-engineering involves the development of biodegradable nanomaterials. Current concerns regarding the persistence of nanoparticles in the environment have prompted research into eco-friendly alternatives that degrade into harmless byproducts after serving their intended purpose. Such biodegradable nanomaterials can be applied safely in agricultural and environmental settings without posing long-term ecological risks.

7.2 Integration of Nanotechnology with Sustainable Practices

To maximize its benefits while ensuring environmental responsibility, nanotechnology must be integrated with sustainable agricultural and land management practices. This integration is crucial for tackling global challenges such as soil degradation, food security, and climate change. Nanotechnology aligns well with the United Nations Sustainable Development Goals (SDGs), particularly those related to sustainable agriculture (SDG 2), clean water and sanitation (SDG 6), and climate action (SDG 13). Nanomaterials have shown potential in enhancing soil water retention, reducing greenhouse gas emissions from fertilizers, and improving crop resilience to climate stressors. For example, silica-based nanoparticles can increase drought tolerance in crops by improving water retention and uptake efficiency. Moreover, nanotechnology supports the principles of a circular economy by facilitating resource recovery and waste minimization. For instance, nanomaterials can be used to extract valuable nutrients from agricultural waste or remediate contaminated sites, thus contributing to the efficient use of resources and reducing environmental pollution. Such applications ensure that nanotechnology remains a sustainable and regenerative tool in soil and land management. The push for sustainable nanomanufacturing is another critical aspect of ensuring long-term viability. Green synthesis techniques, which use natural precursors such as plant extracts or microbial processes to produce nanoparticles, are gaining traction as an alternative to conventional chemical synthesis. This shift toward environmentally friendly production methods reduces the ecological footprint of nanotechnology applications while maintaining their effectiveness.

7.3 Prospects for Policy Development and Implementation

Despite its immense potential, the widespread adoption of nanotechnology in soil and land management requires well-structured policies and regulations to ensure safety, efficacy, and ethical implementation. Currently, regulatory frameworks for nanotechnology remain fragmented, necessitating urgent efforts to establish global standards. Governments and regulatory bodies must work toward comprehensive guidelines that address the production, application, and disposal of nanomaterials. Standardized toxicity assessments, environmental monitoring, and lifecycle evaluations should be implemented to understand the long-term impact of nanoparticles on soil health, biodiversity, and human safety. Additionally, international collaboration is crucial for the development of cohesive policies governing nanotechnology. Countries need to coordinate efforts through initiatives such as the National Nanotechnology Initiative (NNI) or the Organisation for Economic Co-operation and Development (OECD) frameworks to create harmonized safety protocols. Such collaborations can facilitate knowledge sharing, streamline research funding, and ensure that nanotechnology advancements are implemented responsibly across global agricultural landscapes. Finally, stakeholder engagement plays a fundamental role in shaping the future of nanotechnology policies. Farmers, scientists, industry leaders, and policymakers must be actively involved in discussions to balance technological advancements with ethical considerations. Public outreach programs and capacity-building workshops can help demystify nanotechnology, addressing misconceptions and fostering public trust in its applications.

The future of nanotechnology in soil and land management holds transformative potential, with innovations in nano-engineering offering new solutions for soil fertility, land rehabilitation, and environmental sustainability. However, for these technologies to reach their full potential, they must be seamlessly integrated with sustainable practices and supported by comprehensive regulatory frameworks. By fostering interdisciplinary collaboration, investing in green nanomanufacturing, and developing clear policies, nanotechnology can be leveraged to address some of the most pressing global challenges related to land degradation and food security. As research continues to advance, a proactive approach to governance and sustainability will be essential in ensuring that nanotechnology remains a powerful tool for environmental stewardship and agricultural resilience.

8. Acknowledgement

The authors gratefully acknowledge the support of the Department of Agricultural Science & Engineering, College of Agriculture, Tennessee State University, Nashville, USA and Institute of

Plant Protection, Muhammad Nawaz Sharif University of Agriculture, Multan for the support in this publication.

References

Abd-Elsalam, K.A., M.A. Mehmood, M. Ashfaq, T.E. Abdelkhalek, R.K. Hassan and M.J.S.S. Ravichandran. 2024. Liquid nanoclay: Synthesis and Applications to Transform an Arid Desert into Fertile Land. 8(3): 73.

Abid, N., A.M. Khan, S. Shujait, K. Chaudhary, M. Ikram, M. Imran et al. 2022. Synthesis of nanomaterials using various top-down and bottom-up approaches, influencing factors, advantages, and disadvantages: A review. 300: 102597.

Adil, M., S. Bashir, S. Bashir, Z. Aslam, N. Ahmad, T. Younas et al. 2022. Zinc Oxide Nanoparticles Improved Chlorophyll Contents, Physical Parameters, and Wheat Yield Under Salt Stress. 13: 932861.

Alkhaza'leh, H., M. Rahbeh, I. Hamadneh, H. AL-Mashakbeh and Z.J.A. Albalawna. 2025. Nanoparticles in Soil Reclamation: A Review of their Role in Reducing Soil Compaction. 18: 11786221241311725.

Allan, J., S. Belz, A. Hoeveler, M. Hugas, H. Okuda, A. Patri et al. 2021. Regulatory Landscape of Nanotechnology and Nanoplastics from a Global Perspective. 122: 104885.

Amoakwah, E., S.T. Lucas, N.A. Didenko, M.A. Rahman and K.R. Islam. 2022. Impact of Deforestation and Temporal Land-Use Change on Soil Organic Carbon Storage, Quality, And Lability. 17(8): e0263205.

Amutha, C., A. Gopan, I. Pushbalatatha, M. Ragavi and J.A. Reneese. 2024. Nanotechnology and governance: Regulatory framework for responsible innovation. pp. 481–503. Nanotechnology in Societal Development. Springer.

Babu, S., R. Singh, D. Yadav, S.S. Rathore, R. Raj, R. Avasthe et al. 2022. Nanofertilizers for Agricultural and Environmental Sustainability. 292: 133451.

Behera, A., P. Mallick and S. Mohapatra. 2020. Nanocoatings for anticorrosion: An introduction. pp. 227–243. Corrosion Protection at the Nanoscale. Elsevier.

Binns, C. 2021. Introduction to Nanoscience and Nanotechnology. John Wiley & Sons.

Borrelli, P., D.A. Robinson, L.R. Fleischer, E. Lugato, C. Ballabio, C. Alewell et al. 2017. An assessment of the global impact of 21st-Century Land Use Change on Soil Erosion. 8(1): 1–13.

Chen, X., J. Wang, Y. You, R. Wang, S. Chu, Y. Chi et al. 2022. When nanoparticle and microbes meet: The Effect of Multi-Walled Carbon Nanotubes on Microbial Community and Nutrient Cycling in a Hyperaccumulator System. 423: 126947.

Cui, Y.-F., X.-J. Zhou and C.-X. Guo. 2017. Experimental Study on the Moving Characteristics of Fine Grains in Wide Grading Unconsolidated Soil Under Heavy Rainfall. 14(3): 417–431.

Díez-Pascual, A.M. 2021. Carbon-based nanomaterials. MDPI.

El-Saadony, M.T., A.M. Saad, S.M. Soliman, H.M. Salem, E.-S.M. Desoky, A.O. Babalghith et al. 2022. Role of Nanoparticles in Enhancing Crop Tolerance to Abiotic Stress: A Comprehensive Review. 13: 946717.

El-Temsah, Y.S. and E.J.J.C. Joner. 2013. Effects of nano-sized zero-valent iron (nZVI) on DDT degradation in Soil and its Toxicity to Collembola and Ostracods. 92(1): 131–137.

Farooq, M.S., M. Uzair, A. Raza, M. Habib, Y. Xu, M. Yousuf et al. 2022. Uncovering the Research Gaps to Alleviate the Negative Impacts of Climate Change on Food Security: A Review. 13: 927535.

Farooqi, Z.H., R. Begum, K. Naseem, W. Wu and A. Irfan. 2022. Zero valent Iron Nanoparticles as Sustainable Nanocatalysts for Reduction Reactions. 64(2): 286–355.

Garg, R., R. Garg and N.O. Eddy. 2024. Applications of Nanocoatings: Smart and Sustainable Protection. pp. 28–56. Sustainable Approach to Protective Nanocoatings. IGI Global.

Ghadr, S., C.-H. Liu, P. Mrudunayani and C. Hung. 2022. Effects of Hydrophilic and Hydrophobic Nanosilica on the Hydromechanical Behaviors of Mudstone Soil. 331: 127263.

Gramaglia, C., M. Rio, C. Salles and M.G. Tournoud. 2024. Reducing the Imperviousness of Urban Soils to Enhance the Quality of Surface Water: Obstacles and Levers to Implementing Ecological Runoff Management In The South of France. 636: 131168.

Griggs, G. and B.G. Reguero. 2021. Coastal Adaptation to Climate Change and Sea-Level Rise. 13(16): 2151.

Hanemann, T. and D.V. Szabó. 2010. Polymer-nanoparticle composites: from Synthesis to Modern Applications. 3(6): 3468–3517.

Harsh, H., A.A.B. Moghal, R.M. Rasheed and A. Almajed. 2023. State-of-the-art Review on the Role and Applicability of Select Nano-Compounds in Geotechnical and Geoenvironmental Applications. 48(4): 4149–4173.

Hatti, V., M. Raghavendra, Y. Singh, S. Sharma, H.M. Halli and B. Goud. 2020. Application of Nanotechnology in Crop Production to Enhance Resource Use Efficiency with Special Reference to Soil and Water Conservation: A Review. 19(4): 375–381.

He, M., H. Shi, X. Zhao, Y. Yu and B. Qu. 2013. Immobilization of Pb and Cd in Contaminated Soil using Nano-Crystallite Hydroxyapatite. 18: 657–665.

Hristozov, D. and I. Malsch. 2009. Hazards and Risks of Engineered Nanoparticles for the Environment and Human Health. 1(4): 1161–1194.

Huang, X., M. Auffan, M.J. Eckelman, M. Elimelech, J.-H. Kim, J. Rose et al. 2024. Trends, Risks and Opportunities in Environmental Nanotechnology. 5(8): 572–587.

Hull, M.S. and D. Bowman. 2014. Nanotechnology Environmental Health and Safety: Risks, Regulation, and Management. William Andrew.

Ihtisham, M., A. Noori, S. Yadav, M. Sarraf, P. Kumari, M. Brestic et al. 2021. Silver Nanoparticles' Toxicological Effects and Phytoremediation. 11(9): 2164.

Jafari, M., M. Tahmoures, M. Ehteram, M. Ghorbani and F. Panahi. 2022. Agroforestry and its role in soil erosion biological control. pp. 649–700. Soil Erosion Control in Drylands. Springer.

Jenkins, W.A., B.C. Murray, R.A. Kramer and S.P. Faulkner. 2010. Valuing ecosystem services from wetlands restoration in the Mississippi Alluvial Valley. 69(5): 1051–1061.

Kaur, H., A. Kalia, J.S. Sandhu, G.S. Dheri, G. Kaur and S. Pathania. 2022. Interaction of TiO_2 nanoparticles with soil: Effect on Microbiological and Chemical Traits. 301: 134629.

Khan, S.T., J. Musarrat and A.A. Al-Khedhairy. 2016. Countering drug resistance, infectious diseases, and sepsis using metal and metal oxides nanoparticles: current status. Colloids Surf. B Biointerfaces. 146: 70–83.

Krasnopeeva, E.L., G.G. Panova and A.V. Yakimansky. 2022. Agricultural applications of superabsorbent polymer hydrogels. Int. J. Mol. Sci. 23(23): 15134.

Landa, P. 2021. Positive effects of metallic nanoparticles on plants: Overview of involved mechanisms. Plant Physiol. Biochem. 161: 12–24.

Li, R., T. Li and Q. Zhou. 2020. Impact of titanium dioxide (TiO_2) modification on its application to pollution treatment-a review. Catalysts. 10(7): 804.

Liew, S. Types and Impacts of Debris Flow Failure in Tropical Hilly Terrain and Development Areas and its Practical Rehabilitation Solutions.

Liu, R. and D. Zhao. 2007. In situ immobilization of Cu(II) in soils using a new class of iron phosphate nanoparticles. Chemosphere. 68(10): 1867–1876.

Ma, W., Y. Xu, D. Zhou, L. Wang, X. Liang and Y. Sun. 2023. Development and optimization of high–performance nano–biochar for efficient removal of Cd in aqueous: absorption performance and interaction mechanisms. Chem. Eng. Res. Des. 189: 516–529.

Majumdar, S., J.R. Peralta-Videa, J. Trujillo-Reyes, Y. Sun, A.C. Barrios, G. Niu et al. 2016. Soil organic matter influences cerium translocation and physiological processes in kidney bean plants exposed to cerium oxide nanoparticles. Sci. Total Environ. 569: 201–211.

Makvandi, P., C.Y. Wang, E.N. Zare, A. Borzacchiello, L.N. Niu and F.R. Tay. 2020. Metal-based nanomaterials in biomedical applications: antimicrobial activity and cytotoxicity aspects. Adv. Funct. Mater. 30(22): 1910021.

Marella, S., A.N. Kumar and N.P. Tollamadugu. 2021. Nanotechnology-based innovative technologies for high agricultural productivity: Opportunities, challenges, and future perspectives. Res. Dev. Agric. Med. Biochem. pp. 211–220.

Mayel, S., M. Jarrah and K. Kuka. 2021. How does grassland management affect physical and biochemical properties of temperate grassland soils? A review study. Grassl. Sci. 76(2): 215–244.

Medici, S., M. Peana, A. Pelucelli and M.A. Zoroddu. (2021). An updated overview on metal nanoparticles toxicity. Semin. Cancer Biol. 17–26.

Mukhopadhyay, R., B. Sarkar, E. Khan, D.S. Alessi, J.K. Biswas, K. Manjaiah et al. 2022. Nanomaterials for sustainable remediation of chemical contaminants in water and soil. Crit. Rev. Environ. Sci. Technol. 52(15): 2611–2660.

Munir, A., M.A. Salah, M. Ali, B. Ali, M.H. Saleem, K. Samarasinghe et al. 2024. Advancing agriculture: harnessing smart nanoparticles for precision fertilization. Biotechnol. 14(4): 3846–3863

Niroumand, H., L. Balachowski and R. Parviz. 2023. Nano soil improvement technique using cement. Sci. Rep. 13(1): 10724.

Peng, T. and S.-J. Wang. 2012. Effects of land use, land cover and rainfall regimes on the surface runoff and soil loss on karst slopes in southwest China. Catena. 90: 53–62.

Pokrajac, L., A. Abbas, W. Chrzanowski, G.M. Dias, B.J. Eggleton, S. Maguire et al. 2021. Nanotechnology for a sustainable future: Addressing global challenges with the international network4sustainable nanotechnology. ACS Publications.

Rajput, V.D., T. Minkina, S.K. Upadhyay, A. Kumari, A. Ranjan, S. Mandzhieva et al. 2022. Nanotechnology in the restoration of polluted soil. Nanomaterials. 12(5): 769.

Rajput, V.D. and T.M. Minkina. 2025. Nano-Biochar and Nanoparticles for Soil Health Improvement. CRC Press.

Rashid, M.I., G.A. Shah, M. Sadiq, N.U. Amin, A.M. Ali, G. Ondrasek et al. 2023. Nanobiochar and copper oxide nanoparticles mixture synergistically increases soil nutrient availability and improves wheat production. Plants. 12(6): 1312.

Rhodes, C.J. 2014. Soil erosion, climate change and global food security: challenges and strategies. Sci. Prog. 97(2): 97–153.

Sah, M.K., B.S. Thakuri, J. Pant, R.L. Gardas and A. Bhattarai. 2024. The multifaceted perspective on the role of green synthesis of nanoparticles in promoting a sustainable green economy. Sustain. Chem. 5(2): 40-59.

Sharma, A., J. Tracy and P. Panwar. 2022. Ecological restoration of degraded forests for achieving land degradation neutrality. Land Degrad. Neutral. Achiev. SDG 15 For. Manag. Springer, pp. 191-204.

Sharma, S., B.K. Sahu, L. Cao, P. Bindra, K. Kaur, M. Chandel et al. 2021. Porous nanomaterials: main vein of agricultural nanotechnology. Prog. Mater. Sci. 121: 100812.

Sherje, A.P., M. Jadhav, B.R. Dravyakar and D. Kadam. 2018. Dendrimers: A versatile nanocarrier for drug delivery and targeting. Int. J. Pharm. 548(1): 707–720.

Shoukat, A., B. Pitann, M.M. Zafar, M.A. Farooq, M. Haroon, A. Nawaz et al. 2024. Nanotechnology for climate change mitigation: enhancing plant resilience under stress environments. J. Plant Nutr. Soil Sci. 187(5): 604–620.

Tiwari, J.N., R.N. Tiwari and K.S. Kim. 2012. Zero-dimensional, one-dimensional, two-dimensional and three-dimensional nanostructured materials for advanced electrochemical energy devices. Prog. Mater. Sci. 57(4): 724–803.

Vasconcelos, A.R.C. 2024. Self-sensing of soil-cement composites based on carbon nanotubes for civil engineering monitoring.

Vejan, P., T. Khadiran, R. Abdullah and N. Ahmad. 2021. Controlled release fertilizer: a review on developments, applications and potential in agriculture. J. Control. Release. 339: 321–334.

Vincent, W.F., M. Lemay and M. Allard. 2017. Arctic permafrost landscapes in transition: towards an integrated Earth system approach. Anthropocene Sci. 3(2): 39–64.

Xie, Y., Z. Fang, W. Cheng, P.E. Tsang and D. Zhao. 2014. Remediation of polybrominated diphenyl ethers in soil using Ni/Fe bimetallic nanoparticles: influencing factors, kinetics and mechanism. Sci. Total Environ. 485: 363–370.

Xiong, Z., F. He, D. Zhao and M.O. Barnett. 2009. Immobilization of mercury in sediment using stabilized iron sulfide nanoparticles. Water Res. 43(20): 5171–5179.

Xu, Y. and D. Zhao. 2007. Reductive immobilization of chromate in water and soil using stabilized iron nanoparticles. Water Res. 41(10): 2101–2108.

Yang, G.C. and Y.-I. Chang. 2011. Integration of emulsified nanoiron injection with the electrokinetic process for remediation of trichloroethylene in saturated soil. Sep. Purif. Technol. 79(2): 278–284.

Yin, C., W. Zhao and P. Pereira. 2022. Soil conservation service underpins sustainable development goals. Glob. Ecol. Conserv. 33: e01974.

Zafar, S., S. Perveen, M. Kamran Khan, M.R. Shaheen, R. Hussain, N. Sarwar et al. 2022. Effect of zinc nanoparticles seed priming and foliar application on the growth and physio-biochemical indices of spinach (*Spinacia oleracea* L.) under salt stress. PLoS One. 17(2): e0263194.

CHAPTER-15

Effect of Nanoparticles on Soil Microbes

Somasundaram Ambiga,[1] *Raja Suja Pandian,*[2] *Lazarus Vijune Lawrence,*[3] *Arjun Pandian*[4] and *Abdul Bakrudeen Ali Ahmed*[1]

1. Introduction

Nanotechnology shows numerous uses and applications in various domains, particularly in soil sciences. Owing to this specialty, nanosciences and technology has emerged an essential component of developing society. The expansion of the nano-based sector is primarily focused on the manufacture, manipulation and development of advanced nanomaterials for a wide range of applications to resolve present and future issues in various industries (Shang et al. 2019). Numerous estimates indicate that there are numbers of commercial goods based on nanoparticles on the market now, and that number is anticipated to grow over time. Materials with one or more dimensions of 100 nm or less are known as nanomaterials (Kreyling et al. 2010). Due to their specific functional and structural components, either on the inside or the outside, these materials typically exhibit exceptional and unusual biological and physicochemical properties in comparison to their bulk counterparts. Nanomaterials offer unrivalled prospects and applications, but they also carry a significant risk to the environment and to human health (Chen et al. 2017, Gunaki et al. 2025). Figure 1 represents different types of nanoparticles.

2. Effect Of Nanoparticles In Soil

Generally, nanoparticles have encountered or been exposed to agricultural soil through untreated wastewater treatment that is used for irrigation. It is also possible to encounter soil by means of fertilisers. As these particles have much biological potential, they prevail in the soil community for a long time. They may be in free or combined form. Such forms may alter soil fertility, soil microflora population, and even alter plant metabolism. Figure 2 shows the physiological and biochemical responses when the nanoparticles are exposed to soil fertility and foliar uptake. Some of the physiological responses may include stem-root development and fruit-flower

[1] Department of Biochemistry, PRIST Deemed University, Thanjavur, Tamil Nadu, India.
[2] Department of Biochemistry, Rajah Serfoji government college, Thanjavur, Tamil Nadu, India.
[3] Department of Biotechnology, School of Bio Sciences and Technology, Vellore Institute of Technology, Vellore, Tamil Nadu, India.
[4] Department of Biological Sciences, SAVEETHA DEEMED University, Chennai.
* Corresponding author: bakrudeen.ali.ahmed@tdtu.edu.vn

DOI: 10.1201/9781003558354-15

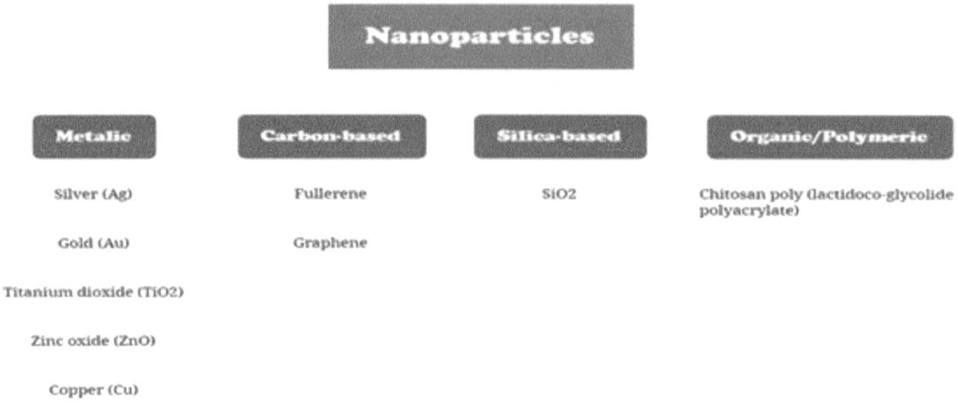

Figure 1. Classification of Nanoparticles.

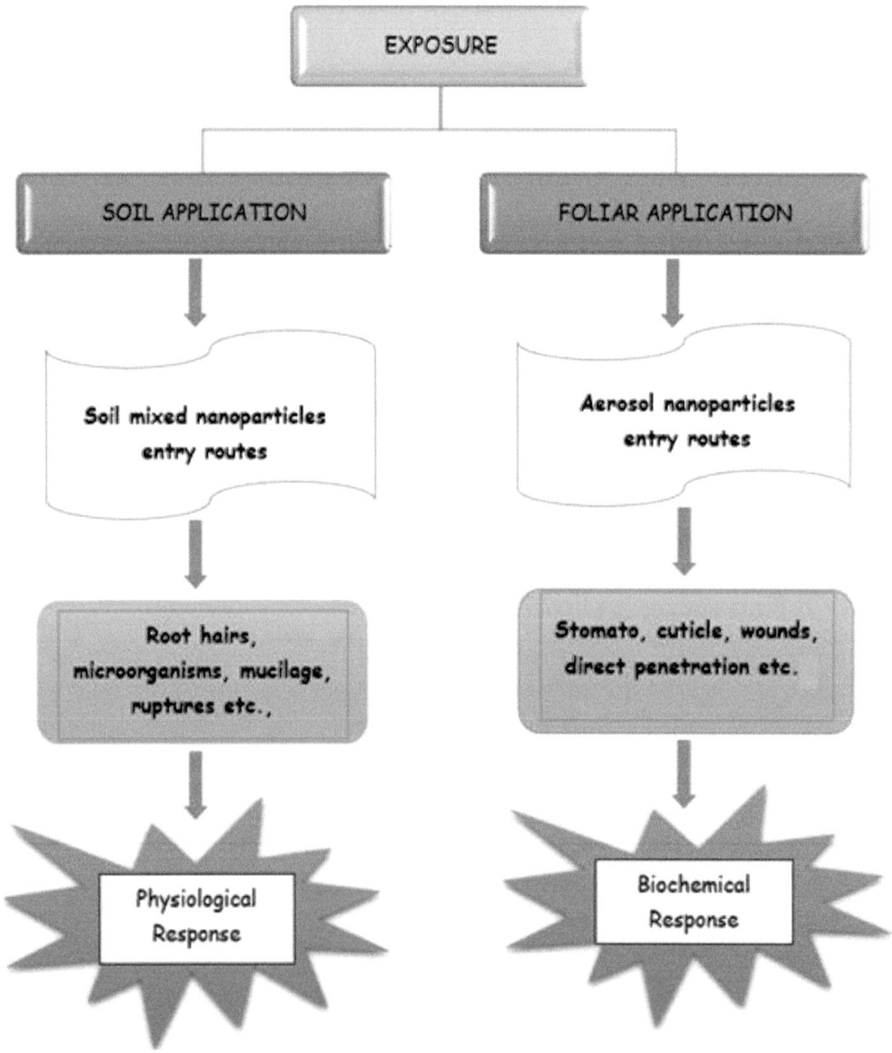

Figure 2. Exposure of Nanoparticles in the Soil Community.

development. Biochemical responses like chlorophyll and nutritional factors include lycopene (Rawat et al. 2018, Shreeshwal et al. 2025).

The ground may also be directly impacted by nanoparticles that have accumulated in soil systems, which can then have an impact on human health. For the rehabilitation of groundwater, certain nanoparticles have been employed to remove both organic and inorganic impurities. Such nanoparticles might also contribute to groundwater nano-pollution. A substantial number of engineered nanoparticles are released into the atmosphere by a variety of industrial activities, including point sources (manufacturing facilities, during transportation, wastewater treatment plants, waste incineration, power plants, and landfilling) and non-point sources. In addition to soil systems and groundwater (vehicle emissions, during washing or abrasion of nano-enabled products). Additionally, the amount of the localised atmospheric concentration may be increased by the unintentional emission of nanoparticles (Mahdi et al. 2018, Peng et al. 2017, Sheershwal et al. 2025).

3. Metallic Nanoparticles (MNPs) in Soil

Numerous MNPs are possible replacements for traditional nutrients and organic matter as an effective, slow-releasing Zn fertiliser, and also zinc oxide NPs (ZnO NPs) have already been investigated as a means of reducing the pervasive Zn deficit in agricultural soils. Due to the slow-release supplication for Cu requirements for the long duration of crops, copper oxide nanoparticles (CuO NPs) have also been researched. MNPs can act as brand-new macronutrient transporters. For example, silicon oxide nanoparticles (SiO_2NPs) have also been created. These SiO_2NPs were created as a slow-release supplication for Si in order to reduce salt stress. They were also created as transporters of some important macronutrients such as nitrogen, potassium, and phosphorus for plants (Mushtaq et al. 2018, Hasan et al. 2025).

The decomposition and mineralisation of soil microbial communities, biogeochemical cycles, and the degradation and transformation of soil pollutants are all directly influenced by the populations of soil microbes. Therefore, the abundance and diversity of soil microorganisms, along with their activities, will vary as a result of their exposure to MNPs. Also, these changes will definitely have an impact on the functions and health of the soil. However, depending on the MNPs, the soils, the dosage and length of exposure, as well as the relevant microorganisms, the reactions of soil microbial mass to MNPs differ (Wenjie et al. 2020, Shamloo et al. 2025).

4. Impact of MNPs on Soil Properties

There have been few studies on how MNPs affect the physical characteristics of soil. The geoenvironmental community, which attempts to alter soil attributes by utilising the relationships among MNPs and soil clay minerals, is where the majority of the literature in this field comes from (Figure 3). For instance, 6% CuONPs or aluminium hydroxide nanoparticles (Al2O3NPs) increased the shrinkage of clay (kaolinite clay) by 8–17%, boosting the capacity of soil to preserve the aggregation of structure under dry conditions. However, their effect on clay contraction was not noticed when the amount of MNPs was decreased to 0.5%. Additionally, soil pore size, as well as its hydraulic properties, is impacted by the shrinking of clay in the soil (Coo et al. 2016, Panwar et al. 2025).

5. Metal-oxide (MO) NPs

By damaging microbial cell membranes, DNA, and mitochondria with reactive oxygen species, nanoparticles and their ions potentially cause microbial cellular death. Measurements of soil aeration and enzymatic activity could be used to assess the effect of nanoparticles on microbial activities (Simonin and Richaume 2015). Metallic NPs may affect changes in the quantity of soil microorganisms and enzymatic activities in the soil. The toxicity of nanoparticles may be

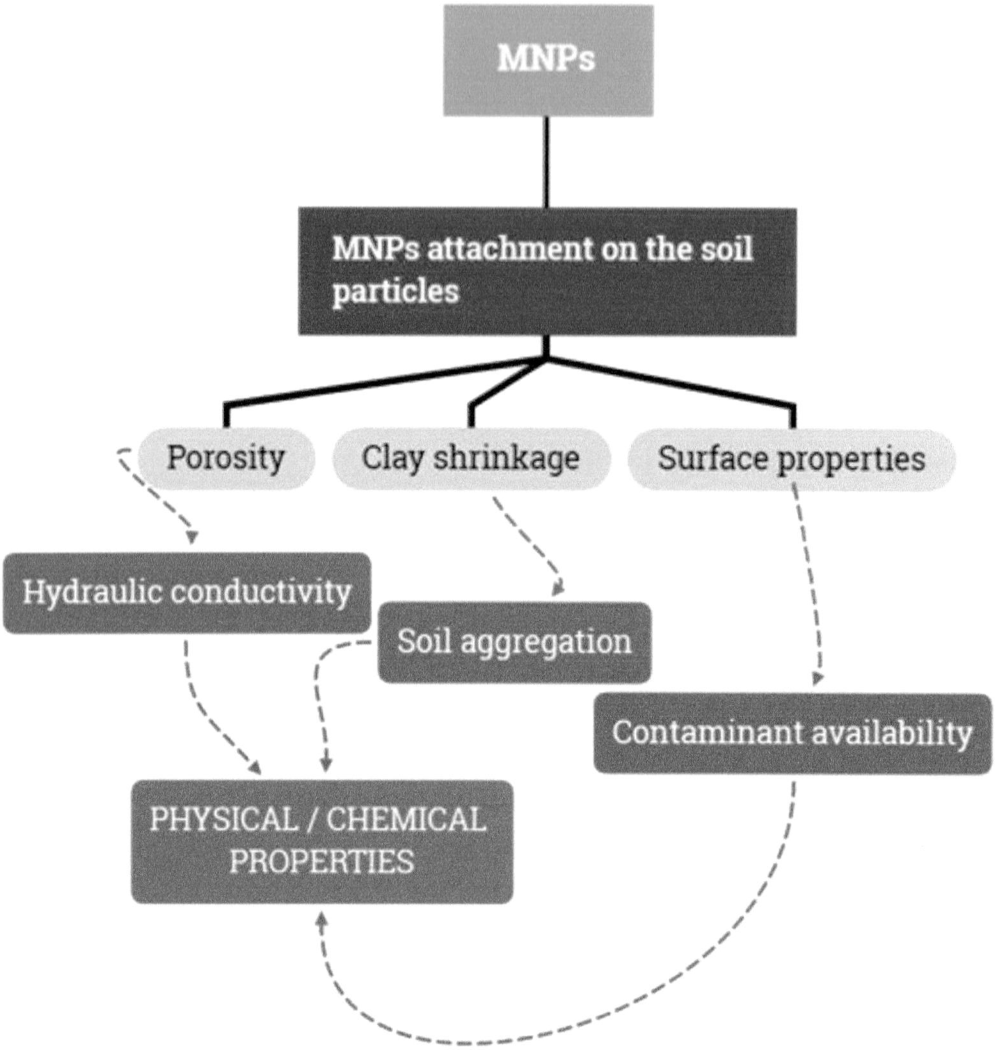

Figure 3. Impact of MNPs on soil's physical properties.

significantly influenced by the types of soil. Additionally, the size and makeup of the soil bacterial communities are significantly influenced by the kind of soil in terms of nanoparticle toxic effects. Ionised nanoparticles are thought to play a crucial role in the lethality of both infectious and helpful microorganisms, as well as being extremely harmful to the soil bacteria that grow there. In addition to affecting rhizobacteria, arbuscular mycorrhiza, nitrifying bacteria, nitrogen-fixing bacteria, as well as other helpful soil microorganisms, the analysis on the hazardous impact of nanoparticles additionally affects other soil microorganisms. Recent research revealed that nanoparticles could have an impact on the diversity and quantity of soil bacteria, as well as enzymatic and metabolic processes, nitrification capability, and colony counts. In order to improve soil health, soil microorganisms are essential for the immobilisation, recycling, and detoxification of nutrients, carbon, and pollutants (Ladhari et al. 2025).

Approximately 15 percent of the total microbial communities in the widely dispersed heterotrophic microflora are made up of bacterial communities from multiple genera that either indirectly or directly encourage plant growth. These bacterial communities that live within the rhizosphere are capable of colonising the root system and are known as plant growth-promoting rhizobacteria (PGPR). *Azotobacter, Pseudomonas, Thiobacillus, Rhizobium, Azospirillum,*

Bradyrhizobium, Acinetobacter, Agrobacterium, Bacillus, Burkholderia, Arthrobacter, Serratia, and other species are prominent PGPRs that promote plant growth. Only 2–5% of rhizosphere bacteria have been identified as powerful PGPRs despite being such a diverse population. According to the rising production of agricultural goods utilizing nanotechnology, including such NPs-based insecticides, fertilizers, and herbicides, it is essential to assess the relationships between NPs and bacteria (Bian et al. 2025) (Table 1).

For instance, direct introduction of TiO2 NPs and Fe NPs is being utilised in water purification and environmental restoration, and also promotes the growth of certain target species, respectively. Azotobacter and P- and K-solubilising bacteria were dramatically decreased by ZnO, CeO_2, and TiO_2 NPs (Chai et al. 2015). Carbon nanotubes, GO/rGO NPs, Fe_2O_3 and Fe_3O_4 nanoparticles,

Table 1. Effect of Different Nanoparticles in Soil.

NPs	Size	Effect of Nanoparticles	References
ZnO, Fe_3O_4	500–1000, 2000	Activity of enzymes were affected. Enzymes include urease, phosphatase, catalase and invertase. Bacterial communities were also affected.	You et al. 2018
ZnO	1000	Acetobacter - plate counts were affected	Chai et al. 2015
CuO	0–1000	Soil community of biomass and enzyme activities were also decreased. Disturbances in the structure of soil community.	Xu et al. 2015
Fe	550	Bacterial groups had effect on individual analysis	Shah et al. 2014
CuO	10	Soil community was affected	Ben-Moshe et al. 2013
Cu	220	Carbon and nitrogen biomass were reduced. Disturbances in the structure of microbial community.	Kumar et al. 2011
ZnO	500–2000	Soil bacterial communities were altered. Bacterial diversity in the soil were decreased	Ge et al. 2011
Ag-NP	0.14	Structure of bacterial communities were modified. Enzymatic activities were also reduced.	Colman et al. 2013
Ag-NP	3.2, 32,320	Soil biomass community was decreased. Metabolic quotient and basal respiration were also increased. Microbial biomass was not effective.	Hänsch and Emmerling 2010
Ag-NP	1.25 µg to 30	Enzymes such as arylsulfatase, leucine-aminopeptidase, phosphomonoesterase and β-D-glucosidase were decreased. These enzymes were decreased when the concentration of Ag was low. Protective role of OM was not observed.	Peyrot et al. 2014
Citrate coated Ag-NPs	1, 10, 100, 1000	Strongly inhibits urease activity and has a negative dosage response. Enzymes such as Arylsulfataseβ glucosidase, acid phosphatase, fluorescein diacetate hydrolase and dehydrogenase were reduced in their activity. The negative effects brought on by the Ag-NP themselves were evident since the impacts of the silver ions that were dissolved from the silver nanoparticle were also not significant.	Shin et al. 2012
CeO_2-NP, SnO_2-NP, Fe_3O_4-NP,	0, 10 and 100	No impact on the microbial biomass of C and N. qCO_2 has increased metabolically in the soil contamination.	Vittori Antisari et al. 2013
Ag-NP, Cu-NP, Si-NP	660	Reduction in the variety of substrate (Biolog) use to 16% in soil treated with Ag NP and to 40% and 52% in soil treated with Cu NP and SiO_2NP. The FAME profiles are unaffected. Ag-NP changed the makeup of the microorganisms (DGGE analysis).	Kumar et al. 2011
Cu-NP, Ag-NP and Si-NP (Mixture)	220	Reduction in microbial biomass in carbon and nitrogen. Changing the organisation of the microbial biomass (using FAME analysis), particularly the species of bacteria (DGGE analysis).	Kumar et al. 2012

Ag, Cu, zinc, and TiO_2 nanoparticles are a few examples of nanomaterials that have been shown to have biologically toxic impacts on both harmful and helpful soil microbes, including *Escherichia coli, Bacillus subtilis,* and *Streptococcus aureus, Pseudomonas chlororaphis, Pseudomonas putida* and *Campylobacter jejuni* (Haris et al. 2017). The effects of TiO_2 NPs and ZnO NPs on substrate-induced respiration showed a decrease with 500, 1000, and 2000 mg kg^{-1} soil following 15 and 60 days (Yi et al. 2025).

5.1 Titanium Dioxide (TIO_2 NPS)

Titanium dioxide (TiO_2-NPs) are artificial nanomaterials which are frequently employed in a variety of industries, including paints, lubricants, UV light attenuation, skincare products, strong pigment coverage, cosmetics, and photocatalytic capabilities. NPs are accidentally discharged into the ecosystem when TiO_2-NP-containing materials are used. Commercial items contain significant amounts of TiO_2-NPs that can enter natural ecosystems, particularly the soil, one of the largest drains for nanomaterials. TiO_2-NPs are subsequently released into agricultural soils either as nanofertilisers or nanopesticides, or passively via irrigation or the use of wastewater sludge, or following an unintentional leak during industrial operations. TiO_2-NPs are being released into the environment at an accelerating rate, which prompts questions about how this can affect the soil's ability to perform crucial ecosystem activities (Figure 4).

Soils are highly permeable structures made up of intricately arranged mineral and organic particle assemblies mixed with liquid and gaseous components. Clay or organic matter level, pH, and other soil characteristics have a significant impact on the activity and bioavailability of typical contaminants such as pesticides, insecticides, heavy metals, or PAHs (polycyclic aromatic hydrocarbons). Plant development and biogeochemical activities of the nitrogen cycle—wherein denitrification and nitrification mechanisms regulate soil inorganic N accessibility and consequent soil quality—largely depend on microbial communities. A coating of TiO_2 NPs on the surface of the root or soil particle base may prevent certain microbes from growing and functioning normally, or it may prevent beneficial soil microbes like mycorrhizal fungi and nitrogen-fixing bacteria from colonising the root. Therefore, it is necessary to look into any potential impacts that TiO_2 NPs may have on soil organisms, plants, and organisms that are linked with plants. Arbuscular mycorrhizal fungi (AMF) are a significant species of soil microbes which interact with two-thirds of all photosynthetic microbes. TiO_2-NPs have the potential to alter the microbial population pattern in soils and reduce microbial productivity (Salama et al. 2025).

According to their adsorption on the cellular membrane with formation of ROS by photocatalytic oxidation, TiO_2-NPs have the potential to be harmful to bacteria. For instance, TiO_2 NPs form ROS when they are exposed to living organisms or UV rays. TiO_2 NPs have been shown to have detrimental impacts on the growth of plants in some investigations, and ROS can harm cellular structure, as well as cell expansion, evaporation, and leaf development (Niu et al. 2025).

The implementation of TiO_2 NPs has also been discovered to possess an impact on communities of crucial plant symbiotic organisms, including arbuscular mycorrhizal fungi which colonise root growth and improve absorption of nutrients. These impacts could have considerable long-term impacts on plant nutrient absorption and efficiency in agro-ecosystems (Figure 5). As opposed to certain other NPs like Ag and ZnO NPs that exhibited severe harmful impacts on plant growth with lower root development and lengthening, several researches have revealed no deleterious impacts of TiO_2-NPs on growing plants, particularly at reduced levels (Table 2). While extensively researched up to this point, the impact of TiO_2-NPs on plants remains not completely clear (David et al. 2015). After being incubated in lysimeters for nine months, older TiO_2-NPs at a concentration of 91 mg kg^{-1} soil increased urease activities while lowering catalase and peroxidase activities (Du et al. 2011, Zafar et al. 2025).

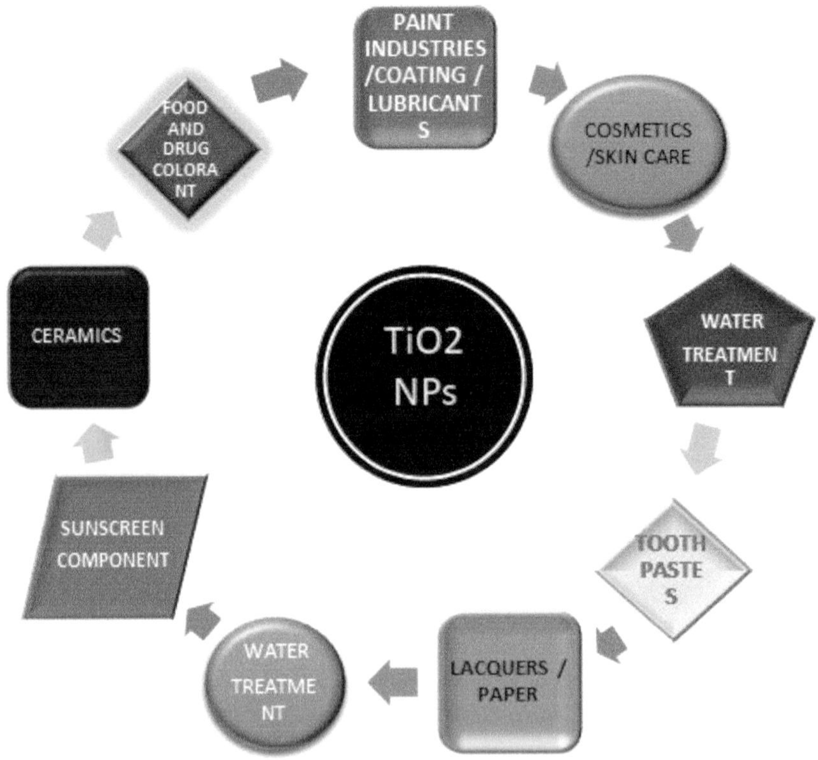

Figure 4. Application of TiO_2-NPs.

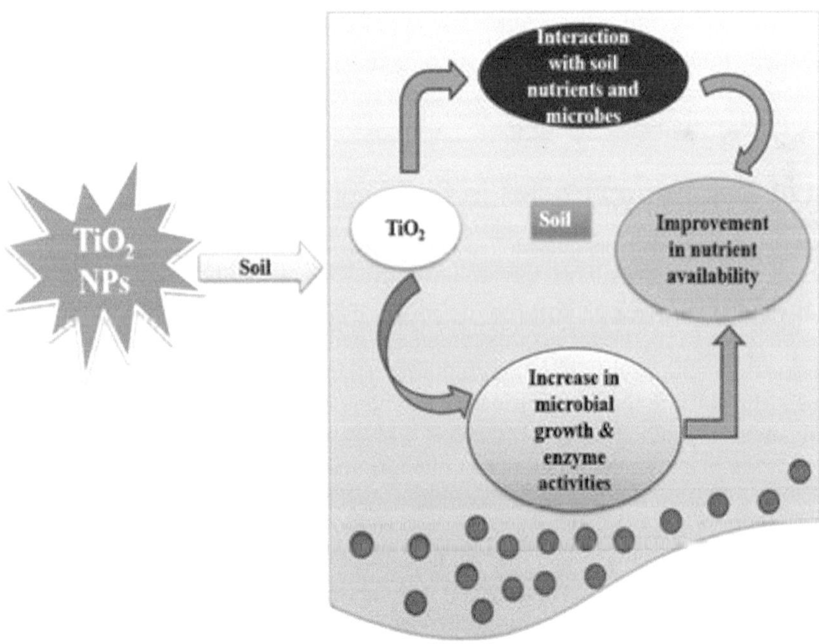

Figure 5. Potential of TiO_2-NPs to alter the microbial population pattern in soils.

Table 2. Effect of microbial community using TiO_2 NPs.

Np	Size	Soil Types	Microbial Community	References
Titanium Dioxide NPs	20 nm	Flooded paddy soil	Decrease the soil microbial biomass. Also, it has no effect on community of microbes	Xu et al. 2015
	15 – 20 nm	Surface grassland soil	Decreased the bacterial diversity and microbial biomass. It shifts the bacterial community	Ge et al. 2011
	21 nm	Acidic soil (weak) and saline-alkaline soil	There are no significant changes of bacterial community.	You et al. 2018
	10 ± 2.5 nm	Farm soil	Decrease the ammonia-oxidizing and nitrate oxidizing activity.	Simonin et al. 2017
	10 – 20 nm	Sandy loam	Increases the P availability.	Zahra et al. 2015
	12 nm	Surface soil	Decreases the P availability.	Taghipour and Jalali 2015

5.2 Iron Oxide or Fe_3O_4

Catalase, invertase, phosphatase, and urease enzyme activity in soil were all adversely influenced by Fe_3O_4 NPs. These nanoparticles are known to inhibit the growth of both Gram-positive and Gram-negative strains of bacteria, including *S. marcescens*, *E. coli*, *S. aureus*, *L. monocytogenes*, and *P. aeruginosa*, based on some recent investigations (Ismail et al. 2015, Ebrahiminezhad et al. 2012, Chatterjee et al. 2011). Iron oxide NPs added to sandy loam soil resulted in an 80% reduction in the number of bacterial colony-forming units. As a result, throughout 180 days, iron oxide NPs decreased N mineralisation and leaf litter degradation by 30% and 60%, respectively. While Fe_3O_4 NPs did not alter enzyme capabilities at elevated levels like 420 mg kg^{-1}, 840 mg kg^{-1}, and 1260 mg kg^{-1} soil, Fe_3O_4 NPs in the soil enhanced urease and invertase activity (He et al. 2011). The modifications found within the bacterial community were used to describe such stimulations and application of nanoencapsulation technology in agriculture for effective and sustainable weed management (Sabarivasan et al. 2025).

6. Effect of Zinc Nanoparticles in Soil Microbes

The previous two decades have seen nearly revolutionary advances in nanotechnology (Roco 2011). Applications of nanomaterials have spread and impacted practically every aspect of human life (Usman et al. 2020). The ecosystem services that soil microbial communities provide, such as soil formation, crop production, waste decomposition, and enhancement of groundwater quality, will be impacted if nanomaterials modify soil microbiota (Sacca et al. 2017). These nanomaterials would be directly added to soil when the nanoparticulate formulations are used as agrochemicals. It is likely that environmental nanomaterial concentrations will eventually rise to harmful levels, given the ongoing increase in manufacturing and use of nanomaterials. At that point, it may be quite difficult to reduce their negative consequences. Therefore, it is vital to research how metal/metal oxide NPs affect soil microbiota (Ren et al. 2024).

Standard zinc fertilisers have a low bioavailability issue because zinc binds to insoluble molecules in the soil, making it a micronutrient with low bioavailability. The first step toward more widespread usage of ZnO NPs in agriculture for plant nutrition and protection could be to improve our understanding of various forms of Zn, as well as Zn uptake and assimilation by higher plants. NP-fixed goods are well utilised in the agricultural industry to increase the fertility of soil as a result of the growing field of nanotechnology. Due to their special characteristics, engineered nanoparticles (NPs) are frequently employed in industrial, agricultural, and consumer manufacturing. The rise in extractable bacterial or fungal groups is evidence that ZnO NPs improved soil microbial populations. With rising ZnO NP concentrations, some vulnerable species, such as Sphingomonadales, exhibited a rise in relative abundance. The ions released from ZnO NPs, which are micronutrients, may help to increase the diversity and richness of the soil microbial community (Gunaki et al. 2025).

6.1 ZnO NPs Have Favourable Effects on Microbial Richness

Zn NPs' released ions are crucial for physiological functions and can be assimilated by bacteria to make up for any possible nutritional deficiencies in soil, particularly red soil (Wyszkowska et al. 2013). Zn takes involvement in the development of protein translation, the enzyme-substrate system, genetic chain multiplication, and duplication of genes (Sekler et al. 2007). Evidence suggests that when ZnO NPs are present, the structure of the microbial community changes. Additionally, at all exposure dosages, the relative abundance of Armatimonadetes decreased when ZnO NPs were applied. As micronutrients, the ions released by ZnO NPs also have the ability to boost the diversity of the soil microbial biomass by increasing the number of soil bacteria or fungi groups as well as the diversity indices. However, at high concentrations, the beneficial effects were compromised.

6.2 Impact of ZnO NPs on Nitrous Oxide Emission

ZnO NPs, or zinc oxide nanoparticles, are often employed and present in soil environments. The most common form of nanomaterial is zinc oxide (ZnO) NPs, which have the ability to affect N_2O generation via at least two different pathways (Rajput et al. 2018). First, a decrease in soil microbial community activity associated with nitrogen (N) cycling (Rashid et al. 2017, Chai et al. 2015, Ge et al. 2011, Du et al. 2011) and nitrogen mineralisation efficiency (Figure 6) can indirectly lower N_2O emissions. This is due to the toxic potential of ZnO NPs or due to the dissolution of Zn^{2+}. Second, ZnO NPs hinder the enzymes involved in the metabolism of glucose, which in turn decreases the energy and electrons that are available for denitrifiers to create N_2O (Ye et al. 2021, 2020, Zheng et al. 2014, Rashid et al. 2017, Sheershwal et al. 2025).

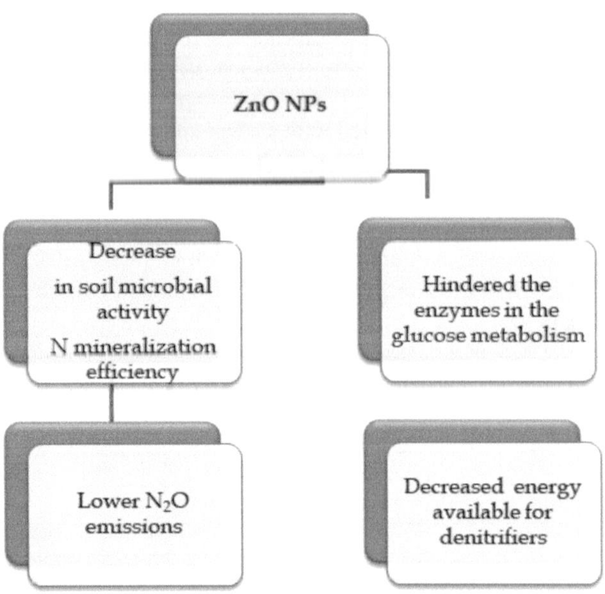

Figure 6. Favourable Effects of Zinc Oxide Nanoparticles in Soil.

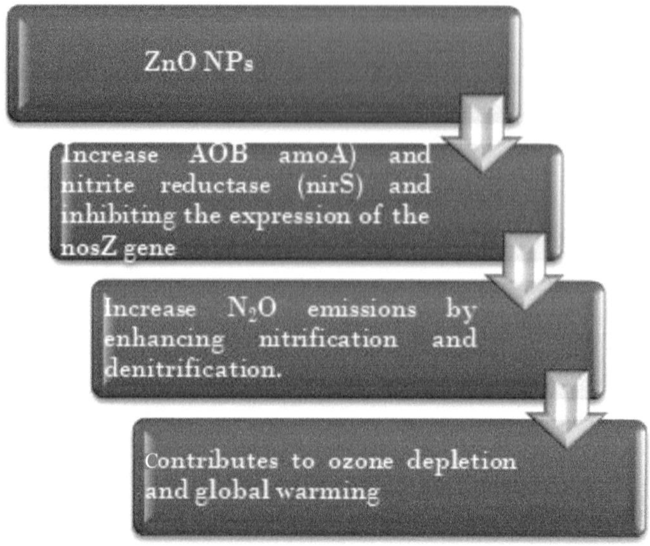

Figure 7. Impact of ZnO NPs on Nitrous Oxide Emission.

ZnO NPs also increased N_2O emissions while largely inhibiting the expression of the *nosZ* gene, which is connected to N_2O reductase in the pathway for denitrification (Zheng et al. 2014). ZnO NPs minimise N_2O emissions by improving N_2O reduction in artificial/synthetic wastewater systems (Ye et al. 2020). Few studies have examined how ZnO NPs affect N_2O emissions in soil environments. ZnO and Ag NPs working together had a negligible effect on N_2O emissions in combinations of soil and sewage sludge (Durenkamp et al. 2016). ZnO NPs decreased the nitrogen mineralisation of leaf litter in the soil (sandy soil), which subsequently most likely decreased the emission of N_2O (Rashid et al. 2017).

The exhaustion of nitrate was caused by ZnO NPs, which also raised the functional genes of ammonia-oxidising bacteria (AOB amoA) and nitrite reductase, but decreased the gene copies of ammonia-oxidising archaea (AOA amoA). The redundancy analysis results also revealed a significant correlation between total N_2O emissions and the AOB amoA and nitrite reductase genes, and that ZnO NPs indirectly impacted N_2O emissions via affecting nitrifiers, denitrifiers, and soil nitrate content (Figure 7). The exposure of ZnO NPs in agricultural fields certainly leads to an increased risk of N_2O emissions when coupled with carbon and nitrogen substrate additions, which contribute to global climate warming. Certain ZnO NPs have increased N_2O emissions by enhancing the nitrification process (AOB amoA) and the denitrification process (nitrite reductase). The activities of nitrifiers and denitrifiers are influenced by the presence of NPs in the soil environment (Parada et al. 2019, Luo et al. 2015, Simonin et al. 2018), which likely affects the release of nitrous oxide (N_2O) from soil content to the atmosphere. This phenomenon contributes to ozone depletion and global warming (Table 3).

7. Acknowledgement

Authors express their deep sense of gratitude to Research Development Cell (RDC), PRIST University, Thanjavur, Tamil Nadu, India. Our sincere thanks to the Director, Dean and faculties of PRIST University at Thanjavur for their support and encouragement.

Table 3. Effect of ZnO NPs on Soil Microbe Species and Their Activities.

Materials	**Soil Microbe Species**	**Effect on Soil Microbes**	**Results**	**References**
ZnO NPs	*Bacillus subtilis*	Growth (*B. subtilis*), FtsZ ring formation, biofilm formation and cytosolic protein activity.	The growth of *B. subtilis* is inhibited by high concentration and reducing biofilm formation.	Hsueh et al. 2015
Mg-doped ZnO NPs	*Bacillus subtilis*	Biofilm formation	Prevents the formation as well as previously formed biofilm.	Auger et al. 2019
ZnO NPs	*Staphylococcus aureus*	Antibacterial and antibiofilm formation	Reduced biofilm formation and have potential against antimicrobes.	Abdelraheem et al. 2021
ZnO NPs	*Klebsiella pneumoniae*	Cytotoxic activities against Hep-2 cell line	Potential for antibacterial use.	Reddy et al. 2014
ZnO NPs	*Pseudomonas aeruginosa*	Quorum sensing inhibitory effect	Reduction in the production of quorum sensing	Saleh et al. 2019
ZnO NPs	*Aspergillus niger*	Antimicrobial activity	Degrading dye	Kalpana et al. 2018
ZnO NPs	*Candida albicans*	Antifungal activity	Alternative for antifungal agent	Hosseini et al. 2020
ZnO NPs	*Escherichia coli*	Antimicrobial activity	Damage the bacterial cells	Liu et al. 2009

References

Abdelraheem, W.M., R.M.M. Khairy, A.I. Zaki and S.H. Zaki. 2021. Effect of ZnO nanoparticles on methicillin, vancomycin, linezolid resistance and biofilm formation in *Staphylococcus aureus* isolates. Ann. Clin. Microbiol. Antimicrob. 20(1): 54.

Al Jabri, H., M.H. Saleem, M. Rizwan, I. Hussain, K. Usman and M. Alsafran. 2022. Zinc oxide nanoparticles and their biosynthesis: Overview. Life. 12: 594.

Asadishad, B., S. Chahal, A. Akbari, V. Cianciarelli, M. Azodi, S. Ghoshal et al. 2018. Amendment of agricultural soil with metal nanoparticles: Effects on soil enzyme activity and microbial community composition. Environ. Sci. Technol. 52(4): 1908–1918.

Auger, S., C. Henry, C. Péchaux, N. Lejal, V. Zanet, M.V. Nikolic et al. 2019. Exploring the impact of Mg-doped ZnO nanoparticles on a model soil microorganism *Bacillus subtilis*. Ecotoxicol. Environ. Saf. 182: 109421.

Ben-Moshe, T., S. Frenk, I. Dror, D. Minz and B. Berkowitz. 2013. Effects of metal oxide nanoparticles on soil properties. Chemosphere. 90: 640–646.

Bian, X., Q. Li, Y. Cai, Y. Lu, H. Wei, T. Zhang et al. 2025. Hollow mesoporous SiO2 nanoparticle-mediated modulation of plant-soil-microbe systems: Case study in *Panax ginseng* cultivation. Ind. Crops Prod. 224: 120297.

Burke, D.J., N. Pietrasiak, S.F. Situ, E.C. Abenojar, M. Porche, P. Kraj et al. 2015. Iron oxide and titanium dioxide nanoparticle effects on plant performance and root-associated microbes. Int. J. Mol. Sci. 16: 23630–23650.

Chai, H., J. Yao, J. Sun, C. Zhang, W. Liu, M. Zhu et al. 2015. The effect of metal oxide nanoparticles on functional bacteria and metabolic profiles in agricultural soil. Bull. Environ. Contam. Toxicol. 94(4): 490–495.

Chatterjee, S., A. Bandyopadhyay and K. Sarkar. 2011. Effect of iron oxide and gold nanoparticles on bacterial growth leading towards biological application. J. Nanobiotechnol. 9: 34.

Chen, M., X. Qin and G. Zeng. 2017. Biodiversity change behind wide applications of nanomaterials? Nano Today. 17: 11–13.

Colman, B.P., C.L. Arnaout, S. Anciaux, C.K. Gunsch, M.F. Hochella Jr., B. Kim et al. 2013. Low concentrations of silver nanoparticles in sewage sludge cause adverse ecosystem responses under realistic field scenario. PLoS ONE. 8: e57189.

Coo, J.L., Z.P. So and C.W. Ng. 2016. Effect of nanoparticles on the shrinkage properties of clay. Eng. Geol. 213: 84–88.

Du, W., Y. Sun, R. Ji, J. Zhu, J. Wu and H. Guo. 2011. TiO_2 and ZnO nanoparticles negatively affect wheat growth and soil enzyme activities in agricultural soil. J. Environ. Monit. 13: 822–828.

Durenkamp, M., M. Pawlett, K. Ritz, J.A. Harris, A.L. Neal and S.P. McGrath. 2016. Nanoparticles within WWTP sludges have minimal impact on leachate quality and soil microbial community structure and function. Environ. Pollut. 211: 399–405.

Ebrahiminezhad, A., Y. Ghasemi, S. Rasoul-Amini, J. Barar and S. Davaran. 2012. Impact of amino-acid coating on the synthesis and characteristics of iron-oxide nanoparticles (IONs). Bull. Korean Chem. Soc. 33(12): 3957–3962.

Feng, Z., Y. Yu, H. Yao and C. Ge. 2021. Effect of zinc oxide nanoparticles on nitrous oxide emissions in agricultural soil. Agriculture. 11: 730.

Ge, Y., J.P. Schimel and P.A. Holden. 2011. Evidence for negative effects of TiO_2 and ZnO nanoparticles on soil bacterial communities. Environ. Sci. Technol. 45(4): 1659–1664.

Gunaki, M.N., S.P. Masti, L.K. Kurabetta, J.P. Pinto, A.A. Hunashyal, N.P. Dalbhanjan et al. 2025. Influence of chitosan-capped quercetin nanoparticles on chitosan/poly(vinyl) alcohol multifunctional films: A sustainable approach for bread preservation. Int. J. Biol. Macromol. 299: 140029.

Hänsch, M. and C. Emmerling. 2010. Effects of silver nanoparticles on the microbiota and enzyme activity in soil. J. Plant Nutr. Soil Sci. 173: 554–558.

Haris, Z. and I. Ahmad. 2017. Impact of metal oxide nanoparticles on beneficial soil microorganisms and their secondary metabolites. Int. J. Life Sci. Sci. Res. 3(3): 1020–1030.

Hasan, A., A. Elumalai and A. Kumar. 2025. Advancements in bioremediation: Harnessing nanomaterials and microorganisms for environmental restoration by degradation of pollutants. Environ. Qual. Manage. 34(3): e70018.

He, S., Y. Feng, N. Gu, Y. Zhang and X. Lin. 2011. The effect of γ-Fe_2O_3 nanoparticles on *Escherichia coli* genome. Environ. Pollut. 159(12): 3468–3473.

Hosseini, S.S., H. Joshaghani, T. Shokohi, A. Ahmadi and Z. Mehrbakhsh. 2020. Antifungal activity of ZnO nanoparticles and nystatin and downregulation of *SAP1-3* genes expression in fluconazole-resistant *Candida albicans* isolates from vulvovaginal candidiasis. Infect. Drug Resist. 13: 385–394.

Hsueh, Y.H., W.J. Ke, C.T. Hsieh, K.S. Lin, D.Y. Tzou and C.L. Chiang. 2015. ZnO nanoparticles affect *Bacillus subtilis* cell growth and biofilm formation. PLoS ONE. 10(6): e0128457.

Ismail, R.A., G.M. Sulaiman, S.A. Abdulrahman and T.R. Marzoog. 2015. Antibacterial activity of magnetic iron oxide nanoparticles synthesized by laser ablation in liquid. Mater. Sci. Eng. C. 53: 286–297.

Kalpana, V.N., B.A.S. Kataru, N. Sravani, T. Vigneshwari, A. Panneerselvam and V.D. Rajeswari. 2018. Biosynthesis of zinc oxide nanoparticles using culture filtrates of *Aspergillus niger*: Antimicrobial textiles and dye degradation studies. OpenNano. 3: 48–55.

Khanna, K., S.K. Kohli, N. Handa, H. Kaur, P. Ohri, R. Bhardwaj et al. 2021. Enthralling the impact of engineered nanoparticles on soil microbiome: A concentric approach towards environmental risks and cogitation. Ecotoxicol. Environ. Saf. 221: 112459.

Kreyling, W.G., M. Semmler-Behnke and Q. Chaudhry. 2010. A complementary definition of nanomaterial. Nano Today. 5: 165–168.

Kumar, N., V. Shah and V.K. Walker. 2011. Perturbation of an arctic soil microbial community by metal nanoparticles. J. Hazard. Mater. 190(1-3): 816–822.

Kumar, N., V. Shah and V.K. Walker. 2012. Influence of a nanoparticle mixture on an arctic soil community. Environ. Toxicol. Chem. 31: 131–135.

Ladhari, S., N.-N. Vu, A. Saidi, A.A. Assadi and P. Nguyen-Tri. 2025. Mechanical and antimicrobial properties of green and photoactive AgTiO2/poly(3-hydroxybutyrate) (PHB) electrospun membranes. Polym. Eng. Sci. 65(2): 714–731.

Liu, Y., L. He, A. Mustapha, H. Li, Z.Q. Hu and M. Lin. 2009. Antibacterial activities of zinc oxide nanoparticles against *Escherichia coli* O157:H7. J. Appl. Microbiol. 107(4): 1193–1201.

Luo, Z., Z. Qiu, Z. Chen, G. Du Laing, A. Liu and C. Yan. 2015. Impact of TiO_2 and ZnO nanoparticles at predicted environmentally relevant concentrations on ammonia-oxidizing bacteria cultures under ammonia oxidation. Environ. Sci. Pollut. Res. Int. 22(4): 2891–2899.

Mahdi, K.N.M., R. Peters, M. van der Ploeg, C. Ritsema and V. Geissen. 2018. Tracking the transport of silver nanoparticles in soil: A saturated column experiment. Water Air Soil Pollut. 229(10): 334.

Mushtaq, A., N. Jamil, S. Rizwan, F. Mandokhel, M. Riaz, G.L. Hornyak et al. 2018. Engineered silica nanoparticles and silica nanoparticles containing controlled release fertilizer for drought and saline areas. IOP Conf. Ser. Mater. Sci. Eng. 414(1): 12029.

Niu, H., S. Shi, S. Zhu, Y. Cai and D. Cao. 2025. Biochars-inlaid nano zero-valent iron reactors: A tool for visualized analysis of soil-nanomaterials micro-interfacial interaction in soil remediation process. Sci. Total Environ. 9581: 177829.

Panwar, A., S. Manna, G. Sahini, V. Kaushik, M. Kumar and M. Govarthanan. 2025. The legacy of endophytes for the formation of bioactive agents, pigments, biofertilizers, nanoparticles and bioremediation of environment. World J. Microbiol. Biotechnol. 41(2): 52.

Parada, J., O. Rubilar, D.Z. Sousa, M. Martínez, M.A. Fernández-Baldo and G.R. Tortella. 2019. Short-term changes in the abundance of nitrifying microorganisms in a soil-plant system simultaneously exposed to copper nanoparticles and atrazine. Sci. Total Environ. 670: 1068–1074.

Peng, C., W. Zhang, H. 2017. Behavior and potential impacts of metal-based engineered nanoparticles in aquatic environments. Nanomaterials. 7(1): 21.

Peyrot, C., K.J. Wilkinson, M. Desrosiers and S. Sauvé. 2014. Effects of silver nanoparticles on soil enzyme activities with and without added organic matter. Environ. Toxicol. Chem. 33: 115–125.

Rajput, V.D., T.M. Minkina, A. Behal, S.N. Sushkova, S. Mandzhieva, R. Singh et al. 2018. Effects of zinc-oxide nanoparticles on soil, plants, animals and soil organisms: A review. Environ. Nanotechnol. Monit. Manage. 9: 76–84.

Rajput, V., T. Minkina, S. Sushkova, A. Behal, A. Maksimov, E. Blicharska et al. 2020. ZnO and CuO nanoparticles: A threat to soil organisms, plants, and human health. Environ. Geochem. Health. 42(1): 147–158.

Rashid, M.I., T. Shahzad, M. Shahid, I.M. Ismail, G.M. Shah and T. Almeelbi. 2017. Zinc oxide nanoparticles affect carbon and nitrogen mineralization of *Phoenix dactylifera* leaf litter in a sandy soil. J. Hazard. Mater. 324(Pt B): 298–305.

Rawat, S., S. Rawat, V.L.R. Pullagurala, I.O. Adisa, Y. Wang, J.R. Peralta-Videa et al. 2018. Factors affecting fate and transport of engineered nanomaterials in terrestrial environments. Curr. Opin. Environ. Sci. Health. 6: 47–53.

Reddy, L.S., M.M. Nisha, M. Joice and P.N. Shilpa. 2014. Antimicrobial activity of zinc oxide (ZnO) nanoparticle against *Klebsiella pneumoniae*. Pharm. Biol. 52(11): 1388–1397.

Ren, Z., X. Xu, J. Liang, C. Yuan, L. Zhao, H. Qiu and X. Cao. 2024. Aging dynamics of polyvinyl chloride microplastics in three soils with different properties. Environ. Sci. Technol. 58(50): 22332–22342.

Roco, M.C. 2011. The long view of nanotechnology development: The National Nanotechnology Initiative at 10 years. J. Nanopart. Res. 13: 427–445

Sabarivasan, R. and P. Murali Arthanari. 2025. Application of nanoencapsulation technology in agriculture for effective and sustainable weed management: A critical review. Commun. Soil Sci. Plant Anal. 56(2): 283–297.

Sacca, M.L., A.B. Caracciolo and M. Di Lenola. 2017. Ecosystem services provided by soil microorganisms. pp. 9–24. *In:* Lukac, M., P. Grenni and M. Gamboni [eds.]. Soil Biological Communities and Ecosystem Resilience. Springer International Publishing.

Salama, D.M. and M.E. Abd El-Aziz. 2025. Nanotechnology and plant probiotics for sustainable agriculture. Agric. Sustain. Nanotechnol. pp. 27–45.

Saleh, M.M., R.A. Sadeq, H. Latif, H.A. Abbas and M. Askoura. 2019. Zinc oxide nanoparticles inhibit quorum sensing and virulence in *Pseudomonas aeruginosa*. Afr. Health Sci. 19(2): 2043–2055.

Sekler, I., S.L. Sensi, M. Hershfinkel and W.F. Silverman. 2007. Mechanism and regulation of cellular zinc transport. Mol. Med. 13(7-8): 337–343. https://doi.org/10.2119/2007-00037.

Shah, V., D. Collins, V.K. Walker and S. Shah. 2014. The impact of engineered cobalt, iron, nickel, and silver nanoparticles on soil bacterial diversity under field conditions. Environ. Res. Lett. 9: 024001.

Shamloo, F., S. Ebrahimi and F.R. Charati. 2025. Effects of biochar modifier and iron nanoparticles on bioremediation of gasoil. Water Air Soil Pollut. 236(2): 132.

Shang, Y., M.K. Hasan, G.J. Ahammed, M. Li, H. Yin and J. Zhou. 2019. Applications of nanotechnology in plant growth and crop protection: A review. Molecules. 24: 2558.

Sheershwal, A., A. Singh, V. Sharma and B. Trivedi. 2025. Microbial synthesis of nanoparticles for sustainable agricultural advancements: A comprehensive review. Nanotechnol. Environ. Eng. 10(1): 16.

Simonin, M. and A. Richaume. 2015. Impact of engineered nanoparticles on the activity, abundance, and diversity of soil microbial communities: A review. Environ. Sci. Pollut. Res. Int. 22(18): 13710–13723.

Simonin, M., J.M. Martins, X. Le Roux, G. Uzu, A. Calas and A. Richaume. 2017. Toxicity of TiO_2 nanoparticles on soil nitrification at environmentally relevant concentrations: Lack of classical dose-response relationships. Nanotoxicology. 11(2): 247–255.

Simonin, M., A.A.M. Cantarel, A. Crouzet, J. Gervaix, J.M.F. Martins and A. Richaume. 2018. Negative effects of copper oxide nanoparticles on carbon and nitrogen cycle microbial activities in contrasting agricultural soils and in presence of plants. Front. Microbiol. 9: 3102.

Sun, W., F. Dou, C. Li, X. Ma and L.Q. Ma. 2020. Impacts of metallic nanoparticles and transformed products on soil health. Crit. Rev. Environ. Sci. Technol.

Taghipour, M. and M. Jalali. 2015. Effect of nanoparticles on kinetics release and fractionation of phosphorus. J. Hazard. Mater. 283: 359–370.

Usman, M., M. Farooq, A. Wakeel, A. Nawaz, S.A. Cheema, H.U. Rehman et al. 2020. Nanotechnology in agriculture: Current status, challenges and future opportunities. Sci. Total Environ. 721: 137778.

Vittori Antisari, L., S. Carbone, A. Gatti, G. Vianello and P. Nannipieri. 2013. Toxicity of metal oxide (CeO_2, Fe_3O_4, SnO_2) engineered nanoparticles on soil microbial biomass and their distribution in soil. Soil Biol. Biochem. 60: 87–94.

Wyszkowska, J., A. Borowik, M. Kucharski and J. Kucharski. 2013. Effect of cadmium, copper and zinc on plants, soil microorganisms and soil enzymes. J. Elem. 18(4): 769–796.

Xu, C., C. Peng, L. Sun, S. Zhang, H. Huang, Y. Chen et al. 2015. Distinctive effects of TiO_2 and CuO nanoparticles on soil microbes and their community structures in flooded paddy soil. Soil Biol. Biochem. 86: 24–33.

Ye, J., H. Gao, J. Wu, Y. Chang, Z. Chen and R. Yu. 2020. Responses of nitrogen transformation processes and N_2O emissions in biological nitrogen removal system to short-term ZnO nanoparticle stress. Sci. Total Environ. 705: 135916.

Yi, L., W. Li, H. Bu, W. Yin, H. Deng and J. Wu. 2025. Effects of humic acid, nitrate and sulfate on hexavalent chromium removal by abiotic and biotic zero-valent iron. Int. J. Environ. Sci. Technol. 22(4): 2409–2420.

You, T., D. Liu, J. Chen, Z. Yang, R. Dou, X. Gao et al. 2018. Effects of metal oxide nanoparticles on soil enzyme activities and bacterial communities in two different soil types. J. Soil Sediment. 18(1): 211–221.

Zafar, S., S. Fatima, F. Asad, M.M. Nazir, S. Batool and A. Ashraf. 2025. Combating lead (Pb) contamination: Integrating biomonitoring, advanced detection, and remediation for environmental and public health. Water Air Soil Pollut. 236(1): 8.

Zahra, Z., M. Arshad, R. Rafique, A. Mahmood, A. Habib, I.A. Qazi et al. 2015. Metallic nanoparticle (TiO_2 and Fe_3O_4) application modifies rhizosphere phosphorus availability and uptake by *Lactuca sativa*. J. Agric. Food Chem. 63(31): 6876–6882.

Zhang, H., X. He, Z. Zhang, P. Zhang, Y. Li, Y. Ma et al. 2011. NanoCeO_2 exhibits adverse effects at environmentally relevant concentrations. Environ. Sci. Technol. 45: 3725–3730.

Zheng, X., Y. Su, Y. Chen, R. Wan, K. Liu, M. Li et al. 2014. Zinc oxide nanoparticles cause inhibition of microbial denitrification by affecting transcriptional regulation and enzyme activity. Environ. Sci. Technol. 48: 13800–13807.

CHAPTER-16

Soil-Nanoparticle Interactions

Fate and Transportation, and Ecological Implications

Ujala Tahir,[1] *Asma Tabassum,*[1] *Ghulam Mustafa,*[2] *Saima Noreen*[1] and *Muhammad Zahid*[1,*]

1. Introduction

Soil is define as "The loose mineral or organic substances that contain solid, liquid, and gas phases, found on the Earth's surface, formed through weathering processes due to the interactions of the lithosphere, atmosphere, hydrosphere, and biosphere, offering a habitat for plants, animals, and micro and macroorganisms, ultimately sustaining human existence and civilization" (Hartemink 2016). Nanoparticles (NPs) are particulate matter with at least one dimension less than 100 nm. Although materials of this size are found in nature, manmade nanoparticles may differ from their natural counterparts in terms of their chemistry and surface characteristics. The size-dependent characteristics and high area-to-volume ratio of manufactured nanoparticles, particularly those smaller than 20 nm, suggest new properties and actions that render them suitable for use in applications involving nanoparticles. That's why they offer versatile usage and applications (Yamini et al. 2023).

The application of fertilizers and plant protection chemicals, biosolids, irrigation with sewage water, and flooding of floodplains are just a few of the ways that nanoparticles can get into soils. After entering the soil, nanoparticles (NPs) undergo interactions with both the mobile and stationary solid phases and the soil solution. Environmental conditions and a variety of processes influence the stability of nanoparticles in soil (Klitzke et al. 2015a).

By affecting several soil processes and characteristics, NMs can improve agricultural productivity and soil quality. By acting as agrochemical nano-carriers or nano-fertilizers, NPs can improve the availability of vital nutrients like potassium, phosphorus, nitrogen, and micronutrients. Zeolites, silica, chitosan, alginate, and polymers are some of the frequently utilized nano fertilizers. They function as slow-releasing fertilizers, lower environmental losses, and boost nutrient efficiency. Both dissolved and colloidal transportation of nanoscale particles is possible, and their behaviors differ from those of conventional solutes that contain comparable particles. Enhanced solubility of active ingredients might hasten degradation and movement with the help of microorganisms that

[1] Department of Chemistry, University of Agriculture, Faisalabad 38040, Pakistan.
[2] Department of Chemistry, University of Okara, Okar, Pakistan.
* Corresponding author: rmzahid@uaf.edu.pk

DOI: 10.1201/9781003558354-16

live in the soil. Because of their greater solubility, NP-based insecticides are thought to provide possibly lower environmental harm (Sharma and Kumar 2024).

Hundreds of thousands of tons of the most common nanoparticles (Ag, Al, C, Ce, Cu, Fe, Si, Ti, and Zn) are released into the environment each year as a result of the production, use, and disposal of different NPs, according to studies based on data and modeling analysis. Most of them are physically introduced into the soil or are disposed of in landfills as sludge or other debris. A disproportionately large fraction is also released into air and water. According to reports, NPs have an impact on the microscopic characteristics of soil, such as the humic acid concentration and the bacterial community. Agro-nanotechnology is becoming more important as a potential solution to address different problems. In order to increase agricultural yield, nanomaterial-based fertilizers, additives, pesticides/herbicides, sensors, and conditioners for the soil have been extensively created and researched in agriculture. For instance, nano-additives improve fertilizer efficiency by allowing a gradual release of nutrients throughout the lifespan of the crop, thereby minimizing nutrient loss from adsorption, decomposition, leaching, and surface runoff. This is made possible by the larger surface area and enhanced absorption capability of these nanosized additives (Rajput et al. 2018).

Given the wide range of uses for nanoparticles, it is likely that these materials will come into contact with the environment at some point throughout their lifetime. The charge, stability, and aggregation state of the nanoparticles are significantly influenced by the nature of the solution and surface functionalization, which in turn affect transport across the receiving matrix. Although agglomerate size, rather than primary size, is associated with transportability, nanoparticle size does have an impact on transport as well. Although a variety of factors affect how nanoparticles move through the environment, the size, charge, and rate of aggregation of the particles in the medium of transport determine how easily the particles will move through soil (Darlington et al. 2009). Although the existence of humic compounds and suspended organic matter can help to stabilize suspension, the availability of adsorbed metals, like copper and lead, which are common in soil minerals, can encourage the aggregation of nanoparticles. Furthermore, it was demonstrated that the establishment of preferential flow routes for colloidal material can result from the heterogeneous pore space structure in natural soils, potentially significantly altering transport properties (Sagee et al. 2012).

2. Sources of Nanoparticles in Soil

It is anticipated that soil has the capacity to absorb nanoparticles. Therefore, it is highly recommended to comprehend nanoparticles' behavior in soil and assess their risk in agricultural soil environments or other actual environmental situations. The most prevalent natural NPs are found in soil, both as basic particles and as aggregates or clusters. Because they may be exceptionally resistant to degradation and have the ability to accumulate in the soil, artificially introducing NPs into the soil could have a significant impact. One of the main concerns in the area of sustainable soil usage is the preservation of soil microbe biomass and diversity. The content, type, and enzyme activity of the soil are all affected by nanoparticles retained in the soil. A few sources of nanoparticles are shown in Figure 1 (Rajput et al. 2018).

Nanoparticles can originate from purposeful, inadvertent, and natural human actions. Since the beginning of time, natural nanoparticles have been a component of the environment. They are frequently and extensively distributed throughout the environment, including in soil, the atmosphere, groundwater, oceans, and living organisms. Hurricanes from deserts, volcanic eruptions, corrosion, creation from clay fragments, wildfires, and soil degradation by air and water are examples of massive biological processes that release nanoparticles into the sky. In a single year, air dust alone is predicted to contain up to several million tons of natural nanoparticles. Ambient nanoparticles that occur naturally are of various sizes, have a short half-life in the environment, and can travel great distances (Yamini et al. 2023).

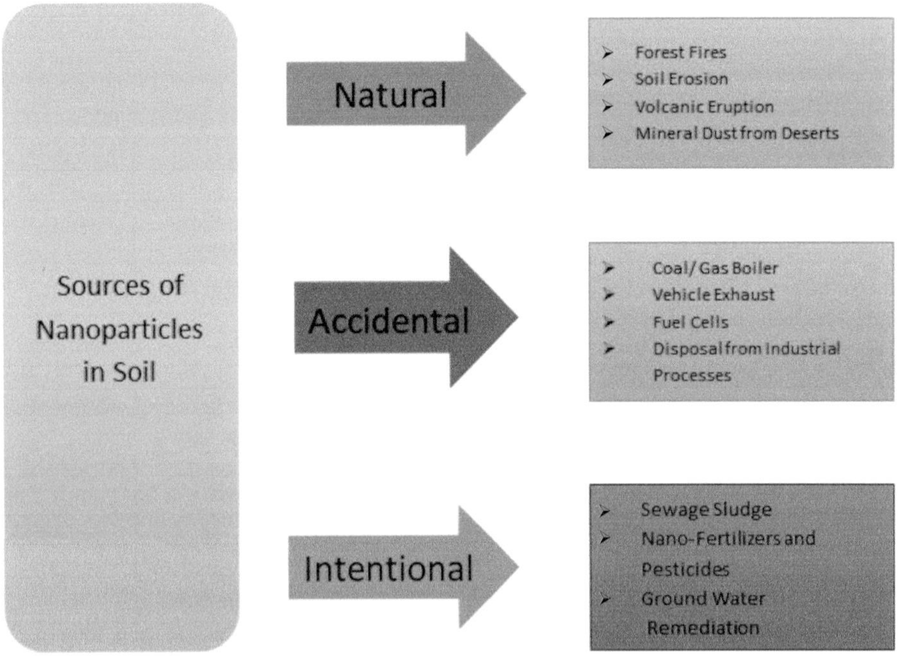

Figure 1. Sources of nanoparticles in soil.

The routes by which NPs enter soil reflect their movement, accumulation (content), and entrance. NPs may enter soil by anthropogenic activity, direct soil absorption of gaseous chemicals, sedimentation in the form of dust and aerosols, atmospheric precipitation, and leaf abscission. In general, there are two types of anthropogenic causes of NP release: point sources and non-point sources. It is possible that free NPs will spread throughout the environment either directly or through the breakdown of NPs (such as surface-bound NPs or nano-sized coatings). The cleansing process using nanotechnology is also expected to release a significant quantity of NPs into the environment. Landfills and wastewater treatment plants also discharge NPs as concentrated sludge or discharged effluent. Other sources of environmental exposure include leaks or spills that occur during the production and shipping of NPs or related goods. In the context of soil and the agro-environment, broad applications such as hydroponic solutions, seed treatment, and nanopesticides and fertilizers are anticipated to create new avenues for NP release into cultivated soils (Rajput et al. 2020).

3. Role of Nanoparticles in Soil

3.1 Improvement in Soil Fertility

In the quest for sustainable solutions, exogenous protectants—also referred to as green technologies—are essential for managing nutrients and protecting crops. Nanotechnology exhibits potential as a green technology for enhancing agricultural productivity and strengthening soil by stimulating energy and nutrients in plant cells. Nanoparticles are employed as nano-fertilizers, nano-insecticides, and nano-fungicides to boost crop yields. A variety of nanoparticles, such as silicon oxide, magnesium oxide, titanium dioxide, zinc oxide, gold, and silver, are being used to improve soil fertility, control nutrients, and boost crop yields (Kale et al. 2024). NPs have garnered increased attention for cleaning up contaminated soils because they mostly exploit soil sorption, reduction, or oxidative processes. Some important nanoparticles in soil along with their functions are shown in Table 1. The nanoparticles' high specific surface area, enhanced chemical reactivity, and specificity make them excellent remediation materials. The primary processes that regulate the activity and toxicity of soil pollutants, which comprise both organic and inorganic compounds, are sorption–desorption, complexation with organic ligands, ionic exchange, and surface precipitation

Table 1. Some Important Nanoparticles in Soil, Sources and Their Functions.

Nanoparticles	Sources	Function	References
ZnO (Zinc Oxide)	Chemical Synthesis Plant Extract	Increases enzyme activity, promotes plant growth and has antibacterial properties	(Rasheed et al. 2022)
FeO (Iron Oxide)	Mineral Ores Green Synthesis	Increases microbial activity, nitrogen availability, and soil aeration	(Khan et al. 2021)
TiO_2 (Titanium Dioxide)	Chemical Synthesis Minerals	Increases soil fertility, aids in the degradation of contaminants, and boosts photocatalysis	(Mgadi et al. 2024)
SiO_2 (Silicon Dioxide)	Chemical Synthesis	Increases a plant's ability to withstand stress and pests and retain water	(Zhang et al. 2024)
Ag (Silver Nanoparticles)	Plant Extracts Chemical Method	Antibacterial properties and lowers the prevalence of soil-borne illnesses	(Javed et al. 2019)
Carbon-Based Nanoparticles	Biochar, Carbon Nanotubes	Enhances microbial development, boosts nutrient retention, and increases soil porosity.	(Alkhaza'leh et al. 2025)
CuO (Copper Oxide)	Chemical Synthesis, Biological Methods	Enhances plant defense systems and functions as a fungicide and bactericide	(Mgadi et al. 2024)
Nano-zeolites	Synthetic Methods	Enhances soil microbial activity, regulates soil pH and improves nutrient retention	(Sharma et al. 2022)
Hydroxyapatite $[Ca_{10}(PO_4)_6(OH_2)]$	Synthetic Methods	Decreases heavy metal pollution and increases soil phosphorus availability	(Dhanapal et al. 2024)

reactions with soil constituents. NPs function as mobilization agents, increasing the bioavailability and mobility of pollutants in the soil so that procedures like soil washing and plant absorption can more easily remove them. NMs are also used strategically to bind contaminants within the soil matrix, serving as immobilizing agents. By reducing the pollutants' mobility and bioavailability, this binding action successfully prevents them from entering the food chain (Sharma and Kumar 2024).

3.2 Enhancement of Soil Micro-Environment

By lowering the surface energy and improving the water repellency of soil particles, as well as by raising soil water conductivity and stabilizing aggregates, nanoparticles may change the structure of soil. These modifications may have an effect on the soil water cycle, affecting elements like evaporation rates and water retention. At the same time, changes in soil composition may also affect how nutrients and pH levels are distributed, which would then affect the rhizosphere's beneficial microbial community. Nanoparticles change the composition, variety, and structure of beneficial microbial communities in plants, which affects how the plant, soil, and microbial community system interacts. Additionally, a healthy soil structure increases the activity and cycling of nutrients by soil microorganisms. By changing the soil's elemental makeup, nanoparticles may promote the growth of microbial community diversity (Jatav et al. 2020). NPs can significantly increase the amount of nitrogen and phosphorus in the soil, creating a favorable environment for enzymes linked to microbial growth, improving soil fertility and health. Biologically produced nanoparticles positively impact the naturally occurring carbon and microorganisms in corn-growing soils, which speeds up plant growth. Novel semi-polymer nanocomposite particles have the potential to significantly increase the levels of organic and active organic carbon, which will support plants' robust growth and development. According to the studies listed above, nanoparticles may improve the pH levels, structure, and other characteristics of soil, which may then affect how plants, soil, and microbes interact (Zhang et al. 2024).

3.3 Suppression of Plant Pathogenic Micro-Organisms

Microorganisms and plants are always interacting in a way that can be advantageous to both. Fungi and bacteria are essential for activities like nitrogen fixation and the roots' assimilation of nutrients from the soil. Viruses, bacteria, fungi, protozoa, and nematodes can all harm and infect plants, impacting their viability and metabolism. Plant disease is characterized by a sequence of reactions in plant tissues and cells to an environmental stressor or pathogen that lead to unfavorable alterations in plant health (Warghane et al. 2024). Plant diseases impair the quantity, quality, color, and flavor of agricultural products, which has a detrimental impact on the economy. Because nanomaterials can manipulate matter at the atomic and molecular level, they can have a variety of properties and uses. They can have a variety of compositions, including inorganic nanostructures of metals, metal oxides, and ceramics; organic nanostructures composed of vesicles like liposomes and polymers; and nanostructures composed primarily of pure carbon like graphene and carbon nanotubes. It is advised to create these nanomaterials using green synthesis since it avoids the use of potentially hazardous chemicals for the environment or human health. In 2006, the first nanomaterial created to treat plant infections was disclosed; it had hydrophilic polymer-coated silver-silica nanoparticles (NPs) to fight fungal diseases (dos Santos et al. 2024). The seeds, leaves, or roots of plants can be directly treated with nanoparticles to protect them from a pathogen invasion. The activities of these nanoparticles may resemble those of chemical insecticides. Nanomaterials can also be used for controlled dispensing of a wide range of substances by serving as carriers for substances like pheromones, drugs that result in systemic acquired resistance, active ingredients in pesticides, etc. It has been demonstrated that pathogens like viruses, fungi, and bacteria are suppressed by nanoparticles. It has been demonstrated that nanoparticles affect pathogen cells in a variety of ways. These characteristics of nanoparticles cause pathogens to die (Hamid and Saleem 2022).

3.4 Water Retention and Soil Resource Management

In nano-remediation applications, metal nanoparticles have shown promise in improving the quality of soil and water in a variety of agricultural systems, including cleaning up pesticide-contaminated soil and purifying irrigation water. Through several procedures, such as adsorption and catalytic degradation, these nanoparticles can be employed to lessen the negative impacts of contaminants, such as pesticides and heavy metals. By lowering pollutant runoff into water sources, titanium dioxide nanoparticles, for instance, have been utilized to remediate pesticide-contaminated soils and thereby enhance irrigation water quality. The long-term productivity and environmental sustainability of agricultural systems can be ensured by integrating nano-remediation technology into agriculture, which can help with sustainable soil and water management (Francis et al. 2024). NPs are presently being investigated as a way to enhance agricultural soil quality and, thus, boost the use of sustainable farming techniques. They have several special qualities (such as improved cation exchange capacity, long-lasting nutrient release, and nutrient delivery) and are highly reactive due to their small size and wide surface area, making them potentially useful for soil restoration issues. Regarding the control of water quality, NMs eliminate unwanted bacteria from the tainted water and prevent the growth of biofilms. As antimicrobial agents, NMs like silver nanoparticles (NPs) and other metallic and metallic oxide NPs including copper, zinc oxide, titanium oxide, magnesium oxide, cadmium telluride, and cadmium selenium are employed. Because of their greater antibacterial capabilities, it is crucial to develop antimicrobial nanocomposites with a larger capacity for immobilization, which lowers the possibility of NPs leaking. We reviewed the function of NMs in managing the quality of soil and water management (Raja and Husen 2020).

4. Fate of Nanoprticles in Soil

Soil becomes more productive due to the interaction between nanoparticles and the microorganisms that are already present. They can regulate mineral transformation, aggregate turnover, biogeochemical reactions, the catalytic creation of humic substances, and the transformation of both organic and inorganic contaminants. Mineral colloid NPs in the soil provide a favorable environment for soil microorganisms if the soil is rich in water, ions, and organic matter. However, because of their NP activity, which increases soil fertility, these microbes encourage mineral weathering. Colloidal NPs are more stable in soil than large-sized particles. Colloidal organic matter aggregates in the soil, forming microstructures that are necessary for microbial activity and nutrient absorption. The fate of NPs depends mostly on the level of mineral solubility in soil (Raut et al. 2024). Agglomeration, fragmentation, and physiological responses are some of the aging mechanisms that affect nanoparticles in the atmosphere. The fate and ecotoxicological potential of nanoparticles are ultimately determined by the interaction of these mechanisms with nanoparticle transit. Because particle characteristics and environmental factors control both aging and transit processes, it can be challenging and occasionally inaccurate to simply transfer knowledge to even vastly different contexts (Barton et al. 2014). Changes in chemical divergence, dispersion, and degradation, along with additional surface modifications brought on by precipitation and adsorption or desorption, are all significant industrial transition mechanisms of nanoparticles. Both terrestrial and marine habitats have been the subject of in-depth research on these activities. From this perspective, the complexation of nanoparticle surfaces, which improves the nanoparticles' qualities for industrial usage, would have a major influence on environmental transformation processes. Therefore, an extensive knowledge of nanoparticle surface chemistry, i.e. the synthesis or reduction of the coatings throughout the period, appears to be necessary to understand nanoparticle fate, despite the fact that it is still very difficult. The current study, however, concentrates on disintegration, volatilization, agglomeration, absorption, precipitation, and accumulation because some of the most important activity takes place in real-world settings (Yamini et al. 2023).

5. Transformation of Nanoparticles in Soil

Nanoparticles and their metal ions interact differently in soils, depending on the aging process and soil conditions; they often persist longer in sedimentation and terrestrial areas. Additionally, plants are unable to absorb nanoparticles due to their aggregation in the soil system. The resilience of the particles is increased by the organic components in the soil, which also have an impact on how well the nanoparticles stick to the soil surface. The soil's nanoparticles are also dissolved by soil pore water. Released ions are more bioavailable than similar nanoparticles, and the kind and physicochemical properties of the soil, as well as the manner in which the nanoparticles are applied (e.g., in powder or solution form), largely dictate how easily they dissolve (Ahmed et al. 2021a). The alteration of nanoparticles' environmental behavior following solute adsorption is a significant transformation process in natural settings. The chemistry of the soil solution, the level of dissolved organic matter, and microorganisms can all affect the nanoparticles' aging process, aggregation, and chemical/biological transformation. When nanoparticles are exposed to soil environments, they affect soil organisms (such as plants, bacteria, nematodes, and earthworms) in both positive and adverse ways (Tao et al. 2023). Through interactions with nearby molecules or physical, chemical, and biological alterations—including aggregation, corrosion, dissolution, and various other processes—nanoparticles attain a thermodynamically more stable state. Figure 2 shows the mechanism of physical and chemical transformation in soil.

5.1 Physical Transformation of Nanoparticles

Physical processes include the production of a double electric layer of counterions and the aggregation of nanoparticles, which is mainly propelled by van der Waals forces and includes the creation of

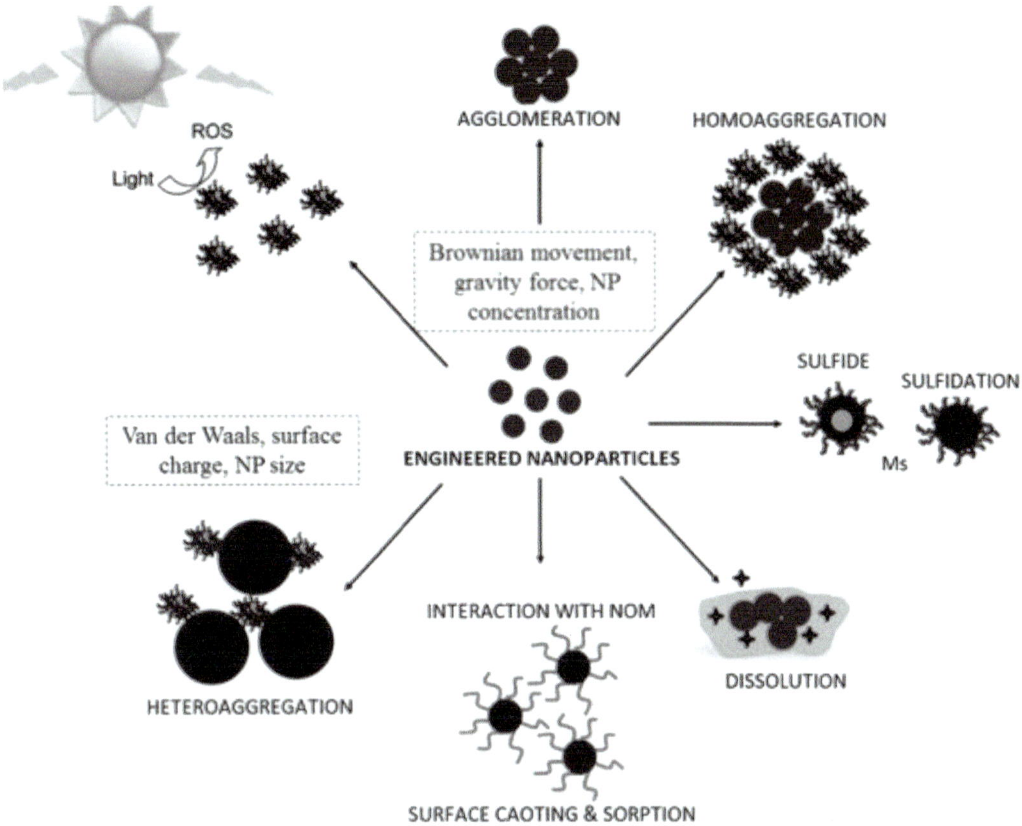

Figure 2. Mechanism of Physical and Chemical Transformation of Nanoparticles in Soil (Abbas et al. 2020).

clusters of varying sizes. Hydrophobic, magnetic, and hydration forces are additional interactions that affect the aggregation of nanoparticles. There are two kinds of aggregation: hetero-aggregation, in which multiple elements interact with NPs to induce aggregation, and homo-aggregation, in which comparable nanoparticles create aggregates (Ahmed et al. 2021a). Environmental conditions are more conducive to hetero-aggregation. Adsorption onto various environmental materials, especially naturally occurring organic matter, is another physical process. This adsorption is primarily controlled by the surface chemistry of nanoparticles and the physicochemical characteristics of natural organic matter. Other elements, including ambient fluids or biomolecules, affect how NPs adsorb onto different surfaces. When environmental components come together to form a corona, commonly referred to as the "protein corona" or "eco-corona," the protein fraction of the environment interacts with nanoparticles to produce this effect. Another way that nanoparticles settle in wet soil environments is by deposition. The amount of deposition depends on NP types, agglomeration levels, and the availability of naturally occurring organic components (Raut et al. 2024).

5.1.1 Agglomeration or Aggregation of Nanoparticles

Accordingly, a lot of research has been done on the fate of metallic trace elements by colloidal transport through soils. The use of wastewater sludge for soil amendment, intentional releases for environmental purposes (such as zero-valent iron nanoparticles), and accidental spills are the three main ways that ENMs get into soils. Furthermore, NPs may be released into the environment through aerosols that settle in soil or aquatic systems and end-of-life disposal in landfills. Many studies have been conducted on the behavior of NPs in environmental matrices, particularly in soils (El Hadri et al. 2018). Agglomeration or aggregation can be classified into two types: (a) homo-aggregation, which occurs when comparable nanoparticles come together, and (b) hetero-aggregation, which occurs when several different factors interact with the nanoparticles to promote aggregation. The more common environmental phenomenon among these is hetero-aggregation. Over extended periods of time, the aggregate size of nanoparticles increases while overall chemical reactivity and bioavailable concentrations frequently decrease. Furthermore, because of their increased collision frequency, the nanoparticles agglomerate quickly at larger concentrations. This phenomenon results in a thermodynamically determined decrease in the surface energy of nanoparticles. This has been verified by a quicker rate of ZnO, TiO_2, and CeO_2 nanoparticle aggregation. Moreover, the coating of nanoparticles may also alter the rate of environmental aggregation (Qiu et al. 2020).

5.1.2 Deposition of Nanoparticles

The process of deposition, which mostly takes place in aquatic environments, is how nanoparticles dispersed in watery environments tend to sink to the bottom. However, depending on the types, level of aggregation, and availability of natural organic materials, the position may vary. Agglomeration, aggregation, and deposition are all connected in general. Deposition rises with increased aggregation, which can be regulated by the characteristics of the nanoparticles and the media's physicochemical characteristics (Amde et al. 2017). The surface area of the porous media and the nanoparticle concentration both affect the deposition rate. The researchers claim that certain types of models fitted to the breakthrough curve produced by the concentrations of nanoparticle effluent escaping the porous media were utilized to capture all known methods of nanoparticle deposition. This tends to restrict our comprehension of how the pore structure of the porous media shapes the many kinds of nanoparticle retention mechanisms (Yusuf et al. 2024).

5.1.3 Adsorption of Nanoparticles

Natural organic matter is one of the many environmental substances that nanoparticles have a propensity to adsorb. Two elements control this adsorption: the physicochemical characteristics of organic materials found in nature and the chemistry of nanoparticle surfaces. If the nanoparticles' surface functionalization is weak, they will be retained by natural organic matter. After adsorption on the surface of nanoparticles, natural organic matter may block oxidation sites, preventing ion release, or fulvic and humic acids may convert the metal ions that have already been released to their zero-valent state. Other factors also affect the adsorption of nanoparticles to various surfaces. These consist of biomolecules or environmental fluids (Nasser et al. 2020).

5.2 Chemical Transformation of Nanoparticles in Soil

Numerous chemical changes, including oxidation, dissolution, sulfidation, complexation, chlorination, and reduction, are experienced by nanoparticles. The physicochemical characteristics of sediments, including pH, redox potential, oxygen content, and natural organic matter (NOM), are critical to this process. Naturally occurring ring silver nanoparticles may release fewer ions if the oxidation is inhibited or slowed down by low oxygen concentration. The environment's pH and NOM have an inverse relationship with this process. The sulfidation of nanoparticles occur and develop a core-shell structure in both aerobic and anaerobic environments. Redox reactions, sulfidation, and dissolution are examples of chemical processes (Sharma and Kumar 2024).

5.2.1 Dissolution of Nanoparticles

Physicochemical characteristics and the chemistry of the surrounding environment affect the dissolution of nanoparticles, which results in the release of soluble metal ions. Relative to their larger counterparts, smaller NPs release more free metal ions; however, coating them slows down the release of metal ions. Sulfate, phosphate, and chlorides are examples of inorganic ligands that are essential to the chemical alteration of NPs (Francis et al. 2024). The breakdown of nanoparticles may release harmful substances into the environment. There have been reports of NP release in a range of sample matrices, such as artificial soil solutions, well-defined aqueous media with varying pH, natural organic matter (NOM) content, and ocean salts. One might infer that NP oxidation in neutral soil fluids has only a minor role based on the findings, which showed that Ag^+ release peaked at low pH (i.e., pH 4) and in the absence of NOM. On the surfaces of NPs, oxidized ions may sorb and leak into the bulk solution (Klitzke et al. 2015b).

5.2.2 Sulfidation of Nanoparticles

The fate of nanoparticles is also impacted by the presence of sulfide in the surrounding medium. Metal ions produced from nanoparticles are reduced during the oxidation of sulfur to sulfate. The concentration of sulfides in the medium determines the process of nanoparticle sulfidation. Occasionally, an overabundance of sulfur causes all of the nanoparticles in a solution to become sulfidated (Ahmed et al. 2021a). The sulfidation process reduces metal ions produced from nanoparticles and transforms sulfide into sulfate. For instance, CuO nanoparticles produce Cu^{2+}, which is reduced to Cu^+ to generate copper sulfate hydroxides. Additionally, after the dissolution of nanoparticles such as CuO and Cu nanoparticles, the high solubility may compete with the sulfidation process (Levard et al. 2013).

5.2.3 Redox Reactions of Nanoparticles

Electrons are transferred between species in redox reactions, which are essential to the transformation of nanoparticles. Oxidation is more prevalent in aerated soils and waterways, whereas reduction is more prevalent in groundwater and carbon-rich sediments. Two significant changes brought about by redox reactions impact the ecological fate of nanoparticles: if the surface becomes active, it facilitates interaction with environmental elements; if the surface becomes insoluble, it increases stability, which boosts persistence (Sharma and Kumar 2024). Redox reactions can provide an insoluble surface layer that improves the NPs' stability and boosts their persistence. However, it can also result in the development of an active surface, which makes it easier for the NPs to react with other elements in the environment and reduces their stability. The dissolving behavior of NPs may also be facilitated by the oxidation process. As a result, it is one of the mechanisms that significantly influences how MeO-NPs behave in the environment (Amde et al. 2017).

5.3 Biological Transformation of Nanoparticles in Soil

The main biologically mediated changes in the environment are biodegradation, macromolecules that help build NPs, and eco-bio corona and nano-bio interactions. The connection of NPs with extracellular enzymes and microorganisms establishes the relative relevance and rate of these processes. Biochar nanoparticles are broken down enzymatically by fungal hyphae to create nanoparticles. In addition to chemical oxidation, iron-oxidizing and iron-reducing bacteria also carry out microbiological activity that oxidizes, preserving the environment's iron biogeochemical cycle. Figure 3 shows the mechanism of biological transformation of nanoparticles in soil (Abbas et al. 2020). Proteins, lipids, nucleic acids, and even biological metabolites are examples of biomolecules that NPs may interact with in biological media due to their nanosize and high surface-to-mass ratio. The environmental macromolecules, including proteins, polysaccharides, and other surfactants, may interact, change the surface, and drastically alter the physicochemical characteristics of NPs. Therefore, the fate of NPs in the environment and their harmful consequences can also be determined by macromolecules (Saptarshi et al. 2013). Numerous mechanisms for interactions between macromolecules and nanoparticles have been proposed, depending on the types and properties of both materials. These consist of Van der Waals interaction, hydrophobic interaction, chelation, ligand exchange, bridging, and hydrogen bonding. Protein adsorption on nanoparticle surfaces is particularly significant. The nanoparticle-protein corona (NP-PC) is the term used to describe the creation of nanoparticle-protein complexes. When proteins adsorb on the NP surface, a number of effects are possible. All things considered, the NP-PC can affect the NP's biological reactivity (Abdelsattar et al. 2021).

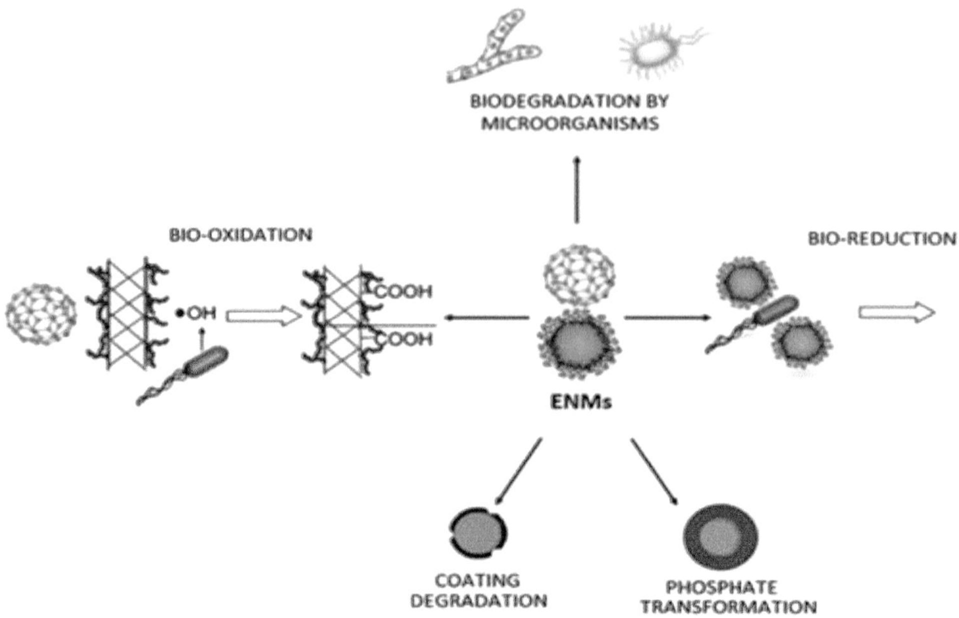

Figure 3. Mechanism of Biological Transformation of Nanoparticles (Abbas et al. 2020).

6. Interaction of Nanoparticles with Soil

Nanoparticles and their metallic ions usually have a longer lifespan in sedimentation and terrestrial settings, but their reactions vary depending on the process of aging and soil characteristics. Furthermore, the soil system's aggregation of nanoparticles prevents plants from absorbing them. For instance, the diffusion of ZnO nanoparticles was impeded by their hetero-aggregation with soil granules. The soil's organic materials also affect how well nanoparticles adhere to the soil surface, increasing their stability. The nanoparticles in the soil are also dissolved by soil pore water. The discharged ions are more accessible than the comparable nanoparticles, whose dispersion is mostly dictated by the kind and physicochemical features of the soil, in addition to the way the nanoparticles are applied in the soil, such as in powder or solution forms. For example, it has been observed that ZnO nanoparticles dissolve in soils to the point that they are not detectable in soil that has been spiked with nanoparticles, as shown in Figure 4 (Ahmed et al. 2021a).

Essential nanomaterials and various soil additions can be used to decontaminate biotic or abiotic soil contaminants. The effectiveness of remediation depends on the characteristics of the nanoparticles and how they interact with the characteristics and materials of the soil. Only laboratory settings are currently being used to test the usage of nanomaterials for soil remediation. The potential for NMs to change soil and plant morphological properties that are fully explained by soil physio-biochemical features and plant morpho-physiological, cellular, and molecular reactions is present when they enter soils. It has been addressed how nanoparticles are absorbed, transported, and have both positive and negative consequences. There have also been reports of nanoparticles' beneficial effects in reducing abiotic stressors and mitigating their phytotoxic effects through detoxifying techniques (Ulhassan et al. 2022). To fully realize metal nanoparticles' promise to increase crop productivity and sustainability, it is essential to comprehend the molecular interactions that occur between them and plants. By elucidating mechanisms, it highlights the possibility for enhancing crop productivity, stress tolerance, and nutrient-use efficiency, all of which promote sustainable agriculture and food security. Quantifying risks and benefits demonstrates significant gains. Metal nanoparticles can reduce disease incidence by up to 50% and boost agricultural productivity by 20% on average when used as antibacterial agents. Additionally, they improve soil carbon sequestration

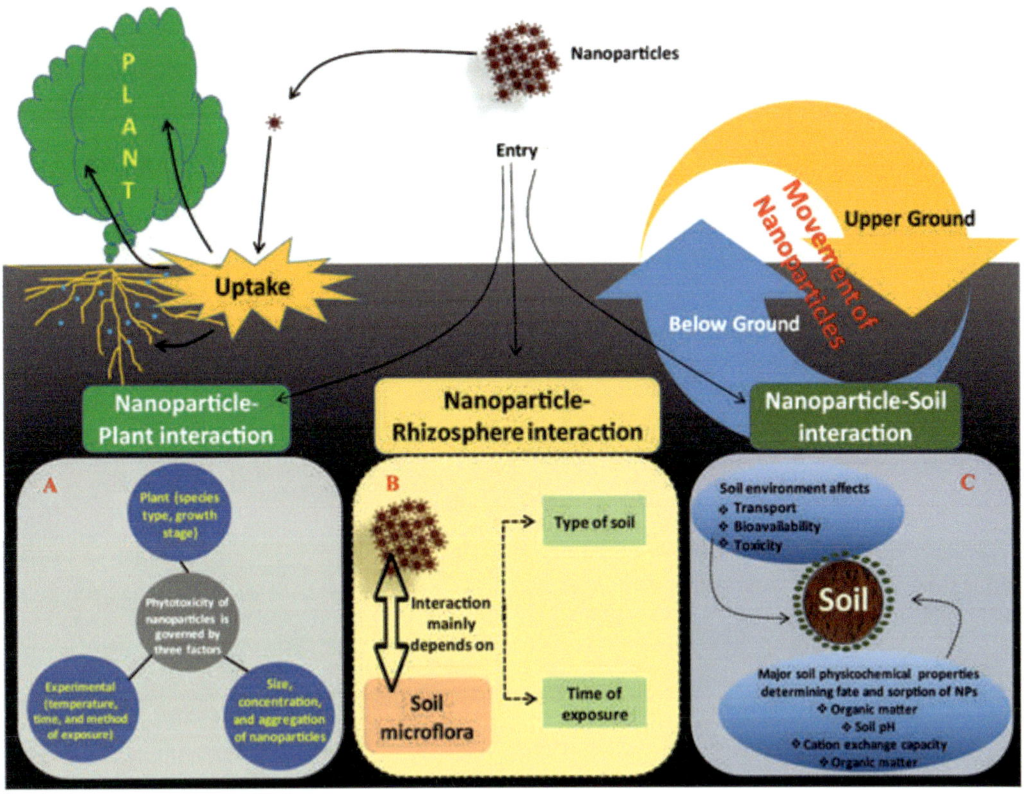

Figure 4. Interaction of Nanoparticles with Soil (Ahmed et al. 2021a).

by 15% and decrease nutrient leakage by 30%; nevertheless, issues with toxicity, negative impacts on non-target organisms, and the buildup of nanoparticles in the food chain need to be addressed. Sensing, signaling, transcription, translation, and post-translational changes are among the biological functions that are impacted by metal nanoparticles. They boost nutrient absorption, strengthen defense mechanisms, activate stress-responsive genes, and function as signaling molecules (Rajput et al. 2020).

7. Bioavailability of Nanoparticles in Plants

A stability-dependent factor is the bioavailability of nanoparticles to plants. As a result, the nanoparticles have low toxicity because their bioavailability decreases with increasing environmental stability. For example, plants were less able to absorb CeO_2 nanoparticles when the soil's Ce content dropped from Ce(IV) to the comparatively stable Ce(III). On the other hand, soil organisms release chelating chemicals and enzymes that change nanoparticles and increase their availability. Furthermore, other parameters such as the particle size of the nanoparticles, the utilization of coated substances, and homo-/hetero-aggregation are also crucial in modulating the accessibility of nanoparticles and therefore must be properly monitored. Moreover, proteins, polysaccharides, fulvic acids, and humic acids are also in charge of the intracellular uptake and surface adsorption of nanoparticles. ZnO nanoparticles were discovered in the roots of the maize plant but not in the shoot system. This could be because the ZnO nanoparticles dissolved in the plant tissues, indicating that different plant parts had varying levels of availability. After dissolving, nanoparticles are known to be readily accessible to plants. For example, it has been found that ZnO nanoparticles are often accessible mostly in their dissolved or ionic state, which indicates that they dissolve quickly. The whole mechanism of availability of nanoparticles is shown in Figure 5 (García-Sánchez et al. 2015).

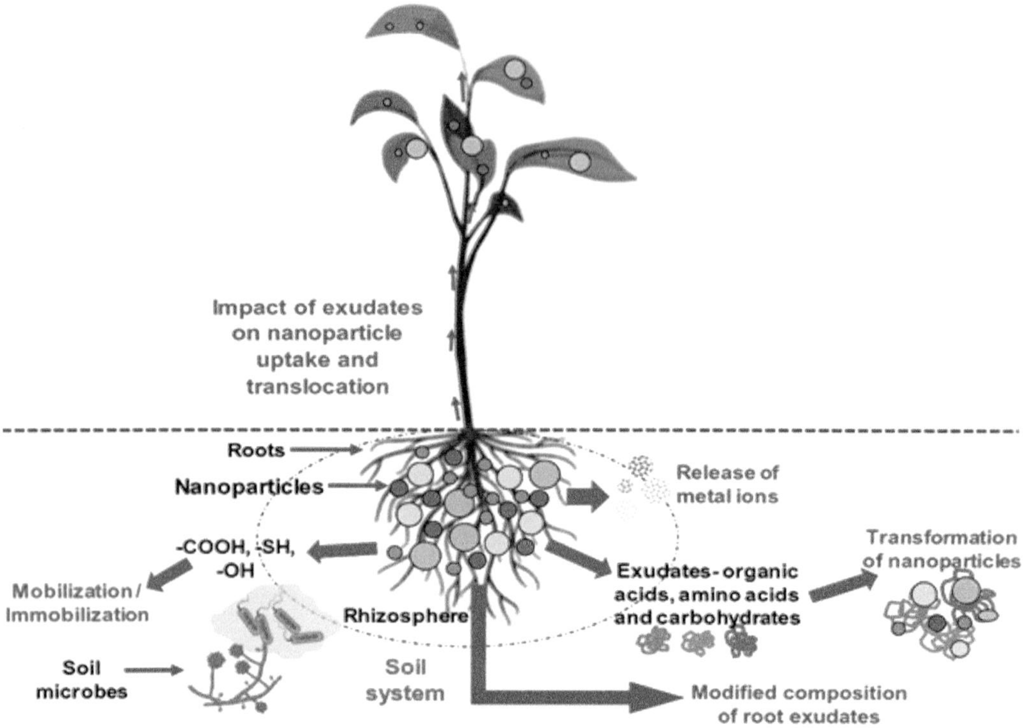

Figure 5. Interaction of Nanoparticles with Soil (Ahmed et al. 2021a).

8. Effect of Nanoparticles on Soil

On the basis of type, concentration and exposure duration, nanoparticles have significant impact on soil. Some important effects are shown below in Figure 6.

8.1 Effect of Nanoparticles on Physical Properties of Soil

NMs affect the physical properties of soil, including bulk density, organization, texture, and porosity. By creating nanoscale pores or channels between soil aggregates, NPs can enhance soil aeration and water drainage. It is said that platinum nanoparticles increase the saturated hydraulic conductivity by decreasing the spacing between soil particles and increasing the stiffness of the soil structure. NMs can also enhance the volume of soil organic matter while lowering the bulk density of the soil by strengthening the connections between soil particles or filling in the spaces between them. Scientists found that the remarkable elastic properties and high aspect ratios of carbon nanotubes increased the aggregation of sandy-textured soil by 35% (Sharma and Kumar 2024).

Physical property studies have demonstrated that NPs such as Fe_3O_4, ZnO, MgO, SiO_2, and TiO_2 can decrease the distance between soil particles and increase hydraulic conductivity and soil porosity. As a result, NPs help create a more rigid matrix, which promotes higher agricultural productivity and results in a safer environment and healthier lives. Additionally, it was discovered that although Fe_3O_4 NPs simply strengthened the links between Fe and soil to increase the aggregates' tensile strength, MgO NPs enhanced soil structure, increased porosity, and decreased bulk density. Furthermore, concentrations and particle sizes of ENPs can have an impact on the physical characteristics of soil. After applying Pt NPs of 3 nm in concentrations of 0.1, 1, and 10 µg per 300 mg^{-1} soil, researchers noticed an increase in the strength of the water molecule bonds and the structural rigidity of the soil. However, they reduced the water content retained in SOM at doses ranging from 100 to 1000 µg per 300 mg^{-1} soil. Similarly, Fe NPs were able to pass through

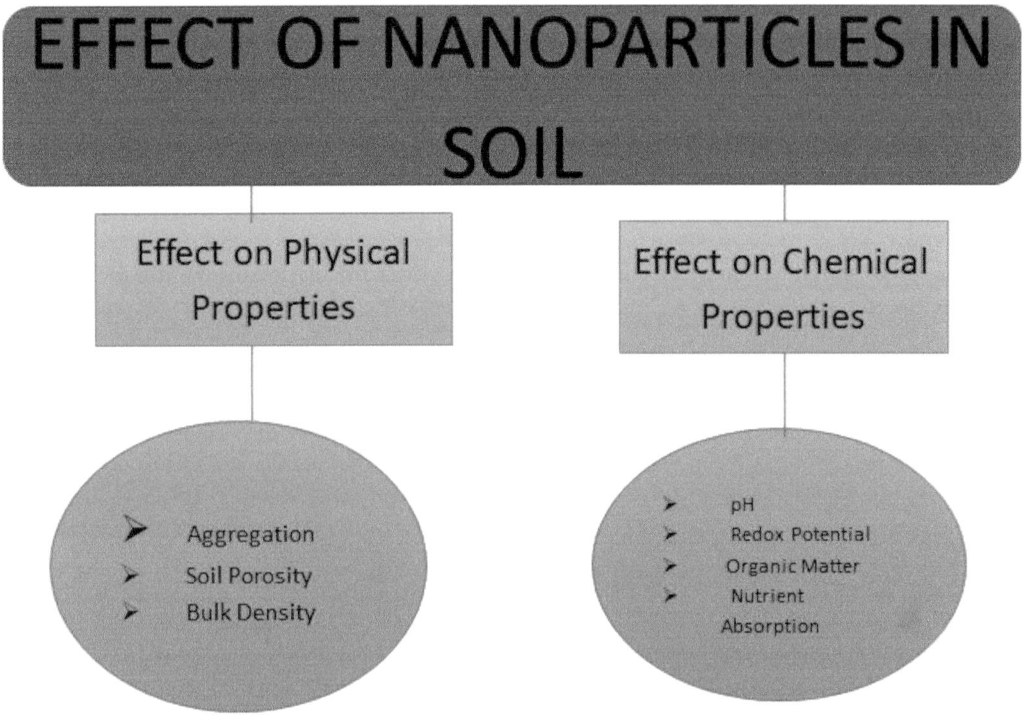

Figure 6. Effects of Nanoparticles in Soil.

clay soil because their particles were smaller than its pores, but they may eventually block the pores and cause aggregates to develop, which would lower the hydraulic conductivity (Suazo-Hernández et al. 2023). By increasing the surface area and hydrophilic qualities of soil particles, NPs can be used to improve soil hydrological regimes. They can also be used to create water-absorbing hydrogels that can help plants cope with drought stress and water loss. Furthermore, it was observed that soil liquefaction and soil settlement had significantly decreased (Arora et al. 2019).

8.2 Effect of Nanoparticles on Chemical Properties of Soil

8.2.1 Effect on pH of Soil

One important characteristic that has a direct bearing on the fertility and health of soil is its pH. It indicates whether the soil is alkaline or acidic. The nutrients available to plants are influenced by the pH of the soil. It has been demonstrated that the composition of the soil determines the type of acidity or basicity of the soil. Generally speaking, plants prefer a pH range of 5–7. It has been noted that the buildup of various nanoparticle types, such as Zn, Ag, Au, Cu, etc., may have an impact on the pH of the soil (Javed et al. 2019). The type of nanoparticles (NPs) and matrix characteristics determine how the presence of nanoparticles alters the pH of the soil. Researchers discovered that the pH of acidic soil rose from about 5.30 to 7.60 after adding Fe NPs, whereas the pH of calcareous soil was almost 8.0 both with and without nanoparticles. Cu_2O nanoparticles have recently been found to selectively alter the pH of rice soil. Cu_2O nanoparticles cause the pH of the soil to rise more, which eventually affects the soil. The pH of the soil influences the uptake of Ag nanoparticles by insects. The pH of the soil seems to be a key factor in regulating the quantities of soluble metals. Furthermore, calcareous soils contain more clay than acidic soils, which increases zinc retention and decreases bacterial availability of zinc since clay is expected to be absorbed through ion-exchange and specific adsorption (Dhankar et al. 2024).

8.2.2 Effect on Redox Potential of Soil

The oxidation-reduction reactions are represented by the soil Eh, which is influenced by temperature, precipitation, organic matter content, and oxygen (O_2) concentration. Because it affects how the soil, plant, and microorganism system functions, as well as how soluble nutrients and pollutants behave, Eh is a crucial agronomic parameter. Since pH and Eh are interdependent and because it can be challenging to replicate, compare, and interpret the results, studies of Eh in soils with and without ENPs have received little attention. Most metal ENPs tend to oxidize in soils. In other words, they lose electrons, which are then absorbed by materials from the external environment, causing the Eh values to change. The scientists concluded that ENP dosages and incubation duration, which were linked to the quantity of reactive mass of ENPs, had a significant impact on the variation in Eh values. Even if the elements in metallic oxide ENPs are oxidized, they can still affect how the soil is modified (Eh). For example, mixed-valence Fe_3O_4 ENPs were introduced to organic-rich natural wetland soil both uncoated and coated with dimercaptosuccinic acid (DMSA) (Van Hoecke et al. 2011). Specifically, ENP dosages were found to have a significant impact on Eh values. The presence of NPs caused the Eh value to decrease during the 30-day period under evaluated flood conditions. The decomposition of organic matter (OM) produced organic reducing chemicals, while microbial respiration reduced the amount of oxygen in the soil, causing this behavior. When paddy soil was flooded, researchers discovered that CuO and ZnO NPs increased Eh values by about 20–30 mV. The ability of ENPs to utilize the reducing chemical (H^+) in flooded conditions provided an explanation for this. Numerous factors that impact soil Eh values have been studied, including the presence of stabilizing agents such as proteins, humic acid, and chloride, as well as the detrimental effects of NPs on soil microorganisms. According to reports, fluctuations in Eh values for Ag NPs, which have antibacterial capabilities, have resulted from a decline in soil microbiology (Suazo-Hernández et al. 2023).

8.2.3 Effect on Organic Matter in Soil

The environment is full of organic matter (OM), which is highly reactive. The fate, transport, bioavailability, and toxicity of engineered nanoparticles (ENPs) to organisms can all be greatly impacted by interactions between OM and ENPs once they are introduced into natural systems. But due to OM's complexity and the lack of practical characterization techniques, there are significant knowledge gaps in our comprehension of how OM and ENPs interact. The highest amount of OM is found in the most productive agricultural soils (Khanna et al. 2021). Numerous investigations have assessed how OM (SOM and DOM) affects the mobility, transformations, and toxicity of ENPs. ENPs' toxicity can be decreased by DOM's ability to be adsorbed on their surface, which increases their stability and stops ions from escaping. This is because of the electrostatic and/or steric repulsion mechanism between ENPs, which reduces their rate of aggregation and duration of residency in soils, increasing their likelihood of migrating to other natural systems like rivers and groundwater (Yu et al. 2018).

8.2.4 Effect on Nutrient Absorption in Soil

Nanoparticles have a lot of potential to improve nutrient uptake, and some of them can increase the efficiency of plant nutrient utilization by using processes including controlled release, sustained release, or directional delivery. Nanoparticles can affect nutrient absorption in two ways while promoting plant nutrient uptake. They can serve as both nutrient carriers and modulators of soil microbes, which can improve the plant's capacity to absorb nutrients. On the other hand, the efficiency of nutrient utilization could be increased by using nanoparticle carriers to specifically transport nutrient elements to different plant areas (Zhang et al. 2024).

Through the accurate administration of chemical fertilizers and plant growth regulators, nanoparticles have enhanced the uptake of nutrients in plants, promoting innovation in the agricultural industry and providing a fresh tactical approach to precision agriculture. Studies have shown that a seed covering made of a zinc and urea hydroxyapatite nanohybrid can improve the efficiency of nutrient utilization and more accurately distribute nutrients to plants. When applied to sunflower crops in acidic soil, novel phosphate hydroxyapatite nanofertilizers—more precisely, hydroxyapatite nanoparticles, or HA-NPs—showed a noticeably quicker and more effective phosphate uptake than traditional phosphate and triphosphate fertilizers. One important direction for further study into the accurate distribution of nanoparticles is the creation of nanocarriers (Riseh et al. 2024).

9. Transportation of Nanoparticles in Soil

Moreover, there are no soil transport parameters specifically for nanomaterials in the literature. Nanomaterials' contact with the soil is based on a flowing liquid, since they are almost dispersed in a liquid solution, through the pores of the soil with a laminar flow regime, being only turbulent in the apparent fractures (Silva et al. 2016). Assuming that the nanomaterials' movement in the soil can occur through advection, diffusion, or mechanical dispersion processes (or a combination of them), this work considers the hypothesis that numerical solutions could be formulated to predict the transportation and impacts by assuming different transport phenomena.

When NPs come into contact with solid surfaces, their travel through soils is dictated. The physicochemical properties of the particles, as well as the surrounding environment, may have an impact on this interaction. Transport qualities are typically influenced by the same factors that affect NPs stability. Gravity and Brownian motion are responsible for the sedimentation and diffusion of NPs, respectively, as has been demonstrated for colloids. Brownian motion governs the movement of NPs in porous media. But when the particles clump together and agglomerate, gravitational forces increase, making it more likely that they will come into contact with the surfaces of soil particles. Because these variables influence agglomeration and electrostatic attraction/repulsion between particles, as well as between particles and soil, the surface charges of metal-based NPs and soil surfaces have an impact on their interaction. When the particle and soil surfaces are equally charged, repulsion between them favors particle mobility over sorption. When the charges of particles and the soil are opposite, the particles will be attracted to the soil, and absorption may occur, reducing mobility (Tourinho et al. 2012).

Since NPs with high diffusivity are more likely to be absorbed by soil grains, the transport of NPs in porous media is regulated by the filtration rate of NPs by the porous media, which is frequently established by diffusion. The mobility of NPs should theoretically decrease as particle size increases. However, because aggregation can make NPs larger and aggregate straining is typical for NPs to be captured in porous media, NPs' mobility is strongly influenced by their aggregation status. Aggregate straining is still difficult to define precisely since the more complex effects, such as changes in particle surface charge and morphology, might actually cause the change in particle size through aggregation. Furthermore, the contact between particles and the soil grain surface has a significant impact on NP capture by the soil matrix. NPs' size, surface charge, and modifications to the surface can all have a significant impact on mobility in porous media. Physical factors of the media, such as size of particles, fluid speed, and grain size, can all play crucial roles in the filtration of NPs (Rajput et al. 2020).

The filtration of NPs is also influenced by soil matrix properties including size and porosity. The mobility of NPs generally decreases when the size of the soil medium grain decreases. The presence of clay in soil can serve as an anionic support material to increase the mobility of NPs and stop them from aggregating. It has also been demonstrated that the mobility and aggregation status of NPs are influenced by the flow rate of the solution. The mobility of NPs often increases as the flow rate of solution increases. Important variables influencing NPs' movement include temperature, ionic strength, and solution pH. The pH of the solution regulates the surface charges of the soil grains and NPs, and consequently, the electrostatic interactions within the system. A higher temperature can cause NPs to move more randomly and improve the medium particles' ability to capture Brownian particles. Dissolved counterions in solution will shield long-range electrostatic interactions in the environmental matrix, reducing NP mobility and stability. Ionic strength rises with increasing dissolved salt content, and NPs will be more attracted to one another than repelled, forming aggregations (Chen 2018).

The transport of nanoparticles, colloids, and the pollutants they are linked with in saturated porous media has garnered increased interest in the last ten years due to the inherent existence of nanomaterials in the environment and the possible hazards they pose to the ecosystem and public health. Colloids migrate in porous media through a variety of physical and chemical processes, including adsorption, convection, advection, and precipitation. In transport processes, adhesion, deposition, and release are all involved in the adsorption of colloids. Essentially, colloidal size, shape, and properties determine how colloid particles move across porous media. It is easier for particles to move through porous material when they are smaller in size. In saturated porous media, the pH and ionic strength of the solution also affect the movement of soil colloids. In natural settings, changes in irrigation and rainfall will affect the ionic strength and pH of the soil. The transport of environmental pollutants adsorbed to the colloidal surface can be improved by the colloids' release brought on by variations in pH and ionic strength. As a result, numerous studies have examined how pH and ionic strength affect the movement of colloids in porous media and have employed quartz sand as a stand-in for soil medium (Jiang et al. 2017).

10. Nanoparticles in Soil Reclamation

In agriculture, nanoparticles have the potential to improve compacted soil. The impacts of NPs on the physical properties of soil must be well documented, nevertheless, particularly in the fields of geological and geotechnical engineering and building. In geotechnical, geological, and engineering applications, nanotechnology (NT), which entails working with materials at the nanoscale (often less than 100 nm), has been created and applied to enhance soil in an economical and ecologically friendly manner. To lessen the impacts of soil compaction, NPs encourage flocculation, which increases soil porosity and decreases bulk density. Nanoparticles have been applied to agriculture for improved productivity, protection, and plant nutrition. The US National Research Council (NRC 2006) emphasized the use of NT in geotechnical materials in 2006, producing 0.002-mm clay-sized particles. Applications of NT for soil improvement have changed over time to conform to sustainable development concepts (Alkhaza'leh et al. 2025).

11. Toxicity of Nanoparticles in Soil

The kind, concentration, and physicochemical features of the nanoparticles involved, as well as the pH and organic matter content of the soil, all affect how poisonous the nanoparticles are. Nanoparticles in soil can be injurious to soil species, as well as microbes and plants, because of their small size, which allows for easy penetration into cells, potentially causing disruption to cellular functions, oxidative stress, and damage to cell membranes, ultimately impacting soil health and ecosystem function. Oxidative stress is brought on by the physicochemical susceptibility of nanoparticles, which can either directly or indirectly trigger oxidative enzymatic pathways to produce free radicals, or reactive oxygen species (ROS), including super oxide radical anions and hydroxyl radicals. In general, oxidative stress can be caused by a variety of circumstances.

- Because nanoparticles have a tendency to produce oxidants and stimulate their formation, when cells react to them, they generate ROS.
- Transition metal-based catalysts or transition metal contaminants are used to create non-metal nanoparticles.
- When nanoparticles combine, redox-active chemicals are created; the reactive surfaces of the particles contain relatively persistent free radical precursors.

About ten years ago, at the start of the scientific area known as "nano-ecotoxicology," researchers hypothesized that nanoparticles could be able to harm the biosphere by producing reactive oxygen species (ROS) that could change biological systems. In contrast, photosystem II was affected by variations in the photosynthetic pigment content of aquatic plants such as Wolffia globosa, Arabidopsis thaliana, Solanum lycopersicum, and others, as well as algae like dinoflagellates, cryptomonads, brown algae, etc. Likewise, some recent analyses have examined the deposition of nanoparticles in terrestrial plants, which could lead to changes in physiology and biochemistry (Yamini et al. 2023).

12. Future Perspective of Nanoparticles in Soil

Future research should focus on design and prediction, novel characterization technologies for tracking nanoparticles' transition in soil ecological systems, and improving their risk evaluation in light of the unresolved problems. It is necessary to look into how plant metabolites affect NP transformation and the complex connection between exudate composition and plant interactions. The function of microbes at plant interfaces and the rate at which NPs change in soil and plants must be investigated. Additionally, it is critical to take into account the long-term impacts of NPs on agricultural ecosystems by examining their linkages to soil quality and climate change (Sharma and Kumar 2024). Furthermore, NPs have the ability to build up in microbial and plant systems, making their way into the food chain and endangering human health. To reduce their buildup in agricultural systems, we should focus on creating biodegradable and biocompatible NPs in the future. To ensure the safe and societal acceptance of NPs for HMs repair, there are important limitations and information gaps. Additional field studies are needed to investigate the application's safety, reliability, effectiveness, consistency, fate, inherent hazard of NPs, and long-term impacts of NPs on HM assimilation and bioavailability in contaminated soils. For NPs to have their potential effects in the agricultural and environmental sectors, future research should focus on safe targeted administration and dose optimization (Ahmed et al. 2021b).

13. Conclusion

The fate and mobility of contaminants in soils are governed by intricate physical, chemical, and biological interactions. The process of desorption, degradation, leaching, volatilization, and plant uptake are all important processes that influence the tenacity and mobility of chemicals in the soil. Soil texture, biological matter concentration, pH, humidity, and microbial activity all have a substantial impact on these dynamics. Understanding these processes is critical for good environmental management, pollution reduction, and sustainable agriculture. Forecasting models and location-specific assessments assist in reducing the hazards associated with pollutant transfer, assuring soil health and the integrity of water. Future research should focus on enhancing monitoring tools with remediation solutions to lessen the adverse consequences of soil contaminants on habitats and human health.

14. Acknowledgement

The authors are thankful to the University of Agriculture Faisalabad, Pakistan for facilitating the accomplishment of this task

References

Abbas, Q., B. Yousaf, Amina, M.U. Ali, M.a.M. Munir, A. El-Naggar, J. Rinklebe et al. 2020. Transformation pathways and fate of engineered nanoparticles (ENPs) in distinct interactive environmental compartments: A review. Environ. Int. 138: 105646.

Abdelsattar, A.S., A. Dawoud and M.A. Helal. 2021. Interaction of nanoparticles with biological macromolecules: A review of molecular docking studies. Nanotoxicology. 15: 66–95.

Ahmed, B., A. Rizvi, K. Ali, J. Lee, A. Zaidi, M.S. Khan et al. 2021a. Nanoparticles in the soil–plant system: a review. Environ. Chem. Lett. 19: 1545–1609.

Ahmed, T., M. Noman, M. Ijaz, S. Ali, M. Rizwan, U. Ijaz et al. 2021b. Current trends and future prospective in nanoremediation of heavy metals contaminated soils: A way forward towards sustainable agriculture. Ecotoxicol. Environ. Saf. 227: 112888.

Alkhaza'leh, H., M. Rahbeh, I. Hamadneh, H. Al-Mashakbeh and Z. Albalawna. 2025. Nanoparticles in soil reclamation: a review of their role in reducing soil compaction. Air Soil Water Res. 18: 11786221241311725.

Amde, M., J.-F. Liu, Z.-Q. Tan and D. Bekana. 2017. Transformation and bioavailability of metal oxide nanoparticles in aquatic and terrestrial environments. A review. Environ. Pollut. 230: 250–267.

Arora, A., B. Singh and P. Kaur. 2019. Performance of nanoparticles in stabilization of soil: a comprehensive review. Mater. Today Proc. 17: 124–130.

Barton, L.E., M. Auffan, M. Bertrand, M. Barakat, C. Santaella, A. Masion et al. 2014. Transformation of pristine and citrate-functionalized CeO_2 nanoparticles in a laboratory-scale activated sludge reactor. Environ. Sci. Technol. 48: 7289–7296.

Chen, H. 2018. Metal-based nanoparticles in agricultural system: behavior, transport, and interaction with plants. Chem. Speciat. Bioavailab. 30: 123–134.

Darlington, T.K., A.M. Neigh, M.T. Spencer, O.T. Guyen and S.J. Oldenburg. 2009. Nanoparticle characteristics affecting environmental fate and transport through soil. Environ. Toxicol. Chem. 28: 1191–1199.

Dhanapal, A.R., M. Thiruvengadam, J. Vairavanathan, B. Venkidasamy, M. Easwaran and M. Ghorbanpour. 2024. Nanotechnology approaches for the remediation of agricultural polluted soils. ACS Omega. 9: 13522–13533.

Dhankar, R., S. Gupta, A. Tomar and P. Dhanker. 2024. Impact of nanotechnology on soil health and plant halobiont. pp. 1–13. *In*: Nanotechnology and Nanomaterials Annual Volume 2024. IntechOpen.

Dos Santos, O.A.L., M.S. Dos Santos, S. Antunes Filho and B.P. Backx. 2024. Nanotechnology for the control of plant pathogens and pests. Plant Nano Biol. 8: 100080.

El Hadri, H., S.M. Louie and V.A. Hackley. 2018. Assessing the interactions of metal nanoparticles in soil and sediment matrices–a quantitative analytical multi-technique approach. Environ. Sci. Nano. 5: 203–214.

Francis, D.V., A.K. Abdalla, W. Mahakham, A.K. Sarmah and Z.F.R. Ahmed. 2024. Interaction of plants and metal nanoparticles: exploring its molecular mechanisms for sustainable agriculture and crop improvement. Environ. Int. 190: 108859.

García-Sánchez, S., I. Bernales and S. Cristobal. 2015. Early response to nanoparticles in the Arabidopsis transcriptome compromises plant defence and root-hair development through salicylic acid signalling. BMC Genomics. 16: 1–17.

Hamid, A. and S. Saleem. 2022. Role of nanoparticles in management of plant pathogens and scope in plant transgenics for imparting disease resistance. Plant Prot. Sci. 58: 173–184.

Hartemink, A.E. 2016. The definition of soil since the early 1800s. Adv. Agron. 137: 73–126.

Jatav, H.S., L.D. Sharma, R. Sadhukhan, S.K. Singh, S. Singh, V.D. Rajput et al. 2020. An overview of micronutrients: prospects and implication in crop production. pp. 1–30. *In*: Plant Micronutrients: Deficiency and Toxicity Management.

Javed, Z., K. Dashora, M. Mishra, V.D. Fasake and A. Srivastva. 2019. Effect of accumulation of nanoparticles in soil health—a concern on future. Front. Nanosci. Nanotechnol. 5.

Jiang, Y., L. Yu, H. Sun, X. Yin, C. Wang, S. Mathews et al. 2017. Transport of natural soil nanoparticles in saturated porous media: effects of pH and ionic strength. Chem. Speciat. Bioavailab. 29: 186–196.

Kale, S.S., R. Chauhan, B. Nigam, S. Gosavi and I.J. Chaudhary. 2024. Effectiveness of nanoparticles in improving soil fertility and eco-friendly crop resistance: a comprehensive review. Biocatal. Agric. Biotechnol. 56: 103066.

Khan, S.T., S.F. Adil, M.R. Shaik, H.Z. Alkhathlan, M. Khan and M. Khan. 2021. Engineered nanomaterials in soil: their impact on soil microbiome and plant health. Plants. 11: 109.

Khanna, K., S.K. Kohli, N. Handa, H. Kaur, P. Ohri, R. Bhardwaj et al. 2021. Enthralling the impact of engineered nanoparticles on soil microbiome: a concentric approach towards environmental risks and cogitation. Ecotoxicol. Environ. Saf. 222: 112459.

Klitzke, S., G. Metreveli, A. Peters, G.E. Schaumann and F. Lang. 2015a. The fate of silver nanoparticles in soil solution—sorption of solutes and aggregation. Sci. Total Environ. 535: 54–60.

Klitzke, S., G. Metreveli, A. Peters, G.E. Schaumann and F. Lang. 2015b. The fate of silver nanoparticles in soil solution—sorption of solutes and aggregation. Sci. Total Environ. 535: 54–60.

Levard, C., E.M. Hotze, B.P. Colman, A.L. Dale, L. Truong, X. Yang et al. 2013. Sulfidation of silver nanoparticles: natural antidote to their toxicity. Environ. Sci. Technol. 47: 13440–13448.

Mgadi, K., B. Ndaba, A. Roopnarain, H. Rama and R. Adeleke. 2024. Nanoparticle applications in agriculture: overview and response of plant-associated microorganisms. Front. Microbiol. 15: 1354440.

Nasser, F., J. Constantinou and I. Lynch. 2020. Nanomaterials in the environment acquire an "Eco-Corona" impacting their toxicity to Daphnia magna—a call for updating toxicity testing policies. Proteomics. 20: 1800412.

Qiu, Y., Z. Mu, N. Wang, X. Wang, M. Xu and H. Li. 2020. The aggregation and sedimentation of two different sized copper oxide nanoparticles in soil solutions: dependence on pH and dissolved organic matter. Sci. Total Environ. 731: 139215.

Raja, M.A. and A. Husen. 2020. Chapter 20 - Role of nanomaterials in soil and water quality management. pp. *In*: Husen, A. and M. Jawaid [eds.]. Nanomaterials for Agriculture and Forestry Applications. Elsevier.

Rajput, V., T. Minkina, M. Mazarji, S. Shende, S. Sushkova, S. Mandzhieva et al. 2020. Accumulation of nanoparticles in the soil-plant systems and their effects on human health. Ann. Agric. Sci. 65: 137–143.

Rajput, V.D., T. Minkina, S. Sushkova, V. Tsitsuashvili, S. Mandzhieva, A. Gorovtsov et al. 2018. Effect of nanoparticles on crops and soil microbial communities. J. Soils Sediments. 18: 2179–2187.

Rasheed, A., H. Li, M.M. Tahir, A. Mahmood, M. Nawaz, A.N. Shah et al. 2022. The role of nanoparticles in plant biochemical, physiological, and molecular responses under drought stress: A review. Front. Plant Sci. 13: 976179.

Raut, S.S., R. Singh and U.M. Lekhak. 2024. Naturally occurring nanoparticles (NONPs): A review. Next Sustain. 3: 100037.

Riseh, R.S., M. Vatankhah, M. Hassanisaadi and R.S. Varma. 2024. A review of chitosan nanoparticles: Nature's gift for transforming agriculture through smart and effective delivery mechanisms. Int. J. Biol. Macromol. 129522.

Sagee, O., I. Dror and B. Berkowitz. 2012. Transport of silver nanoparticles (AgNPs) in soil. Chemosphere. 88: 670–675.

Saptarshi, S.R., A. Duschl and A.L. Lopata. 2013. Interaction of nanoparticles with proteins: relation to bio-reactivity of the nanoparticle. J. Nanobiotechnol. 11: 1–12.

Sharma, R. and V. Kumar. 2024. Nano enabled agriculture for sustainable soil. Waste Manag. Bull. 2: 152–161.

Sharma, V., B. Javed, H. Byrne, J. Curtin and F. Tian. 2022. Zeolites as carriers of nano-fertilizers: From structures and principles to prospects and challenges. Appl. Nano. 3: 163–186.

Suazo-Hernández, J., N. Arancibia-Miranda, R. Mlih, L. Cáceres-Jensen, N. Bolan and M.D.L.L. Mora. 2023. Impact on some soil physical and chemical properties caused by metal and metallic oxide engineered nanoparticles: A review. Nanomaterials. 13: 572.

Tao, Z., Q. Zhou, T. Zheng, F. Mo and S. Ouyang. 2023. Iron oxide nanoparticles in the soil environment: Adsorption, transformation, and environmental risk. J. Hazard. Mater. 459: 132107.

Tourinho, P.S., C.A. Van Gestel, S. Lofts, C. Svendsen, A.M. Soares and S. Loureiro. 2012. Metal-based nanoparticles in soil: Fate, behavior, and effects on soil invertebrates. Environ. Toxicol. Chem. 31: 1679–1692.

Ulhassan, Z., A.R. Khan, Y. Hamid, W. Azhar, S. Hussain, M.S. Sheteiwy et al. 2022. Chapter 14 - Interaction of nanoparticles with soil–plant system and their usage in remediation strategies. *In*: Aftab, T. and Hakeem, K. [eds.]. Metals Metalloids Soil Plant Water Systems. Academic Press.

Van Hoecke, K., K.A. De Schamphelaere, P. Van Der Meeren, G. Smagghe and C.R. Janssen. 2011. Aggregation and ecotoxicity of CeO2 nanoparticles in synthetic and natural waters with variable pH, organic matter concentration and ionic strength. Environ. Pollut. 159: 970–976.

Warghane, A., R. Saini, M. Shri, I. Andankar, D.K. Ghosh and B.A. Chopade. 2024. Application of nanoparticles for management of plant viral pathogen: Current status and future prospects. Virology. 592: 109998.

Yamini, V., V. Shanmugam, M. Rameshpathy, G. Venkatraman, G. Ramanathan, H. Al Garalleh et al. 2023. Environmental effects and interaction of nanoparticles on beneficial soil and aquatic microorganisms. Environ. Res. 236: 116776.

Yu, S., J. Liu, Y. Yin and M. Shen. 2018. Interactions between engineered nanoparticles and dissolved organic matter: A review on mechanisms and environmental effects. J. Environ. Sci. 63: 198–217.

Yusuf, E.O., I. Amber, S. Officer and G.F. Oluyemi. 2024. Transport of nanoparticles in porous media and associated environmental impact: A review. J. Eng. Res. 12: 275–284.

Zhang, H., T. Zheng, Y. Wang, T. Li and Q. Chi. 2024. Multifaceted impacts of nanoparticles on plant nutrient absorption and soil microbial communities. Front. Plant Sci. 15: 1497006.

CHAPTER-17

Nanotechnology for Soil Carbon Sequestration and Climate Change Mitigation

Najma Majeed,[1] *Muhammad Imran,*[1,*] *Akhtar Hameed,*[2] *Waseem Hassan,*[1] *Shakeel Ahmad*[1] and *Mudassar Yasin*[3]

1. Introduction

Global warming is one of the principal global concerns of the 21st century, characterized by the emission of increasing volumes of greenhouse gases (GHG) into the atmosphere. Carbon dioxide (CO_2) is the most common of these gases, released mainly by the combustion of fossil fuels, forest destruction, and many other industrial processes (Arya et al. 2017). Currently, there is no greater need for efficient carbon sequestration methods than now. This chapter aims to highlight nanotechnology as a new frontier in the strategic fight against climate change, proposing innovative pathways to enhance soil carbon sequestration (Malhi et al. 2021, Smith et al. 2007). Actually, soil—the Earth's largest carbon reservoir—is critically important in the fight against climate change. However, conventional efforts in soil management may not reach the desired level of scale and intensity to adequately address global emissions of CO_2. There is potential for nanotechnology to change the future of this field through its ability to alter the soil's capability to capture and store carbon by upgrading the soil's properties, nutrient retention, and microbial activity (Patle et al. 2024, Rout and Aggarwal 2024).

This chapter reviews the state of knowledge regarding the application of nanotechnology to carbon storage in soil and the role of such science in climate change solutions. This paper explains the current practices in carbon sequestration, describes the process of carbon storage in soils, and presents the basic concepts concerning nanotechnology in this context. The chapter also explores some case studies regarding nanomaterials used in soil management, outlines risks and challenges, and offers a critical review of cross-cutting themes and future research.

[1] Department of Soil and Environmental Sciences, MNS-University of Agriculture, Multan, Pakistan.
[2] Institute of Plant Protection, MNS-University of Agriculture, Multan, Pakistan.
[3] Department of Agri Extension, MNS-University of Agriculture, Multan, Pakistan.
* Corresponding author: m.imran@mnsuam.edu.pk

DOI: 10.1201/9781003558354-17

2. Soil Carbon Sequestration: An Overview

Soil carbon sequestration refers to the process of capturing and storing atmospheric carbon dioxide in soils through biological, chemical, and physical processes (Jose and Bardhan 2012). The two main forms of carbon stored in soil are organic carbon (derived from decomposed plants and microbial residues) and inorganic carbon (from minerals). Healthy soils act as carbon sinks, capturing CO_2 through plant photosynthesis and converting it into organic matter that is stored in the soil (Bernadett 2013). However, human activities such as deforestation, overgrazing, and unsustainable farming practices have led to soil degradation and significant carbon loss (Lal 2004).

2.1 Mechanisms of Soil Carbon Storage

The ability of soil to sequester carbon depends on various factors, including soil texture, organic matter content, microbial activity, and climate (Reetika et al. 2024). There are two primary mechanisms of carbon storage:

- Soil Organic Carbon (SOC): This is carbon that is sequestered in the organic residues resulting from plant and microbial organic matter. Stimulated organic carbon (SOC) is mobile and constantly moving between the atmosphere and soil (Pandey et al. 2024).
- Soil Inorganic Carbon (SIC): This carbon is associated with minerals in the soil, especially in arid and semi-arid areas, that is relatively stable as compared to SOC (Naorem et al. 2022).

2.2 Challenges in Soil Carbon Sequestration

Even though the potential of soils to act as carbon sinks is high, various challenges limit the effectiveness of current practices:

Saturation Limits: Soils can sequester carbon only up to a limit, beyond which they cannot hold additional carbon (Poeplau et al. 2024).

Decomposition: Organic compounds in soil undergo microbial decomposition, returning stored carbon to the atmosphere (Mattila and Vihanto 2024).

Soil Degradation: Unsustainable and modern agricultural practices erode soils, thereby limiting their carbon storage capacity (Morya et al. 2023).

Some of these limitations may be addressed by nanotechnology, which creates ways of improving soil structure, nutrient retention, and microbial activity that lead to stable organic carbon formation (Bai and Cotrufo 2022, Zhang et al. 2011).

3. Nanotechnology in Soil Carbon Sequestration

Nanotechnology is the process of controlling matter on the scale of nanometers (1–100 nm), where it exhibits distinct characteristics (Amenta et al. 2015). Nanotechnology in agriculture and soil management is still in its infancy but has shown promising potential in the improvement of carbon sequestration (Duhan et al. 2017).

3.1 Types of Nanomaterials Used in Soil Management

Various nanomaterials have been explored for their potential to improve soil carbon sequestration, including:

Biochar-based Nanocomposites: Biochar is a charcoal-like substance that is obtained from the pyrolysis of biomass. It has been blended with nanomaterials to form carbon cenosites for enhanced carbon capture. The enhancement of biochar with nanotechnology has resulted in a highly efficient method of stabilizing carbon in the soil. Biochar itself is highly effective in this regard (Sarma et al. 2024).

Nanoclays: These are nanoparticles made from silicates. Their application enables them to help promote the conservation of organic matter and nutrients in the soil structure. They form coherent bodies of sedimentation that minimize the breakdown of organic carbon as practicably achievable, with the ultimate goal of enhancing the long-term sequestration of carbon (Patle et al. 2024).

Carbon-based Nanomaterials: This includes the use of carbon nanotubes (CNTs) and graphene oxide and how they enhance the capacity of soil to sequester carbon through reactions with the soil organic matter, and how they immobilize carbon within the soil (Kale et al. 2024).

Metallic Nanoparticles: The engineered nanoscale iron and zinc both can alter atmospheric CO_2 to competent inorganic forms in the soil, which in turn improves soil carbon sequestration (Rout and Aggarwal 2024).

3.2 Mechanisms of Nanomaterial Action

Nanomaterials influence soil carbon sequestration through several mechanisms:

Correcting Soil Structure: Nanomaterials such as nanoclay improve the structure, and hence aggregation, of the soil to reduce erosion and increase the potential of the soil to retain organic matter. More recent soil structural conditions facilitate water management—a crucial factor for plant growth and microbial function that further contributes to carbon sequestration (Das et al. 2023).

Enhancing Microbial Activity: Some nanomaterials can transport enzymes or nutrients that increase microbial quantity and hence facilitate degradation of organic compounds into stable carbon forms (Pagano et al. 2023).

Stabilizing Organic Matter: Nanomaterials including biochar-based composites and nanoclays control the aggregation of organic matter by linking it to structural units of soil particles (Wang et al. 2016).

Promoting Carbon Mineralization: Mineral particles, especially ferrous oxide nanoparticles, are capable of promoting reactions that cause carbon in the organic matter to become fixed in a stable inorganic form in the soil matrix (Mukherjee et al. 2016).

4. Applications and Case Studies

Nanotechnology has been proven to benefit soil carbon sequestration through the enhancement of soil carbon minerals. This next part focuses on describing a number of useful case studies, which explain how nanotechnology can be used for managing soil carbon sequestration issues (Jimenez-Castaneda et al. 2022, Youns et al. 2023). Nanotechnology-enhanced soil carbon sequestration is advantageous over conventional methods, which is compared in Table 1.

4.1 Nanoclays for Enhanced Carbon Storage

According to a study by Patle et al. (2024), nanoclays prove useful in soil aggregation and stabilization of organic carbon in degraded soil. Soil samples taken in this experiment and treated with nanoclay had 30% better capacity for holding carbon than the samples without the nanoclay. This enhancement was found to result from the formation of stable aggregates that shield organic matter from microbial deterioration (Ramadass et al. 2020).

Table 1. Comparison of Traditional and Nanotechnology-Enhanced Soil Carbon Sequestration Methods.

Method	Carbon Sequestration Potential	Longevity of Carbon Storage	Effect on Soil Properties	References
Traditional Soil Amendments	Moderate	Short to Moderate	Minimal improvement in soil structure	(Cristea et al. 2020)
Biochar	High	Long	Enhances water retention	(Han et al. 2024, Tan et al. 2016)
Nanoclays	High	Long	Significantly improves soil aggregation	(Calabi-Floody et al. 2011)
Biochar-Nanocomposites	Very High	Very Long	Improves nutrient retention, reduces erosion	(Verma et al. 2014)

4.2 Biochar-Nanoparticle Composites

In another study, biochar-based nanocomposites were added to agricultural soils with a consequent enhancement of carbon stocks. Biochar combined with zinc and iron nanoparticles favored soil organic matter due to the slow rate of decomposition of the components (Alam et al. 2020). After three years, the nanocomposite also enabled the sequestration of 40% more carbon in treated soils than conventional biochar (Khan et al. 2022).

5. Challenges and Risks

As has been seen, nanotechnology offers great potential for enhancing practices of soil carbon sequestration, but the issues and prospects have their limitations which must be overcome.

5.1 Environmental Risks

An outstanding cutting-edge issue involves the implications of introducing nanomaterials into soils. To date, there is still scarce information on the impacts of nanoparticles, especially when presented in different concentrations, on soil typology, microbial flora and fauna, and general plant health. Other negative side effects include nanoparticle migration into the water table with possible negative effects on human health and the environment. Moreover, the production and use of nanomaterials in large quantities may be expensive (Auffan et al. 2009).

5.2 Economic Viability

While nanotechnology helps increase the carbon sequestration capacity of plants, it is expensive, so its application may be restricted by costs in areas that often suffer most from climate change. While there are advantages to improved carbon storage through the use of nanomaterials, there is a downside that has been touched on: cost. The generally high costs of nanotechnology may compromise the practical application of this technology in areas that need it most developing countries (Dai et al. 2020).

5.3 Regulatory and Ethical Considerations

In some areas, there is a need for regulation regarding the use of nanotechnology in food production and the environment. Furthermore, the issues of possible emergent effects of nanomaterials on the environment and food safety should be taken into consideration as ethical concerns (Dai et al. 2020).

6. Nanotechnology and Soil Health

Among the workable benefits of nanotechnology for enhanced carbon sequestration in the soil is the potential to enhance the physical health of the soil. Soil health doesn't only involve enhanced production capacity but also increased capacity for carbon sequestration. This section reviews how nanotechnology affects soil health under better nutrient control, improved water-holding capacity, and encouragement of the growth of beneficial microorganisms (Dhanapal et al. 2024).

6.1 Nanoparticles for Nutrient Management

Leaching loss contributes to the depletion of soil nutrient values and increases emissions of greenhouse gases in agricultural lands. For example, nitrogen is an important nutrient for plant growth but can be easily leached from soils by means of denitrification and leaching (Khan et al. 2021). Nanotechnology provides options to decrease the loss of nutrients and increase the fertility of the soil by preparing nanofertilizers to release nutrients more efficiently (Faizan et al. 2024, Ghormade et al. 2011).

It can also be used to regulate the rate at which nitrogen, phosphorus, potassium, or other nutrients are made available for plant uptake without being washed off by rainwater (Gogos et al. 2012, Yasmin et al. 2021). Research indicates that nanofertilizers can increase nutrient use efficiency by as much as 30%, thereby enhancing crop yields while minimizing the emissions of nitrous oxide (N_2O) (Kumari et al. 2024).

Besides, it has a positive effect on improving nutrient status and immobilizing toxic heavy metals and other analogous substances in the soil (He et al. 2007). Nanoparticles such as nano-iron and nano-alumina have also been reported to reduce the solubility of heavy metals (Wang et al. 2009), thus limiting their chances of being absorbed by plants and polluting the environment.

6.2 Enhancing Water Retention

For plant growth, as well as for carbon sequestration, water retention is a significant factor. Waterlogged soils have a higher ability to store carbon and support microbial growth and productivity compared to soils with low water-holding capacity. Nanomaterials, especially nanoclays and nano-SiO_2, can enhance the water retention ability of the soil by improving soil porosity and particle aggregation (Srivastava et al. 2024). For instance, nano-silica can increase the soil's water retention while improving aeration, allowing roots easier access to water. Besides improving plant growth, it contributes to the increase of organic matter in the soil, which supports long-term carbon storage (Nthebere et al. 2022). In situations where carbon sequestration potential is severely constrained by water limitation, such as in arid and semi-arid lands, the ability to manage water is transformative (Sarma et al. 2024). Nanotechnology can enhance the soil's moisture retention, thus lengthening the growing period and increasing the feasibility of using otherwise infertile lands to produce food and feed, in the process sequestering more carbon through plant growth and decomposition of soil organic matter.

6.3 Stimulating Beneficial Microbial Communities

Microorganisms have a significant role in the carbon cycle, recycling organic carbon and packaging it as soil organic matter. The changes elicited in microbial consortia by nanotechnology are either through direct carriers of enzymes and nutrients or through altered substrate properties of soils used to foster microbial growth (Santoyo et al. 2021). Some of these nanomaterials include biochar nanoparticles, which have been identified to offer "microbial niches" that support healthy microbial life. These materials improve the richness and metabolism of microbial populations involved in carbon cycling and the stabilization of organic carbon (Luo et al. 2023). Also, using nanomaterials, it is possible to control the dispersion of nutrients and enzyme substances to stimulate the activity of microorganisms. For example, magnetite nanoparticles made of iron and magnesium increase the activity of nitrogen-fixing soil bacteria, which is essential for soil fertility and plant growth (Mahapatra et al. 2024, Upadhayay et al. 2024). The increase in microbial activity is not only beneficial to soil aeration but also enhances efficiency in carbon storage.

7. Role of Nanotechnology in Climate-Smart Agriculture

Climate-smart agriculture (CSA) has been defined as addressing specific systemic challenges by intervening in one or more of the four domains, namely croplands, pastoralists' livestock, forest, and fisheries (Harvey et al. 2014, Karbo and Crentsil 2021). Since it enhances soil and carbon stocks and union, nanotechnology can be said to be compatible with CSA. Specifically, this section demonstrates the position of nanotechnology within the context of CSA and its ability to offset the consequences of further climate change while maintaining agricultural productivity (Garg et al. 2024, Raza et al. 2023). A summary of the role of nanotechnology in climate-smart agriculture is provided in Table 2.

Table 2. The Role of Nanotechnology in Climate-Smart Agriculture.

CSA Goal	Nanotechnology Contribution	Outcome	References
Reducing GHG emissions	Nanofertilizers, nanopesticides, water retention	Lower emissions of CO_2, CH_4, and N_2O	(Gupta et al. 2018)
Enhancing crop resilience	Nanoseed coatings, nanonutrient delivery systems	Increased tolerance to drought, heat, and nutrient stress	(Zhao et al. 2022)
Improving soil carbon storage	Biochar nanocomposites, nanoclays, nano-silica	Enhanced soil organic matter and long-term carbon storage	(Don et al. 2023)

7.1 *Reducing Greenhouse Gas Emissions*

Agricultural operations contribute to the emission of the main greenhouse gases—methane, nitrous oxide, and carbon monoxide. Nanotechnology unveils new approaches to minimizing such discharges through improved nutrient use, optimized water consumption, and carbon storage in the soil. For example, controlled-release nanofertilizers reduce N_2O emissions, and nanoparticles that enhance water retention in the soil can decrease CH_4 emissions from flooded soils, such as those used for rice cultivation (Santoyo et al. 2021).

7.2 *Enhancing Crop Resilience to Climate Change*

Impacts of climate change include more frequent and severe weather patterns, such as droughts, floods, heat waves, and others. These events pose potential food security threats because they reduce yields and lower soil fertility through land degradation. This section establishes that nanotechnology can increase the system's ability to resist stress arising from water and nutrient deficits, as well as biotic and abiotic stress (Zhao et al. 2023). For instance, nanocoatings can be used to protect seeds against drought and heat by coating the seeds with a protective element. These coatings regulate water throughput and prevent heat damage to the seed, consequently enhancing germination and early plant development. Similarly, nanomaterials can be employed to convey growth-enhancing constituents, such as plant growth regulators, to the desired area within the root zone to enhance crop resilience to unfavorable weather conditions.

7.3 *Improving Soil Carbon Storage in Agroecosystems*

Another aim of CSA is to strengthen the ability of agricultural systems to store carbon. Nanotechnology offers unique methods through which physical rearrangement of the soil promotes enhanced accumulation of organic carbon, greater stability of organic matter, and improved formation of stable soil aggregates (Bera et al. 2024). For instance, applications of biochar-based nanocomposites have enhanced the physical properties of various soils in temperate and tropical agroecosystems by raising the organic carbon content. These nanocomposites improve soil fertility by immobilizing nutrients, thereby decreasing the rate of loss of organic carbon from the soil and enhancing carbon sequestration.

8. Nanotechnology and Policy Frameworks

Regulatory, economic, and ethical implications of nanotechnology application for the management of soil and carbon sequestration are crucial. To address the benefits of applying nanotechnology, collaboration between governments, international organizations, and other private actors is crucial as risk groups (Renn and Roco 2006).

8.1 Regulatory Challenges

There are no general and broad regulatory provisions concerning the application of nanomaterials in agricultural processes and soil treatment. Based on the potential environmental implications of the release of nanoparticles into the environment, there is a need for standards to regulate the use, handling, and disposal of nanomaterials (Morya et al. 2023). Since nanomaterials are novel, the government and regulatory fronts should emulate the precautionary principle in this reckoning to ensure that nanomaterials are adequately evaluated for risks before they are allowed into agricultural systems.

8.2 Economic Considerations

Nanotechnology has great potential for enhancing carbon sequestration in the soil, but this depends on the prices of the materials and technologies used. Some nanomaterials can be expensive to produce, and even if farmers in developing countries embrace the new technologies, they may not be able to afford them. To overcome this difficulty, governments and international organizations should support research and offer financial incentives in the form of subsidies or concessional loans to promote the adoption of nanotechnology in soil management (Jose and Bardhan 2012).

8.3 Ethical Implications

While nanotechnology may changes into the agricultural sphere, it also raises crucial ethical questions related to environmental justice, food security issues, and other potential consequences on the ecosystems. Policymakers simply have to guarantee that the process of nanotechnology innovation and application in agriculture does not become one-sided and closed to public and stakeholder participation of farmers, environmental groups and the public (Amenta et al. 2015).

9. Conclusion and Future Directions

Nanotechnology, therefore, offers an emergent and potentially revolutionary mechanism to increase organic carbon in soil and combat climate change. Through the incorporation of nanomaterials, including nanoclays, biochar-based nanocomposites, and nanofertilizers, the physical properties of the soil can be improved, microbial activity enhanced, and organic carbon stabilized—thereby increasing carbon sequestration (Basavegowda and Baek 2021).

It is evident that there are challenges facing nanotechnology, including environmental safety, economic constraints, and, most critically, the absence of enabling policies. More studies are required to better understand the consequences of nanomaterials on soil ecosystem functioning and to develop effective large-scale remediation strategies. Furthermore, policymakers must establish suitable policy frameworks to facilitate the safe and equitable use of nanotechnology in agriculture (Buerkle et al. 2023). Therefore, we can appreciate the fact that nanotechnology can greatly assist the world in its struggle to fight climate change by improving sequestration of carbon in the soil.

As research in this field advances, it is expected that nanotechnology will play an increasingly significant role in fight against global warming and promoting sustainable agriculture (Newell et al. 2010).

10. Conclusion

Nanotechnology is recognized as having potential to improve soil carbon storage and management, offering new approaches to address challenges associated with conventional soil management practices. Applying nanomaterials alters the physical structure of soil, enhances microbial activity, and facilitates more effective carbon storage. Nonetheless, further scientific studies are needed to investigate the long-term effects of nanomaterials on soil ecosystems and to develop sustainable and commercially viable technologies. In conclusion, nanotechnology should not be seen as a standalone solution to climate change, but rather as a critical component in the broader toolbox of strategies aimed at mitigating climate change impacts due to rising global temperatures. This technology is still in the research and development phase; therefore, increased investment and the creation of appropriate policies are necessary to enable nanotechnology to contribute meaningfully to the fight against climate change.

11. Acknowledgement

We (Authors) are highly thankful to the University of Agriculture Faisalabad and Yangzhou University for their support in this publication. Authors also want to thank Professor Dr Abdul Wakeel from Soil Fertility and Plant Nutrition Laboratory, University of Agriculture Faisalabad, Pakistan and Dr. Akhtar Hameed, Assistant Professor (Plant Pathology), Institute of Plant Protection for providing us this opportunity. We are highly thankful to our families who sacrificed their good times for the accomplishment of this milestone.

References

Alam, M.S., B. Bishop, N. Chen, S. Safari, V. Warter, J.M. Byrne et al. 2020. Reusable magnetite nanoparticles–biochar composites for the efficient removal of chromate from water. Sci. Rep. 10: 19007.

Amenta, V., K. Aschberger, M. Arena, H. Bouwmeester, F. Botelho Moniz, P. Brandhoff et al. 2015. Regulatory aspects of nanotechnology in the agri/feed/food sector in EU and non-EU countries. Regul. Toxicol. Pharmacol. 73: 463–476.

Arya, A., S.S. Negi, J.C. Kathota, A.N. Patel, M.H. Kalubarme and J.K. Garg. 2017. Carbon sequestration analysis of dominant tree species using geo-informatics technology in Gujarat State (India). Int. J. Environ. Geoinformatics. 4. Article 2.

Auffan, M., J. Rose, J.-Y. Bottero, G.V. Lowry, J.-P. Jolivet and M.R. Wiesner. 2009. Towards a definition of inorganic nanoparticles from an environmental, health and safety perspective. Nat. Nanotechnol. 4: 634-641.

Bai, Y. and M.F. Cotrufo. 2022. Grassland soil carbon sequestration: Current understanding, challenges, and solutions. Science. 377: 603–608.

Basavegowda, N. and K.-H. Baek. 2021. Current and future perspectives on the use of nanofertilizers for sustainable agriculture: The case of phosphorus nanofertilizer. 3 Biotech. 11: 357.

Bera, T., S. Samui, A. Dey and J. Ankireddypalli. 2024. Soil carbon sequestration in the context of climate change. pp. 63–106. *In*: Pathak, H., D. Chatterjee, S. Saha and B. Das [eds.]. Climate Change Impacts on Soil-Plant-Atmosphere Continuum. Springer Nature, Singapore.

Bernadett, L. 2013. Agricultural soil carbon sequestration offset programs: Strengths, difficulties, and suggestions for their potential use in AB 32's cap and trade program. UCLA J. Environ. Law Policy. 31: Article 1.

Calabi-Floody, M., J.S. Bendall, A.A. Jara, M.E. Welland, B.K.G. Theng, C. Rumpel et al. 2011. Nanoclays from an Andisol: Extraction, properties and carbon stabilization. Geoderma. 161: 159–167.

Cristea, O.D., V. Vlăduţ, N. Ungureanu, D.L. Popa, S. Isticioaia, I. Dumitrescu et al. 2020. Research on methods, techniques and technologies of carbon sequestration in soil. E3S Web Conf. 180: 03021.

Dai, X., G. Yang, D. Liu and R. Wan. 2020. Vegetation carbon sequestration mapping in herbaceous wetlands by using a MODIS EVI time-series data set: A case in Poyang Lake Wetland, China. Remote Sens. 12: Article 18.

Das, C.G.A., V.G. Kumar, T.S. Dhas, V. Karthick and C.M.V. Kumar. 2023. Nanomaterials in anticancer applications and their mechanism of action-A review. Nanomedicine. 47: 102613.

Dhanapal, A.R., M. Thiruvengadam, J. Vairavanathan, B. Venkidasamy, M. Easwaran and M. Ghorbanpour. 2024. Nanotechnology approaches for the remediation of agricultural polluted soils. ACS Omega. 9: 13522–13533.

Don, A., F. Schneider, H. Heinemann, D. Seitz, N. Begill and C. Poeplau. 2023. What limits carbon sequestration in soils? https://doi.org/10.5194/egusphere-egu23-13898.

Duhan, J.S., R. Kumar, N. Kumar, P. Kaur, K. Nehra and S. Duhan. 2017. Nanotechnology: The new perspective in precision agriculture. Biotechnol. Rep. 15: 11–23.

Faizan, M., A. Singh, A. Eren, H. Sultan, M. Sharma, I. Djalovic et al. 2024. Small molecule, big impacts: Nano-nutrients for sustainable agriculture and food security. J. Plant Physiol. 301: 154305.

Garg, S., N.P. Rumjit and S. Roy. 2024. Smart agriculture and nanotechnology: Technology, challenges, and new perspective. Adv. Agrochem. 3: 115–125.

Ghormade, V., M.V. Deshpande and K.M. Paknikar. 2011. Perspectives for nano-biotechnology enabled protection and nutrition of plants. Biotechnol. Adv. 29: 792–803.

Gogos, A., K. Knauer and T.D. Bucheli. 2012. Nanomaterials in Plant Protection and Fertilization: Current State, Foreseen Applications, and Research Priorities. J. Agric. Food Chem. 60(39): 9781–9792. https://doi.org/10.1021/jf302154y.

Gupta, M., T. Mondal, E. Lokesha, A. Singh, K. Ashwin, S. Perween and J.K. Sahoo. 2018. Methane Emission by Ruminants and Its Measurement- A Review. IJCMAS. https://doi.org/10.20546/ijcmas.2018.707.364.

Han, L., C. Lu, L. Chen, F. Wang, Q. Chen, K. Gao et al. 2024. Carbon sequestration potential of biochar in soil from the perspective of organic carbon structural modification. Appl Soil Eco. 198: 105389. https://doi.org/10.1016/j.apsoil.2024.105389.

Harvey, C.A., M. Chacón, C.I. Donatti, E. Garen, L. Hannah, A. Andrade et al. 2014. Climate-Smart Landscapes: Opportunities and Challenges for Integrating Adaptation and Mitigation in Tropical Agriculture. Conser Lett. 7(2): 77–90. https://doi.org/10.1111/conl.12066.

He, F., Zhao, D., Liu, J. and Roberts, C.B. 2007. Stabilization of Fe−Pd Nanoparticles with Sodium Carboxymethyl Cellulose for Enhanced Transport and Dechlorination of Trichloroethylene in Soil and Groundwater. Ind. Eng. Chem. Res. 46(1): 29–34. https://doi.org/10.1021/ie0610896.

Jose, S. and S. Bardhan. 2012. Agroforestry for biomass production and carbon sequestration: An overview. Agrofor Syst. 86(2): 105–111. https://doi.org/10.1007/s10457-012-9573-x.

Kale, S.S., R. Chauhan, B. Nigam, S. Gosavi and I.J. Chaudhary. 2024. Effectiveness of nanoparticles in improving soil fertility and eco-friendly crop resistance: A comprehensive review. Biocatal. Agric. Biotechnol. 56: 103066.

Karbo, R.T. and A.O. Crentsil. 2021. Climate-smart agriculture (CSA) adaptation strategies of farmers against climate change in Lawra Municipality, Upper West Region, Ghana. J. Sci. Res. Rep. 27: 10–19.

Khan, M.N., M. Siddique, N. Mirza, R. Khan, M. Bilal, N. Riaz et al. 2022. Synthesis, characterization, and application of Ag-biochar composite for sono-adsorption of phenol. Front. Environ. Sci. 10: 823656.

Khan, S.T., S.F. Adil, M.R. Shaik, H.Z. Alkhathlan, M. Khan and M. Khan. 2021. Engineered nanomaterials in soil: Their impact on soil microbiome and plant health. Plants. 11: 109.

Kumari, P., S. Kumari, T. Risha, P. Bhagat, P. Kumari, P. Beck et al. 2024. Unveiling the potential scope of nanoparticles in the current era of crop productivity. International Journal For Multidisciplinary Research, 6(3): 20616. https://doi.org/10.36948/ijfmr.2024.v06i03.20616.

Lal, R. 2004. Soil carbon sequestration to mitigate climate change. Geoderma. 123: 1–22.

Luo, L., J. Wang, J. Lv, Z. Liu, T. Sun, Y. Yang et al. 2023. Carbon sequestration strategies in soil using biochar: Advances, challenges, and opportunities. Environ. Sci. Technol. 57: 11357–11372.

Jimenez-Castaneda M. E., R. Lal, F.F.C. Mello, K. Witkowski, C. St. Martin and F. Villarreal. 2022. An overview of carbon sequestration in agricultural soils of Latin America and the Caribbean. Critical Reviews in Plant Sciences, 41: 6, 391–405, DOI: 10.1080/07352689.2022.2148923.

Mahapatra, A., B. Misra, D. Banik, V.K. Gupta, D.S. Rout and C. Sahu. 2024. Role of nanoparticles on soil microbial community and functionality. pp. 43–56. *In*: Rajput, V.D. [ed.]. Advances in Environmental Engineering and Green Technologies. IGI Global.

Malhi, G.S., M. Kaur and P. Kaushik. 2021. Impact of climate change on agriculture and its mitigation strategies: A review. Sustainability.

Mattila, T.J. and N. Vihanto. 2024. Agricultural limitations to soil carbon sequestration: Plant growth, microbial activity, and carbon stabilization. Agric. Ecosyst. Environ. 367: 108986.

Morya, R., A. Bhargava, G.R. Lalitha, M. Vamshi, S. Verma, M. Rastogi et al. 2023. Soil management practices to enhance carbon sequestration rates- A review. Int. J. Environ. Clim. Change. 13: 3762–3776.

Mukherjee, A., S. Majumdar, A.D. Servin, L. Pagano, O.P. Dhankher and J.C. White. 2016. Carbon nanomaterials in agriculture: A critical review. Front. Plant Sci. 7.

Nanotechnology, soil biology, and nutrient cycling-ST Biologicals. 2024. Available at: https://stbiologicals.com/nanotechnology-soil-biology-and-nutrient-cycling/.

Naorem, A., S. Jayaraman, R.C. Dalal, A. Patra, C.S. Rao and R. Lal. 2022. Soil inorganic carbon as a potential sink in carbon storage in dryland soils-A review. Agric. 12: Article 8.

Nthebere, K., S.H.K. Sharma, Ch.P. Kumari and A.A. Qureshi. 2022. Long-term effect of different cropping systems on carbon sequestration in a sandy loam soil of Telangana. Agric. Rev.

Pagano, L., R. Rossi, J.C. White, N. Marmiroli and M. Marmiroli. 2023. Nanomaterials biotransformation: In planta mechanisms of action. Environ. Pollut. 318: 120834.

Pandey, P., V.J. Swaroopa, S. Kushal, P.L. Sahu and B. Parmar. 2024. A comprehensive review on carbon sequestration potential and addition of organic carbon to soil. Int. J. Environ. Clim. Change. 14: 724–733.

Patle, T., B. Tomar, S.S. Parihar, S.S. Tomar and P. Singh. 2024. Nanotechnology-based soil improvement and conservation for enhancement of crop production. pp. 217–236. *In*: Rajput, V.D., A. Singh, K. Ghazaryan, A. Alexiou and A.R.M. Said Al-Tawaha [eds.]. Advances in Environmental Engineering and Green Technologies. IGI Global.

Poeplau, C., R. Dechow, N. Begill and A. Don. 2024. Net primary production rather than saturation of mineral surfaces limits soil carbon sequestration. Available at: https://doi.org/10.5194/egusphere-egu24-9216.

Ramadass, K., C.I. Sathish, S. MariaRuban, G. Kothandam, S. Joseph, G. Singh et al. 2020. Carbon nanoflakes and nanotubes from halloysite nanoclays and their superior performance in CO_2 capture and energy storage. ACS Appl. Mater. Interfaces. 12(10): 11922–11933.

Raza, A., S. Charagh, H. Salehi, S. Abbas, F. Saeed, G.E.J. Poinern et al. 2023. Nano-enabled stress-smart agriculture: Can nanotechnology deliver drought and salinity-smart crops? J. Sustain. Agric. Environ. 2(3): 189–214.

Reetika, G., I. ingle, S. Kumar and S. Kaushal. 2024. Soil management strategies for increasing carbon sequestration: A review. Int. J. Res. Agronomy. 7(4): 496–499. https://doi.org/10.33545/2618060X.2024.v7.i4g.596.

Rout, C. and D. Aggarwal. 2024. Nanotechnology in restoration of degraded soil fertility. pp. 1–23. *In*: Rajput, V.D. [ed.]. Advances in Environmental Engineering and Green Technologies. IGI Global. https://doi.org/10.4018/979-8-3693-1471-5.ch001.

Santoyo, G., P. Guzmán-Guzmán, F.I. Parra-Cota, S. de los Santos-Villalobos, M.C. Orozco-Mosqueda and B.R. Glick. 2021. Plant growth stimulation by microbial consortia. Agronomy. 11(2): Article 2. https://doi.org/10.3390/agronomy11020219.

Sarma, H.H., S.K. Borah, O. Das, S. Sonowal, N. Talukdar and B.C. Das. 2024. Nano biochar: Properties, preparation, key features and its impact on soil and crop ecosystem. J. Adv. Biol. Biotechnol. 27(7): 66–78. https://doi.org/10.9734/jabb/2024/v27i7967.

Smith, P., D. Martino, Z. Cai, D. Gwary, H. Janzen, P. Kumar et al. 2007. Greenhouse gas mitigation in agriculture. Philos. Trans. R. Soc. B Biol. Sci. https://doi.org/10.1098/rstb.2007.2184.

Srivastava, S., C.M. Mehta, M. Rana, D. Banik, V.D. Rajput and E. Konstantinova. 2024. Nanotechnology and soil health: Current updates and future perspective. pp. 24–42. *In*: Rajput, V.D. [ed.]. Advances in Environmental Engineering and Green Technologies. IGI Global. https://doi.org/10.4018/979-8-3693-1471-5.ch002.

Tan, X., Y. Liu, Y. Gu, Y. Xu, G. Zeng, X. Hu et al. 2016. Biochar based nano composites for the decontamination of wastewater: A review. Bioresour. Technol. 212: 318–333. https://doi.org/10.1016/j.biortech.2016.04.093.

Upadhayay, V.K., M.K. Chitara, A. Sharma, P. Chaudhary, A. Jaiswal, G. Kumari et al. 2024. Role of nanomaterials on soil microbial community and functionality: Nanomaterial impact - Unraveling the dynamics of soil microbial communities and functions. pp. 333–353. *In*: Rajput, V.D. [ed.]. Advances in Environmental Engineering and Green Technologies. IGI Global. https://doi.org/10.4018/979-8-3693-1471-5.ch017.

Verma, M., N. M'hamdi, Z. Dkhili, S.K. Brar and K. Misra. 2014. Thermochemical transformation of agro-biomass into biochar: Simultaneous carbon sequestration and soil amendment. pp. 51–70. *In*: Brar, S.K., G.S. Dhillon and C.R. Soccol [eds.]. Biotransformation of Waste Biomass into High Value Biochemicals. Springer New York. https://doi.org/10.1007/978-1-4614-8005-1_3.

Wang, H., R.L. Wick and B. Xing. 2009. Toxicity of nanoparticulate and bulk ZnO, Al2O3 and TiO2 to the nematode *Caenorhabditis elegans*. Environ. Pollut. 157(4): 1171–1177. https://doi.org/10.1016/j.envpol.2008.11.004.

Wang, Z., L. Zhang, J. Zhao and B. Xing. 2016. Environmental processes and toxicity of metallic nanoparticles in aquatic systems as affected by natural organic matter. Environ. Sci. Nano. 3(2): 240–255. https://doi.org/10.1039/C5EN00230C.

Yasmin, H., J. Mazher, A. Azmat, A. Nosheen, R. Naz, M.N. Hassan et al. 2021. Combined application of zinc oxide nanoparticles and biofertilizer to induce salt resistance in safflower by regulating ion homeostasis and antioxidant defence responses. Ecotoxicol. Environ. Saf. 218: 112262. https://doi.org/10.1016/j.ecoenv.2021.112262.

Youns, Y.T., Manshad, A.K. and Ali, J.A. 2023. Sustainable aspects behind the application of nanotechnology in CO_2 sequestration. Fuel. 349: 128680. https://doi.org/10.1016/j.fuel.2023.128680.

Zhang, H.-L., Lal, R., Zhao, X., Xue, J.-F. and Chen, F. 2014. Chapter One-Opportunities and Challenges of Soil Carbon Sequestration by Conservation Agriculture in China. pp. 1–36. *In*: Sparks, D.L. [ed.]. Advances in Agronomy, Vol. 124. Academic Press, San Diego. https://doi.org/10.1016/B978-0-12-800138-7.00001-2.

Zhao, J., Liu, D. and Huang, R. 2023. A Review of Climate-Smart Agriculture: Recent Advancements, Challenges, and Future Directions. Sustainability. https://doi.org/10.3390/su15043404.

Zhao, L., Bai, T., Wei, H., Gardea-Torresdey, J.L., Keller, A. and White, J.C. 2022. Nanobiotechnology-based strategies for enhanced crop stress resilience. Nature Food. 3(10): 829–836. https://doi.org/10.1038/s43016-022-00596-7.

CHAPTER-18

Regulatory and Environmental Considerations of Soil Nanotechnology

Wazir Ahmed, Umair Riaz, Waseem Hassan, Muhammad Imran, Shakeel Ahmed* and *Hasan Riaz*

1. Introduction

With a rich history spanning several decades and involving the convergence of disciplines like physics, chemistry, biology, and materials science, nanotechnology is the manipulation of matter at the atomic and molecular scale (Santos et al. 2024). Japanese scientist Norio Taniguchi originally used the term "nanotechnology" to refer to nanoscale precision manufacturing. Through his book "Engines of Creation," American engineer Eric Drexler popularized the idea of molecular nanotechnology (MNT) in the 1980s, imagining a world where machines could replicate themselves. The STM, created in 1981 by Gerd Binnig and Heinrich Rohrer, allowed for the visualization and manipulation of individual atoms and earned them the 1986 Nobel Prize in Physics (Swati and Hemant 2021).

Interest in carbon-based nanomaterials was piqued in the 1980s and 1990s by the discovery of carbon nanotubes and buckminsterfullerene (C_{60}). To advance nanoscience and nanotechnology research, the United States established the National Nanotechnology Initiative (NNI) in 2000. Nanotechnology has improved many areas in the modern era, including drug delivery, semiconductor design, renewable energy technologies, and more. But even as the field develops, worries about health, environmental hazards, and moral ramifications remain. The advancement of nanotechnology has moved from theoretical concepts to real-world uses that have a significant influence on many different industries (Ariga and Azzaroni 2022, Prasad et al. 2022).

In the 21st century and beyond, the continued advancement of nanotechnology, especially in the field of molecular engineering, has the potential to completely transform society and technology (Tegart 2004, Aithal and Aithal 2022). Although nanotechnology holds great promise, it also poses moral, health, and environmental issues that need to be resolved as it continues to influence technology and society in the future (El-Ramady et al. 2022, Majeed et al. 2024). A subfield of nanotechnology known as "soil nanotechnology" is dedicated to studying and improving soil systems, especially

Department of Soil and Environmental Sciences, MNS-University of Agriculture, Multan, Pakistan.
* Corresponding author: m.imran@mnsuam.edu.pk

DOI: 10.1201/9781003558354-18

in agriculture and environmental management, by using materials and techniques at the nanoscale. This field emerged from the convergence of soil science, environmental engineering, and nanoscience, among other scientific and technological advances. Although its specific applications in soil systems have emerged more recently in response to agricultural and environmental challenges, its development is consistent with the broader historical progression of nanotechnology.

Although soil nanotechnology is a relatively new field, its theoretical roots trace back to broader advancements in nanotechnology and soil science (Patle et al. 2024). As early as the 1960s and 1970s, soil scientists recognized that soil contains naturally occurring nanoparticles (e.g., clay particles, minerals, and organic matter) that affect soil properties such as structure, water retention, and nutrient availability (Kansotia et al. 2024). The general field of nanotechnology, especially with Richard Feynman's visionary 1959 lecture, *There's Plenty of Room at the Bottom*, and the development of essential tools like the scanning tunneling microscope (STM) in the early 1980s, created a foundation for the understanding and manipulation of matter at the nanoscale. This laid the groundwork for future applications in soil systems, though soil-specific applications did not emerge until later (Prasad et al. 2022).

In the early 1990s, researchers identified organic colloids, clay minerals, and iron oxides as nanoparticles and started to explore their potential for improving physicochemical properties of soils, roles in nutrient cycling, and remediation. Initially, scientists explored the nanoscale characteristics and behavior of these soil nanoparticles using advanced techniques such as X-ray diffraction and electron microscopy (Patle et al. 2024). Owing to significant achievements in natural soil nanoparticles, efforts were made to develop engineered nanoparticles for soil nanotechnology in the beginning of the 21st century, which gained significant attention because of extensive advancements in nanotechnology and its growing significance in the fields of engineering, environmental science, and agriculture (Aithal and Aithal 2022). In this era, extensive research started to evaluate the potential use of engineered nanoparticles for improving nutrient use efficiency, water use efficiency, soil characteristics, crop productivity, and addressing environmental problems. Although researchers have reported many beneficial impacts of these soil nanoparticles, particularly of engineered nanoparticles, less work has been done regarding their fate in the environment and soil, detrimental impacts on ecosystems, and legislation and quality control aspects of soil nanoparticles (Ariga and Azzaroni 2022). To ensure sustainable and safe use of these soil nanoparticles in soil and environmental sciences, there is a need to address issues regarding regulatory and environmental considerations in soil nanotechnology.

2. Soil Nanotechnology's Prospects

The world's population in 2024 (8 billion) is projected to cross the figure of 9 billion by 2050 (Sands and Suttles 2022). Approximately a 35 to 56% increase in global food demand is projected by 2050 (van Dijk et al. 2021), which is further projected to increase up to 62% under the prevailing threat of climatic changes, based on a meta-analysis of projected global food demand and population from 2010 to 2050, with a significant increase in the severity of environmental problems (van Dijk et al. 2021, Sands and Suttles 2022). Additionally, natural resources, particularly land and water, are being exhausted at an alarming rate. According to reports, global fertilizer demand will increase by 70% to feed 8 to 10 billion people (Sands and Suttles 2022).

Nanotechnology has provided innovative solutions to old problems through an enormous shift in agricultural and environmental technologies (Kansotia et al. 2024). Without the application of nanotechnology, it seems impossible to feed 9 billion people, especially under vulnerable climatic conditions and the continuous global decline in land and water resources (Singh et al. 2024). Currently, soil nanoparticles are being used in soil science for crop production, soil conservation, soil health, soil remediation, environmental monitoring, and sustainable agriculture (Alsharef et al. 2016, Patle et al. 2024). Precision farming, healthy soil and plants, and higher yields with minimal utilization of chemical inputs are the revolutionary aspects of soil nanotechnology (Singh et al. 2024). To achieve the UN Sustainable Development Goal "Zero Hunger," the application of nanotechnology principles is of great importance (van Dijk et al. 2021) because soil nanotechnology has shown great potential for enhancing soil productivity and environmental protection and has the potential to revolutionize agriculture by inducing remarkable transformations in technology and farming practices (Gondal et al. 2023). It is included among one of the six "Key Enhancing Technologies" identified for sustainable breakthroughs (Merzbacher 2020). Some prospects of soil nanoparticles and the scope of soil nanotechnology are mentioned in Table 1.

3. Soil Nanoparticles Categories

The properties and impacts on the environment and soil depend solely on the nature and composition of soil nanoparticles (Kumari and Sarkar 2021, Khan and Hossain 2022, Damm et al. 2024). Soil nanoparticles can be primarily natural, engineered, or a combination of both (hybrid). A comparative overview of natural and anthropogenic nanoparticles is shown in Figure 1.

4. Importance of Regulatory and Environmental Considerations

Regulatory and environmental considerations in soil nanotechnology ensure the safe use of soil nanoparticles and help to identify hazardous effects associated with their use. These considerations outline the standards and rules that must be followed, as well as the strategies that must be implemented to mitigate potential risks (Allan et al. 2021). While soil nanotechnology has demonstrated significant benefits for soil health, crop production, and environmental sustainability—including advancements in the fertilizer sector and the reduction of soil and water pollution—researchers have also highlighted potential negative effects of soil nanoparticles on plants, animals, and the ecosystem through bioaccumulation (Karthigadevi et al. 2021, Mohammadi and Jafari 2021, Dhanapal et al. 2024). Therefore, establishing regulatory frameworks for soil nanotechnology is essential to ensure the safe and sustainable use of these nanoparticles in agriculture and environmental protection (Karim and Munir 2015). Here are some key factors that should be considered before adopting soil nanotechnology principles.

5. Nanoparticle Behavior in Soil

(a) In soil, particle adsorption or desorption is solely associated with particle size, surface area, and the adhering and cohesive forces of soil colloids and soil organic matter. As nanoparticles possess nanoscale size with significantly higher surface area, the behavior of nanoparticles regarding adsorption or desorption is totally different from that of soil particles (Mandal et al. 2020). Therefore, principles of ion exchange and exchange-based models cannot be applied to soil nanoparticles.

(b) The dynamics of nanoparticles in soil purely depend on interactions of these particles with the mineralogical features of soil particles. For example, 1:1 type clay behaves totally differently from 2:1 type clay with respect to surface charge. Nutrient interactions with mineral surfaces and soil organic colloids determine the soil's capacity to hold nutrients. It is essential to study soil nanoparticles' behavior after their interactions with mineral surfaces and organic colloids (Kurganskaya 2024).

(c) As nanoparticles have minute sizes and larger surface areas compared to soil colloids, these nanoparticles can alter the mechanism of soil aggregation, resulting in unpredictable changes in soil structure and nutrient mobility within the soil profile. These alterations in soil structure and ion mobility due to the application of soil nanoparticles are expected to regulate ionic strength up or down, leading to salinity or detrimental effects on soil physicochemical characteristics and plant physiology. According to Kurganskaya (2024), nanoparticles in soil can modify adhesive

Table 1. Prospects and scope of nanotechnology.

Soil nanotechnology's prospects	**Examples/Nanoparticles used**	**Source**
Restoration of polluted soil and water	Carbon nanomaterials, iron(III) oxide, titanium oxide, zinc oxide, nZVI, nanocomposites.	(Dhanapal et al. 2024, Majeed et al. 2024, Nouri 2024, Rajput et al. 2022)
Remediation of heavy metals (e.g., Pb, Cd, As)	Nano-Fe3O4, SiO2, TiO2, ZnS and MoS2, ZnO, CeO2, TiO2, HAP Bio-Se, Ast-Au, Mel-Au, Multiwall carbon nanotubes (MWCNTs), CNTs, Graphene oxide, ZnO + Bacillus cereus and Lysinibacillus macrolides, SiO2, TiO2, ZnS and MoS2, Nzvi, Ag NPs, Silicon dioxide nanoparticle	(Ali et al. 2023, Aliyari Rad et al. 2023, Chaudhary et al. 2023)
Nano bioremediation	Metal nanoparticles (Pt, Rh, Pd, Ir, Ag, Au, Cu, Co, Ni, FeNi, Cu3Au, CoNi, CdTe, CdSe, ZnS), Cylindrical nanotube (SWNT, MWNT) fullerenes, ZnO, Fe2O3, Fe3O4, MgO, BaCO3, BaSO4, TiO2 nanowire of polypyrrole, polyaniline, poly (3,4-ethylenedioxythiophane) dendrimers (PAMAM), Bionanomaterials (Viruses, plasmids, and protein NPs)	(Rizwan et al. 2014, Vázquez-Núñez et al. 2020)
Degradation of xenobiotic pollutants	Nanocomposite (polyethylene oxide and polyethyleneimine, CNT epoxy composites include hydrocarbon polymer composites, conjugated polymer composites, CNTs with polycarbonates, fluoropolymers, polyethylene glycol, polyester polyamides)	(Chaudhary et al. 2023, Karthigadevi et al. 2021)
Restoring of metalloids (e.g., Sb, As, B, Ge, Te, Si) contaminated soils	nAg, nAl, nCeO2, nTiO2, nZVI, nZnO, Nano-CuO	(Chowardhara et al. 2024, Martínez-Fernández et al. 2017, Mohanty et al. 2022)
Sustainable Agriculture/Precision Agriculture		(Ardra Lekshmi et al. 2024, Parameswari et al. 2024, Santos et al. 2024, Aggrawal et al. 2024)
Restoration of polluted and degraded soils	Carbon tubes, graphene	(Dhanapal et al. 2024, Majeed et al. 2024, Nouri 2024, Rajput et al. 2022)
Commercial nanofertilizers	Biozar Nano-Fertilizer - Iran Hero Super Nano - Thailand Master Nano Chitosan Organic Fertilizer - Thailand Nano-Ag Answer - USA Nano Calcium (Magic Green) – Germany Nano Fertilizer (Eco Star) – India Nano Green – India Nano-Gro – USA Nano Max NPK Fertilizer – India Nano Micro Nutrient (Eco Star) – India Nano Ultra-Fertilizer – Taiwan Plant Nutrient Powder (Green Nano) – Thailand PPC Nano – Malaysia TAG NANO - India	(Abdelghany et al. 2022, Alekhya et al. 2024, Channab et al. 2024, Dimkpa and Bindraban 2018, Guardiola-Márquez et al. 2024, Santos et al. 2024)
Soil Health Monitoring (pH, nutrients, organic matter)	Soil nanoparticle (carbon nanodots, carbon nanotubes, graphene) based sensors for real time assessment of soil health parameters Metal nanoparticles (Au, Ag) for pH and salinity Metal oxide nanoparticles (TiO2, ZnO) for salinity and microbial activity Quantum dots for nutrients and organic matter	(Alsharef et al. 2016, Bharati et al. 2023, Das et al. 2024, Mandal et al. 2020, Parameswari et al. 2024, Elahi et al. 2018)
Moisture and nutrient retention in soils	Soil nanoparticle-based sensors like carbon nanotubes, graphene, nano-functionalized hydrogels	(Das et al. 2024, El-Ramady et al. 2022, Singh et al. 2024, Parameswari et al. 2024)
Crop Yield Enhancement	Multi-walled carbon nanotubes, metal nanoparticles	(Abdelghany et al. 2022, Alekhya et al. 2024, Channab et al. 2024, Guardiola-Márquez et al. 2024, Maity et al. 2018)
Carbon Sequestration	Allophanenano-structure, Fe oxide,	(Aggrawal et al. 2024, Filimonova et al. 2016, Pramanik et al. 2020)
Environmental protection and monitoring	Nanopesticides, nanoherbicides, and nanobactericides	(Biswas and Rai 2024, Singh et al. 2024)

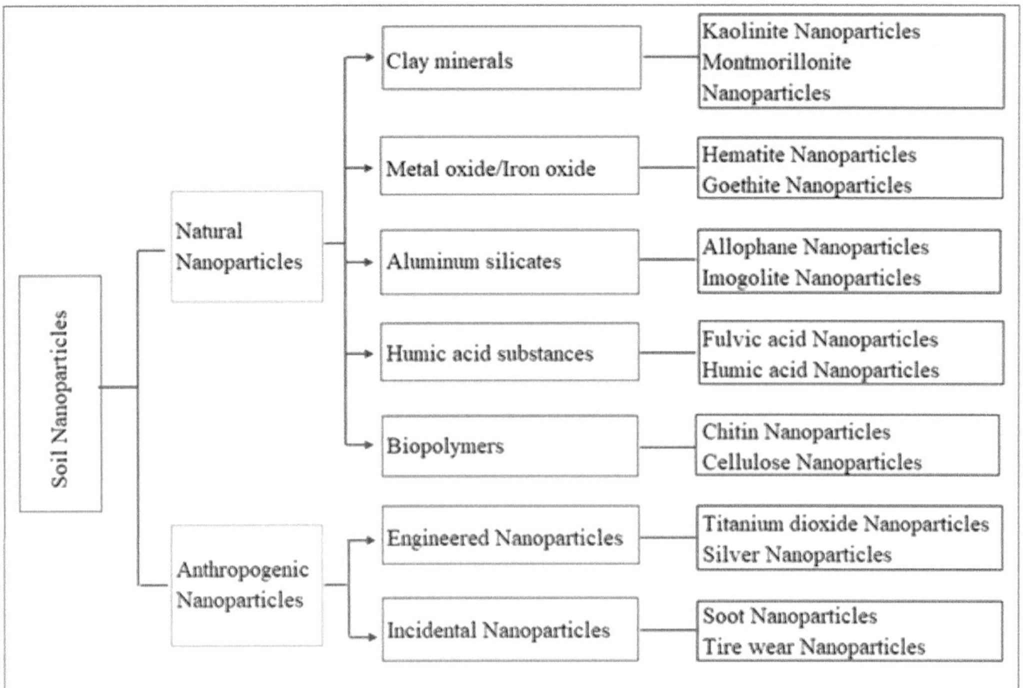

Figure 1. Flow chart overview of natural and anthropogenic nanoparticles.

and cohesive forces responsible for aggregation or disaggregation, resulting in changes in surface charge and stability of these nanoparticles (Bharati et al. 2023).

(d) The applied soil nanoparticles, particularly engineered nanoparticles, have great potential to cross different soil layers and contaminate underground water due to their nanoscale size and high mobility in soils. Moreover, due to the lack of assessment of pH-dependent nanoparticle transformations, these transformed nanoparticles can influence mobility and transport through advection, diffusion, or preferential flow pathways and bioaccumulate in the food chain (Katzourakis and Chrysikopoulos 2024).

6. Nanoparticle Fate/Processes in Soil

The fate processes of soil nanoparticles, especially metallic and hybrid nanoparticles in the soil environment, must be considered in soil nanotechnology because these particles' fate in soils differs from that of soil organic and mineral colloids (Chowardhara et al. 2024). Already established principles cannot be implemented in soil nanotechnology. Key processes responsible for the conversion of the fate of applied nanoparticles into toxic ones are as follows:

(a) Transformation into toxic substances is possible due to higher reactivity potential.

(b) Excessive dissolution is expected because of unpredictable oxidation or reduction, leading to excessive uptake by plants.

(c) Redox reactions can alter the soil-associated chemical structure and characteristics of applied nanoparticles, resulting in unknown impacts on the environment.

(d) Nanoparticle degradation in soil is possible due to the metabolic activity of microorganisms. Without studying microbial responses to applied nanoparticles, biodegradation and remediation cannot be achieved (Chaudhary et al. 2023).

(e) The size and contents of clay and organic colloids determine soil porosity. Soil porosity is responsible for soil health and nutrient and water holding. Due to the nanoscale size of soil particles and their unknown interactions with other soil colloids, the impact of soil nanoparticles on soil porosity must be evaluated. Otherwise, adsorption, aggregation, and mobility of nanoparticles will be unpredictable, thereby leading to detrimental effects on biota and the environment (Sabarinath et al. 2024).

(f) Contrary to mineral and organic colloids, soil nanoparticles induce buffering capacity, pH changes, and CEC that cannot be predicted without assessment studies. Without assessing these factors, the fate and transport of nanoparticles in the soil environment cannot be predicted (Kaur 2021, Mubeen et al. 2023).

(g) The colloidal characteristics of natural nanoparticles can be associated with fulvic and humic acids, but engineered nanoparticles must be studied before their use in soil nanotechnology. Soil nanoparticles have colloidal properties and can affect the fate of the environment, allowing toxic or xenobiotic compounds to bind with them and transporting these xenobiotics in soil environments, leading to bioaccumulation in biota (Karthigadevi et al. 2021).

7. Nanoparticle Sequestration and Accumulation

Sequestration and bioaccumulation of a nutrient or pollutant depend on its soil solution concentration, mechanism of uptake, and other environmental factors (Nouri 2024).

8. Regulatory Considerations in Soil Nanotechnology

Regulatory considerations in nanotechnology involve laws and guidelines that regulate factors responsible for the hazardous effects of soil nanoparticles (Mohammadi and Jafari 2021), summarized in Table 2. They ensure enforcement and compliance to safeguard sustainable practices of these nanoparticles in agriculture and environmental sustainability (Biswas and Rai 2024). These regulatory considerations involve creating rules to ensure the safe and responsible use of materials and technologies at the nanoscale and assessing health and environmental effects, setting guidelines for consumer protection, labeling, and manufacturing. Effective laws help manage risks and maximize the benefits of nanotechnology in environmental and agricultural practices. See Table 1 for key regulatory issues in soil nanotechnology.

9. Definitions and Classification of Nanomaterials

Natural, artificial, or accidental materials having at least one exterior or interior dimension, usually between 1 and 100 nm, are called nanoparticles (Sabarinath et al. 2024). Nanoparticles have the following five essential properties,

a. Size, between approximately 100 nm and less than 0.1 nm.
b. A variety of technologies, such as those for imaging, measuring, modelling, and matter manipulation.
c. Multidisciplinary, including intentionally created physical, chemical, and biological elements.
d. Novel features and functions that depend on size.
e. The deliberate control and manipulation of matter at the atomic level.

Size-based classification of nanomaterials is an important regulatory factor in nanotechnology. Specialized monitoring and risk assessment are required due to the distinct physicochemical characteristics of nanomaterials in this size range. Nanomaterials can be classified as organic, inorganic, metal-based, carbon-based, or a combination of these based on their chemical makeup. This method is used by regulatory agencies to assess possible hazards and safety issues for various kinds of nanomaterials. Table 3 summarizes the definitions of nanoparticles by different regulatory bodies (Allan et al. 2021). Figure 1 shows how soil nanoparticles are categorized according to their chemical makeup and place of origin.

Table 2. Significant regulatory considerations for soil nanotechnology.

Sr#	Regulatory considerations	Sub-components of considerations
1	Implementation of uniformity in nanomaterials	Implementation of nanomaterial classification Size based classification Composition based classification
2	Nanomaterial Registration and Reporting	Chemical composition of nanoparticles Size and shape characterization of nanoparticles Nanoparticle inventory tracking
3	Regulatory Frameworks	International Guidelines OECD Guidelines ISO Standards National / Regional Regulations United States EPA Nanomaterial Policies FDA Regulations TSCA Regulations OSHA Standards European Union REACH Regulations Nano-specific Directives Asia Pacific Japan Chemical Regulations China Nanomaterials Standards
4	Product Approval Regulations	Pre-market Evaluation Toxicological test Environmental Risk Assessment Labelling and Disclosure Consumer Awareness Supply Chain Traceability Post-marketing Surveillance Adverse Event Reporting Product Irascibility Monitoring and Surveillance
5	Health and Safety Regulations/ Guidelines	Safety Guidelines Exposure Monitoring Contaminant Measures Emission Control Risk Assessment/ Hazard identification Size Distribution, Surface Area and Reactivity Acute Toxicity, Chronic Toxicity or Genotoxicity Exposure Assessment Persistence in the Environment Bioaccumulation in Biota
6	Risk Characterization Regulations	Dose Response Modelling No Observed Adverse Effect Level Benchmark Dose Calculation
6	Risk Estimation Regulations	Probabilistic Modelling Deterministic Approaches
7	Risk Management Regulations	Contaminant Strategies Engineering Control Personal Protective Equipment Waste Disposal Nano-waste Characterization Contaminant Procedures

Table 2 contd. ...

...Table 2 contd.

Sr#	Regulatory considerations	Sub-components of considerations
8	Ethical Considerations	Public Perception Management Nanomaterial Risk Perception Transparency Initiatives Risk Communication Strategies Societal Impact Assessment Equity and Access Concerns Environmental Justices Issues
9	Intellectual Property Rights	Patenting Nanomaterials Invention Disclosure Process Patent Infringement Disputes Data Exclusivity Regulations Test Data Protection Data Sharing Requirements
10	Regulatory Homogenization Efforts	International Standards Development ISO Nanotechnology Standards OECD Testing Guidelines
11	Regulatory Cooperation Framework	EU-US Trade Agreements Asia Pacific Economic Cooperation

Table 3. Overview of nanomaterials/nanoparticles as discussed by Allan et al. (2021).

Regulatory agency	Nanoparticles' definition
European Union	Natural, incidental or manufactured materials with particle size range 1 nm–100 nm.
FDA	Engineered materials or end-products having at least one dimension in the nanoscale, either externally or internally. OR engineered particles exhibiting physiochemical and biological properties similar to nanoparticles, even if their nanoscale is up to 1 um.
US-EPA	Chemical substances being solid at 25°C with particles of nanoscale ranges 1–100 nm (at least one dimension).
ISO	Materials with nanoscale dimensions (at least).
REACH	Natural or manufactured substance, size range 1 nm–100 nm.

10. Nanomaterial Registration and Reporting

10.1 International Guidelines for Regulatory Framework

The Environmental Protection Agency was given the power to control the entry of new or existing chemical substances, including nanomaterials, in 2018 under the Toxic Substances Control Act (TSCA) regulations in the United States (Allan et al. 2021). The Occupational Safety and Health Administration in the U.S. created the OSHA standards, which are rules pertaining to occupational health and safety. They seek to shield employees from risks, such as those that nanomaterials might present. With special provisions for nanomaterials, the REACH regulations are a body of EU legislation that mandates the registration, assessment, approval, and limitation of chemicals made or imported into the EU. Representatives of 36 developed nations from North and South America, Europe, and the Asia-Pacific region, as well as the European Commission, meet at the Organisation for Economic Co-operation and Development (OECD), an intergovernmental organization, to coordinate and harmonize policies, exchange concerns, and work together on global issues. The OECD recommendations provide best practices for the safe and responsible use of nanotechnology,

Table 4. Classification of nanoparticles based on size and composition.

Nanoparticles' Category	**Nanoparticle examples**
Classification based on composition	
Elemental Nanoparticles	Carbon Nanoparticles Carbon Nanotubes Graphene Fullerenes Metallic Nanoparticles Iron Nanoparticles Gold Nanoparticles Silver Nanoparticles Copper Nanoparticles Ceramic Nanoparticles Titanium Dioxide Nanoparticles Silicon Dioxide Nanoparticles Aluminum Oxide Nanoparticles
Composite Nanoparticles	Polymer Nanocomposites Polymer-Clay Nanocomposites Polymer-Carbon Nanotubes Nanocomposites Polymer-Graphene Nanocomposites Ceramic-Metal Nanocomposites Alumina-Nickle Nanocomposites Zirconia-Copper Nanocomposites Silicon Carbide – Aluminum Nanocomposites Organic Nanomaterials Liposomes Dendrimers Micelles
Biological Nanomaterials	Protein Nanostructures Virus like particles Ferritin Nanoparticles DNA and RNA Nanostructures DNA Origami RNA Nanotechnology
Hybrid Nanomaterials	Inorganic-Organic Hybrids Silica-Polymer Hybrids Metal-Organic Frameworks Bio-Inorganic Hybrids Enzyme-Nanoparticle Conjugates Peptides-Functionalized Nanoparticles

Source: (Ba-Abbad et al. 2012, Rizwan et al. 2014, Ijaz et al. 2020, Kumari and Sarkar 2021, Joudeh and Linke 2022, Khan and Hossain 2022, Damm et al. 2024)

focusing on protecting the environment, human health, and safety (Mohammadi and Jafari 2021). Guidelines given by the organizations should be considered in soil nanotechnology.

a. World Standards Cooperation (WSC)
b. International Electrotechnical Commission (IEC)
c. International Telecommunication Union (ITU)
d. International Organisation for Standardisation (ISO)
e. European Telecommunications Standards Institute (ETSI)
f. European Committee for Electrotechnical Standardisation (CENELEC)
g. European Committee for Standardisation (CEN)
h. German Institute for Standardization (DIN)
i. International Organisation for Standardisation (ISO)

11. Environmental Considerations in Soil Nanotechnology

Environmental considerations are a set of scientific guidelines that address characteristics and special behaviors of nanoparticles in the soil, which should be considered when assessing their human health and ecological risks (Ijaz et al. 2020). Implementation of these guidelines is mandatory for the safe use of nanoparticles in soil and environmental sustainability because recent literature has reported the retention and mobility of soil-applied nanoparticles in soils and their bioaccumulation in ecosystems. These facts enforce that these aspects should be considered during the testing of the nanoparticle before use in soil and the environment.

One of the key factors that plays a critical role in the risk assessment of nanoparticles is the mobility and transport of nanoparticles in the system of media, e.g., soil and the environment (US EPA 2014, Mohammadi and Jafari 2021, OECD 2021). Generally, the mobility and transport fate of nanoparticles is governed by:

a. Physiochemical characteristics of nanoparticles such as particle size, shape, contents, sorption and desorption behavior, surface area, colloidal properties and potential for aggregation or degradation
b. Physiochemical characteristics of bulk soil such as particle size distribution, surface interactions, clay type and contents, soil organic matter, iron oxides, hydroxides, oxygen, surfactants, and microorganisms
c. Soil pore water chemistry, e.g. base saturation percentage, ionic strength, pH, CEC, EC, presence of dissolved organic matter (DOM) and soil colloids, hydraulic conductivity, infiltration rate, and runoff potential.

The summary of processes monitoring the fates of nanoparticles in soils are as follows:

a. Leaching (nanoparticles easily cross soil layers and contaminate soil)
b. Homoagglomeration (nanoparticles show self-aggregation under controlled conditions)
c. Heteroagglomeration (nanoparticles potentially aggregate with other soil colloids)
d. Deposition (nanoparticle attachment with colloidal surfaces after transport or mobility)
e. Straining (nanoparticles' physical entrapment in pore space)

The following conditions must be addressed in soil nanotechnology.

11.1 Specific Guidelines/Considerations on Nanoparticles

a. Selection of right analytical techniques and strategies for testing nanoparticles or risk assessment
b. Labelling of nanoparticles / information on a test substance
c. Use of right reference substances for assessment of nanoparticles' transport or fate in the environment.
d. Adaptation of standardized definition and units associated with nanoparticles' transport and fate
e. Enforcement of quality criteria for test conduction to achieve maximum recovery of nanoparticles

11.2 Considerations on Test Method

a. Use glass leaching columns as test system and laboratory equipment that have potential to prevent environmental contamination.
b. Regulatory bodies' approved procedures and conditions should be followed during leaching column of test particles.

c. The projection models developed for testing nanoparticles' fate processes should be adopted for assessment of nanoparticles' transport and mobility in soils as given below:
   i. STANMOD (STudio of ANalytical MODels)
   ii. MNMs 2021 (Micro-and Nanoparticle transport, filtration and clogging Model – Suite)
   iii. Hydrus-1D
   iv. ColloidFit

Understanding the physiological transformation mechanisms, multi-compartmental transport, and environmental fate modeling can help assess the severity of nanoparticle toxicity in the atmosphere, lithosphere, hydrosphere, and biosphere. Additionally, modeling the environmental fate of nanoparticles can reveal correlations and interactions among various environmental components (Kaur 2021, Kumari and Sarkar 2021).

Exposure risk assessment of nanoparticles is also one of the main components of risk assessment, in addition to environmental fate risk assessment modeling. Timely assessment of nanomaterials' content in environmental components such as air, water, and soil could be effective in avoiding potential exposures and risks associated with engineered nanomaterials in the environment (US EPA 2014, Ijaz et al. 2020).

Researchers have identified some types of engineered barriers and remediation technologies to reduce the risk of nanoparticles on the environment and biota. One of the most important tools of environmental consideration is assessing the potential of engineered barriers and remediation technologies. Based on nanoparticles' exposure risk assessment, different strategies should be adopted to enforce these barriers and remediation techniques. OECD guidelines for analyzing leachate and soil nanomaterials are illustrated in Figure 2.

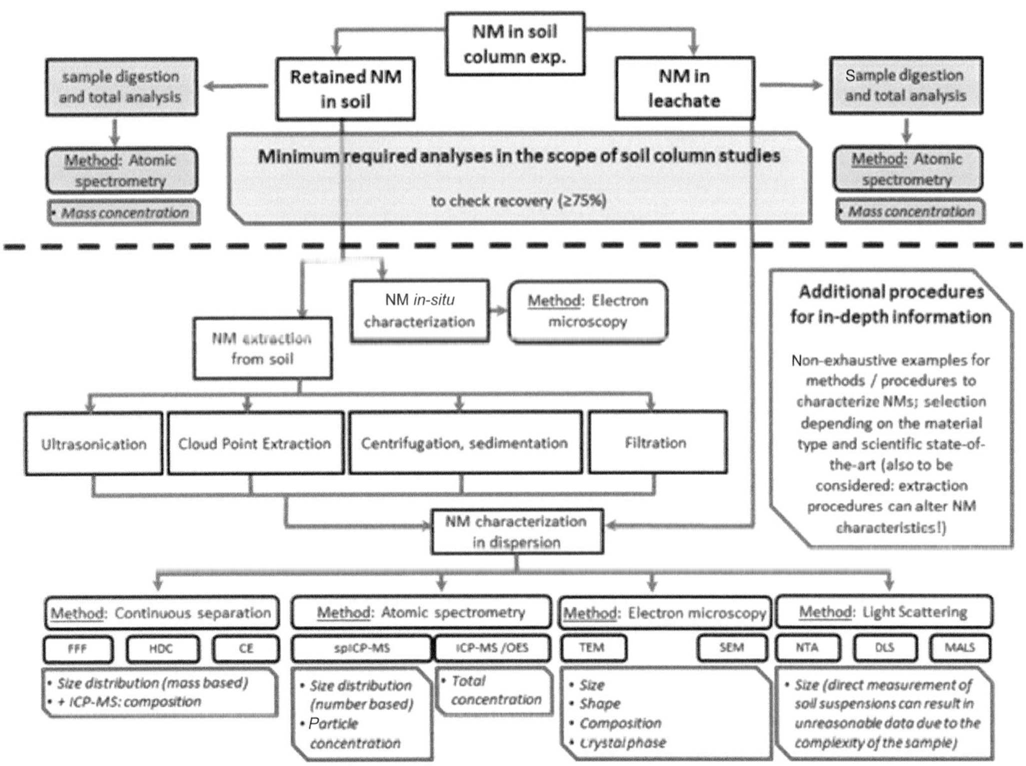

Figure 2. OECD guidelines for analyzing leachate and soil nanomaterials (OECD 2021) .

Table 5. Overview of guidelines or consideration for fate of nanoparticles (US EPA 2014, Ayaz et al. 2023, Macko et al. 2023).

Topic	**Details of considerations**
Applicability domain	Use only those nanoparticles for which analytical method (approved by regulatory body) is available.
Information on nanoparticles	Information related to nanoparticles' chemical composition, particle size distribution, shape, physicochemical properties, possible toxicity, synergism and antagonism with other parties or test substance, etc., must be labelled on nanoparticles' product or packaging.
Reference nanoparticle	Always use a reference nanoparticle with known high leaching behavior during risk assessment of a test nanoparticles in *in situ* or *ex situ* studies.
Quality criteria	The recommended recovery efficiency of test non-labelled nanoparticles from the system or environment ranges from 70 to 110 % while radio-labelled nanoparticles range from 90 to 110% depending on repeatability and sensitivity of the analytical method.
Enrichment factors (EF)	EF used for assessing the degree of pollution and health status of soils. Always use a reference nanoparticle with known quantification of engineered nanoparticle pollution levels and whose threshold value is given by international regulatory agencies such as ISQG, ISO, ETSI, CENELEC, DIN, etc. (Ayaz et al. 2023).
Geo-accumulation Indices (Igeo)	Igeo is essential to assess the degree of soil pollution due to engineered particles. Use nanoparticles with known threshold values described by international regulatory agencies.
Human Health Risk Assessment Accumulation index	The US EPA guidelines for developmental toxicity risk assessment should be applied to assess human health risks associated with engineered nanoparticles. EPA guidelines for health risk assessment should be adhered to for non-carcinogenic and carcinogenic analysis of nanoparticles, such as heavy metals (Gnonsoro et al. 2021, US EPA 2014)
Heavy metal Pollution Index (HPI)	
Metal Index (MI)	
Hazard Quotient (HQ)	
Hazard Index (HI)	
Carcinogenic Risk (CR)	
• Non-carcinogenic Analysis • Hazard Quotients (HQ) • Hazard Index (HI) • Incremental Lifetime Cancer Risk (ILCR) • Chronic Daily Intake (CDI) • Chronic Daily Intake (CDI)	
Implementation of environmental regulatory framework	For marketing and use in soil and environmental sustainability, an environmental regulatory framework for the registration of nanoparticle-based products should be implemented. Compliance and enforcement of these regulatory considerations are the backbone of soil nanotechnology.

12. Conclusion

Without the application of nanotechnology, it seems impossible to feed 9 billion people, especially under vulnerable climatic conditions and the continuous global decline in land and water resources. Precision farming, healthy soil and plants, and higher yields with minimal utilization of chemical inputs are the revolutionary aspects of soil nanotechnology. Regulatory and environmental considerations in soil nanotechnology ensure the safe use of soil nanoparticles and help to identify hazardous effects associated with their use. These considerations outline the standards and rules that must be followed, as well as the strategies that must be implemented to mitigate potential risks.

13. Acknowledgement

We (Authors) are highly thankful to the University of Agriculture Faisalabad and Yangzhou University for their support in this publication. Authors also want to thank Professor Dr Abdul Wakeel from Soil Fertility and Plant Nutrition Laboratory, University of Agriculture Faisalabad Pakistan and Dr. Akhtar Hameed, Assistant Professor (Plant Pathology), Institute of Plant Protection for providing us this opportunity. We are highly thankful to our families who sacrificed their good times for the accomplishment of this milestone.

References

Abdelghany, A.M., A.A.A. El-Banna, E.A.A. Salama, M.M. Ali, A.A. Al-Huqail, H.M. Ali et al. 2022. The individual and combined effect of nanoparticles and biofertilizers on growth, yield, and biochemical attributes of peanuts (*Arachis hypogea* L.). Agron. 12: 398.

Aggrawal, K., V. Dixit, A.K. Pal, K.K. Chaubey, S. Barman, S. Pandey et al. 2024. Recent research on the use of carbon nanomaterials in plant growth and development. pp. 123–144. *In*: (Joshi) Bachheti, A., R.K. Bachheti and A. Husen [eds.]. Carbon-Based Nanomaterials, Synthesis, Agricultural, Biomedical, and Environmental Interventions. Springer Nature.

Aithal, P.S. and S. Aithal. 2022. Opportunities and challenges for green and eco-friendly nanotechnology in the twenty-first century. pp. 31–50. *In*: Pathak, Y.V., G. Parayil and J.K. Patel [eds.]. Sustainable Nanotechnology. 1st ed. Wiley.

Alekhya, V., B. Rajalakshmi, S. Kochar, I. Khan, S. Paul, Z. Alsalami et al. 2024. Nanofertilizers, transforming agriculture for sustainable food production. In, Proc. E3S Web Conf. 507: 01062.

Ali, Q., M.A. Zia, M. Kamran, M. Shabaan, U. Zulfiqar, M. Ahmad et al. 2023. Nanoremediation for heavy metal contamination, a review. Hybrid Adv. 4: 100091.

Aliyari Rad, S., K. Nobaharan, N. Pashapoor, J. Pandey, Z. Dehghanian, V. Senapathi et al. 2023. Nano-microbial remediation of polluted soil, a brief insight. Sustainability. 15(1), Article 1.

Allan, J., S. Belz, A. Hoeveler, M. Hugas, H. Okuda, A. Patri et al. 2021. Regulatory landscape of nanotechnology and nanoplastics from a global perspective. Regul. Toxicol. Pharmacol. 122: 104885.

Alsharef, J.M.A., M.R. Taha, A.A. Firoozi and P. Govindasamy. 2016. Potential of using nanocarbons to stabilize weak soils. Appl. Environ. Soil Sci. 2016, 1-9.

Ardra Lekshmi, A., S. Lekshmi, S.H., K. Gopal, S.R. Sambhudevan and S. Smitha Chandran. 2024. Nanofertilizers for the sustainable management of agricultural resources, a pathway to meet U.N. Millennium Sustainable Development Goals (MDGs). pp. 129–147. *In*: Abd-Elsalam, K.A. [ed.]. Nanofertilizer Delivery, Effects and Application Methods.

Ariga, K. and O. Azzaroni. 2022. What is nanoarchitectonics, origin and task. pp. 1–28. *In*: Azzaroni, O. and K. Ariga [eds.]. Concepts and Design of Materials Nanoarchitectonics. The Royal Society of Chemistry.

Ayaz, H., R. Nawaz, I. Nasim, M.A. Irshad, A. Irfan, I. Khurshid et al. 2023. Comprehensive human health risk assessment of heavy metal contamination in urban soils, insights from selected metropolitan zones. Front. Environ. Sci. 11: 1260317.

Ba-Abbad, M.M., A.A.H. Kadhum, A.B. Mohamad, M.S. Takriff and K. Sopian. 2012. Synthesis and catalytic activity of TiO_2 nanoparticles for photochemical oxidation of concentrated chlorophenols under direct solar radiation. Int. J. Electrochem. Sci. 7(6): 4871–4888.

Bharati, N.K., D.D. Kadam, A. Samanta, A. Kumar, B.T. Bhushan and E. Rajeswari. 2023. Role of nanomodification and nanofertilizers in crop production and soil health. pp. 381–408. *In*: Malik, J.A. and M.J. Sadiq Mohamed [eds.]. Modern Nanotechnology, Volume 1, Environmental Sustainability and Remediation. Springer International Publishing.

Biswas, J.K. and M. Rai. 2024. Nanotechnology for environmental management. CRC Press.

Channab, B.E., A.E. Idrissi, A. Ammar, O. Dardari, S.E. Marrane, A. el Gharrak et al. 2024. Recent advances in nano-fertilizers, synthesis, crop yield impact, and economic analysis. Nanoscale. 16(9): 4484–4513.

Chaudhary, P., L. Ahamad, A. Chaudhary, G. Kumar, W. J. Chen and S. Chen. 2023. Nanoparticle-mediated bioremediation as a powerful weapon in the removal of environmental pollutants. J. Environ. Chem. Eng. 11(2): 109591.

Chowardhara, B., B. Saha, J.P. Awasthi, B.B. Deori, R. Nath, S. Roy et al. 2024. An assessment of nanotechnology-based interventions for cleaning up toxic heavy metal/metalloid-contaminated agroecosystems, potentials and issues. Chemosphere. 359: 142178.

Damm, C., D. Long, J. Walter and W. Peukert. 2024. Size and shape selective classification of nanoparticles. Powders. 3(2): 255–279.

Das, H., S.K. Singh, E. Priyadarshini, C.L. Maurya, A.K. Tiwari, S. Gulaiya et al. 2024. Nanotechnology facilitated real time soil monitoring for optimized crop production. Powders. 10(2): Article 2.

Dhanapal, A.R., M. Thiruvengadam, J. Vairavanathan, B. Venkidasamy, M. Easwaran and M. Ghorbanpour. 2024. Nanotechnology approaches for the remediation of agricultural polluted soils. ACS Omega. 9(12): 13522–13533.

Dimkpa, C.O. and P.S. Bindraban. 2018. Nanofertilizers, new products for the industry? J. Agric. Food Chem. 66(26): 6462–6473

Elahi, N., M. Kamali and M.H. Baghersad. 2018. Recent biomedical applications of gold nanoparticles, A review. Talanta. 184: 537–556.

El-Ramady, H., E.C. Brevik, Z.F. Fawzy, T. Elsakhawy, A.E.-D. Omara, M. Amer et al. 2022. Nano-Restoration for Sustaining Soil Fertility, A Pictorial and Diagrammatic Review Article. Plants. 11(18), Article 18.

Filimonova, S., S. Kaufhold, F.E. Wagner, W. Hausler and I. Kögel-Knabner. 2016. The role of allophane nano-structure and Fe oxide speciation for hosting soil organic matter in an allophanic Andosol. Geochimica et Cosmochimica Acta. 180: 284–302.

Gnonsoro, U.P., Y.E.D.A. Assi, N.S. Sangare, Y.U. Kouako and A. Trokourey. 2021. Health risk assessment of heavy metals (Pb, Cd, Hg) in hydroalcoholic gels of Abidjan, Cote d'Ivoire. Biol. Trace Elem. Res. 200(5): 2510.

Guardiola-Márquez, C.E., M. del, C. Martínez-Ballesta, C.V. García-Sánchez, E.M. Bojorquez-Rodríguez, D.A. Jacobo-Velázquez et al. 2024. Effects of foliar and root application of Zn and Fe bio-nanofertilizers on the glucosinolate, Zn and Fe contents of Pak choi (*Brassica rapa* Subsp. Chinensis) grown under hydroponic and pot cultivation. Sci. Hortic. 338: 113654.

Ijaz, I., E. Gilani, A. Nazir and A. Bukhari. 2020. Detail review on chemical, physical and green synthesis, classification, characterizations and applications of nanoparticles. Green Chem. Lett. Rev. 13(3): 223–245.

Joudeh, N. and D. Linke. 2022. Nanoparticle classification, physicochemical properties, characterization, and applications, A comprehensive review for biologists. J. Nanobiotechnol. 20(1): 262.

Kansotia, K., R. Naresh, Y. Sharma, M. Sekhar, P. Sachan, K. Baral et al. 2024. Nanotechnology-driven solutions, Transforming agriculture for a sustainable and productive future. J. Sci. Res. Rep. 30(3), Article 3.

Karim, M.E. and A.B. Munir. 2015. Nanotechnology in Asia, A preliminary assessment of the existing legal framework (SSRN Scholarly Paper 2589995). Soc. Sci. Res. Netw.

Karthigadevi, G., S. Manikandan, N. Karmegam, R. Subbaiya, S. Chozhavendhan, B. Ravindran et al. 2021. Chemico-nanotreatment methods for the removal of persistent organic pollutants and xenobiotics in water-A review. Bioresour. Technol. 324: 124678.

Katzourakis, V.E. and C.V. Chrysikopoulos. 2024. Aggregating nanoparticle transport with nonlinear attachment, Modeling and experimental validation. Adv. Water Resour. 193: 104819.

Kaur, T. 2021. Fate and transport of engineered nanoparticles as an emerging agricultural contaminant. pp. 283–308. *In*: Singh, V.K., R. Singh and E. Lichtfouse [eds.]. Sustainable Agriculture Reviews 50, Emerging Contaminants in Agriculture. Springer Int. Publ.

Khan, S. and M.K. Hossain. 2022. Classification and properties of nanoparticles. pp. 15–54. *In*: Nanoparticle-Based Polymer Composites. Elsevier.

Kumari, S. and L. Sarkar. 2021. A review on nanoparticles, Structure, classification, synthesis and applications. J. Sci. Res. 65.

Kurganskaya, I. 2024. Dissolution mechanisms and surface charge of clay mineral nanoparticles, Insights from kinetic Monte Carlo simulations. Minerals. 14(9): 900.

Macko, M., J. Antos, F. Bozek, J. Konecny, J. Huzlik, J. Hegrova et al. 2023. Development of new health risk assessment of nanoparticles, EPA health risk assessment revised. Nanomaterials. 13(1), Article 1.

Maity, A., N. Natarajan, D. Vijay, R. Srinivasan, M. Pastor and D.R. Malaviya. 2018. Influence of metal nanoparticles (NPs) on germination and yield of oat (*Avena sativa*) and berseem (*Trifolium alexandrinum*). Proc. Natl. Acad. Sci. India Sect. B Biol. Sci. 88(2), 595–607.

Majeed, M., W. Noor, A. Rasool, M. Saleem, I. Bashir, R. Fatima et al. 2024. Nanotechnology for soil remediation and land restoration. pp. 60–77. *In*: Harnessing Nano-Omics and Nanozymes for Sustainable Agriculture. IGI Global.

Mandal, N., S. Adhikary and R. Rakshit. 2020. Nanobiosensors, Recent developments in soil health assessment. pp. 285–304. *In:* Rakshit, A., S. Ghosh, S. Chakraborty, V. Philip and A. Datta [eds.]. Soil Analysis, Recent Trends and Applications. Springer.

Martínez-Fernández, D., M. Vítková, Z. Michálková and M. Komárek. 2017. Engineered nanomaterials for phytoremediation of metal/metalloid-contaminated soils, Implications for plant physiology. pp. 369–403. *In*: Ansari, A.A., S.S. Gill, R. Gill, G.R. Lanza and L. Newman [eds.]. Phytoremediation. Springer Int. Publ.

Mohammadi, Z. and S.M. Jafari. 2021. Regulatory principles on food nanoparticles legislated by Asian and Oceanian countries. pp. 201–238. *In*: Jafari, S.M. [ed.]. Safety and Regulatory Issues of Nanoencapsulated Food Ingredients. Vol. 7. Academic Press.

Mohanty, S., S. Chakraborty, M. Das and S. Paul. 2022. Role of nanomaterials in phytoremediation of tainted soil. pp. 329–353. *In*: Kumar, V., M.P. Shah and S.K. Shahi [eds.]. Phytoremediation Technology for the Removal of Heavy Metals and Other Contaminants from Soil and Water. Elsevier.

Mubeen, B., A. Hasnain, J. Wang, H. Zheng, S.A.H. Naqvi, R. Prasad et al. 2023. Current progress and open challenges for combined toxic effects of manufactured nano-sized objects (MNO's) on soil biota and microbial community. Coatings 13(1), Article 1.

Nouri, M. 2024. Transformative nanotechnology applications in soil remediation and ecosystem restoration. pp. 126–147. *In*: Harnessing NanoOmics and Nanozymes for Sustainable Agriculture. IGI Global.

OECD. 2021. Guidance document on testing nanomaterials using OECD TG No. 312 "Leaching in soil columns". OECD.

Parameswari, P., N. Belagalla, B.V. Singh, G.J. Abhishek, G.M. Rajesh, D. Katiyar et al. 2024. Nanotechnology-based sensors for real-time monitoring and assessment of soil health and quality, A review. Asian J. Soil Sci. Plant Nutr. 10(2), Article 2.

Atle, T., B. Tomar, S.S. Parihar, S.S. Tomar and P. Singh. 2024. Nanotechnology-based soil improvement and conservation for enhancement of crop production. In Harnessing NanoOmics and Nanozymes for Sustainable Agriculture (pp. 217–236). IGI Global.

Patel, B.V., K. Yadav, R. Desai, S. Patel, A. Amari, N. Choudhary et al. 2024. Bacteriogenic synthesis of morphologically diverse silver nanoparticles and their assessment for methyl orange dye removal and antimicrobial activity. PeerJ, 12: e17328.

Pramanik, P., P. Ray, A. Maity, S. Das, S. Ramakrishnan and P. Dixit. 2020. Nanotechnology for Improved Carbon Management in Soil. pp. 403–415. *In*: Ghosh, P.K., S.K. Mahanta, D. Mandal, B. Mandal and S. Ramakrishnan [eds.]. Carbon Management in Tropical and Sub-Tropical Terrestrial Systems. Springer, Singapore.

Prasad, S.V.S., S.B. Prasad and S. Singh. 2022. Introduction and origin to nano manufacturing. In, Nanomanufacturing and Nanomaterials Design. CRC Press.

Rajput, V.D., T. Minkina, S.K. Upadhyay, A. Kumari, A. Ranjan, S. Mandzhieva et al. 2022. Nanotechnology in the Restoration of Polluted Soil. Nanomater. 12(5), Article 5.

Rizwan, M., M. Singh, C.K. Mitra and R.K. Morve. 2014. Ecofriendly Application of Nanomaterials, Nanobioremediation. J. Nanoparticle Res. 2014(1): 431787.

Sabarinath, S., V. Raj, V.R. Kurup and S.P. Rajeev. 2024. Enhancing material properties with nano clay, Applications in bioimplants and epoxy composites. In, Nanoclay-Based Sustainable Materials. pp. 403–428. Elsevier.

Sands, R.D. and S.A. Suttles. 2022. World agricultural baseline scenarios through 2050. Appl. Econ. Perspect. Policy. 44(4): 2034–2048.

Santos, P.A., X. Biraku, E. Nielsen, A.C. Ozketen, A.A. Ozketen and E.E. Hakki. 2024. Agricultural nanotechnology for a safe and sustainable future, Current status, challenges, and beyond. J. Sci. Food Agric.13922.

Singh, A., S. Rawat, V.D. Rajput, T. Minkina, S. Mandzhieva, A. Eloyan et al. 2024. Nanotechnology products in agriculture and environmental protection, Advances and challenges. Egypt. J. Soil Sci. 4(4): 1355–1378.

Swati, A.P. and H.S. K. Hemant. 2021. Nanotechnology-Finding proofs for its ancient origin. Asian J. Res. Pharm. Sci., 11(1): 65–70.

Tegart, G. 2004. Nanotechnology, The technology for the twenty-first century. Foresight. 6(6): 364–370.

US EPA, O. 2014, July 21. Human Health Risk Assessment [Collections and Lists]. https,//www.epa.gov/risk/human-health-risk-assessment.

van Dijk, M., T. Morley, M.L. Rau and Y. Saghai. 2021. A meta-analysis of projected global food demand and population at risk of hunger for the period 2010–2050. Nat. Food, 2(7): 494–501.

Vázquez-Nunez, E., C.E. Molina-Guerrero, J.M. Peña-Castro, F. Fernández-Luqueno and M.G. de la Rosa-Alvarez. 2020. Use of nanotechnology for the bioremediation of contaminants, A review. Processes, 8(7), Article 7.

CHAPTER-19

Future Perspectives and Challenges in Soil Nanotechnology Research

Waseem Hassan,[1] *Sundas Riaz,*[1] *Umair Riaz,*[1] *Shakeel Ahmad,*[1] *Muhammad Imran*[1,*] and *Hasan Riaz*[2]

1. Introduction

The goal of soil nanotechnology research is to employ nanotechnology to improve soil fertility, health, and overall agricultural output. Through materials manipulation at the nanoscale, scientists hope to solve a number of issues pertaining to environmental sustainability, nutrient management, and soil quality. Soil may be made more water-retaining by adding nanoparticles, such as nano-clays, which is especially useful in dry areas. With better water retention, soil moisture is maintained, promoting plant development and lowering the frequency of watering. Interdisciplinary cooperation between soil scientists, agronomists, chemists, and environmental scientists is necessary for the future orientations of soil nanotechnology research. Researchers want to create novel approaches that improve agricultural output and soil health while maintaining environmental sustainability by fusing knowledge from various domains (Rajput et al. 2022).

1.1 Nanotechnology

Soil nanotechnology is the use of nanotechnology and nanoparticle principles to increase soil properties and enhance agricultural practices by addressing environmental challenges such as soil contamination and degradation (Bhushan and Bharat 2017).

1.2 Importance of Soil Nanotechnology Research

In biology, nanomaterials are an adaptable tool for improving plant output, growth, and development. By increasing input efficiency and lowering relevant losses, nanotechnology makes it easier to boost agricultural productivity. Agrochemicals have a greater and more targeted surface area thanks to nanomaterials, which also serve as unique carriers that improve nutrient bioavailability and nutrient element accumulation in crop plants. Nanomaterials are employed in agriculture for the production of biosensors, nano-fertilizers, nano-insecticides, and nano-genetically modifying plants and animals to produce desired traits. They are also utilized in the development of nano plant growth regulators

[1] Department of Soil and Environmental Sciences, Muhammad Nawaz Shareef University of Agriculture, Multan.
[2] Institute of Plant Protection, Muhammad Nawaz Shareef University of Agriculture, Multan.
* Corresponding author: m.imran@mnsuam.edu.pk

DOI: 10.1201/9781003558354-19

(Singh and Gujar 2022). Nanodevices have been developed and used in fields to improve existing plant breeding and genetic transformation technology (Mukhopadhyay and Sharma 2013, Prasad et al. 2017). Researchers have reported the encouraging applications of nanosensors in precision farming (Dasgupta et al. 2017). In order to create a sustainable and environmentally friendly environment, nanodevices have also been used for the monitoring and assessment of growth, crop yield, soil characteristics, insect and microbial diseases, and fertilizer distribution (Chen et al. 2016). Nano-barcodes have been designed for monitoring the quality of agricultural products (Li et al. 2005). Pandey (2018) has reported that nanotechnology can increase the 25% growth rate of the agri-market. The agricultural system's integration of nano-based techniques has the potential to boost current economic growth by up to $3 trillion USD. The environment is seriously threatened by the inefficient use of agrochemicals. As a result, researchers, producers, and decision-makers are always looking for fresh approaches to solve pressing problems. Growing evidence points to nanotechnology as the new sustainable agricultural savior. Nanosensors have been used in agri-fields for purposes other than precision farming, such as the identification of chemically hazardous analytes and crop diseases. Furthermore, research has demonstrated that the use of nanorobotics and nano-barcodes can significantly improve agricultural operations and increase crop yields. Additionally, there are several remarkable uses of nanotools in agriculture, including bioimaging, sensing, photocatalysis, and the delivery of agrochemicals. Comparison of conventional, modern and nano based practices in agricultural productivity highlighting the importance of nanotechnology is provided in Table 1.

1.3 Purpose of the Chapter

This chapter looks at upcoming trends, potential uses, and obstacles in the field of soil nanotechnology in order to offer insights that help guide future research efforts and aid in the development of sustainable and useful nanotechnological solutions for problems related to soil.

2. Current State of Soil Nanotechnology Research

Another important area of study for soil nanotechnology is nutrient management. In order to produce a more effective substitute for conventional fertilizers, scientists are developing nanofertilizers that release nutrients gradually and under control, giving plants a continuous supply of vital nutrients over time. This method increases crop yield while lowering the risk of nutrient runoff, which can contaminate the environment (Usman et al. 2020).

2.1 Overview of Current Research Areas

The synthesis of nanoenzymes has also changed how plants respond to stress. These enzymes are potent antioxidants that have recently been used extensively to combat salt tolerance. The capacity of nanotechnology to transfer genetic material efficiently for genetic engineering and gene editing procedures has also tremendously benefited crop development. By using methods like nanobioremediation and nanophotocatalysis, hazardous pollutants may also be removed from the environment (Hasan et al. 2020). It is clear that the nanotechnology-driven agri-food sector is expected to grow rapidly. The pre-Green Revolution period revealed famine, but the post-Green Revolution era has greatly enhanced the practical use of pesticides and fertilizers to meet the expanding food demand by the growing population. But it also results in the loss of organic matter, the development of disease resistance, the irreversible eradication of microorganisms, and the humus of the soil—all of which present a significant risk in the modern world. However, smart agriculture might undergo a revolution due to the broad variety of uses of nanotechnology in delivery systems, nutrition, remediation, and nanosensor instruments (Zain et al. 2023).

Table 1. Comparison of conventional, modern and nano based practices used for increasing agriculture productivity.

Aspects of agriculture	Conventional agriculture practices	Modern agriculture practices	Nano-based agriculture practices	References
Methods used	These practices are based on the use of distinct types of chemicals	These practices involve mixed cropping, crop rotation, companion planting and organic compost.	These practices are based on the use of active ingredients in the form of distinct types of nanomaterials.	(Babu et al. 2022)
Fertility of soil	Poor fertility	Augmented the fertility of soil by improving the nutrients content.	Increased the nutrients' content of soil, thereby significantly augmented the soil fertility.	(Singh and Singh 2017)
Types of fertilizers	Fertilizers used are mainly inorganic.	These practices use organic fertilizers.	Both inorganic as well as organic fertilizers are used in these practices.	(Lowry et al. 2019)
Concentration of active ingredients	Agrochemicals are required in very high amount.	These practices require moderate number of agrochemicals.	These practices require very low number of agrochemicals.	(Lowry et al. 2019)
Nutrient content	Soil contain scarce amount of nutrients.	Soil contain all nutrients required for growth of crops.	Soil is very rich in nutrients.	(Duhan et al. 2017)
Efficiency	These practices show poor efficiency.	These practices show high efficiency.	These practices show exceptional high efficiency.	(Mittal 2020)
Targeted/non-targeted	Agrochemicals used in these practices are non-targeted.	Agrochemicals used in these practices are also non-targeted.	Agrochemicals used in these practices are targeted.	(Lowry et al. 2019)
Stability	Stability of spraying of agrochemicals is very less.	Stability of spraying of agrochemicals is better than conventional practices.	Stability of spraying of agrochemicals is good.	(Mittal 2020)
Cost	These practices use very costly agrochemicals in large quantity.	These practices use cost-effective.	These practices use cost-effective as they are used in minimum quantity.	(Babu et al. 2022)
Toxicity	Very hazardous, consequently increases the risk of cancer, reproductive toxicity and neuronal ailments. These also increase the risk of eutrophication in water bodies.	These practices minimum toxicity.	These practices show moderately but green nanomaterials do not cause toxicity.	(Duhan et al. 2017)
Effect on soil	Fertile but soon become infertile due to exploitation.	These practices keep soil fertile but may increase the prevalence of deforestation and soil erosion.	Nanomaterials based approaches significantly enhance fertility of soil.	(Mittal 2020)
Eco-friendly	Non-sustainable as conventional agrochemicals kill beneficial insecticides that cause bio-diversity loss.	These practices are sustainable.	Green nano based approaches are sustainable as they are compatible with living system.	(Shang et al. 2019)

2.1.1 Soil Remediation

Engineered nanomaterials (ENMs) and nanotechnology, with their special qualities of high reactivity, selectivity, and flexibility, have shown great promise in solving challenging problems and creating new approaches to soil remediation. The intricate and dynamic nature of the soil-water interface has posed significant financial and technical challenges for the remediation of polluted groundwater and soil in a variety of environmental media. The areas of in situ monitoring, getting the right mixing, and promoting interactions between the remedial agents and the target pollutants are the areas of challenge (Usman et al. 2020). In addition to conventional *ex situ* remediation approaches like removing contaminated soils or recapping new soils with equal contents, engineering-type remediation techniques including thermal desorption, soil vapor extraction (SVE), and surfactant flushing have been applied in the past 20 years. Recently, new methods for in situ cleaning have been developed, such as bioremediation, sequestration, immobilization, and chemical oxidation and reduction (using Fenton reagent or persulfate, for example). Most of these technologies are still time- and energy-consuming, inefficient economically, or detrimental to the environment. The goals of innovation in this field should include enabling simultaneous monitoring and consideration of emergent pollutants in addition to the frequently targeted contaminants—heavy metals (As, Cr, Pb, Zn, etc.); organic compounds (pesticides, PAHs, polychlorinated biphenyls or PCBs); polyfluoroalkyl substances (PFASs); and combined contaminants (Linley et al. 2021).

2.1.2 Nutrient Delivery Systems

Through the application of strategies such as targeted dispersion and controlled or slow release, nutrient-use efficiency may be enhanced using nanostructured fertilizers. In exact response to environmental signals and biological demands, they may release their active molecules. Through controlled or gradual release and targeted dispersion, the use of nanostructured fertilizers can improve nutrient efficiency. In exact response to environmental signals and biological demands, they may release their active molecules (Achari et al. 2018). Through controlled or gradual release, they can precisely release their active substances in response to indications from the environment and biological demands. The mineral components in the soil must be in a condition that is accessible for the plant to easily absorb them (McClements et al. 2015). Which nutrients are accessible to the plants depends on their quantity, kind, and interaction with other nutrients in the solid phase. The capacity of the soil-plant system to provide and absorb nutrients—including the plant's movement and absorption of nutrients upon their release from the solid phase into solution—is another way to characterize it. The mineral components in the soil must be in a condition that is accessible for the plant to easily absorb them. Which nutrients are accessible to the plants depends on their quantity, kind, and interaction with other nutrients in the solid phase. The capacity of the soil-plant system to provide and absorb nutrients—including the plant's movement and absorption of nutrients upon their release from the solid phase into solution—is another way to characterize it (Kanjana and Duraisamy 2017).

Plant nutrition is a crucial component of sustaining production. It has a significant role in determining the amount and quality of agriculture. A lack of any one of the 17 essential nutrients might hinder the development of plants. The use of fertilizers is crucial to raising plant production. However, the low nutrient utilization efficiency of conventional fertilizers (CFs) has been related to various losses, including widespread multi-nutrient deficiencies in the soil and lower output. Furthermore, the excessive use of these substances has caused severe harm to the environment and soil. The recent increase in the use of nanotechnology in plant nutrition has made it feasible to manufacture fertilizers triggered by nanoparticles. Thus, in sustainable agriculture, nano fertilizers are an especially excellent alternative to CFs. The components of nano fertilizers are macro- and micronutrient-rich nanoparticles that are managed in their delivery to the plant's rhizosphere (Arora et al. 2022). This increases nutrient utilization efficiency It is also possible to provide nutrients as nanoparticles or emulsions in nanotubes and nanoporous materials covered in a thin polymer coating. These represent only a handful of the numerous ways fertilizers may be enclosed in nanoparticles. Smart delivery methods may be used to provide fertilizers that contain nanoparticles gradually and under control due to their special qualities, which include size, shape, solubility, and crystal state, as well as the materials and doses that make them up. By halting their chemical breakdown, runoff, and leaching losses, this reduces their harmful impacts on the ecosystem (Solanki et al. 2015).

2.1.3 Soil Monitoring

Scientists describe chemical sensors as devices that identify the observable presence, concentration, or amount of a specific analyte. Despite this potential, chemical sensors are costly and controlled analytical equipment or instruments for environmental monitoring and soil analysis in particular. The necessity and diversity of chemical measurements have surpassed the large and consistently growing capability of chemical sensing technologies during the past few decades. A large amount of research is anticipated to be directed towards the detection of various compounds in the regulation of manufacturing processes, as well as toxins in groundwater, car emissions, and biotoxins in public areas. It has not been possible to obtain real-time data for soil microbial identification using conventional culture-dependent methods. Biosensors' efficiency can be raised by using nanomaterials because of their distinct physicochemical characteristics. Soil Observation and Perception The understanding of the intricate interactions between microorganisms and nanomaterials and the ensuing biological effects has been sparked by this field (Nocita et al. 2015).

2.1.4 Crop Enhancement

Agrochemicals have a larger and more focused surface area thanks to nanoparticles, which act as special carriers that boost nutrient bioavailability and accumulation of nutrient components in crop plants. Crop protection is another use for nanomaterials. Through the activation of nutrients and energy in plant cells, nanotechnology is developing as a viable green technology for improving agricultural productivity and fortifying soil (Singh et al. 2024). Crop yields are increased by the use of mineral components in the soil that must be in a condition accessible for the plant to easily absorb. The nutrients accessible to the plants depend on their quantity, type, and interaction with other nutrients in the solid phase. The capacity of the soil-plant system to provide and absorb nutrients, including the plant's movement and absorption of nutrients upon their release from the solid phase into solution, is another way to characterize it. Nanoparticles as fungicides, insecticides, and fertilizers. The mineral nutrients present in the soil must be in bioavailable form, so that the plant can take them up easily (Liu et al. 2020). The availability of nutrients to plants depends on their amount, nature, and their association with other nutrients in the solid phase. It can also be explained as the capacity of the soil-plant system to supply/absorb nutrients, which includes the release of nutrients from the solid phase to solution. A plant's ability to receive and utilize nutrients is determined by the amount, type, and relationship of the mineral nutrients it contains with other nutrients in the solid phase. It can also be defined as the soil-plant system's ability to supply and absorb nutrients, which includes the movement and absorption of nutrients by the plant as well as their release from the solid phase into solution (Younis et al. 2021).

2.2 Recent Advances in Soil Nanotechnology

Common remediation techniques include physicochemical adsorption, (photo) chemical redox reactions, and filtration. Recently, a vast array of customized inorganic and organic nanocomposites has been created. Membrane-like materials, embedding structures, and composites with remarkable physicochemical properties and outstanding immobilization or removal abilities from polluted areas are among them. Emerging strategies include the synthesis of bio-nanocomposites and nanomaterials with both adsorptive and oxidative characteristics (Qian et al. 2020).

3. Future Perspectives in Soil Nanotechnology Research

Research on soil nanotechnology has a bright future ahead of it, with several important areas of focus. For specific uses like soil remediation, pest control, and nutrient delivery, advanced nanomaterials will be created. In order to reduce ecological risks, sustainable practices will prioritize the use of biodegradable and environmentally friendly nanoparticles. Through real-time data and intelligent delivery systems, integration with precision agriculture will improve monitoring of soil health, maximize resource use, and increase crop productivity (Hasan et al. 2020).

3.1 Emerging Trends and Technologies

One of the main aspects of nanotechnology is the evaluation of nanoparticles. The usage of nanomaterials has ushered in a new era known as the "nano-revolution," which involves moving particles with at least one dimension at the nanoscale (Syed et al. 2016). Considered the particles of the century, nanomaterials have attracted a lot of attention (Zhao et al. 2014). These nanoparticles have advanced in many domains and are currently regarded as the most exciting materials in the technical world because of their physicochemical properties that rely on size and structure (Porchezhiyan and Noorjahan 2016). Tiny nanomaterials have tremendous reactivity because of their large surface plasmon resonance, which significantly improves their analytical capabilities (Baker et al. 2015). In the context of farming, sustainability remains a challenge despite the exponential progress of nanotechnology. The agriculture sector must make sure that the world's rapidly expanding population gets enough food in the current environment (Mba et al. 2012). Unfortunately, it is estimated that around one-third of global food production is lost each year during the transition from production to consumption. Diverse technological approaches are being developed to tackle this issue and improve standards after harvest. The use of products based on nanomaterials in contemporary agricultural reform is one such trend (Sekhon 2014). The application of nanoparticles in agriculture has received a lot of attention lately, and it is expected to completely transform the way that agriculture is done today (Fraceto et al. 2016). According to Pouratashi and Iravani (2012), the agricultural community is currently dealing with a wide range of issues, including soil contamination from pollutants in the environment that have caused significant loss, yield stagnation, crop destruction from pests, loss of soil fertility, insufficient water, and unpredictable weather patterns. Overcoming these obstacles requires careful design and innovative thinking (Parisi et al. 2015). define nanoagroparticles as nanosized particles designed to alleviate problems related to the agriculture sector.

These nanoagroparticles are not only efficient insecticides, fungicides, herbicides, and pesticides; they may also promote plant development. Nonetheless, one of the primary problems is how to synthesize basic nanoparticles without utilizing any dangerous components. Most traditional techniques sometimes have several disadvantages, such as producing excessive heat, requiring hazardous ingredients for synthesis, or requiring sophisticated equipment (Kavitha et al. 2013). As a result, research teams are trying to develop straightforward, user-friendly procedures for producing nanoparticles with the required properties. One such area is the synthesis of nanoparticles from biogenic sources. Microorganisms and plants, or their byproducts, are the biogenic sources. They facilitate the synthesis, stabilize the nanoparticles, and function as reductive agents.

3.2 Nanomaterials for Sustainable Agriculture

Nanotechnology may significantly increase production through nutrient control in addition to monitoring pesticides and water quality for the sustainable growth of agriculture (Gruère 2012, Mukhopadhyay 2014, Prasad et al. 2014). Given the wide diversity of characteristics and activities of nanoparticles, it is hard to offer a comprehensive evaluation of the health and environmental problems connected with them (Prasad et al. 2014). In addition to size, additional properties of NPs that influence toxicity include their chemical composition, form, surface charge, surface structure, behavior, and level of clumping or disaggregation. Manufactured NPs are associated with these features (Ion et al. 2010). As a result, nanomaterials with the same chemical composition but various sizes or shapes can still exhibit a range of toxicity. Research on studying the effects of nanotechnology on agriculture is increasingly vital to sustainable development, if not necessary. The agri-food industries have been seen to employ nanotubes, fullerenes, biosensors, controlled delivery systems, nanofiltration, etc., in pertinent ways (Ion et al. 2010, Sabir et al. 2014). It has been demonstrated that this technique works well for maintaining soil fertility, administering medications to plants, and managing agricultural resources. It is also regularly evaluated in terms of risk assessment, food processing, food packaging, and the use of agricultural waste and biomass (Floros et al. 2010). Because of its ability to identify environmental pollution in soils and water rapidly and accurately, nanosensors are being employed in agriculture on a large scale these days. The main instruments for detecting heavy metals in the trace range will be a spectrum of sensors based on nano-detection technologies, such as biosensors, electrochemical sensors, optical sensors, and devices (Ion et al. 2010).

In addition to directly catalyzing the breakdown of waste and hazardous materials, nanomaterials also help microorganisms degrade waste and harmful materials more efficiently. Utilizing living organisms, bioremediation eliminates or degrades toxins and other dangerous materials from agricultural soil and water. Specifically, additional terminology like phytoremediation (plants), mycoremediation (fungi and mushrooms), and bioremediation (beneficial bacteria) are also frequently employed. Thus, microorganisms may effectively and environmentally eliminate heavy metals from soil and water through bioremediation (Dixit et al. 2015). Consequently, agricultural bioremediation aids in the resolution and restoration of the soil's natural state using sustainable remediation processes. When one considers how nano-nano interaction can eliminate harmful elements from agricultural soil and create sustainable soil, it is an intriguing phenomenon. The main instruments for detecting heavy metals in the trace range will be a spectrum of sensors based on nano-detection technologies, such as biosensors, electrochemical sensors, optical sensors, and devices (Ion et al. 2010, Dixit et al. 2015). Here is the illustration of application of nanotechnology in agriculture in Figure 1.

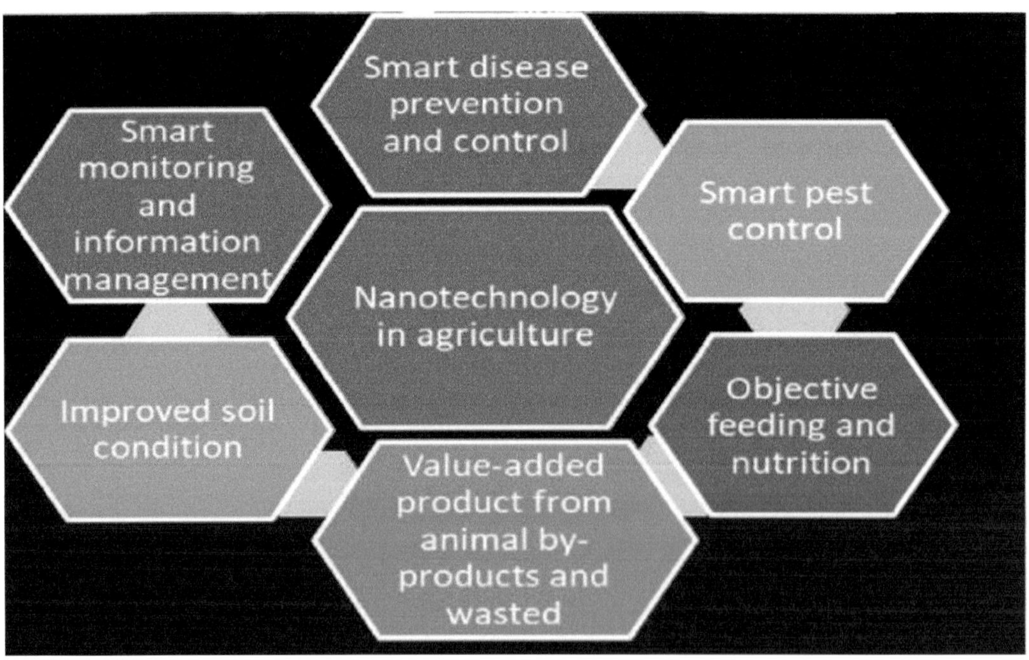

Figure 1. Application of nanotechnology in agriculture.

3.3 Precision Agriculture and Smart Farming

Global agriculture is changing as a result of precision agriculture's potential benefits for crop quality, environmental preservation, sustainability, food safety, economic growth, profitability, and productivity. Precision farming uses information technology (IT) to manage commercial agriculture, with the main objective being to reduce inputs (fertilizers, herbicides, and insecticides) and increase crop production. Precision farming and agriculture employ a variety of technologies, such as GPS, remote sensing, and geographic information systems (GIS). With a focus on the integration of precision agriculture and nanotechnology, this study underscores the significance of diverse nanosensors in soil-plant systems. We looked at the advantages and disadvantages of high-tech farming with biosynthesized nanoparticles (NPs). We concluded by outlining the main fields of study for nanotechnology and offering a variety of alternative policies and projects to ensure secure and sustainable agricultural production (Zain et al. 2023).

3.4 Potential Applications and Impacts

Nanoparticles have applications in a wide range of sectors, such as industry, agriculture, health, energy, and environmental cleaning. The introduction of management strategies has resulted from agricultural development for quicker growth due to resource constraints; among these, fertilizer management has proven to be extremely significant. The use of nanoparticles in fertilizer application has improved the soil system and encouraged the growth characteristics of plants.

The slow-release method of nutrients boosts the efficacy of fertilizer application, even if the major variables are the field conditions, soil pH, fertilizer changes, and other factors. The plant root system's activity was also improved by the use of nanoparticles. Although chemical fertilizers also reduced soil fertility, scientists looked at them as the main factor influencing crop yield. But the careful application of nano fertilizer has improved agricultural plants' physio-chemical composition in all respects, from metabolism to synthesis (Farooq et al. 2022).

Precision farming has enhanced and increased the efficiency of agricultural practices via the use of technologies such as pH release, ultrasonic release, heat release, moisture release, quick release, and specific release. Although nanotechnology has obvious advantages in agriculture, there are also a number of risks to any population. Bangladesh and other developing countries must act quickly to take advantage of the immense potential of nanotechnology in order to implement the expansion of the agriculture sector (Ali et al. 2021). Variable Rate Technology (VRT) is a crucial element of precision agriculture. It helps with the precise application of various input resources, including fertilizers, herbicides, and pesticides, based on soil types and crop health. By utilizing resources only when required, farmers are able to optimize crop growth and yield while cutting costs and waste through the use of VRT. Furthermore, by serving as a transporter of these resources, nanoparticles can boost the effectiveness of VRT.

Using VRT-based nanoparticles, herbicides, fertilizers, pesticides, and other compounds may be accurately supplied to a specific plot of land. This is particularly helpful in cases where crop health and soil type vary. Moreover, VRT can be helpful in figuring out the right seeding rate for a certain kind of field. At the end of the day, employing nanocarriers in VRT has several advantages. For example, it makes it possible to apply inputs accurately in the first place. Secondly, the possibility of environmental contamination is decreased since it only supplies nutrients where they are needed. Thirdly, it significantly increases crop growth and productivity (Basavegowda et al. 2021).

3.5 Impact

In an era where climate conditions are continually changing, a range of biotic (viruses, bacteria, fungi, nematodes, etc.) and abiotic (long-term floods, drought, soil salinization, etc.) stress factors constitute a major danger to food production, quality, and ultimately food security. They also harm the microorganisms in the soil. Therefore, innovative new technology must be adopted in order to control the tense environment. With its enormous potential, nanotechnology has the ability to revolutionize agri-production and food security. Cutting-edge substances called nanoparticles (NPs) have enormous potential to support the present agriculture system. NP-based nanomaterials, including nanobiofertilizers, nanoclay, nanoagrosensors, nanocarriers, nanohybrids, nanobarcodes, nanoherbicides, and fungicides, are being extensively researched and applied in contemporary agriculture to control the health of plants and soil (Basavegowda et al. 2021).

4. Challenges and Limitations

There are a number of obstacles and restrictions with nanotechnology in soil science and agriculture. The long-term environmental effects of nanomaterials are poorly understood, and their potential toxicity to humans, plants, and soil organisms is a serious concern. The absence of safety standards and regulatory frameworks makes it more difficult to govern the application of nanotechnology. Accurately identifying and detecting nanoparticles in soil, as well as guaranteeing their stability and controlled release, are technical challenges (Singh and Gurjar 2022).

4.1 Environmental Concerns and Risks

Nanotechnology is widely used in the contemporary fields of agriculture, food processing, food safety, packaging, dairy, transportation, and agricultural product quality control. With the current pace of research and invention, it is predicted that the worldwide market for nanotechnology might reach a valuation of up to \$100 billion by 2020. By 2050, the globe will need to deal with the difficult problem of feeding and sustaining almost 9.7 billion people. New research projects aimed at comprehending the environmental effects of these nanoparticles have been launched as a result of growing awareness of the possible, unanticipated, negative effects of generated nanoparticles in the ecosystem. Human activity can create nanoparticles, both intentionally and accidentally. Natural nanoparticles have been in the environment from the beginning of time (Yao et al. 2013). They are widely and often dispersed across the ecosystem, the oceans, the soil, the atmosphere, the groundwater, and the creatures. Nanoparticles are released into the atmosphere by massive biological processes including wildfires, volcanic eruptions, corrosion, and synthesis from clay particles, soil depletion by air and water, or sandstorms from deserts. It is estimated that in a single year, air dust alone may include several million tons of natural nanoparticles. Naturally occurring ambient nanoparticles can move across long distances, appear in a wide variety of sizes, and have a brief floating lifespan in the environment. Man-made engineered nanoparticles (ENPs) can be released into the environment accidentally or on purpose as a result of a variety of physical and economic activities. The number of nanoparticles produced is constantly increasing, indicating the vast range of industrial and domestic uses for which they are suitable. Owing to their environmental presence, these economic achievements could negatively impact natural systems. The biological activities and consequences of nanoparticles are dependent on how they enter the body and on their unique properties, including size, shape, charge, area, and chemical composition. Certain properties allow for the classification of particles as either safe or dangerous. Both organic and inorganic pollutants that have an impact on aquatic and terrestrial systems continue to be absorbed by nanoparticles. Since nanoparticles are so prevalent in the environment, beneficial bacteria start to pose a threat to their own cells, which inhibits the manufacture of bioactive chemicals by preventing molecular alterations and ultimately leads to a decrease in the population of germs (Usman et al. 2020).

4.2 Regulatory Issues and Safety Standards

In response to these concerns, international regulatory bodies have developed rules and policies to ensure the safe use of nanotechnology in agricultural products. These are a few of the principal laws and regulations that deal with agricultural products that use nanotechnology.

4.2.1 Regulatory Oversight

The laws regulating agriculture based on nanotechnology vary from country to country. For example, the Environmental Protection Agency (EPA) is in charge of pesticides across the nation, while the U.S. Department of Agriculture (USDA) is in charge of agricultural biotechnology products. The European Chemicals Agency (ECHA) and the European Food Safety Authority (EFSA) oversee regulations in the European Union pertaining to nanotechnology in food and pesticides (Mishra et al. 2017).

4.2.2 Risk Assessment

Before regulatory bodies will authorize any agricultural product based on nanotechnology, risk evaluation is a must. This involves evaluating the toxicity of the used nanomaterials, their potential for release into the environment, and their impact on human health (Singh et al. 2022).

4.2.3 Labeling

Products made in agriculture that incorporate nanomaterials need to be labelled in compliance with laws. This helps consumers make informed product selections and uses (Prasad et al. 2017).

4.3 International Standards

International standards based on nanotechnology have been developed to ensure the safety and quality of agricultural goods. The International Organization for Standardization (ISO) has developed a variety of standards pertaining to nanotechnology, including ISO/TS 80004-1, which defines and uses terminology unique to nanoparticles (Mishra et al. 2017).

4.4 Research and Development

Regulatory bodies fund research and development in this field to ensure that agriproducts based on nanotechnology are safe for the environment and human health. All things considered, collaboration between scientists, producers, authorities, and end users is required for the responsible deployment of nanotechnology in agricultural commodities. It is essential to monitor and assess the risks associated with nanotechnology-based agriproducts in order to ensure their efficacy and security. Several countries have developed a range of regulations and guidelines to ensure the safe application and development of agri-products based on nanotechnology (Supplementary Table S6). Here are a couple of such examples:

The United States of America: Based on nanotechnology, a number of government organizations oversee agricultural goods in the nation, such as the United States Department of Agriculture (USDA), the Environmental Protection Agency (EPA), and the Food and Drug Administration (FDA). A number of nanoparticles have received approval from the US Food and Drug Administration to be used in food applications. Here are some such examples:

- Distilled titanium dioxide: This material is a common food ingredient that is used to whiten and brighten a range of foods, such as powdered sugar, gum, and candy. Nanoscale versions of titanium dioxide are now allowed in food items.
- Silica: Powdered coffee creamers are one food product that uses nano-sized silica as an anti-caking ingredient (US Food and Drug Administration 2020).
- Zinc oxide: The US Food and Drug Administration (2020) states that food coloring and nutritional supplements containing nanoscale zinc oxide are allowed.
- Iron oxide: The use of nanoscale iron oxide as a food colorant has been allowed by the US Food and Drug Administration (2020).

- Generally speaking, these organizations seek to ensure that agricultural products utilizing nanotechnology are safe and effective, and that they also comply with applicable laws and regulations.

4.5 Technological Barriers and Limitations

Soil nanotechnology is the application of nanotechnology principles to soil science, and it faces a number of technological challenges.

4.5.1 Soil Settings

One of the primary issues facing modern science at the moment is understanding the motion and fate of nanoparticles in soil settings. The way that soil constituents react and move may be influenced by the different interactions that nanoparticles have with them. For some applications, it is essential to regulate the motion of the nanoparticles inside the soil matrix.

4.5.2 Soil Heterogeneity

The uniform distribution and effectiveness of nanoparticles are challenged by soil heterogeneity because of the significant variation in soil composition and properties. The way that nanoparticles interact with different types of soil can have an impact on their efficacy and possible environmental impacts.

4.5.3 Nanoparticle Stability

In soil environments, nanoparticles can undergo changes due to a range of parameters such as pH, organic matter concentration, and microbial activity. Maintaining the stability and longevity of nanoparticles in soil is essential for their continued efficacy.

4.5.4 Risk Assessment and Environmental Impact

Concerns have been raised regarding the potential impacts of soil-applied nanoparticles on human health and the environment. Understanding the long-term effects of nanoparticle exposure on plant development, soil organisms, and ecosystem functioning is essential for safe deployment.

4.5.5 Enhanced Economic Viability and Scale

Two challenges that come up when transferring laboratory-scale research to field applications are scalability and cost-effectiveness. The development of scalable production procedures and economically viable solutions for soil nanotechnology is essential for wider application.

4.5.6 Regulatory Frameworks

The set of regulations pertaining to the use of nanotechnology in soil is currently evolving. To guarantee compliance and safety, rules and regulations pertaining to the use of soil technologies based on nanoparticles must be developed.

4.5.7 Restricted Knowledge of Processes

Despite significant progress, much remains unknown about the processes underlying the interactions between soil components and nanoparticles. To completely understand these mechanisms and maximize the effectiveness of nanoparticles in soil applications, more study is required.

5. Strategies for Overcoming Challenges

Solving problems in soil science nanotechnology calls for an all-encompassing strategy. To fully comprehend the toxicity and ecological effects of nanoparticles, extensive environmental and safety research is essential. Clear safety standards and regulatory frameworks will aid in regulating the application of nanotechnology, and technical advancements should concentrate on enhancing the characterization, detection, stability, and controlled release mechanisms of nanoparticles. It is crucial to lower costs and improve scalability through research and public-private partnerships.

5.1 Risk Assessment and Mitigation Measures

After a specific danger has been determined, the second step in any risk analysis is to examine the risk related to nanomaterials. Risk assessment will provide the information needed for a realistic risk management plan. Nanomaterials are often relevant in the context of chemical risk assessment (Coll et al. 2015). Here, there are two topics to talk about. The substance must be recognized, and any potential risks must be investigated under controlled conditions, as in a laboratory, before anything more is done. While the majority of nanomaterials' biological and environmental dispersion pathways are still unclear, new research is starting to clarify things (Wang et al. 2016). The advantages of NMs and the potential of nanotechnology to address a wide variety of difficulties in the agri-food business are highlighted in several research and reviews.

Nanoparticle concentration is important for evaluating the cytotoxic and genotoxic effects of seed treatments, which might vary depending on the kind of agricultural seed. Particles of varying sizes can reach the nucleus through nuclear pores, indicating that the size of the particles influences how they interact with various cell components. Furthermore, nanoparticles can bypass cell cycle checkpoints, interrupt cell division pathways by attaching to proteins and preventing protein synthesis, or enter cells through chemical or mechanical interactions with enzymes that generate reactive oxygen species (ROS). When creating seed treatments, it is essential to consider the characteristics of the nanomaterial in order to regulate toxicity. Size, doses, exposure durations, surface chemistry, structures, immunological reactions, accumulations, and other effects are some of these characteristics (Sharma et al. 2022).

There is a stalemate in the field of nanotoxicology on the best method to assess the risk of agrochemicals based on nanomaterials for environmental monitoring and human health (Ray et al. 2009). Toxicology testing can be done in vitro on cell cultures or in vivo (on living organisms like fish, mice, other animals, microcrustaceans, and plant models), according to Bahadar et al. (2016). It is difficult to compare the results and reach an agreement about the toxicity of agrochemicals based on nanomaterials since there are no set procedures for doing so (Chaud et al. 2021). NMs may be readily produced with the use of top-down methods like grinding and ball milling. They are also consistently produced chemically by bottom-up methods. Modern, state-of-the-art technology has simplified the process of their physicochemical characterization for further applications (Baek et al. 2011). Soil organisms play a critical role in the functioning of the ecosystem by influencing the physical properties and biological activities of the soil. These organisms participate in the cycling of nutrients, carbon, and energy fluxes, which affect the biological condition of the soil in addition to its chemical and physical properties and processes. The size of the soil fauna is closely tied to the physical properties of the soil. The soil biota includes microorganisms and macrofauna (such as ants) as well as microfauna (bacteria, fungi, and actinomycetes). It is difficult to assess the effects of new technologies, like nanotechnology, in developing countries like China and India because there is a dearth of trustworthy historical data, most research is still in its early phases, and findings are only reported in publications and patents with promises of further investigation (Sastry et al. 2011, Kostoff et al. 2007)

5.2 Development of Sustainable Nanomaterials

Nanotechnology may significantly increase production through nutrient control in addition to monitoring pesticides and water quality for the sustainable growth of agriculture (Gruère 2012, Mukhopadhyay 2014, Prasad et al. 2014). Because nanomaterials have so many different characteristics and applications, it is challenging to give a comprehensive evaluation of the health and environmental risks they pose (Prasad et al. 2014). In addition to size, additional properties of NPs that influence toxicity include their chemical composition, form, surface charge, surface structure, behavior, and level of clumping or disaggregation. Manufactured NPs are associated with these features (Ion et al. 2010). As a result, nanomaterials with the same chemical composition but various sizes or shapes can still exhibit a range of toxicity. Studying the effects of nanotechnology on agriculture is increasingly vital to sustainable development, if not necessary. The agri-food industries have been shown to employ fullerenes, nanotubes, biosensors, controlled delivery systems, nanofiltration, etc. in pertinent ways (Ion et al. 2010, Sabir et al. 2014). It has been demonstrated that this technique works well for maintaining soil fertility, administering medications to plants, and managing agricultural resources. It is also regularly evaluated in terms of risk assessment, food processing, food packaging methods, and the use of agricultural waste and biomass (Floros et al. 2010). Nanosensors have been widely used in agriculture recently because of their efficiency and speed in detecting environmental pollution in soils and water (Ion et al. 2010). The main instruments for detecting heavy metals in the trace range will be a spectrum of sensors based on nano-detection technologies, such as biosensors, electrochemical sensors, optical sensors, and devices (Ion et al. 2010).

Nanomaterials assist microorganisms in more effectively decomposing garbage and hazardous materials in addition to directly catalyzing their disintegration. Bioremediation is the process of removing or breaking down toxins and other hazardous substances from agricultural soil and water by using living organisms. More specific terms like "bioremediation" (beneficial bacteria), "mycoremediation" (fungi and mushrooms), and "phytoremediation" (plants) are also commonly used. Therefore, through a process called bioremediation, microorganisms may safely and sustainably remove heavy metals from soil and water, as stated by Dixit et al. (2015). Thus, by employing sustainable remediation techniques, agricultural bioremediation helps resolve and return the soil to its original state. It is an interesting topic to consider how hazardous materials may be removed from agricultural soil and sustainable soil can be created through nano-nano interaction—an area with enormous potential for the creation and use of new technologies (Ion et al. 2010, Dixit et al. 2015).

Nanotechnology has already proven to have many positive social effects; however, in addition to its benefits, nanotechnology research and development also has inherent risks. During revolutions, unexpected consequences are typical, and sometimes they inspire countermovement (such as the usage of genetically modified organisms in agriculture). Our understanding of the effects of nanotechnology on the environment and human health is still lacking (Heng et al. 2009). A poorly-thought-out commercialization strategy lowers the likelihood of commercialization success in several ways, such as inadequate protection of intellectual property, poor target market selection, early alliance partner selection gone wrong, inexperienced founding team, and restricted capital raising capabilities.

5.3 Stakeholder Engagement and Communication

The urgent need for new models of nanotechnology innovation and governance is being recognised by a wide range of stakeholder groups, including academia, governmental organisations, environmentally concerned businesses, and citizen advocacy groups (Renn and Roco 2006). While several scenarios about nanotechnologies have been created by academics in academia and industry, few of them examine alternative strategies for creating and controlling nanotechnologies. Predictive governance (Guston 2008), responsible innovation (Roco et al. 2011, von Schomberg 2013), and sustainable governance (Lang and Siegrist 2008) offer compelling guiding concepts. The investigation of such alternative models may benefit from these ideas. Nevertheless, local governments, corporations, company owners, and individuals are overburdened by the whirlwind of everyday operations and lack the time and energy to look into such alternative models. This study poses the following question: what would alternative models look like, and how might future versions of the current innovation models appear? And how well-suited are those different approaches to lowering concerns about sustainability and improving living standards? In order to do this, the research focuses on nanotechnologies in urban settings, including systems for energy generation, transmission, and storage; genetically based security applications; multifunctional surface coatings; and nano-enhanced building materials, to mention a few. The project aims to provide a distinct set of diverse scenarios that consider the possible future impacts of different models of nanotechnology innovation, particularly in an urban environment (Wiek et al. 2009, 2013).

5.4 Policy Recommendations and Guidelines

Although nanotechnology has made significant strides in many fields, its agricultural applications have lagged behind global demand, mostly due to a lack of understanding and biosafety concerns. Geographically limited regulatory standards, legislative frameworks, and negative public opinion are the main reasons for the lack of commercial growth of nanotechnology in agriculture (Arts et al. 2014). The economic benefits of items based on nanotechnology for the agriculture industry are highlighted by the growing difficulties of climate change and global food security. According to (Rossi et al. 2014), it is essential to thoroughly investigate the risk assessment and risk management components associated with the application of nanoparticles in the agricultural sector prior to enacting regulatory requirements (FAO/WHO, 2013). Several regulatory agencies, including the US Food and Drug Administration (USFDA), the International Standard Organization (ISO), and the Organisation for Economic Co-operation and Development (OECD), are tackling this issue. The ISO and OECD merely provide guidelines and suggestions to the regulatory agencies; the USFDA, in particular, is responsible for enforcing regulations on the ground. OECD and non-OECD countries are regulating nanotechnology in the food, feed, and agricultural sectors using different approaches.

REACH (Registration, Evaluation, Authorization and Restriction of Chemicals) is the main EU law that addresses the use of nanomaterials in plant protection products, food additives and supplements, and materials that come into contact with food (European Commission 2013). With regard to the food, feed, and agricultural sectors, only the European Union and Switzerland have successfully enacted regulations relevant to nanotechnology; other non-EU countries have non-binding frameworks that offer non-legal guidance (OECD 2013). Nonetheless, it is notable that products based on nanoparticles for agricultural application are struggling to find a market due to murky legal frameworks and conflicting views around the globe. Prioritising risk assessment and risk management when creating regulatory regulations to address biosafety concerns should be the first priority. Furthermore, exchanging ideas and opinions on global public forums will be beneficial for handling effective regulatory actions.

- Increase funding and resources for soil nanotechnology research: Governments and funding agencies should prioritize funding for soil nanotechnology research to support the development of innovative solutions for soil health and agriculture.
- Encourage interdisciplinary collaboration: To promote a comprehensive approach to tackling soil health concerns, encourage collaboration between academics from different fields, such as soil science, nanotechnology, and agriculture.
- Establish standard operating procedures and guidelines: Establish standard operating procedures and guidelines for soil nanotechnology research to ensure consistency and repeatability of findings.
- Promote transparency and communication: To promote collaboration and information sharing among scientists, encourage researchers to openly share their findings and data.
- Handle moral and legal concerns: Ensure that moral and legal considerations are made throughout the development and use of soil nanotechnology solutions.
- Involve stakeholders: To ensure that solutions based on soil nanotechnology meet the needs and desires of final consumers, such as farmers, policymakers, and other participants in the research and development process.

- Monitor and evaluate the results of soil nanotechnology applications: Establish mechanisms for monitoring and evaluating the long-term impacts of soil nanotechnology solutions on soil health, environmental sustainability, and agricultural yield.
- Promote education and outreach: Disseminate information about the possible benefits and challenges of soil nanotechnology research to the general public and the scientific community through outreach and education initiatives.

By implementing these principles and recommendations for policy, we may overcome the challenges in soil nanotechnology research and promote the creation of a more resilient and sustainable agricultural system.

6. Future Directions and Recommendations

Future developments in soil nanotechnology ought to be directed toward strengthening soil health and promoting sustainable farming methods. To reduce ecological impact, research should focus on creating biodegradable and environmentally friendly nanomaterials. Increasing the accuracy of nanoparticle delivery systems can boost productivity and cut down on waste. Working across disciplines is crucial to filling in knowledge gaps and developing novel applications. Assessing long-term effects and effectiveness will be made easier with the implementation of strong monitoring and evaluation frameworks.

6.1 Research Priorities and Areas for Focus

Priorities for soil nanotechnology research should concentrate on a few important areas. Comprehending the behavior and environmental fate of nanoparticles in soil, along with their interactions and potential accumulation, is of utmost importance. It is crucial to look into the toxicity and long-term ecological effects on plants, soil organisms, and the larger ecosystem. The creation of safer, more efficient nanomaterials with specialized uses in agriculture can maximize gains and reduce hazards (Bottero et al. 2015). The importance of nanotechnology in agriculture is highlighted in Figure 2.

6.1.1 Nanotechnology to Control Plant Diseases

Every year, between 20% and 40% of the world's crops are destroyed by viruses and pests. The primary instrument used in contemporary farming is a pesticide, which might include herbicides, fungicides, and insecticides. Pesticides that are affordable and environmentally benign must be created. Improved shelf life, toxic-free insecticides, and increased solubility of poorly water-soluble pesticides may all have favorable environmental consequences because of cutting-edge ideas like nanotechnology (Elmer et al. 2018). Agricultural nanotechnology is significant for disease control and safety. Copper and zinc oxide compositions on a nanoscale can be effectively used to manage bacterial and fungal issues in tomato plants. It has also been reported that silver nanoparticles were effective in preventing plant diseases in rice. NPs are a sprayable insecticide that may be used as a carrier.

6.1.2 Nanotechnology to Improve the Quality of Soil and Fertilizer Distribution

Using nanotechnology to manage crops is essential to raising agricultural production. Researchers examine soil quality and fertilizer dispersion using nanostructures and materials, such as carbon nanotubes, quantum dots, and nanofibers. NPs seek to improve quality and output through efficient nutrient administration, minimize nutrient loss, and slow the spread of pollutants. In ecological agriculture, the use of zeolite, vermiculite, and nanoclay may increase crop productivity and fertilizer efficacy. In order to reduce NH_4-N passage and boost N fertilizer production, sandy loam soils can benefit from the addition of inorganic amendments in ecological agriculture. According to,

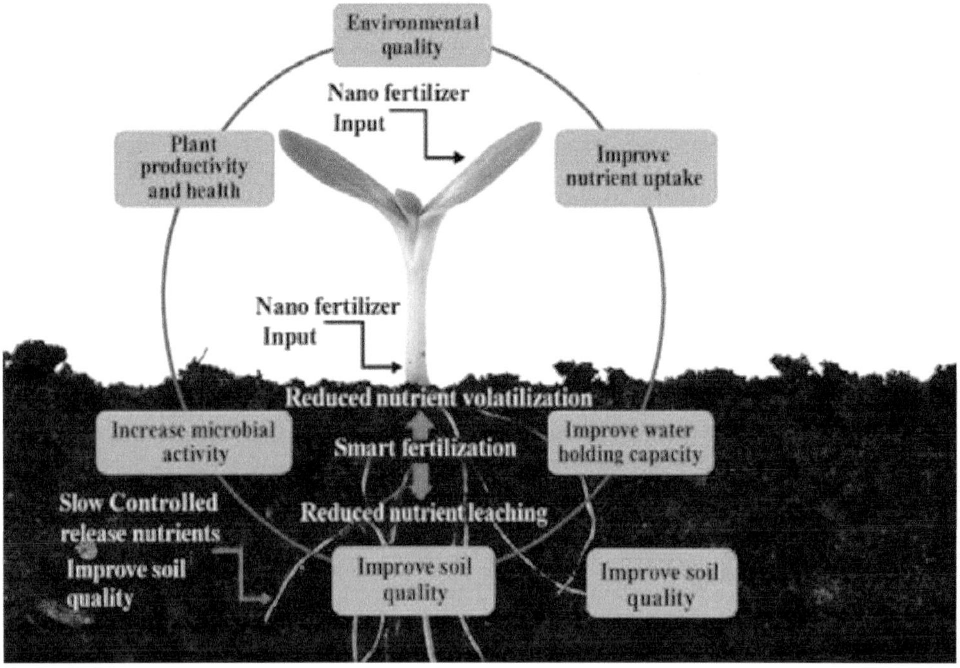

Figure 2. Importance and benefits of nano fertilizer.

NH_4-N leaching was reduced by 32% and N fertilizer efficiency was raised by 21% in sandy loam soils after applying nano-hydroxyapatite (Kerry et al. 2017).

6.1.3 Nanotechnology to Reduce Post-Harvest Loss

After harvest, food losses in wealthy nations often exceed 40%. In the least developed countries, more than 40% of losses occur in the post-harvest and processing stages. Unpreserved, highly damp crops that have just been harvested can be rapidly ruined by microorganisms. Using more modern and cutting-edge technologies, including nanotechnology, might reduce post-harvest losses. This may lessen such losses by developing functional packaging components with minimal concentrations of bioactive chemicals, enhanced mechanical and gas barrier properties, and no effect on the sensory qualities of fruits and vegetables. In a study, the addition of silver NPs (AgNPs) to chitosan films using nanotechnology extended the shelf life of strawberries (Neme et al. 2021).

6.1.4 Nanotechnology to Enhance Food Processing

The largest food corporations in the world use a range of methods to improve the quality, health, and nutritional value of their products. In the food industry, newer technology is required to boost productivity, quality, and market competitiveness. A number of applications of nanotechnology in the food production and processing sector include nano-based food additives, nanosensors, nanoencapsulation, NP-based smart delivery systems, nanotechnology enabled packaging, pharmaceuticals, and healthcare. A rising number of people are interested in food texture modification, flavor encapsulation, novel gelation or viscosity-increasing agents, and other applications of nanotechnology in commercial food processing (Thiruvengadam et al. 2018). In a study, Mustafa et al. developed a nanosensor to detect dangerous pathogens, such as Escherichia coli, in food products. The nanosensor's high sensitivity, fast reaction time, and selectivity made it an effective tool for assessing the safety and quality of food. In food processing, food nanosensing, mechanical strength and stability fillers, antibacterial agents, nutrient-delivery transporters, and anti-caking chemicals are utilized to enhance food value and safety when nanoparticles are used as food additives (Moraru et al. 2009).

6.2 Education and Training Initiatives

The usage of diverse materials by humans has been a defining feature of their lives across generations, making material transformation the cornerstone of global technological growth. The Stone Age, Bronze Age, Iron Age, Steel Age, Semiconductor Age, Advanced Materials (ceramic, polymer, and metal matrix composites), and Nanomaterials/Nanocomposites are among the divisions that follow. The manufacturing of radios, televisions, and other electronic devices using analog inputs began between 1900 and 1950. Between 1951 and 2000, semiconductor microchips were introduced, resulting in the production of increasingly sophisticated radios, televisions, small-sized computers, and internet systems. Nanowires and nanodevices have been used since 2000 to characterize stronger and more efficient products (Wong et al. 2012).

A plethora of novel materials have emerged in recent times, including nanomaterials, nanotubes, nanowires, nanocrystals, and nanocatalysts (e.g., carbon nanotubes or CNTs). These materials are stronger, lighter, and more resistant to chemicals. A department or ministry of nanotechnology should be established under the Ministry of Science and Technology to oversee the nation's overall nanotechnology initiatives and focus on developing human capital through student researcher support programs. This is one of the initiatives that must be taken to raise awareness of the use of nanotechnology in the country. It will be helpful to establish a strong cooperative relationship between nations possessing considerable nanotechnology capabilities, such as South Africa, India, and the European Union, to offer assistance on several aspects of nanotechnology operations, including funding (Fingeret and Arlene 1984).

The nation should make nanoscience and nanotechnology a priority in policy development, set clear goals, and ensure that relevant government ministries allocate budgets for nanotechnology projects and programs within their respective sectors (Newell et al. 2010). Stated differently, all levels of government—federal, state, and local—should collaborate with developed countries on training and human resource development. For instance, over the next 10 years, at least three PhD students in nanoscience and technology should receive annual funding for fellowships or training abroad (Newell et al. 2010). It is recommended that academic institutions and governmental bodies initiate a comprehensive campaign that emphasizes the critical role that nanotechnology will play in the upcoming industrial revolution through symposiums and seminars (Newell et al. 2010).

Businesses in the private sector should be encouraged to collaborate with the government to promote nanotechnology and strengthen the national economy. Both short- and long-term strategies for nanotechnology should be implemented to encourage the development of novel products and services as well as state-of-the-art materials (Newell et al. 2010). It is imperative that states and feasible local administrations create their own autonomous infrastructure. In order to ensure technological progress and bolster the nation's economy, scientists and the government can establish centers for nanoscience or form nanotechnology-focused enterprises that produce and market specific nanoproducts (Newell et al. 2010). This will promote employment and labor connected to nanotechnology, especially in the area of research and development. The potential health risks associated with nanoproducts might be one of the main barriers to poor countries' engagement in nanotechnology. Thus, it is essential that more independent research on the overall health risk associated with nanoproducts be carried out and made publicly available. In light of this, governments everywhere ought to consistently finance health risk research related to nanotechnology, giving enough resources to provide objective results within a suitable and impartial legal framework (Newell et al. 2010).

6.3 Long-term Sustainability Goals

The extensive use of chemical fertilisers in traditional agriculture has a detrimental effect on the environment and public health. Long-term soil health maintenance necessitates the intelligent, lossless delivery of nutrients to crop plants. Utilising nanomaterials has the potential to greatly increase agrochemical efficiency (50%) and agricultural production (10–20%), as well as soil health. The agro-ecosystem is under tremendous strain due to a number of factors, including the world's rapidly growing population, growing food demand, rising energy and freshwater withdrawals, excessive food waste, inefficient use of agrochemicals, deteriorating environmental conditions, and climate change (Mirti et al. 2018). Enhanced membranes and sorbents for distributed water treatment and resource recovery; novel materials for the targeted and timed delivery of agrochemicals; improved sensors for tracking physical, chemical, or biological properties and processes; pathogen-controlling technologies to maximise food safety and reduce food waste; and new materials for tracking and improving animal health are just a few of the ways that nanotechnology offers opportunities to make food production more sustainable (Buerkle et al. 2023).

7. Conclusion

To sum up, soil nanotechnology offers a practical answer to a range of issues in environmental and agricultural management. An extensive review of soil nanotechnology, including its present status, potential applications, difficulties, and solutions, has been given in this chapter. We have discussed the many uses of soil nanotechnology, from crop improvement to soil cleanup, with an emphasis on current developments and trends. The importance of nanomaterials in addressing concerns about global food security has been emphasised by their ability to assist agriculture sustainably and enable precision farming. These prospects do come with a number of noteworthy drawbacks and difficulties, though, including technological constraints, legal difficulties, and environmental concerns. Proactive risk assessment, sustainable material development, and stakeholder involvement are essential to overcoming these challenges. It is also critical to offer strong legislative suggestions and guidelines in order to guarantee the moral and secure application of soil nanotechnology. In the future, more investigation and creativity will be required to fully achieve soil nanotechnology's potential. Closing large knowledge gaps should be the primary goal of research, and efforts to improve the scientific community's capacities must include teaching and training. Furthermore, maximising the advantages of soil nanotechnology while preserving the environment and public health will depend on achieving long-term sustainability goals.

In conclusion, it is impossible to exaggerate the significance of ongoing research and development in soil nanotechnology. We may potentially realize the revolutionary promise of nanotechnology in transforming agriculture and environmental stewardship for future generations if we embrace multidisciplinary collaboration and take a holistic strategy.

8. Acknowledgement

We (Authors) are highly thankful to the University of Agriculture Faisalabad and Yangzhou University for their support in this publication. Authors also want to thank Professor Dr Abdul Wakeel from Soil Fertility and Plant Nutrition Laboratory, University of Agriculture Faisalabad, Pakistan, and Dr. Akhtar Hameed, Assistant Professor (Plant Pathology), Institute of Plant Protection for providing us this opportunity. We are highly thankful to our families who sacrificed their good times for the accomplishment of this milestone.

References

Achari, G.A. and M. Kowshik. 2018. Recent developments on nanotechnology in agriculture: plant mineral nutrition, health, and interactions with soil microflora. J. Agric. Food Chem. 66: 8647–8661.

Ali, S.S., R. Al-Tohamy, E. Koutra, M.S. Moawad, M. Kornaros, A.M. Mustafa et al. 2021. Nanobiotechnological advancements in agriculture and food industry: applications, nanotoxicity, and future perspectives. Sci. Total Environ. 792: 148359.

Alidokht, L., I. Anastopoulos, D. Ntarlagiannis, P. Soupios, B. Tawabini, D. Kalderis et al. 2021. Recent advances in the application of nanomaterials for the remediation of arsenic-contaminated water and soil. J. Environ. Chem. Eng. 9: 105533.

Arora, S., G. Murmu, K. Mukherjee, S. Saha and D. Maity. 2022. A comprehensive overview of nanotechnology in sustainable agriculture. J. Biotechnol. 355: 21–41.

Awasthi, G., T. Singh, Y. Tiwari, A. Awasthi, R.D. Tripathi, S. Shrivastava et al. 2020. A review on nanotechnological interventions for plant growth and production. Mater. Today: Proc. 31: 685–693.

Baker, S., T. Volova, S.V. Prudnikova, S. Satish and N.P. MN. 2017. Nanoagroparticles emerging trends and future prospect in modern agriculture system. Environ. Toxicol. Pharmacol. 53: 10–17.

Basavegowda, N. and K.H. Baek. 2021. Current and future perspectives on the use of nanofertilizers for sustainable agriculture: the case of phosphorus nanofertilizer. 3 Biotech. 11: 357.

Bhushan, B. 2017. Introduction to nanotechnology. pp. 1–19. *In*: Springer Handbook of Nanotechnology. Springer, Berlin.

Bottero, J.Y., M. Auffan, D. Borschnek, P. Chaurand, J. Labille, C. Levard et al. 2015. Nanotechnology, global development in the frame of environmental risk forecasting. A necessity of interdisciplinary researches. Comptes Rendus. Géoscience. 347: 35–42.

Buerkle, A., A. O'Dell, H. Matharu, L. Buerkle and P. Ferreira. 2023. Recommendations to align higher education teaching with the UN sustainability goals-A scoping survey. Int. J. Educ. Res. Open. 5: 100280.

Elmer, W. and J.C. White. 2018. The future of nanotechnology in plant pathology. Annu. Rev. Phytopathol. 56: 111–133.

Farooq, M.A., F. Hannan, F. Islam, A. Ayyaz, N. Zhang, W. Chen et al. 2022. The potential of nanomaterials for sustainable modern agriculture: present findings and future perspectives. Environ. Sci. Nano. 9: 1926–1951.

Gari, G.N., S. Saritha, A. Thattantavide and A. Kumar. 2022. Nanotechnology - Big impact: How nanotechnology is changing the future of agriculture? J. Agric. Food Res. 10: 100457.

Kale, S.S., R. Chauhan, B. Nigam, S. Gosavi and I.J. Chaudhary. 2024. Effectiveness of nanoparticles in improving soil fertility and eco-friendly crop resistance: A comprehensive review. Biocatal. Agric. Biotechnol. 56: 103066.

Kanjana, D. 2017. Advancement of nanotechnology applications on plant nutrients management and soil improvement. pp. 209–234. *In*: Nanotechnology: Food and Environmental Paradigm. Elsevier, Amsterdam.

Kerry, R.G., S. Gouda, G. Das, C.N. Vishnuprasad and J.K. Patra. 2017. Agricultural nanotechnologies: Current applications and future prospects. pp. 3–28. *In*: Microbial Biotechnology: Volume 1. Applications in Agriculture and Environment. Springer, Berlin.

Liu, J., A.R. Fernie and J. Yan. 2020. The past, present, and future of maize improvement: domestication, genomics, and functional genomic routes toward crop enhancement. Plant Commun. 1: 1.

Linley, S. and N.R. Thomson. 2021. Environmental applications of nanotechnology: nano-enabled remediation processes in water, soil, and air treatment. Water, Air, Soil Pollut. 232: 59.

McClements, D.J. 2015. Nanoscale nutrient delivery systems for food applications: improving bioactive dispersibility, stability, and bioavailability. J. Food Sci. 80: N1602-N1611.

Mishra, S., C. Keswani, P.C. Abhilash, L.F. Fraceto and H.B. Singh. 2017. Integrated approach of agri-nanotechnology: challenges and future trends. Front. Plant Sci. 8: 471.

Miteu, G.D., A.A.E. Irenosen Addeh, O. Ojeokun, T. Olayinka, J.S. Godwin, O.I. Adeyemo et al. 2023. Nanoscience and technology as a pivot for sustainable agriculture and its One Health approach awareness. Sci. One Health. 2: 100020.

Moraru, C., Q. Huang, P. Takhistov, H. Dogan and J. Kokini. 2009. Food nanotechnology: current developments and future prospects. pp. 369–399. *In*: Global Issues in Food Science and Technology. Elsevier, Amsterdam.

Neme, K., A. Nafady, S. Uddin and Y.B. Tola. 2021. Application of nanotechnology in agriculture, postharvest loss reduction and food processing: food security implication and challenges. Heliyon. 7: e08510.

Newell, M.L., B.K. Nastasi, C. Hatzichristou, J.M. Jones, G.T. Schanding Jr. and G. Yetter. 2010. Evidence on multicultural training in school psychology: Recommendations for future directions. School Psychol. Quart. 25: 249.

Nocita, M., A. Stevens, B. van Wesemael, M. Aitkenhead, M. Bachmann, B. Barthès et al. 2015. Soil spectroscopy: an alternative to wet chemistry for soil monitoring. Adv. Agron. 132: 139–159.

ounis, S.A., K.H. Kim, S.M. Shaheen, V. Antoniadis, Y.F. Tsang, J. Rinklebe et al. 2021. Advancements of nanotechnologies in crop promotion and soil fertility: Benefits, life cycle assessment, and legislation policies. Renew. Sustain. Energy Rev. 152: 111686.

Pandey, G. 2018. Challenges and future prospects of agri-nanotechnology for sustainable agriculture in India. Environ. Technol. Innov. 11: 299–307.

Qian, Y., C. Qin, M. Chen and S. Lin. 2020. Nanotechnology in soil remediation– applications vs. implications. Ecotoxicol. Environ. Saf. 201: 110815.

Rajput, V.D., T. Minkina, S.K. Upadhyay, A. Kumari, A. Ranjan, S. Mandzhieva et al. 2022. Nanotechnology in the restoration of polluted soil. Nanomaterials. 12(5): 769.

Sashidhar, P., M.K. Dubey and M. Kochar. 2019. Sensing soil microbes and interactions: how can nanomaterials help? pp. 111–130. *In*: Prasad, R. [ed.]. Microbial Nanobionics. Nanotechnology in the Life Sciences. Springer, Cham.

Singh, A., V.D. Rajput, A. Varshney, R. Sharma, K. Ghazaryan, T. Minkina et al. 2024. Revolutionizing crop production: Nanoscale wonders-current applications, advances, and future frontiers. Egypt. J. Soil Sci. 64(1).

Singh, D. and B.R. Gurjar. 2022. Nanotechnology for agricultural applications: Facts, issues, knowledge gaps, and challenges in environmental risk assessment. J. Environ. Manag. 322: 116033.

Shanmugam, V., M. Rameshpathy, G. Venkatraman, G. Ramanathan, H.A. Garalleh, A. Hashmi et al. 2023. Environmental effects and interaction of nanoparticles on beneficial soil and aquatic microorganisms. Environ. Res. 236: 116776.

Solanki, P., A. Bhargava, H. Chhipa, N. Jain and J. Panwar. 2015. Nano-fertilizers and their smart delivery system. pp. 81-101. In: Nanotechnologies in Food and Agriculture. Springer, Cham.

Thiruvengadam, M., G. Rajakumar and I.M. Chung. 2018. Nanotechnology: current uses and future applications in the food industry. 3 Biotech. 8: 1–13.

Usman, M., M. Farooq, A. Wakeel, A. Nawaz, S.A. Cheema, H. ur Rehman et al. 2020. Nanotechnology in agriculture: Current status, challenges and future opportunities. Sci. Total Environ. 721: 137778.

Wong, B.M., W. Levinson and K.G. Shojania. 2012. Quality improvement in medical education: current state and future directions. Med. Educ. 46(1): 107–119.

Yadav, N., V.K. Garg, A.K. Chhillar and J. Singh. 2023. Recent advances in nanotechnology for the improvement of conventional agricultural systems: A review. Plant Nano Biol. 4: 100032.

Yao, D., Z. Chen, K. Zhao, Q. Yang and W. Zhang. 2013. Limitation and challenge faced to the researches on environmental risk of nanotechnology. Procedia Environ. Sci. 18: 149–156.

Yadav, R.K., N.B. Singh, A. Singh, V. Yadav, C. Bano, S. Khare et al. 2020. Expanding the horizons of nanotechnology in agriculture: recent advances, challenges and future perspectives. Vegetos. 33: 203–221.

Zain, M., H. Ma, S. Chaudhary, M. Nuruzaman, I. Azeem, F. Mehmood et al. 2023. Nanotechnology in precision agriculture: Advancing towards sustainable crop production. Plant Physiol. Biochem. 108244.

Zain, M., H. Ma, S.U. Rahman, M. Nuruzzaman, S. Chaudhary, I. Azeem et al. 2024. Nanotechnology in precision agriculture: Advancing towards sustainable crop production. Plant Physiol. Biochem. 206: 108244.

Index

Editors' Biography

Dr. Akhtar Hameed is currently serving as an Assistant Professor at the Institute of Plant Protection, MNS University of Agriculture Multan (MNSUAM), Pakistan. He is an accomplished plant pathologist with a strong academic background and extensive research contributions in the field of plant disease management. Dr. Hameed earned his Ph.D. in Plant Pathology from the Department of Plant Pathology, University of Agriculture Faisalabad (UAF), Pakistan, during the period of 2016 to 2020. After successfully completing his doctoral studies, he joined MNSUAM as an Assistant Professor in 2021, where he has been actively engaged in research, teaching, and mentorship. His primary research interests lie in the molecular characterization of plant pathogens, plant-microbe interaction, quorum sensing and quorum quenching, as well as the management of plant diseases through innovative approaches such as biopesticides, nanoparticles, and quantum dots. Over the years, Dr. Hameed has made significant contributions to the field, authoring 12 internationally recognized book chapters and publishing more than 72 peer-reviewed journal articles, which collectively hold an impressive cumulative impact factor of over 150. His research has not only been recognized nationally but has also gained international acknowledgment, solidifying his reputation as a leading expert in plant pathology. He has successfully supervised six MPhil research students, providing them with the necessary guidance to contribute meaningfully to the field. He has also served as a visiting researcher at the DNA Zoo Lab, University of Western Australia. Dr. Hameed has been a key participant in various research projects funded by both national and international agencies. These include prestigious organizations such as Grand Challenges Canada, the United States-Pakistan Center for Advanced Studies in collaboration with Dr. Douglas Cook, the Feed the Future Innovation Lab at the University of California, Davis (UC Davis), and the Higher Education Commission (HEC) of Pakistan. His work in these projects has primarily focused on the development of novel bio-based solutions for the sustainable management of plant diseases. Furthermore, his research also extends to studying the impact of climate change on plant diseases, CRISPR/Cas9-based genome editing for disease resistance, and microbial applications for improving soil health. Recognizing the importance of biosafety and risk management in scientific research, Dr. Hameed has obtained an internationally recognized certification in "Biorisk Management" from the International Federation of Biosafety Associations (IFBA), Canada. His expertise and contributions to the scientific community are further highlighted by his active involvement as a peer reviewer for several high-impact journals published by leading academic publishers. Additionally, he holds the prestigious position of Editor-in-Chief at the Journal of Agriculture and Biology. Dr. Hameed's scholarly contributions extend beyond research papers and journal editorial roles; he is also an active participant in global scientific discussions. As a multilingual academic, he has presented his research findings at numerous international conferences and has undergone specialized training programs in China, Australia, Turkey and Saudi Arabia. He is an esteemed member of several professional organizations, including the American Phytopathological Society (APS), the American Society for Microbiology (ASM), the Pakistan Phytopathological Society (PPS), and the Pakistan Biological Safety Association (PBSA).

Dr. Muhammad Zahid is currently serving as a Professor at the Department of Chemistry, University of Agriculture Faisalabad (UAF) Pakistan. He completed his PhD during 2007–2011 in Chemistry from the Institute of Physical and Theoretical Chemistry, Graz University of Technology, Austria. He joined UAF Pakistan in 2011 as an Assistant Professor after completing his PhD degree. His research interests include the synthesis of advanced nanomaterials such as graphene, graphene oxide, graphitic carbon nitride, carbon quantum dots, semiconductor nanoparticles, nanocomposites, etc., and their applications in environmental remediation, sensors, energy materials, etc. Dr. Zahid has published 04 books, 30 book chapters (05 National and 25 International) and more than 170 peer-reviewed journal articles (with a cumulative impact factor of more than 700), having above 5500 citations and h-index > 43 (SCOPUS). He has supervised 07 PhD, and more than 120 MPhil/MS research students. He has won several research projects, from national and international funding agencies for developing new materials for various applications in environment, energy, sensors, etc. He also secured a position in **Stanford - Elsevier's Top 2% Scientist** (published in the years 2022, 2023, 2024 and 2025). He is involved as a reviewer for several peer-reviewed journals of high repute published by leading publishers. He served as a **Guest-Editor** for "Carbon-based Materials: Powering the Future of Energy and Environmental Progress" Frontiers in Chemistry.